THE ROUTLEDGE HANDBOOK
OF SCIENCE AND EMPIRE

The Routledge Handbook of Science and Empire introduces readers to important new research in the field of science and empire. This compilation of inquiry into the inextricably intertwined history of science and empire reframes the field, showing that one could not have grown without the other.

The volume expands the history of science through careful attention to connections, exchanges, and networks beyond the scientific institutions of Europe and the United States. These 27 original essays by established scholars and new talent examine: scientific and imperial disciplines, networks of science, scientific practice within empires, and decolonised science. The chapters cover a wide range of disciplines, from anthropology and psychiatry to biology and geology. There is global coverage, with essays about China, Southeast Asia, the Pacific, Australia and New Zealand, India, the Middle East, Russia, the Arctic, and North and South America. Specialised essays cover Jesuit science, natural history collecting, energy systems, and science in UNESCO.

With authoritative chapters by leading scholars, this is a guiding resource for all scholars of empire and science. Free of jargon and with clearly written essays, the handbook is a valuable path to further inquiry for any student of the history of science and empire.

Andrew Goss is Professor of History at Augusta University, Georgia.

THE ROUTLEDGE HANDBOOK OF SCIENCE AND EMPIRE

Edited by Andrew Goss

LONDON AND NEW YORK

First published 2021
by Routledge
2 Park Square, Milton Park, Abingdon, Oxon OX14 4RN

and by Routledge
605 Third Avenue, New York, NY 10158

Routledge is an imprint of the Taylor & Francis Group, an informa business

© 2021 selection and editorial matter, Andrew Goss;
individual chapters, the contributors

The right of Andrew Goss to be identified as the author of the
editorial material, and of the authors for their individual chapters, has been
asserted in accordance with sections 77 and 78 of the Copyright,
Designs and Patents Act 1988.

British Library Cataloguing-in-Publication Data
A catalogue record for this book is available from the British Library

Library of Congress Cataloging-in-Publication Data
A catalog record has been requested for this book

ISBN: 978-0-367-22125-6 (hbk)
ISBN: 978-1-032-02653-4 (pbk)
ISBN: 978-0-429-27336-0 (ebk)

Typeset in Bembo
by Newgen Publishing UK

CONTENTS

Contents

Contents

FIGURES

CONTRIBUTORS

Casper Andersen is Associate Professor at the University of Aarhus in Denmark. His field of research is history of knowledge, science, and technology in imperial and global contexts with a main focus on Africa. His publications include *British Engineers and Africa 1876–1914* (2011), a co-edited five-volume collection on *British Governance and Administration in Africa 1880–1939* (2013), and numerous research articles. His current research focuses on UNESCO and the decolonisation of academic knowledge in Africa.

Timothy P. Barnard is Associate Professor in the Department of History at the National University of Singapore, where he specialises in the environmental and cultural history of island Southeast Asia. His most recent books include *Nature's Colony* (2016) and *Imperial Creatures* (2019).

Antonio Barrera-Osorio is Associate Professor of History at Colgate University, Hamilton, NY. Barrera-Osorio received his BA from Universidad de los Andes, Bogotá, Colombia, and his PhD in history from the University of California, Davis. His book, *Experiencing Nature: The Spanish American Empire and the Early Scientific Revolution* (2006), explores the emergence of empirical practices in the Spanish American empire. Other publications include: "Experts, Nature, and the Making of Atlantic Empiricism," in *Osiris* 25 (2010); "Experiencia y Empiricismo en el Siglo XVI: Reportes y Cosas del Nuevo Mundo," in *Memoria y Sociedad* 27 (2009); "Knowledge and Empiricism in the Spanish Atlantic World," in *Science, Power, and the Order of Nature in the Spanish and Portuguese Empires* (2008); "Empiricism in the Spanish Atlantic World," in *Science and Empire in the Atlantic World* (2007); and "Empire and Knowledge: Reporting from the New World" in *Colonial Latin American Review* (2006). He is currently working on a book that explores relationships between empires, science, and local knowledge.

James Beattie is Associate Professor at Victoria University of Wellington, New Zealand. He has written 12 books and over 70 articles and chapters on aspects of British imperial history, including its environmental, conservation, scientific, gardening, and ecological histories. He is Founding Editor of *International Review of Environmental History*, as well as Founding Co-editor of both *Palgrave Studies in World Environmental History*, and *Routledge Research on Gardens in History*. He is currently co-editing, with Edward Melillo and Ryan Jones, *Migrant*

Ecologies: Environmental Histories of the Pacific World and working on a book manuscript, *China's Pacific: Migration and Environmental Change, 1790s–1920s.*

Dorit Brixius is a postdoctoral researcher at the German Historical Institute Paris. She received her PhD in history from the European University Institute in Florence. Brixius is author of several articles on the history of the construction of plant knowledge in the early-modern French Empire and co-editor (with S. Kroupa and S. Mawson) of the special issue "Science and Islands in the Indo-Pacific Worlds" of *The British Journal for the History of Science* (2018). She is currently completing a manuscript about plant knowledge in eighteenth-century Mauritius while working on her second book on gender and medicine in seventeenth-century France.

Hugh Cagle is Associate Professor in the Department of History at the University of Utah, where he is also Director of the International Studies program. He specialises in the history of science, technology, and colonialism in the early-modern world, with a particular emphasis on Portugal and its colonies. His first book, *Assembling the Tropics: Science and Medicine in Portugal's Empire, 1450–1700* (2018), won the 2019 Leo Gershoy Award from the American Historical Association.

Pratik Chakrabarti is Chair in History of Science and Medicine and Director of the Centre for the History of Science, Medicine, and Technology (CHSTM) at the University of Manchester, UK. He has contributed widely to the history of science, medicine, and imperialism from the eighteenth to the twentieth century. His publications include *Western Science in Modern India: Metropolitan Methods, Colonial Practices* (2004); *Materials and Medicine: Trade, Conquest and Therapeutics in the Eighteenth Century* (2010); *Bacteriology in British India: Laboratory Medicine and the Tropics* (2012); and *Medicine and Empire 1600–1960* (2014). His most recent monograph is *Inscriptions of Nature: Geology and the Naturalization of Antiquity* (2020).

Maria Pia Donato is Centre National de la Recherche Scientifique (CNRS) Research Professor at the Institut d'Histoire Moderne et Contemporaine in Paris, France. She teaches the history of science and medicine at the École Normale Supérieure Paris. She is the author of numerous articles and books on cultural history and the history of science and medicine, especially in the early-modern Catholic context, including *Sudden Death: Medicine and Religion in 18th-Century Rome* (2014) and *Medicine and the Inquisition in the Early Modern World* (2019).

James Flowers is a Brain Pool Program research fellow at Kyung Hee University, South Korea. He is funded by the Ministry of Science and ICT through the National Research Foundation of Korea. He works as a historian of science and medicine in Korea and the East Asian region. He received his PhD in the history of medicine in 2019 from the Johns Hopkins University School of Medicine, with his dissertation titled "Koreans Building a New World: The Eastern Medicine Renaissance in Japan-ruled Korea, 1910–1945."

Andrew Goss is Professor of History at Augusta University, Georgia. His research expertise is the history of science and politics in Indonesia and Asia. He is the author of *The Floracrats: State-Sponsored Science and the Failure of the Enlightenment in Indonesia* (2011).

Matthew M. Heaton is Associate Professor in the Department of History at Virginia Tech. His research interests are in the history of health and illness, migration, and globalisation in Africa with particular emphasis on Nigeria. He is the author of *Black Skin, White Coats: Nigerian*

Psychiatrists, Decolonization, and the Globalization of Psychiatry (2013) and co-author with Toyin Falola of *A History of Nigeria* (2008).

Nanna Katrine Lüders Kaalund is a postdoctoral research associate at the Scott Polar Research Institute and Darwin College, University of Cambridge, where she works as part of the European Research Council-funded "Arctic Cultures" project. She received her PhD in science and technology studies from York University, Canada. She is the author of the forthcoming book *Explorations in the Icy North: How Travel Narratives Shaped Science in the Arctic*, which will be published by the University of Pittsburgh Press in 2021.

Nathan Kapoor is Visiting Assistant Professor of History at Grand Valley State University, Allendale, MI. He specialises in the history of electric power technologies in the British Empire. He received his PhD in the history of science, technology, and medicine in 2019 from the University of Oklahoma. His current book project examines the relationship between electrification and colonialism in New Zealand between the nineteenth and twenty-first centuries.

Ulrike Kirchberger is a research fellow at the University of Kassel, Germany. She has published on different aspects of global and colonial history from the eighteenth to the twentieth centuries. Her current research deals with ecological networks and transfers between Australia, South Asia, and Africa, 1850–1920. Recent publications include the edited collection *Environments of Empire: Networks and Agents of Ecological Change* (2020), and the article "Temporalising Nature: Chronologies of Colonial Species Transfer and Ecological Change across the Indian Ocean in the Age of Empire," in *International Review of Environmental History* (2020).

Anna Kuxhausen is Associate Professor of History and Russian Language and Area Studies at St. Olaf College, Northfield, MN, where she is also Chair of the History Department. In addition to articles on the history of medicine, the body, and reproductive practices, she is the author of *From the Womb to the Body Politic: Raising the Nation in Enlightenment Russia* (2013). Her current project explores the intersections of self-fashioning, masculine identities, and scientific vocation on the eighteenth-century Russian frontier.

Martin Mahony is Lecturer in Human Geography at the University of East Anglia, UK. He works on the historical and contemporary geographies of atmospheric science and technology, and is co-editor of *Cultures of Prediction in Atmospheric and Climate Science* (2017) and *Weather, Climate, and the Geographical Imagination* (2020). He is currently working on two book projects, one which examines the intersections of atmospheric science, aviation technology, and empire in interwar Britain, and another which explores the impact of the concept of the "Anthropocene" on the discipline of geography.

Ruth A. Morgan is an environmental historian and historian of science at the Australian National University, where she is Director of the Centre for Environmental History. She has published widely on the climate and water histories of Australia and the British Empire, including her award-winning book, *Running Out? Water in Western Australia* (2015). Her current project, on environmental exchanges between British India and the Australian colonies, has been generously supported by the Australian Research Council and the Alexander von Humboldt Foundation. She is co-author of the upcoming *Cities in a Sunburnt Country: Water and the Making of Urban Australia*, and a lead author in Working Group II of the Intergovernmental Panel on Climate Change's Assessment Report 6.

Jennifer R. Morris received her PhD in history from the National University of Singapore. Her research explores the history of museums and collecting in Southeast Asia, and she is currently working on a book examining the development of the Sarawak Museum in Malaysian Borneo and its scientific networks under the rule of the Brooke Rajahs. She is also serving as an academic advisor for the new Borneo Cultures Museum in Kuching, Sarawak.

Michael A. Osborne is Professor Emeritus of History of Science at Oregon State University and Research Professor of Environmental Studies and History at the University of California, Santa Barbara. He is also President of the International Union of History and Philosophy of Science and Technology through 2021. He is the author of *The Emergence of Tropical Medicine in France* (2014) and other studies on the history of medicine, natural history, and popular zoology in France and the French Empire.

Sabina Pavone is Associate Professor in Early-Modern History at the University of Macerata, Italy. Her research area is situated at the intersection of religious history and cultural history. She has written several books on Jesuits in Russia, anti-Jesuitism, conversions, and Jesuit missions in India. Among her works are *The Wily Jesuits* (2005, 2000) and *Una strana alleanza. La Compagnia di Gesù in Russia dal 1772 al 1820* (2010), and she is co-editor of *Missioni, saperi e adattamento tra Europa e imperi non cristiani* (2015) and *Compel People to Come In: Violence and Catholic Conversions in the Non-European World* (2018).

Hans Pols is Professor in the School of History and Philosophy of Science at the University of Sydney. He is interested in the history of medicine in the Dutch East Indies and Indonesia, and focuses in particular on the history of mental health. In 2018, Cambridge University Press published his book *Nurturing Indonesia: Medicine and Decolonisation in the Dutch East Indies.*

James Poskett is Assistant Professor in the History of Science and Technology at the University of Warwick, UK. He completed his PhD at the University of Cambridge and is the author of *Materials of the Mind: Phrenology, Race, and the Global History of Science, 1815–1920* (2019).

Megan Raby is Associate Professor of History at the University of Texas at Austin. She is the author of *American Tropics: The Caribbean Roots of Biodiversity Science* (2017), which examines the relationship between field ecology, the emergence of the modern concept of biodiversity, and the expansion of US hegemony in the circum-Caribbean during the twentieth century. Her work has appeared in the journals *Isis, Environmental History*, and *History of Science.*

Thomas Simpson is a research associate on the "Making Climate History" project at the University of Cambridge. His research interests span the history of field sciences, imperial history, and historical geography, with a particular focus on the British Empire in Asia. His publications include *The Frontier in British India: Space, Science, and Power* (2021) and articles on maps, borders, and the history of mountain spaces in *The Historical Journal* and *History of Science.*

Charu Singh holds the Adrian Research Fellowship in the History of Science at Darwin College, University of Cambridge. She received her PhD in history from Jawaharlal Nehru University and previously taught at Shiv Nadar University in India. Her research focuses on the problems and labours of translating scientific knowledge for vernacular reading publics in South Asia. She explores the transformations in colonial scientific subjectivity through the Hindi print archive.

Daniel A. Stolz is the Kemal H. Karpat Assistant Professor of History at the University of Wisconsin-Madison. He is the author of *The Lighthouse and the Observatory: Islam, Science, and Empire in Late Ottoman Egypt* (2018). He is currently working on a book that examines the late Ottoman bankruptcy as part of a global history of public finance.

Cameron B. Strang is Associate Professor of History at the University of Nevada, Reno, and the author of *Frontiers of Science: Imperialism and Natural Knowledge in the Gulf South Borderlands, 1500–1850* (2018). He is currently working on a history of Indigenous explorers in North America.

Fenneke Sysling is Assistant Professor of Colonial and Global History at the University of Leiden, the Netherlands. She specialises in the history of science and Dutch colonialism, and her interests include colonial heritage, museum objects, race, the body, and natural history. She is the author of *Racial Science and Human Diversity in Colonial Indonesia* and the editor of a special issue on "Measurement, Self-Tracking and the History of Science" in *History of Science*.

Andreas Weber is Assistant Professor of Science, Technology, and Culture at the University of Twente in the Netherlands. Andreas received his PhD in history from Leiden University. He is the author of various articles and books on the history of science and natural history collections in the nineteenth-century Dutch Empire, including *Hybrid Ambitions: Science, Governance, and Empire in the Career of C.G.C. Reinwardt, 1773–1854* (2012), *Laborious Transformations: Plants and Politics at Bogor Botanical Gardens* (2018), and *Collecting Colonial Nature: European Naturalists and the Netherlands Indies in the Early Nineteenth Century* (2019). Andreas is co-editor of the Brill book series *Emergence of Natural History*.

1

INTRODUCTION

An imperial turn in the history of science

Andrew Goss

For the last three decades, empire—alongside related terms such as colonialism and global history—has been a productive category of analysis in the history of science.[1] Science and its history is no longer just a narrative of Western civilisation, with historians only occasionally glancing to China, India, and the Middle East. The history of science can only be understood, or even told, with close attention to the connections, exchanges, and networks that extended beyond the learned societies, universities, gardens, laboratories, and research institutions of Europe and the United States. What previously had been thought of as a footnote—the diffusion of science from the metropoles of Europe to the colonial territories of Asia, Africa, and the Americas—has emerged as a vibrant and exciting sub-discipline. Since the 1990s, an entire new historiography has sprung up with an interest in the conditions, practices, and findings of imperial, colonial, and global science. This volume brings together 27 essays by experts in this field, and surveys multiple ways in which careful attention to the imperial, colonial, and global contexts of science enriches our understanding of the development of early-modern and modern science.

The purpose of this handbook is to introduce the reader to the most important new research and results in the field of science and empire. A rich literature has been developed by historians of science as well as scholars of empire demonstrating the numerous ways science and empire grew together, from the fifteenth century until now. Starting three decades ago, research about science in colonial settings began to reveal that neither science nor empire could have existed without each other. A mutually beneficial, symbiotic relationship developed between science and empire, and this spawned complex systems, institutions, and networks which were not only interwoven, but supported, nurtured, and sustained each other. This insight has led to a new way of thinking, an imperial turn, in the history of science. In particular, since the imperial turn, historians of science have investigated the creation of new disciplines of expertise in colonial and imperial contexts. They have explicated the contribution of local Indigenous systems of knowledge to imperial expertise. They have pointed to the importance of networks of knowledge that crossed political, social, and cultural boundaries and which entangled knowledge brokers, material objects, and imperial agents. Finally, they have shown that decolonisation was not a simple matter of newly independent nations and their scientists taking over the reins of formerly colonial efforts, but it was a process in which imperial knowledge was slowly reshaped into new systems, institutions, and disciplines. This handbook is a compilation of the results of this new way of thinking.

Making and sustaining empire required systems and procedures for organising increasingly complex information about nature, land, and people. Governing elites and officials sought ways to channel information from its imperial territories to the centre, and employed those who could provide this in an organised fashion. An emphasis on ordering imperial knowledge in turn created multifaceted relationships between scholars, explorers, rulers, Indigenous experts, and imperialist officials. Generating practical, imperial knowledge was rarely done at court, but instead in the provinces and territories of empire, where scholars, local officials, and Indigenous knowledge brokers created and invented the methods of imperial science. The scholarly relationships and knowledge networks then spawned their own histories and narratives, sometimes overlapping with European political and culture systems, sometimes not. One finding of the new literature is that these networks of expertise existed independently from European institutions of science; they happened in contexts that little resembled the traditional Western scholarly environment. This shaped both the new disciplines of science, including natural history, cartography, anthropology, and meteorology, all of which thrived in the imperial environment, and the political regimes that consumed and sometimes funded this knowledge. Different disciplines, regions, and empires created different forms of expert knowledge and science, and it has proven impossible to write this into a single narrative of global science.

As the essays in this volume attest, a lot more is known about the individuals, research programmes, and goals that shaped imperial science than even just 20 years ago. Just as importantly, this literature has moved past the narrower guideposts which had earlier defined research about science in imperial and global contexts. Empire has not traditionally been at the centre of the historiography of science. Textbook histories of the scientific revolution, from Herbert Butterfield's *The Origins of Modern Science, 1300–1800* (1949) to Steven Shapin's *The Scientific Revolution* (1996), look only rarely past the European scholarly settings.[2] And when empire is invoked, in studies about natural history for example, imperial politics or economics is little remarked on. In much of this literature, empire is context only, an expanding global arena for scientists to gather data and develop theories.

While empire may not have traditionally been a category of analysis, writing about naturalists and scientists as explorers, collectors, and discoverers in imperial territories has a long tradition. Whether it is the observation of the transit of Venus during Cook's first voyage, Humboldt's expedition to Latin America, Darwin's voyage of the *Beagle*, or Eddington's observations of the solar eclipse in 1919, examples of scientists travelling the globe, to collect and observe, have long been recognised by historians as important for the development of some sciences. But with the focus on the history of ideas, on the development, refinement, and improvement of natural explanations for observed phenomena, the context of empire was secondary to the science produced. Classic treatments of the intellectual history of scientific theories did not hide the importance of global and imperial travel, but have not asked how empire shaped, formed, and altered scientific thinking and practice.[3] Although there is a healthy tradition of biographical treatment of and by European imperial scientists, this genre usually framed scientific achievements as happening despite the imperial context, not because of it. The story was of the pioneering and heroic scientists who prepared the imperial territories for the expansion of Western scientific culture. Building on this literature, the historian of science Lewis Pyenson wrote three books that seriously examined the work and lives of some of these scientists. He argued that innovative scientific researchers produced scientific knowledge, in what he called the exact sciences, even when their colonial colleagues ruled and exploited the colonial population ruthlessly. In Pyenson's argument, exact sciences such as astronomy and physics—unlike anthropology and medicine—transcended the brutality and opportunism of colonialism, and these sciences provided cultural lustre to empire, which was then passed on to colonial subjects, thus redeeming at least some of science even in the face of empire's racism and economic exploitation.[4] Pyenson was one of the first scholars to write a collective biography of colonial scholars, but his insistence that exact science was untainted by empire or politics has not held up to closer scrutiny; a critical review in the flagship journal *Isis* in 1993 by Paolo Palladino and Michael Worboys signalled the arrival of empire as a problematic in the discipline.[5]

Even before the imperial turn in the history of science, a small group of scholars in the 1970s and 1980s, most of them not trained in the history of science, had begun to lay the groundwork for understanding the political and economic context of imperial science. They were inspired by a series of new approaches to knowledge and empire—including early efforts by anthropologists to reflect on their discipline's historical connection to empire. Scholars demonstrated how some imperial scientists were directly involved in the economic and political exploitation of the colony. Naturalists and others at botanical gardens did research which helped European imperial powers exploit natural resources more effectively. As the book-length study by Lucille Brockway demonstrated in the context of Kew Botanic Gardens' network of scientists, these imperial scientists were embedded within the culture of mainstream metropole science.[6] Other landmark studies were Daniel Headrick's two books about science, technology, and imperial economies, and James E. McClellan's 1992 study arguing that eighteenth-century science in Saint-Domingue facilitated colonial development.[7] Headrick and McClellan concluded that science was a tool of utility in European colonisation, with science a handmaiden of the colonial state.

The articles in this volume review the scholarship since the insight that global networks of science and knowledge were enmeshed with empire. Early efforts, produced by younger scholars in the years around the turn of the century, focused more squarely on colonial science, that is, science as practised in the European colonies, especially through histories of colonial scientific institutions. Although this approach later showed its limitations, it was methodologically and logistically an appropriate place to start empirical research into the practices of imperial scholarship. In 2000, Roy MacLeod, one of the pioneers whose early work on Australian colonial science had previewed the imperial turn, brought together in one volume the fruits of the early efforts to understand knowledge and power within the colonial enterprise.[8] Colonial science produced numerous studies that examined how science in the European colonies in Asia and Africa was intertwined with colonial politics, and developed practices and knowledge that were distinct from metropole science.[9] Nonetheless, the narrowness of these studies, usually defined by a single colony and discipline, was that they failed to account for the ways science transcended colonial boundaries and categories. They did not sufficiently examine the bridging of imperial borders or the inclusion of Indigenous knowledge practices as the basis for expertise. And with the focus on the practices of European-directed science, there was insufficient attention to the ways in which knowledge regimes were resisted and altered by non-Western intermediaries, guides, and go-betweens. In the last decade or two, a multitude of studies have remedied this, and have explored the complexity and ambiguity of imperial science.[10] And we now know that knowledge creation did not neatly conform to the dominant colonial politics. This handbook brings together the findings from a generation of scholarship examining imperial networks of scientists, imperial ways of knowing, and knowledge entangled with colonialism.

The imperial turn has not produced a single, predominant method or narrative of science and empire. Study subjects have multiplied, so while there continues to be interest in naturalists, anthropologists, and cartographers in the eighteenth-century Atlantic World and the European colonies of late-nineteenth- and early-twentieth-century Asia, the number of scientists and sciences, and the number of locations, has grown immensely. No single framework encompasses the richness of the questions about how empires collected, processed, described, and printed expert knowledge of imperial spaces. As Chapter 21 by James Beattie and Ruth A. Morgan argues, science as a category of analysis does not always assist in understanding the creation of imperial expertise. What is clear is that the empires of the early-modern and modern world produced official and unofficial knowledge of the natural world, and these imperial histories are inseparable from the development of modern science.

In Chapter 2 Pratik Chakrabarti examines why the histories of science and empire have remained separate and marginal from mainstream histories of science. He shows how the discipline of the history of science is built on the history of European exceptionalism, in which European science is a unique set of practices not found elsewhere. The authority of the discipline rests upon scholars continuing to reinscribe the singularity of science's historical location in European culture, even when they do so critically. And the result is a history in which imperial interactions might hybridise science, complicating its global spread, but does not threaten the European origin story of the fundamentals of science. One way forward, Chakrabarti argues, is for historians of science to reconceive of science itself as an imperial knowledge. He suggests that, after a generation of exploring hybrid research done in colonial contact zones, future research must do more, by squarely investigating science as an imperial epistemology.

Many of science's most potent and long-lasting categories, theories, and approaches grew from imperial roots. The essays by Hugh Cagle (Chapter 14) and Antonio Barrera-Osorio (Chapter 22) examine how even before the end of the fifteenth century, long before the so-called Scientific Revolution, science was being revolutionised in the Portuguese and Spanish Empires. As studies in this book attest, numerous scientific concepts such as race, energy, geological time, biodiversity, and climate, to name only a few, are imperial creations. A challenge of ruling empire, how to create cultural certainty that is clear and consistent across land and peoples defined by complexity and difference, incentivised ruling officials and scholars to create practical data and knowledge of the imperial worlds. What resulted were new imperial cultures, disciplines, and systems of expertise, which later came to be labelled as science. Racial science, as James Poskett reminds us in Chapter 4, is not just to be understood as an intellectual history leading to modern racism, but was a set of scientific practices shaped by imperial sites, technologies, power dynamics, and practices. In Chapter 6, Matthew M. Heaton shows how colonial psychiatry defined normal and abnormal behaviour through imperial experiences, and that these categories were resisted and transformed by colonial subjects. Chapter 3 by Thomas Simpson illustrates how maps and cartography have, as has long been understood, shaped the ability of empires to exert control over people and resources. But he shows how, from the very beginning, maps were not just the creation of imperial elites, but were formed by non-Westerners, many of whom used maps for their own purposes.

After the end of the European empires in the middle of the twentieth century, most scholars, and in particular scientists, downplayed the imperial origins of science. This survives even now in some of the traditions of the history of science. But prior to the fall of European empires that was not the case. It is clear from research into science's historical archives that disciplines of science that were born inside empires rarely hid their imperial natures. Scientists and statesmen alike saw the need for knowledge which encompassed the vastness of different territories and peoples but also revealed their distinctness. They created new methods and practices. Key scientific methods such as questionnaires, taxonomies, fieldwork, and of course measuring, surveying, and mapping natural phenomena, originated in imperial contexts. Imperial scholars took the data and sorted, classified, and published it as imperial and scientific knowledge. So, new disciplines and methodologies of sciences were created. As Martin Mahony's Chapter 5 shows, meteorology started in efforts during the nineteenth century to measure, predict, and understand weather not just locally, but across the oceans and lands that comprised the European empires in the Atlantic and Indian Oceans. And subsequent interest in micro-climates started in the mountains of the Austro–Hungarian Empire. Another example is archaeology, with imperial scholars creating the world's earliest heritage sites.[11] While these methods may have started in the reaches of empire, they circulated freely, and found practitioners and audiences outside the imperial contexts that created them.

The increasing awareness that European empire and science were intertwined from the beginning has also forced a re-evaluation of non-European imperial science. Ironically, older arguments for what distinguished non-Western knowledge from the scholarly traditions of Europe, and explained why science arose in Europe and not in China, India, or the Middle East, stressed the independence of European science from power and politics. This autonomy supposedly distinguished European science from the system of close ties between scholars, bureaucrats, and statesmen outside Europe, which in a traditional interpretation stifled the emergence of science in China and elsewhere. As the studies by James Flowers in Chapter 18 and Daniel A. Stolz in Chapter 16 argue, science in the Chinese and Ottoman Empires has much in common with European science. These empires also promoted knowledge and science that revealed their land and people, what Stolz calls the "sciences of worldmaking." Sciences such as astronomy and cartography assisted the Ottoman court in maintaining the empire and strengthening the central governing apparatus. Much like in other empires, this led to complex and varied efforts by provincial and rival scholars to interact with court scholars. This insight helps us to understand the growth of science as culturally authoritative, in the Ottoman lands as well as in Europe and elsewhere. While historians have long been interested in Western science in China, the article by Flowers shows that what we thought of as European knowledge systems were joint efforts by Chinese and European scholars, who shared much more than previously thought.

Anna Kuxhausen's study in Chapter 17 demonstrates how it was in the context of imperial expansion and colonial settlement that science was planted in the Russian soil. By the nineteenth century, empires around the world were enthusiastic consumers and creators of expert knowledge, and "science" was a system of knowledge and practices ready-made for their imperial politics. Although I was unable to secure an essay for this volume which synthesises the scholarship on the intertwined impetus toward science and imperialism in Japan, that scholarship is illuminating. During the Meiji era, when new Japanese elites were bent on building a modern nation-state, sciences not only made rapid inroads, but led to the creation of a new discipline of seismology in Japan.[12] And after 1920, science in Japan gained legitimacy and authority at the same time that Japan's political and imperial culture stressed its uniqueness and anti-Western nature.[13]

One of the most influential methodologies for historians of imperial science has been drawn from cultural anthropology, namely Nicholas Thomas's theory of entangled objects, entanglement, and exchange. Thomas argued, with special reference to colonial and imperial interactions, that Pacific cultures have been entangled with colonial and capitalist systems for hundreds of years, and this history pervades meaning into material artifacts and exchanged goods.[14] Thomas in particular showed how Europeans appropriated Indigenous objects during Cook's voyages. This approach has inspired numerous new and productive approaches to science and empire, especially as it relates to conceptualising networks of exchange and circulation. In Chapter 12, Dorit Brixius argues that a focus on local networks of knowledge production, and their connection to regional systems of exchange, can reframe our understanding of imperial and global histories of science. In a similar way, James Beattie and Ruth A. Morgan in Chapter 21 show how the study of non-Western knowledge systems in colonial Australasia allows researchers to escape the constraints of the history of science traditions in order to make sense of knowledge making by Indigenous and non-Indigenous people.

From the eighteenth century onwards, imperial scientists cast themselves as researchers inscribing order and objectivity upon a blank slate. This was part of a larger effort by European imperial powers to emphasise stark differences between coloniser and colonised.[15] Within imperial science this practice, modelled by such imperial scientific celebrities as Alexander von Humboldt, has erased the intellectual and other contributions of local experts and scientists. This larger ideology—of a European scientific culture that rationalised the world without assistance—then became embedded in colonial institutions in the colonies of Asia and Africa. The predominant genre of scientific publishing removed the labour, knowledge, and analysis of local scholars, collectors, and technicians. More careful historical research, by scholars such as Sujit Sivasundaram, shows that from the beginning imperial methodology relied on absorbing knowledge from local inhabitants. Although mediated by colonial officials and scholars, it was true and valuable because it was the result of local ways of knowing.[16] In Chapter 20, Charu Singh demonstrates that, in British India in the nineteenth century, not only did local knowledge brokers provide the practical knowledge that enabled colonial rule, but Indian colonial subjects crafted scientific selves as professionals and consumers within the colonial system. And, as Jennifer R. Morris shows in her study of non-Western collectors (Chapter 9), these practices continued long after formal scientific institutions were built in the nineteenth century. Although there are methodological challenges in uncovering the agency of Indigenous knowledge brokers, new methodologies and approaches, in particular through careful study of the natural history collections and museums, mean scholars can today reveal the expertise of non-Western collectors. In Chapter 8 Andreas Weber explains how other innovative techniques, in particular digitalisation efforts, now allow scholars to examine natural history collections and archives as products of complexly entangled imperial interactions.

Science was conceived in Europe as a hierarchical system, with powerful metropole institutions directing the work of subordinate scholars in the imperial field. Michael A. Osborne in this volume addresses the unequal co-evolution of science, technology, and medicine in France and its colonies. Studies of, for example, the Botanic Gardens in Europe show how these institutions oversaw networks of collectors operating in and beyond their empire, responsible for providing specimens, data, and observations to their scientific masters.[17] And while scientific leaders in the capitals and universities of Europe tried well into the nineteenth century to manage communities of scholars who were to answer back to Europe, studies in this book show this was not the sum total of imperial science. Imperial scientists participated in a multitude of networks—local, regional, and global—which transcended and upended imperial hierarchies. Almost all imperial scientists, once they were in the field, found they could not generate new and meaningful knowledge without local alliances. This included not only collaborating and cooperating with other imperial officials and scholars, but also relying upon the knowledge and expertise of local intermediaries and knowledge brokers.

One important conclusion, which a number of studies in this book emphasise, is that imperial scholars who created their own networks, cutting across local, regional, imperial, and global contexts, could have success cultivating scientific and imperial authority. Best known perhaps are Jesuits, whose networks are examined by Maria Pia Donato and Sabina Pavone in Chapter 11. Jesuits created the conditions for imperial science by acting as agents of complexly interconnected structures of governance. Ulrike Kirchberger's study of German networks within the British Empire (Chapter 13) shows how neither colonial nor imperial borders defined the scientific networks of German scientists, which created its own specific dynamics of knowledge production.

The earliest research into imperial science already asked the question of scientists' participation in the exploitation and violence of imperial rule. As Fenneke Sysling's study in Chapter 7 shows, while in the late 1960s anthropologists examined their discipline's imperial echoes, much of the imperial anthropology research was ignored or discarded by colonial administrators and rulers. Nonetheless, anthropological ideas helped justify and sustain imperialism, and anthropologists were rarely critical of the violence of colonialism. Moreover, certain fields of expertise were critically important to colonial exploitation. This was particularly the case for natural history and biology. In Chapter 19, Timothy P. Barnard explores the role of imperial botanic gardens in producing knowledge and procedures to grow plants and profits in the European colonies of Southeast Asia. In Chapter 10, Nathan Kapoor demonstrates that imperial coal extraction and then electrification created energy regimes which continue to shape energy infrastructure today, and retain the political values, categories, and power dynamics of their imperial origins.

There is a rich historiography of science in the Atlantic World, highlighting the importance of oceanic networks of exchange between Europe, Africa, North America, the Caribbean, and South America during the early-modern period.[18] Cameron B. Strang's essay in Chapter 23 extends this literature by examining how the five empires in North America in the seventeenth and eighteenth centuries produced rival knowledge networks, built in similar ways. In Chapter 25, Nanna K.L. Kaalund's essay about Arctic science focuses on the British imperial context of nineteenth-century scientific practice in the North, emphasising the way the people and places of the Arctic were critical in shaping what we think of as Arctic science. Megan Raby's Chapter 24 examines scientific connections between the United States and Latin America, which show how imperial ambitions and expansions in the Americas shaped US scientific institutions, while Latin American scholars subverted the hierarchical nature of those connections.

The new research showing that science is an imperial discipline has also complicated the historiography of decolonised science. New research has undercut older assumptions which characterised decolonisation as a time of new beginnings, with opportunities for national and independent scientists to finally begin to practise Western science, starting with imitation of European and colonial models. Long an overlooked history, it is now clear that decolonization was not a break or definitive watershed. To be sure, empire could not be removed from science, as shown in the now classic study by Joseph Hodge of postcolonial careering by former colonial scientists in Africa; in the 1960s, former colonial scientists in Africa and Asia led development programmes that had been first developed during the late imperial era.[19] Chapter 26 by Casper Andersen examines the initiatives taken by UNESCO to facilitate the creation of a global international science infrastructure in the era of decolonisation. Hans Pols's Chapter 27 shows how Indonesian scientists and physicians continue to experience international participation in science through the legacies of imperial science.

How should you use this book? I encourage all readers to read Pratik Chakrabarti's chapter, to orient themselves to the challenges and opportunities in the field of science and empire. The book is roughly divided into three sections, where the earlier chapters are thematically organised around disciplines, while the middle chapters focus on networks of science, followed by essays with a geographical focus. The final two essays are dedicated to decolonised science, but numerous other essays, especially in the first section, take up the issue of decolonising science. I encourage all readers to examine chapters from each of these sections, as they feature different approaches and methods, often bringing different perspectives to similar questions. For example, anyone interested in the history of race, science, and empire might start with the chapter "Racial science," and then look at "Anthropology and empire" and "Colonial psychiatry." The reader could then read the chapters about Jesuit ethnography or the nineteenth-century

French Empire, as well as chapters specific to a geographical region, perhaps the Americas. Finally, the chapter about decolonisation and UNESCO provides a review of the place race science played in the 1950s and 1960s. Similarly, readers interested in imperial mapping should start with the chapter "Cartography and empire from early modernity to postmodernity," and then, depending on their geographical or disciplinary interest, can follow it up in subsequent entries, as mapping and surveying figures in many subsequent chapters. And for those interested in a geographical region, I encourage them to start in the relevant chapter, but also be aware that earlier chapters about disciplines and networks may have relevant information.

Notes

1 I want to thank everyone who helped bring this book to publication, including at Taylor & Francis my editor Max Novick, as well as Jennifer Morrow, Robert Langham, and Eve Setch, and at Newgen Publishing, Faye Gardner and Liz Williams. Many thanks to Pax Bobrow and Sage Bobrow-Goss for their editing assistance and wisdom.

2 Herbert Butterfield, *The Origins of Modern Science, 1300–1800*, rev. ed. (New York: Collier, 1962); Steven Shapin, *The Scientific Revolution* (Chicago: University of Chicago Press, 1996).

3 There is no index entry for empire in the standard textbook treatment of the history of evolution, still in print, Peter J. Bowler, *Evolution: The History of an Idea*, rev. ed. (Berkeley: University of California Press, 1989).

4 Lewis Pyenson, *Cultural Imperialism and Exact Sciences: German Expansion Overseas, 1900–1930* (New York: Peter Lang, 1982); Lewis Pyenson, *Empire of Reason: Exact Sciences in Indonesia, 1840–1940* (Leiden: Brill, 1989); Lewis Pyenson, *Civilizing Mission: Exact Sciences and French Overseas Expansion, 1830–1940* (Baltimore, MD: Johns Hopkins Press, 1993).

5 Paolo Palladino and Michael Worboys, "Science and Imperialism," *Isis* 84, no. 1 (1993): 91–102.

6 Lucille Brockway, *Science and Colonial Expansion: The Role of the Royal British Botanic Gardens* (New York: Academic Press, 1979).

7 Daniel Headrick, *The Tools of Empire: Technology and European Imperialism in the Nineteenth Century* (New York: Oxford University Press, 1991); Daniel Headrick, *The Tentacles of Progress: Technology Transfer in the Age of Imperialism, 1850–1940* (New York: Oxford University Press, 1988); James E. McClellan III, *Colonialism and Science: Saint Domingue in the Old Regime* (Baltimore, MD: Johns Hopkins Press, 1992).

8 Roy MacLeod, ed., "Nature and Empire: Science and the Colonial Enterprise," special issue, *Osiris* 15 (2000).

9 See for example William Kelleher Storey, *Science and Power in Colonial Mauritius* (Rochester, NY: University of Rochester Press, 1997); Andrew Goss, *The Floracrats: State-Sponsored Science and the Failure of the Enlightenment in Indonesia* (Madison, WI: University of Wisconsin Press, 2011); Pratik Chakrabarti, *Bacteriology in British India: Laboratory Medicine and the Tropics* (Rochester, NY: University of Rochester Press, 2012).

10 Brett M. Bennett and Joseph M. Hodge, eds., *Science and Empire: Knowledge and Networks of Science across the British Empire, 1800–1970* (Houndmills, Basingstoke: Palgrave Macmillan, 2011).

11 Marieke Bloembergen and Martijn Eickhoff, "A Moral Obligation of the Nation-State: Archaeology and Regime Change in Java and the Netherlands in the Early Nineteenth Century," in *Empire and Science in the Making: Dutch Colonial Scholarship in Comparative Global Perspective, 1760–1830*, ed. Peter Boomgaard (New York: Palgrave Macmillan, 2013), 185–205.

12 Gregory Clancey, *Earthquake Nation: The Cultural Politics of Japanese Seismicity, 1868–1930* (Berkeley: University of California Press, 2006).

13 Hiromi Mizuno, *Science for the Empire: Scientific Nationalism in Modern Japan* (Stanford, CA: Stanford University Press, 2009). James R. Bartholomew, "Japan," in *The Cambridge History of Science*, vol. 8, *Modern Science in National, Transnational, and Global Context*, ed. Hugh Richard Slotten, R.L. Numbers, and D.N. Livingstone (Cambridge: Cambridge University Press, 2020), 555–76.

14 Nicholas Thomas, *Entangled Objects: Exchange, Material Culture, and Colonialism in the Pacific* (Cambridge, MA: Harvard University Press, 1991).

15 Jane Burbank and Frederick Cooper, *Empires in World History: Power and the Politics of Difference* (Princeton, NJ: Princeton University Press, 2010).

16 Sujit Sivasundaram, *Islanded: Britain, Sri Lanka, and the Bounds of an Indian Ocean Colony* (Chicago: University of Chicago Press, 2013).

17 Richard Drayton, *Nature's Government: Science, Imperial Britain, and the "Improvement" of the World* (New Haven, CT: Yale University Press, 2000).

18 For two collections of essays that brought this research together, see James Delbourgo and Nicholas Dew, eds., *Science and Empire in the Atlantic World* (New York: Routledge, 2008); Daniela Bleichmar, Paula DeVos, Kristin Huffine, and Kevin Sheehan, eds., *Science in the Spanish and Portuguese Empires* (Stanford, CA: Stanford University Press, 2009). See also Londa Schiebinger, *Plants and Empire: Colonial Bioprospecting in the Atlantic World* (Cambridge, MA: Harvard University Press, 2014).

19 Joseph Hodge, "British Colonial Expertise, Postcolonial Careering and the Early History of International Development," *Journal of Modern European History* 8, no. 2 (2009): 24–46.

2

SITUATING THE EMPIRE IN HISTORY OF SCIENCE

Pratik Chakrabarti

Empire remains peripheral to history of science. As a discipline, history of science emerged due to the distinctiveness of the new world of nature, based on experimental, observational, and empirical methods that appeared in Europe from the seventeenth century. Therefore, conventional histories of science are about European savants who created that vision. These define histories of geology, natural history, physical sciences, and ecology from the perspectives of European intellectuals, institutions, and epistemological moments. The birth of natural history, Linnaean classification, Lavoisier's Chemical Revolution, Darwinian evolutionary theories, the discovery of deep time and prehistory, the rise of European and North American environmentalism, and discourses of environmental apocalypse form the basis of these narratives. While these processes have been shown to be complex and often tentative, these form the general outline.[1] Empire provides instances of travel, diversions, and nuances along with the material specimens to this archetypal narrative of the natural sciences. Consequently, imperial history of science appears as derivative of the normative one. This essay investigates why the historiography of science and empire has remained distinct from that of the history of science. It suggests that the problem lies in the conceptualisation of history of science as a discipline.

I will refer to a particular instance of the emergence of this dominant European vision of nature in history of science. The birth of natural history from the seventeenth century established that nature had its own secular history; of reproduction, dispersion, and life cycle, which is not governed by the Divine mind or any celestial design. The subsequent discovery of deep time from the eighteenth century by European savants such as Horace Bénédict de Saussure, Georges Cuvier, William Hamilton, and Charles Lyell entrenched that specific historicity of nature. Their study of fossils and landscapes aligned the history of the earth with that of nature and humans and placed these beyond the timescale of textual or religious traditions. In the nineteenth century, the palaeontological discoveries of human and animal fossils and deliberations on the earth's strata provided the basis for the emergence of the new secular understanding of time in Europe.[2] These gradually replaced Biblical notions of time and the sacred history of nature and the earth.[3] These constitute some of the core characteristics of the emergence of European secular ideas of nature and of natural history. These have also been *the* template for the general history of geology or geohistory, while drawing almost exclusively from European and North American contexts. These do not usually refer to imperial history or even to historical and intellectual developments beyond Europe and North America. The colonial history of geology on the other hand has focused either on its deployment in the exploitation of natural resources and territorial expansion or the unfolding of this deep historical vision in the empire.[4] The intellectual deliberations on deep time generally remain outside this purview.

This does not mean that there are no works on deep history in India, Australia, or South Africa. Precisely the opposite. There are robust traditions of deep history here, connected with much more eclectic intellectual traditions, such as aboriginality, indigeneity, memory, tribal politics, folklore, art, and tribal geomyths.[5] Here historians have sought to understand the deep time of nature from aboriginal or tribal perspectives along with the social and cultural contexts of colonialism, settler histories, and displacement. These demonstrate how colonialism fundamentally transformed discussions on the historicity of the land and the earth. Yet, this intellectual tradition has remained outside conventional histories of geology. How can historians of earth sciences engage with this multitude of visions of nature and time?

Such a trend can be observed in several other scientific disciplines or themes, such as the history of meteorology, biology, laboratory medicine, biomedicine, and zoology.[6] Several of the European scientific experiences were concurrent with European imperialism, such as Darwin's travels to the Pacific islands, Humboldt's expeditions to the Andes, the explorations and collection of imperial flora and fauna, James Cook's Pacific voyages, or the colonial extraction of coal and diamond. Yet, imperial history and colonial epistemologies have remained marginal to the narration of such histories. Scholarly histories of "modern science" usually draw almost exclusively from European intellectual traditions.[7] The non-West here tend to appear as histories of "translations."[8] These give the impression that modern science is a European phenomenon.[9]

This is despite the fact that efforts at developing non-European perspectives in the history of modern science have been as old as the discipline of history of science itself. These two traditions have often remained parallel to each other with distinct intellectual and historical trajectories. Starting with George Sarton and Joseph Needham in the 1930s and 1960s respectively, scholars have highlighted the non-Western, imperial, and global influences in the making of modern science. The intellectual deliberations have undergone three distinct phases. In the first phase, the main emphasis, particularly in the works of Sarton and Needham, was to highlight the influence of not only ancient or medieval but also of Arabic and Chinese science on modern science.[10] Their efforts were defined by their belief in the universality of science. Sarton and Needham viewed modern science as an ecumenical discipline formed by the various streams of "ethno-sciences" hailing from China, Europe, India, and the Islamic world. While these provided a global scope to the evolution of modern science, they did not question its core European doctrines.

The next phase reflected attempts at situating imperial or colonial practices of science within the history of modern science, starting with George Basalla, and subsequently taken up by others such as Roy MacLeod, Robert A. Stafford, and Deepak Kumar.[11] These scholars derived from the postcolonial writing of institutional, intellectual, and nationalist histories of colonialism in Asia, Australia, South America, and Africa and argued that colonialism played a critical role in the global dispersion of modern science. They also highlighted the role of science in the exploitation of colonial resources. There has been another parallel intellectual tradition, which has highlighted the role of science in Western cultural and political dominance.[12] These often developed alongside the general critique of science as hegemonic and part of Western cultural and imperial dominance.[13]

The last phase of the debates has centred on the global history of science. Here scholars have rejected both the *universalist* predispositions of Needham and Sarton as well as the diffusionist model of Basalla and others to suggest that modern science and even European Enlightenment had a global inheritance.[14] These have proposed that what is often known as modern science was both formulated through global cultural exchanges from the eighteenth century as well as absorbed within a global scope through various translations and interlocutions.[15] In doing so, scholars have stressed the "polycentric" nature of the history of science, derived from various non-European traditions.[16]

Global and imperial histories have provided new dimensions and perspectives to modern scientific traditions. In particular, in the history of natural history and botany, the global history of trade and commerce has been seen to have made significant contributions.[17] We are now aware that non-European princes, physicians, and pundits played significant roles as cultural and intellectual interlocutors and relocated European enlightenment through various "intercultural exchanges."[18] We also know that European scholars such as Charles Darwin, Carl Linnaeus, and Alexander von Humboldt had derived their ideas and facts from various local traditions beyond Europe.[19] There has also been considerable scholarship on how, in the processes of colonialisation, Europeans incorporated insights from Indigenous forms of knowledge within Western sciences such as anthropology, biomedicine, and agriculture, which Helen Tilley has defined as "vernacular science."[20] At the same time, scholars have relocated Darwinian evolutionary ideas to late-nineteenth-century Arabic culture and polity.[21] Others have identified the Ottoman roots of modern natural history.[22] Collectively, these have provided wider and deeper intellectual and cultural scopes to the major tenets of Enlightenment or European science—for example, the birth of natural history or deep evolutionary thinking.

These reorientations of the history of science have emerged mainly from beyond Europe, from historical experiences in Asia, Africa, and South America. Despite this strong tradition there remains a lack of alignment between imperial and normative histories of science, which is reflected in the problems of terminologies and frames. In these different phases of global and imperial histories of science, historians have adopted and used frames and concepts such as "global," "colonial," "vernacular," "circulation," "networks," and "translational" to incorporate non-European voices and contributions to modern science. At the same time, it has remained difficult to identify any single theme to define these diverse histories. For example, historians working on the non-European history of science have found George Basalla's phases and frames of "colonial science" limited.[23] Yet, as Warwick Anderson has recently pointed out, while Basalla's model appears problematic and outdated, no single frame that defines the history of globalisation of science has emerged.[24] The phrase "colonial science," shorn of Basalla's diffusionist model, appears useful in situating modern science within social and economic contexts as created by colonialism.[25] "Colonial science" identifies unique or particular locations of modern science, particularly in the empire.[26] As social history has provided important context to the history of European science, "colonial science" provides a similar social, political, and economic setting of modern science as it operated in European empires from the eighteenth century.

The "global" frame has appeared to be problematic as well. Fa-ti Fan has highlighted the problem with the assumption of the apparently "smooth" movement of materials and ideas through a circulation narrative of global science.[27] While the global history of science has subsumed the history of colonialism, it has often failed to highlight the histories of violence, marginalisation, and rejection inherent in such global connections.[28] The tendency to see intermediaries and agents as interstitial brokers of the global scientific heritage can ignore the role these groups played as agents of the European trading companies in the backdrop of expanding imperial control and conquest, which in turn became critical in the expansion of modern science.

Why has it remained difficult to develop any alternative frame for non-European histories of modern science? One suggestion is that the global history of science is not a "singular" story. It engenders diverse connections, disconnections, and multiple narratives.[29] Significantly, "global" here is constituted exclusively of the non-West: Asia, Africa, the Arab world, and South America.[30] This reflects the problem in this search for alternative categories or multiple narratives. If European or North American science continues to be regarded as just "science" and the rest as "global" or "colonial," then these categories will continue to appear as derivative and contingent.[31] Why should historians describe the knowledge, particularly scientific knowledge, in the empire as one either in "transition" or "circulation," or in terms of "networks"?[32] Similarly, why should themes such as "transmission," "contact zones," or intermediaries, which historians have developed persuasively over the last two decades, be particularly (and almost exclusively) relevant when science leaves the shores of Europe or North America?[33]

The problem is not necessarily with categories such as colonial/global/transnational but with the Eurocentrism of the history of science. Whatever terms or adjuncts those who work on the history of science in the Global South come up with for their vocation, these run into problems of contingency. While historians have deliberated upon whether to refer to science in the Global South as "colonial," "translational," or "global," they have not debated whether the history of European and North American science should be referred to as *just* science. It is only when this assumption is questioned that it is possible to decolonise the discipline and conceptualize history of science in the true sense of the term.

The Eurocentrism of history of science

Why has the history of science remained Eurocentric? There is an epistemological problem with history of science. As I have suggested earlier, the discipline is historically predetermined to draw primarily from Western (northern European and American) intellectual, social, cultural, and institutional experiences, no matter how global or diverse those experiences have then been shown to be. The fact that historians tend to depend on Humboldt, Darwin, and Linnaeus to trace the global trajectories of natural sciences is an indication of this epistemic problem of science. Here the non-West finds its space as "ethno," "vernacular," "imperial," and "global" experiences; hybridising, globalising, nuancing, and punctuating those central narratives, rather than dismantling it.

Moreover, there is still discernible scientism in history of science, particularly in the history of environmental science, deep history, history of geology, palaeontology, evolution, and prehistory, in which the research agendas and tones are often set by scientists, or scholars with strong backgrounds in science, rather than humanities. The recent emphasis on the Anthropocene, climate change, and genomics has made this scientism even more evident. For example, in the current literature on Anthropocene, climate scientists have often set the agendas and even their historical scopes, which are usually at a planetary scale, disregarding the regional, historical, economic, cultural, and political layers of lived climatic and geohistorical experiences.[34] Scholars have highlighted the need to engage the idea of the Anthropocene with the colonial and postcolonial exploitation of natural resources.[35] Discussions on genomics and human antiquity have similarly taken a broad "species" approach.[36] Between 1990 and 2000, the Human Genome Project mapped the complete DNA sequence of the human genes.[37] These show that variations among humans in physical features have a connection to the geographical origins of ancestors. These have generated new discussions on the role of science in defining the relationship between race and environment, which started with colonialism in the eighteenth century. These discussions highlight the need to align the search for human prehistory with the history of imperialism; that the scientific discovery of the human prehistory and the colonial discovery of aboriginality took place at the same time and often simultaneously. Modern cultural and social imagination of race, aboriginality, and prehistory is thereby entrenched in colonial histories of settlement and displacement, "discoveries" of Indigenous races and tribes, and their associations with prehistoric humans.

Yet, in the discussions on climate change, the Anthropocene, and human genomics the empire remains conspicuously absent. Here scientific facts, achievements, and viewpoints, which for historical reasons too tend to be Eurocentric, drive the narrative.[38] In their historical deliberations, the problematics of history of science are often conflated with those of science itself. This particular representation of the fundamental challenges for or of history of science is reminiscent of the discipline in its formative years in the early twentieth century. This reflects the broader issue that science itself has remained a metropolitan and Eurocentric discipline. As a combination of both these factors, the history of non-Western science appears not the conventional one, only its punctuation.

One of the ways to address the asymmetries between the imperial and histories of science is by recognising the essential fragmented nature of the discipline and its implications for discussions on its core theme: nature and time. Carla Nappi has urged the need to move away from the mainstream "genealogy"-centric histories of science, which treats the Scientific Revolution of Europe as both its site of origin and its telos. Through the study of the fragmented nature of early-modern Manchu texts (which are a complex mesh of texts comprising letters, diaries, and other records compiled and composed by Jesuits and local Manchu intellectuals) Nappi makes the case for narrating layered and ephemeral stories set in their contexts. It is also essential, she stresses, to do so for *all* stories, not just those beyond Europe. In such a truly global history of science, which is not "a subset of area studies," the category of "global," as it is currently being used, might become redundant and "drop out of the picture." She suggests that such a fragmented history will be the asset of history of science:

> What I am proposing here is a reorientation that takes that fragmentariness as a strength and a raw material to work with, rather than seeing it as a necessary draw-back. It places terms or sounds or texts or ideas next to each other and pays attention to see what might emerge in-between.[39]

It is important to recognise the fragmented and incommensurable nature of the histories of science as a whole; that not all histories or sciences connect, travel, or cognise at a global or planetary scale.

In recent years, several scholars have explored how European naturalists collected exotic plants and other curiosities from remote parts of the world from the seventeenth century, named and sketched them, or placed them inside their field diaries, or laced them in their distinct yet diverse orders of classification.[40] They have also juxtaposed various texts of early-modern natural history, determining the intricate classificatory differences in one from the other.[41] At the same time, the emergence of the global history of natural history has brought new dimensions to these discussions. Historians have established that non-European princes, physicians, traders, and scholars played significant roles in collecting and channelling these plants, herbs, and animals through complex geographical and epistemological routes in the making of the global natural historical heritage.[42]

There are histories of fragments here, which show that the same plants, drugs, practices of collecting and naming, and institutions such as botanical gardens or museums had different histories in distinct contexts. From the seventeenth century, Europeans were engaging with a complex world of nature in the colonies in Asia, Africa, and America. They were simultaneously imposing their own spiritual and liminal ideas on these. That history is often lost in the conventional narratives of the natural history. This history of nature was not just based on the observation, collection, and relocation of nature, but also on various conceptions of nature. This explains why in colonial South Asia, for example, Darwinian ideas, rather than establishing more secular traditions of nature, galvanised ideas of sacred Hindu evolutionism.[43]

This is also not just a matter of appreciating the role of vernacular or indigenous systems of knowledge in Western science. The need is to understand the various metaphysical, physical, and mythological encounters that shaped the modern study of nature away from Europe, which often provides insights into Europe's own multiple selves. For example, the Orientalist study of deep Indian time in the eighteenth century started with attempts to align Hindu scriptures and with Genesis, to find in the former references to the Deluge, which in turn led to the emergence of Hindu notions of deep time.[44]

The history of poisons and antidotes provides an important insight into this fragmented history of science and medicine as shaped by colonialism. Several Asian medicinal substances such as mercury, hemlock, opium, datura, and nux vomica have been known and used as poisons and drugs. Bezoar stones, used for antidotes to poisons, were part of European therapeutics since the seventeenth century. Some of the earliest mentions of the "bezoar" in European texts was in Garcia d'Orta's *Colloquies on the Simples and Drugs*. D'Orta was a Portuguese Jewish physician and professor at Lisbon, who came to India in 1534 and stayed there until his death. Here he studied Indian plants and drugs and in 1563 wrote *Colloquies*. The book became popular across Europe and was translated into Latin and other European languages. This was the first modern European printed text which provided first-hand knowledge of the diseases and drugs of India. D'Orta explained that the main use of the bezoars was to counteract all kinds of poisons.[45] The word "bezoar" itself was derived from the Arabic word *badzahr*, which means antidote or counter-poison. Through d'Orta and other Portuguese physicians, the bezoar stone became popular in Europe in the seventeenth century. This Oriental medicinal object has a South American history as well. The Spanish, as part of their bioprospection for medicinal plants in South America, found Peruvian bezoar stones. Here these were part of a different history of European cultural imperialism, as they became associated with Spanish fears of and attempts at exterminating Andean idolatry.[46]

Several colonial plants and medicines never became part of European natural history or pharmacopoeia. Londa Schiebinger has studied the rejection of some Caribbean treatments in Europe. The peacock flower used as an abortifacient by slave women in the West Indies failed to receive any acceptance in Europe. Schiebinger explains this rejection in the attitude of the scientific elites of Europe towards gender and female bodies.[47]

Colonial practices of poisons and antidotes reflected this fractured nature of natural history, shaped in the colonies, in particular by histories of resistance. In the exploitative plantation systems in the West Indies and southern Americas, the native Amerindian populations in South America and the African slaves regularly practised poisoning as a form of resistance. They also cultivated knowledge of their antidotes. They used plants such as Jamaican blood flower, *contrayerva*, or the *curare*, each carrying in its nomenclature and usage histories of resistance and violence, as poisons and antidotes. European physicians often adopted these plants and these medicinal practices with fear and anxiety, as was the case with the Spanish fear of the Andean bezoar stones. While vital in the colonial contexts, these therapeutics often did not become part of European materia medica, mostly due to their negative cultural and political associations.

Abena Dove Osseo-Asare has studied the colonial bioprospection of African medicinal herbs from the nineteenth century. She has shown that imperial expansion in Africa took place at the time of the rise of pharmaceutical chemistry in Europe. Europeans were simultaneously introducing new pharmaceuticals in African markets and looking for ingredients and plants. This led to European interest in African medicinal practices as well as the appropriation of African ethnobotanical knowledge for modern pharmaceuticals.[48] African healers had used the *Strophanthus*, which was used to produce the strophanthin drug in Europe, as both a poison and medicine traditionally. The plant was "discovered" by the British during David Livingstone's Zambezi expedition. The drug, which Africans used in their poisoned arrows against the British on the Gold Coast, entered the *British Pharmacopoeia* in 1898 following a history of European distrust, anxieties, as well as interest in African plants and medicinal practices. Following investigation and experimentation in Edinburgh, British pharmaceutical companies similarly procured the *Strophanthus kombe* in eastern and southern Africa to produce a drug for cardiac diseases.[49]

Here this history becomes fragmented. By the early twentieth century, as the British established their military presence in West Africa, they outlawed African use of the plant in poisoned arrows. At the same time, the drug became vital to modern pharmaceutical companies. International pharmaceutical demand for *Strophanthus* seeds increased, leading to its large-scale export from the Gold Coast during the First World War. The use of the plant became simultaneously marginalised within African medicine.

Conclusion

In a review essay in 1993, Michael Worboys and Paolo Palladino made the point that "the history of science and imperialism *is* the history of science."[50] They argued that science's modes of operating within the colonial context, in situ with histories of exploitation and the innovative use of resources, reflect some of the key features of the history of modern science. Historians have explored the ways to integrate colonial and global histories into the European history of science, towards a more pluralistic vision of science.[51] These deliberations are ultimately linked with the question about what modern science is and what it represents in the modern world. In Britain, the public debates on the Rhodes statue at Oxford University and the toppling of Edward Colston's statue at Bristol, of UCL's supposed eugenic past and Cambridge University's presumed links with slavery have emerged as key questions of decolonisation.[52] These have been part of wider discussions around decolonising science. That discussion has focused on the need to deconstruct science as a Western/European discipline and adopt a more cosmopolitan approach to acknowledge non-Western and non-European contributions to it. However, as this essay has argued, it is not enough to acknowledge the historical contribution of other cultures and communities or of the empire to Britain's past. That trajectory may well lead to an idea of science (or even Britain's imperial past) as ecumenical and pluralistic, enriched by global contributions. While we acknowledge the global contributions to science, we need to engage with the fact that, from the perspectives of race, gender, and geography, which too are connected to the history of imperialism, science itself has been an elitist, exclusionary, and exploitative discipline. Therefore, is it even possible to decolonise science? The relationships of power and biases engendered within science are innate to the writing of its past as well. Decolonisation of science requires not merely an awareness of Europe's or European science's hidden colonial past. It necessitates seeing science itself as an imperial episteme. The need is neither to "provincialise" nor to "globalise" history of science but to recognise or decolonise the provinciality and marginalisation inherent to it.

Notes

1 For the unfolding of this relatively "coherent" vision of the world of *naturalia* see Nicholas Jardine and Emma Spary, "Introduction: Worlds of Natural History," in *Worlds of Natural History*, ed. H.A. Curry, N. Jardine, J.A. Secord, and E.C. Spary (Cambridge: Cambridge University Press, 2018), 3–16.

2 See for example Martin Rudwick's *magnum opus*, *Bursting the Limits of Time: The Reconstruction of Geohistory in the Age of Revolution* (Chicago: University of Chicago Press, 2005) and his *Worlds before Adam: The Reconstruction of Geohistory in the Age of Reform* (Chicago: University of Chicago Press, 2008); Paolo Rossi, *The Dark Abyss of Time: the History of the Earth & the History of Nations from Hooke to Vico* (Chicago: University of Chicago Press, 1984); Noah Heringman, *Sciences of Antiquity; Romantic Antiquarianism, Natural History, and Knowledge Work* (Oxford: Oxford University Press, 2013); A. Bowdoin Van Riper, *Men among the Mammoths: Victorian Science and the Discovery of Human Prehistory* (Chicago: University of Chicago Press, 1993); Stephen Jay Gould, *Time's Arrow, Time's Cycle: Myth and Metaphor in the Discovery of Geological Time* (Cambridge, MA: Harvard University Press, 1987); Mott T. Greene, *Geology in the Nineteenth Century: Changing Views of a Changing World* (Ithaca, NY: Cornell University Press, 1982).

3 Some notable examples are Rudwick, *Worlds before Adam* and Charles Coulston Gillispie's classic book, *Genesis and Geology: A Study in the Relation of Scientific Thought, Natural Theology, and Social Opinion in Great Britain, 1790–1850* (New York: Harper Torch Book, 1959), Vybarr Cregan-Reid, *Discovering Gilgamesh: Geology, Narrative and the Historical Sublime in Victorian Culture* (Manchester: Manchester University Press, 2011).

4 James A. Secord, "Global Geology and the Tectonics of Empire," in Curry, Jardine, Secord, Spary, *Worlds of Natural History*, 401–17; Robert A. Stafford, "Geological Surveys, Mineral Discoveries, and British Expansion, 1835–71," *The Journal of Imperial and Commonwealth History* 12 (1984): 5–32; Jim A. Secord, "King of Siluria: Roderick Murchison and the Imperial Theme in Nineteenth-Century British Geology," *Victorian Studies* 25, no. 4 (July 1, 1982): 413–44; Robert A. Stafford, "Annexing the Landscapes of the Past," in *Imperialism and the Natural World*, ed. John M. MacKenzie (Manchester: Manchester University Press, 2017), https://doi.org/10.7765/9781526123671.00008; Orlando Bentancor, *The Matter of Empire: Metaphysics and Mining in Colonial Peru* (Pittsburgh, PA: University of Pittsburgh Press, 2017); Suzanne Zeller, "The Colonial World as Geological Metaphor: Strata(gems) of Empire in Victorian Canada," *Osiris* 15 (2000): 85–107; Cameron Strang, *Frontiers of Science: Imperialism and Natural Knowledge in the Gulf South Borderlands, 1500–1850* (Chapel Hill: University of North Carolina Press, 2018); Tim Bonyhady, *The Colonial Earth* (Melbourne: Miegunyah Press, 2001); Deepak Kumar, "Science, Resources, and the Raj: A Case Study of Geological Works in the Nineteenth Century India," *Indian Historical Review* 10 (1983–84): 6–89. For a general exposition of this thesis of science and colonization, see Robert A. Stafford, *Scientist of Empire, Sir Roderick Murchison, Scientific Explorations and Victorian Imperialism* (Cambridge: Cambridge University Press, 1989).

5 Sumathi Ramaswamy, *The Lost Land of Lemuria; Fabulous Geographies, Catastrophic Histories* (Berkeley: University of California Press, 2004); Anne McGrath and Mary Anne Jebb, *Long History, Deep Time: Deepening Histories of Place* (Acton: Australian National University Press, 2015); Libby Robin, "Perceptions of Place and Deep Time in the Australian Desert," in *Thinking through the Environment: Green Approaches to Global History*, ed. Timo Myllyntaus *et al.* (Cambridge: White Horse Press, 2011), 81–99; Denis Byrne, "Deep Nation: Australia's Acquisition of an Indigenous Past," *Aboriginal History* 20 (1996): 82–107; Tom Griffiths, "Travelling in Deep Time: La Longue Durée in Australian History," http://australianhumanitiesreview.org/2000/06/01/travelling-in-deep-timela-longue-dureein-australian-history/.

6 The recent book traces the emergence of the general idea of the "environment" entirely as a post-World War II development based almost entirely from the intellectual traditions and political discourses in North America and Europe; see Paul Warde, Libby Robin, and Sverker Sörlin, *The Environment: A History of the Idea* (Baltimore, MD: Johns Hopkins University Press, 2018).

7 Peter J. Bowler and Iwan Rhys Morus, *Making Modern Science: A Historical Survey* (Chicago: University of Chicago Press, 2005); H. Floris Cohen, *How Modern Science Came into the World: Four Civilizations, One 17th-Century Breakthrough* (Amsterdam: Amsterdam University Press, 2010).

8 Bernard V. Lightman, *A Companion to the History of Science* (Chichester: Wiley, 2016). See chapters by Kapil Raj, Marwa Elshakry, and Carla Nappi.

9 In the wonderfully informative and illustrated *Oxford Illustrated History of Science*, for example, the only chapters with any geographical specifications are those on East Asia and the Middle East. The rest on

more generic themes of the history of science and, without regional themes (under "Doing Science"), cover only Europe and North America. Iwan Rhys Morus, *The Oxford Illustrated History of Science* (Oxford: Oxford University Press, 2017).

10 George Sarton, *The History of Science and the New Humanism* (New York: Holt, 1931); Joseph Needham and Ling Wang, *Science and Civilisation in China,* vol. 1, *Introductory Orientations* (Cambridge: Cambridge University Press, 1954). A much shorter and more accessible book by Needham is *The Grand Titration: Science and Society in East and West* (London: Allen & Unwin, 1969).

11 Roy MacLeod, "On Visiting the 'Moving Metropolis': Reflections on the Architecture of Imperial Science," *Historical Records of Australian Science* 5 (1980): 1–16; Stafford, *Scientist of Empire*; John Gascoigne, *Science in the Service of Empire: Joseph Banks, the British State, and the Uses of Science in the Age of Revolution* (Cambridge: Cambridge University Press, 1998); Deepak Kumar, *Science and the Raj, 1857–1905* (New Delhi: Oxford University Press, 1995). For a general review of this scholarship see Pratik Chakrabarti and Michael Worboys, "Science and Imperialism since 1870," in *The Cambridge History of Science,* vol. 8, *Modern Science in National, Transnational, and Global Context,* ed. Hugh Richard Slotten, Ronald L. Numbers, and David N. Livingstone (Cambridge: Cambridge University Press, 2020), 9–31.

12 Ashis Nandy, ed., *Science, Hegemony and Violence: A Requiem for Modernity* (Delhi: Oxford University Press, 1988).

13 Michael Adas, *Machines as the Measure of Men: Science, Technology, and Ideologies of Western Dominance* (Ithaca, NY: Cornell University Press, 1989). For a review of this literature, see Agusti Nieto-Galan, "Antonio Gramsci Revisited: Historians of Science, Intellectuals, and the Struggle for Hegemony," *History of Science* 49, no. 4 (2011): 453–78.

14 The literature is too vast to cite under one footnote. These are referred to subsequently. A useful overview is in the *Isis* special issue edited by Sujit Sivasundaram, "Global Histories of Science," special issue, *Isis* 101, no. 1 (2010): 95–158.

15 See for example Alper Bilgili, "An Ottoman Response to Darwinism: Ismail Fennî on Islam and Evolution," *The British Journal for the History of Science* 48, no. 4 (2015): 565–82; Savithri Preetha Nair, *Raja Serfoji II: Science, Medicine and Enlightenment in Tanjore* (London: Routledge, 2014).

16 Jonardon Ganeri, "Well-Ordered Science and Indian Epistemic Cultures: Toward a Polycentered History of Science," *Isis* 104, no. 2 (June 1, 2013): 348–59.

17 Harold J. Cook, *Matters of Exchange: Commerce, Medicine, and Science in the Dutch Golden Age* (New Haven, CT: Yale University Press, 2008).

18 Lissa Roberts, "Situating Science in Global History: Local Exchanges and Networks of Circulation," *Itinerario* 33 (2009): 9–30, here p. 10. There is a vast range of literature that generally conforms to that view; see for example, Anna Winterbottom, *Hybrid Knowledge in the Early East India Company World* (Basingstoke: Palgrave, 2016); Kapil Raj, *Relocating Modern Science: Circulation and the Construction of Scientific Knowledge in South Asia and Europe, 17th–19th Centuries* (Delhi: Permanent Black, 2006); Nair, *Raja Serfoji*; Simon Schaffer, Lissa Roberts, Kapil Raj, and James Delbourgo, eds., *The Brokered World: Go-Betweens and Global Intelligence, 1770–1820,* Uppsala Studies in History of Science (Sagamore Beach, MA: Science History Publications, 2009).

19 Neil Safier, *Measuring the New World: Enlightenment Science and South America* (Chicago: University of Chicago Press, 2008); Linda Andersson Burnett, "An Eighteenth-Century Ecology of Knowledge: Patronage and Natural History," *Culture Unbound: Journal of Current Cultural Research* 6, no. 7 (2014): 1275–97; C.M. Brown, *Hindu Perspectives on Evolution: Darwin, Dharma and Design* (Abingdon: Routledge, 2012); Janet Browne, "Missionaries and the Human Mind: Charles Darwin and Robert Fitzroy," in *Darwin's Laboratory: Evolutionary Theory and Natural History in the Pacific,* ed. R. MacLeod and P.F. Rehbock (Honolulu: University of Hawaii Press, 1994), 263–82.

20 Helen Tilley, "Global Histories, Vernacular Science, and African Genealogies; or, Is the History of Science Ready for the World?" *Isis* 101, no. 1 (March 2010): 110–19.

21 Marwa Elshakry, *Reading Darwin in Arabic, 1860–1950* (Chicago: The University of Chicago Press, 2013).

22 Harun Küçük, *Science Without Leisure: Practical Naturalism in Istanbul, 1660–1732* (Pittsburgh, PA: University of Pittsburgh Press, 2020).

23 George Basalla, "The Spread of Western Science," *Science* 156 (1967): 611–22. For such critiques of Basalla, see Dhruv Raina, "From West to Non-West? Basalla's Three-Stage Model Revisited," *Science as Culture* 8, no. 4 (1999): 497–516; Zaheer Babar, *The Science of Empire: Scientific Knowledge, Civilization, and Colonial Rule in India* (Albany, NY: SUNY Press, 1996), 10; Ian Inkster, "Scientific

Enterprise and the Colonial 'Model': Observations on Australian Experience in Historical Context," *Social Studies of Science* 15 (1985): 677–704; Mark Harrison, "Science and the British Empire," *Isis* 96 (2005): 56–63.

24 Warwick Anderson, "Remembering the Spread of Western Science," *Historical Records of Australian Science* 29, no. 2 (2018): 73–81.

25 See Londa Schiebinger's definition of "colonial science" as "any science done during the colonial era that involved Europeans working in a colonial context." "Forum Introduction: The European Colonial Science Complex," *Isis* 96 (2005): 52–5, here p. 52.

26 Pratik Chakrabarti, *Medicine and Empire, 1600–1960* (Basingstoke: Palgrave Macmillan, 2014), xiii–xiv.

27 Fa-ti Fan, "The Global Turn in the History of Science," *East Asian Science, Technology and Society* 6, no. 2 (January 2012): 249–58.

28 Pratik Chakrabarti, *Materials and Medicine: Trade, Conquest and Therapeutics in the Eighteenth Century* (Manchester: Manchester University Press, 2010), has highlighted the fragility of understanding science through networks and circulation, as much of the history is of rejection, violence, and marginalisation.

29 Sujit Sivasundaram, "Introduction," *Isis* 101, no. 1 (2010): 95–7.

30 I am referring particularly to the focus issue in *Isis*: Sivasundaram, ed. "Global Histories of Science."

31 By "global history," I mean how that discipline is represented in Western Europe and North America. Global history in China or South Asia would mean quite the opposite, i.e., non-Chinese or non-South Asian history, respectively.

32 Bernard Lightman, Gordon McOuat, and Larry Stewart, eds., *The Circulation of Knowledge between Britain, India and China: The Early-Modern World to the Twentieth Century* (Leiden: Brill, 2013).

33 Raj, *Relocating Modern Science* and Roberts, "Situating Science in Global History."

34 Atmospheric scientist Paul J. Crutzen and biologist Eugene F. Stoermer coined the term "Anthropocene" in 2000 to refer to the current epoch of geological and historical time in which humans have become the most dominant geological force shaping global ecology, geology, and atmosphere. Paul J. Crutzen, "Geology of Mankind," *Nature* 415, no. 6867 (2002): 23. Historians have adopted this frame and projected it backwards into writing histories of Anthropocene; see Will Steffen, Jacques Grinevald, Paul Crutzen, and John McNeill, "The Anthropocene: Conceptual and Historical Perspectives," *Philosophical Transactions of the Royal Society A* 369, no. 1938 (2011): 842–67; Dipesh Chakrabarty, "The Climate of History: Four Theses," *Critical Inquiry* 35, no. 2 (2009): 197–222; Naomi Oreskes, "The Scientific Consensus on Climate Change," *Science* 306, no. 5702 (2004): 1686.

35 The point is made by Kathryn Yusuff, *A Billion Black Anthropocenes or None* (Minneapolis: University of Minnesota Press, 2018). Pratik Chakrabarti has traced the colonial origin of the idea of Anthropocene in the dual frame of enchantment with the deep history of nature and the disenchantment with its concomitant economic quotient, *Inscriptions of Nature; Geology and the Naturalization of Antiquity* (Baltimore, MD: Johns Hopkins University Press, 2020), 190.

36 For example, Kumarasamy Thangaraj et al. and Vincent Macaulay et al. conducted DNA research among tribal populations in the Andaman islands and the Malayan peninsula respectively, to identify the routes of the earliest human migration out of Africa; see K. Thangaraj et al., "Reconstructing the Origin of Andaman Islanders," *Science* 308 (2005): 996; Vincent Macaulay et al., "Single, Rapid Coastal Settlement of Asia Revealed by Analysis of Complete Mitochondrial Genomes," *Science* 308 (2005): 1034–36.

37 Sadie Bergen, "What Are You? Historians Confront Race, Genealogy, and Genetics," *Perspectives on History*, February 2018, www.historians.org/publications-and-directories/perspectives-on-history/february-2018/what-are-you-historians-confront-race-genealogy-and-genetics.

38 For example, in the recent volume, almost all the examples of contemporary challenges of trust, empiricism, violence, and climate risk emerge from West European and North American experiences, Naomi Oreskes, *Why Trust Science?* (Princeton, NJ: Princeton University Press, 2019). Similarly, see David Sepkoski's recent book for a similar geographical scope for deliberations on the general intellectual and social roots of planetary climate change and ecological disaster, *Catastrophic Thinking; Extinction and the Value of Diversity from Darwin to the Anthropocene* (Chicago: University of Chicago Press, 2020).

39 Carla Nappi, "Paying Attention: Early Modern Science beyond Genealogy," *Journal of Early Modern History* 21, no. 5 (October 2017): 459–70.

40 Ken Arnold, *Cabinets for the Curious: Looking Back at Early English Museums* (Aldershot: Ashgate, 2006). Susan Scott Parrish, *American Curiosity: Cultures of Natural History in the Colonial British Atlantic World* (Chapel Hill: University of North Carolina Press, 2006); James Delbourgo, "Divers Things: Collecting

the World under Water," *History of Science* 49 (2011): 149–85; Pamela H. Smith and Paula Findlen, eds., *Merchants and Marvels: Commerce, Science, and Art in Early Modern Europe* (New York: Routledge, 2002). Also see Kay Dian Kriz, "Curiosities, Commodities, and Transplanted Bodies in Hans Sloane's 'Natural History of Jamaica,'" *The William and Mary Quarterly* 57 (2000): 35–78.

41 Matthew Cobb, "Malpighi, Swammerdam and the Colourful Silkworm: Replication and Visual Representation in Early Modern Science," *Annals of Science* 59 (2002): 111–47.

42 Winterbottom, *Hybrid Knowledge*; Raj, *Relocating*; Nair, *Raja Serfoji*; and Roberts, *The Brokered World*.

43 Brown, *Hindu Perspectives*.

44 Pratik Chakrabarti and Joydeep Sen, "'The World Rests on the Back of a Tortoise'; Science and Mythology in Indian History," *Modern Asian Studies* 50 (2016): 808–40.

45 Luis Millones Figueroa, "The Bezoar Stone: A Natural Wonder in the New World," *Hispanófila* 171 (2014): 139–56.

46 Marcia Stephenson, "From Marvelous Antidote to the Poison of Idolatry: The Transatlantic Role of Andean Bezoar Stones during the Late Sixteenth and Early Seventeenth Centuries," *Hispanic American Historical Review* 90 (2010): 3–39.

47 Londa Schiebinger, *Plants and Empire: Colonial Bioprospecting in the Atlantic World* (Cambridge, MA: Harvard University Press, 2004), 150–93.

48 Abena Dove Agyepoma Osseo-Asare, *Bitter Roots: The Search for Healing Plants in Africa* (Chicago: University of Chicago Press, 2014).

49 Markku Hokkanen, "Imperial Networks, Colonial Bioprospecting and Burroughs Wellcome & Co.: The Case of *Strophanthus Kombe* from Malawi (1859–1915)," *Social History of Medicine* 25, no. 3 (August 2012): 589–607.

50 Michael Worboys and Paolo Palladino, "Science and Imperialism," *Isis* 84 (1993): 84–102, here p. 102.

51 James Delbourgo, "The Knowing World: A New Global History of Science," *History of Science* 57, no. 3 (September 2019): 373–99.

52 "How Do You Decide When a Statue Must Fall?" *BBC Magazine* 9 November 2017, www.bbc.co.uk/news/magazine-41904800; Anna Fazackerley, "UCL Launches Inquiry Into Historical Links with Eugenics," *Guardian* 6 December 2018, www.theguardian.com/education/2018/dec/06/ucl-launches-inquiry-into-historical-links-with-eugenics; Sean Coughlan, "Cambridge Investigates its Slavery Links," 30 April 2019, www.bbc.co.uk/news/education-48097051.

3

CARTOGRAPHY AND EMPIRE FROM EARLY MODERNITY TO POSTMODERNITY

Thomas Simpson

Scholars of the past 40 years have expanded our definitions of "a map" and, to a lesser extent, "an empire." Whereas their predecessors confined themselves to "plane figure[s] representing the surface of the earth, or part of it,"[1] most historians of cartography now work with Brian Harley and David Woodward's famous redesignation of maps as "graphic representations that facilitate a spatial understanding of things, concepts, conditions, processes, or events in the human world."[2] What counts as an empire has been subject to a more diffuse array of challenges and provocations. Whether reminders of the ubiquity and centrality of empires in world history,[3] attempts to decouple empire from centralised states and sedentarised subjects,[4] or efforts to move beyond a "hub-and-spoke" model of imperial metropoles and colonial peripheries,[5] these interventions fragment a unitary model of empire based on the modern maritime powers of Western Europe. I work with this global history approach to empire to show that mapping is better understood as a broadly imperial activity rather than a narrowly European one.

Pluralising our basic definitions carries some risks, not least the loss of a coherent and compelling explanation of the relation between cartography and empire. Brian Harley opted to tell a straightforward tale of the complicity of spatial representation and imperial power: "as much as guns and warships, maps have been the weapons of imperialism."[6] Although heavily indebted to him, scholars today tend to draw more complicated conclusions. This chapter's broadly chronological overview of imperial cartographies from the fifteenth century to the present demonstrates that there are fruitful ways of understanding the co-constitution of cartographies and empires without resorting to a simplistic equation of knowledge and power.[7] Cartography relied not just on surveying imperialists but also the knowledge of empires' subjects, extracted through various encounters—often violent, sometimes more mutual. Maps shaped how empires governed, grew, and declined; empires shaped what maps displayed, their material forms, and who used them. But these processes took diverse forms not only between but also within empires, often confounding the expectations of the many people—colonisers and colonised, producers and viewers—who had a stake in maps. Imperial mapmakers were not necessarily masters of all they surveyed, and imperial subjects were not uniformly disempowered through cartography.

Imperial world maps, c.1400–c.1600

The fifteenth and sixteenth centuries saw intense activity from a host of Asian powers—including Ming China and the Ottomans, Mughals, and Safavids—and from upstart maritime empires in Western Europe. Pushing against the notion that European global hegemony stretched back to Spanish and Portuguese voyaging either side of the turn of the sixteenth century, historians in recent years have shown that this imperial explosion was polycentric and involved overlaps and borrowings between empires.[8] Cartography was no exception. As well as reshaping the world in a political sense, the imperial dynamics of this era reshaped how it was known and represented. This shift of cartographic consciousness is among the developments that allow us to talk of an early-modern era distinct from the medieval period. Older explanations of early-modern global perceptions exaggerated the role of "discoveries" of new lands and routes, especially European contact with the Americas.[9] Less celebrated and more gradual processes, especially the intensification of communications along already-established Eurasian trade networks, were at least as significant.

The most conspicuous changes in spatial representations during this era were in world maps. The first such map produced in East Asia, known as the Kangnido map, was completed in 1402 (Figure 3.1). Although the creation of the recently minted Chosŏn dynasty in Korea, it relied on cartographic traditions in neighbouring Ming China—an influence apparent in its depiction

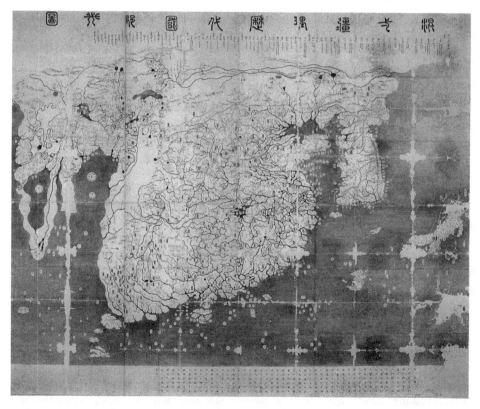

Figure 3.1 Kangnido map of East Asia, originally from 1402. "Honkōji" version rendered c. 1560. Public domain, via Wikimedia Commons.

of this polity as a preponderant, universal empire.[10] The map's novel representation of Europe and Africa possibly derived from the *Geography* of the Greek astronomer of the second century CE Claudius Ptolemy, which was rediscovered and reproduced (and probably embellished) in thirteenth-century Byzantium.[11] Elements of Ptolemy's work could have spread to East Asia through long-running intellectual and material exchanges with the Middle East.[12] Chinese cartography, however, was a vibrant tradition capable of incorporating external influences on its own terms, not a product of the eastward diffusion of European science. In fact, Ptolemy substantially emerged in Western European cartography only later in the fifteenth century, and through the mediation of Sultan Mehmed II's Ottoman Empire (1451–1481), which occupied a central position in the thriving traffic of cartographic objects throughout the Mediterranean.[13] Alongside collecting Islamic geographical works from Western and Central Asia,[14] Mehmed commissioned maps based on the copies of Ptolemy's *Geography* that he obtained having conquered Constantinople in 1453.[15] In Christian Europe, unearthing Ptolemy combined with Iberian imperial expansion and the development of print technology to generate major changes in the comprehension and depiction of terrestrial space among a specialist elite. Particularly significant was that, contrary to medieval *mappae mundi*, European world maps from the later fifteenth century were no longer structured around religious iconography.

Maritime exploration went hand in hand with cartographic innovation in Ming China and the Iberian and Ottoman Empires. Information on the Indian Ocean littoral from voyages during the early fifteenth century led by Zheng He was rendered in Chinese charts.[16] The Ottoman naval commander Piri Reis, the leading agent of imperial expansion into the eastern Mediterranean and Persian Gulf, drew on his own observations and on Portuguese and Indian nautical charts to supplement a Ptolemaic framework and produce a world map in 1513.[17] And Portuguese navigators found that surveying and representational techniques used in the compact Mediterranean basin were inadequate for long-distance oceanic voyaging as they journeyed ever further south along the western coast of Africa during the second half of the fifteenth century. In place of the embodied "shipboard perspective" of plotting rhumb lines that enabled sailors to follow a fixed bearing relative to magnetic north, they employed instruments developed in the Islamic world such as astrolabes and quadrants to determine latitude, based on the abstract division of the entire planet.[18] Within 50 years, calculating longitude had also become a pressing if intractable issue as a result of Spanish and Portuguese efforts to delineate separate hemispheres of activity for their trading empires.[19]

This shift to global-scale abstraction was one way in which, having been rooted in and reliant upon cartographic developments that spanned Eurasia, early-modern Iberian Empires developed distinctive ways of knowing and representing space in tandem with their particular form of maritime trade. Another was the advent in the early sixteenth century of centralised agencies for accumulating and combining survey data and distributing maps and charts to sailors, merchants, and administrators. The operations of these institutions were, however, far from seamless. Data remained patchy and ships' captains proved to be variably reliable informants.[20] Mapmakers' status as independent craftsmen rather than salaried officials meant that cartographic information also leaked out across Western Europe rather than remaining confined within these "centres of calculation."[21] And many traders and colonisers continued throughout the sixteenth century to use traditional representations of space, such as rutters (written maritime navigation manuals), rhumb-line charts, and itinerary maps, rather than the official charts with latitude and longitude graticules.[22] In terms of the limited penetration of innovative means of configuring global space and the continued importance of older forms of representation for most practical purposes, the Iberian Empires were similar to their land-based counterparts across Eurasia during the sixteenth century.[23] Cartographic practices changed in much-celebrated ways between 1400 and 1600.[24] But these shifts were incremental and uneven, and a host of prior traditions continued to structure how most imperial agents understood space.

Mapping early-modern possessions

European maritime powers' establishment of overseas settler colonies and Eurasian land empires' expansion into continental hinterlands generated strong imperatives for mapping land and coasts at scales below the global. Although survey and representational techniques varied substantially over time and between and within empires, recent scholarship suggests that some common features remained surprisingly persistent from the sixteenth to the twentieth centuries. Although they often erased or underplayed the presence of colonised populations, imperial maps were thoroughly dependent on various people beyond colonising elites. There were also major material and political challenges in the field and variable practices of map publication and distribution, all of which meant that depictions of imperial space were highly diverse and bore signs of the specific circumstances of their production. While some mapmakers evinced aspirations to consistent or even comprehensive spatial knowledge, the products of their ventures were invariably more piecemeal.

Early-modern Spanish and Portuguese cartographers relied on local knowledge for their maps and charts. Whereas the Portuguese were prepared to admit this fact and celebrate their "hybrid cultural artefacts," the Spanish sought more uniform data on their growing possessions in the Americas.[25] As Barbara Mundy details in *The Mapping of New Spain*, among the most famous—and most troubled—attempts to map the "New World" was a questionnaire devised in the 1570s by the cartographer Juan López de Velasco at the behest of the Spanish king Philip II. De Velasco intended to collate responses from colonial officials in New Spain (later Mexico) into a unified map and text, but the results thoroughly confounded his aims. Since administrators seemed ill equipped—in terms of knowledge of their domains, cartographic expertise, and commitment to maps as a representational form—to render the requested images, they instead contracted Indigenous painters. Many of the painters drew on older representational modes, expressing "ideas of community" rather than focusing on the built and topographical environments that de Velasco sought. Although they bore signs of Spanish colonial dispossession through land grants and the imposition of new place names, the images' unmanageable array of styles and details rendered them unusable as cartographic instruments of imperial power.[26]

During the seventeenth century, some empires developed more successful strategies for incorporating local information into cartographic products that suited their form of expansion. Although lacking central coordination, maps of English settler colonies of North America tended to inscribe European modes of land ownership and place naming that overturned the spatial logics of the same Indigenous communities upon whose knowledge they relied.[27] The trade-focused Dutch Empire, comprising the United East India Company (VOC, established 1602) in the Indian Ocean and the Chartered West India Company (WIC, established 1603) in the Atlantic Ocean, developed a somewhat distinct cartographic culture. Mapmakers attached to these companies built upon the surveying practices of their Iberian predecessors, using de Velasco's questionnaire as a model for acquiring information on the hinterlands beyond Dutch outposts along the eastern coast of the Americas. Aware of the limitations of Spanish attempts at cartographic centralisation, both Dutch companies allowed for greater local flexibility and improved incorporation of Indigenous knowledge by establishing a hub for mapping in the colonies—the VOC in Batavia, the WIC in Recife, Brazil.[28]

The Russian Empire, which expanded rapidly during the same period but across contiguous land rather than distant littoral *entrepôts*, also made extensive use of local knowledge

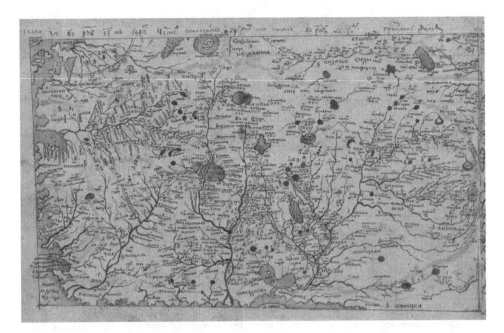

Figure 3.2 Semyon Remezov, manuscript map of Siberia, Khorograficheskaya kniga, 1697.
MS Russ 72 (6), f. 162v. Houghton Library, Harvard University.

for mapping. Semyon Remezov, a soldier-administrator based east of the Urals in the later seventeenth century, rendered some of the first Russian representations of its recently acquired empire using a self-taught cartographic practice that incorporated Siberian traditions. Forgoing a graticule of longitude and latitude, he instead structured his depiction of Siberia around the river networks that were critical to empire building and trading in the region (Figure 3.2).[29]

So, despite a shared reliance on Indigenous knowledge, these three divergent seventeenth-century imperial formations—the English establishing agricultural colonies across the Atlantic, the Dutch comprising a series of coastal outposts, the Russian dominating a continuous but varied landmass—developed distinctive cartographies. The English generally tried to occlude Indigenous presence and depict *terra nullius*—emptying the land symbolically as an adjunct to its being emptied by physical violence and legal dispossession. By contrast, its loose network of far-flung colonial *entrepôts* was part of the reason that, by 1700, in Benjamin Schmidt's words, "Dutch geography tolerated—encouraged even—variety."[30] As Valerie Kivelson shows, maps and geographical texts emanating from Russia's peripheries, such as Remezov's, were crucial components in the seventeenth-century imperial ideology of Russia as an "in-between" empire, spanning the fringe regions of Europe and Asia.[31]

At the southern outskirts of Russia's newly mapped Siberian territories, rapidly changing cartographies played a central role in a key region of imperial contestation. As the seventeenth century drew to a close, it became evident to agents of the expansionary Russian, Chinese, and Mongol Zunghar Empires that they were competing with each other for control over populations, resources, and land at the heart of Asia.[32] Along with contemporaneous contests such as the tussle between the English, French, and Dutch in eastern North America,[33] this clash of empires generated what Laura Hostetler terms "a flurry of map-making activity," leading to a new phase in "the global integration of space."[34] Every bit as much as Portugal and Spain's hemispheric division of the globe approximately 200 years earlier, this era of great empires making and breaking shared borders supercharged the significance of mapmakers and their images. Even the nomadic Zunghars, without a prior cartographic tradition, depicted their region in order to compete with Russia and China. The Zunghar Khan supposedly rendered one map himself, staking territorial claims by clearly labelling his realm but also incorporating features appropriate to an empire based on fluidity, including the omission of defined political boundaries and a clear frame at the map's edges.[35]

Competition in this region was a key factor in an unprecedented shift in Qing China's cartographic practices. Like his Russian contemporary Peter the Great, the Kangxi emperor (r. 1661–1722) employed international personnel to execute state-of-the-art maps of his expanding, multi-ethnic dominions.[36] In the Qing case these outsiders were French Jesuit missionaries.[37] Their decade-long surveys, executed in collaboration with Qing officials and local informants, resulted in an atlas, *Huangyu quanlan tu* (*Overview Maps of the Imperial Territories*), produced in multiple editions from 1717.[38] European methods did not simply override East Asian traditions. Instead, there was a "convergence of interests" among the key actors, with the Jesuits' instrumental surveying fitting with the Chinese literati's turn towards *kaosheng* (evidential learning).[39] As of the mid eighteenth century, diverse traditions of survey and representation remained vital elements of imperial cartographies, especially those used for day-to-day administrative tasks. But the publication and popularity across Europe during the 1730s of only slightly amended versions of the maps from the *Huangyu quanlan tu* is one indication that, at least for overview maps, cartographic practices were increasingly shared across the leading imperial states.[40]

Cartographic power in the age of European imperialism

The century from 1650 to 1750 can be broadly characterised as a period in which the making and breaking of inter-imperial borders drove a limited form of global cartographic integration. The following hundred years or so saw the making and breaking of empires more broadly.[41] The great Eurasian empires—Qing, Ottoman, and Mughal—continued or began to decline, while European empires faced challenges from settler communities in the Americas and reconfigured their global dominions and trade interests. Cartographies reflected and shaped these trends. Expanding European empires, especially the French and British, innovated mapping technologies of unprecedented accuracy and scope that became widely recognised as essential elements of modern government. Imperial maps generally became less pictorial and more consistent in suppressing local variation and Indigenous presence. But spatial representations of this era were far from uniform or infallible weapons of empire. Practices of production in the field and in the cartographer's office continued to be highly diverse, with continued reliance on various local and non-elite actors that often remained visible in published maps. Many survey data and representations laboriously collected and collated in the service of empire were ignored, contested, or reworked in unexpected ways by unexpected people.

While they flourished throughout much of the Indian and Pacific Ocean worlds and their continental hinterlands, during the century from 1750 European empires and their cartographic projects faltered in the Americas. Britain was unable to prevent its 13 American colonies from achieving independence in 1783, and Spain faced a widespread and prolonged crisis in the face of discontent from its colonial subjects. As Magali Carrera shows, maps were a crucial element in both anti-imperial assertions and attempts to reconstitute imperial authority. From the 1750s, the government in New Spain attempted to impose spatial order through sophisticated surveys. These projects prompted elites in the colony to undertake counter-mapping efforts in a very different form—narrative itineraries that instead conceived of territory as affinity with the physical environment.[42] In New Spain's northern reaches, colonial authorities faced another oppositional spatial logic from a regional power that Pekka Hämäläinen classes as an "indigenous empire." The Comanches—nomadic groups operating in the region that now forms the southwestern United States—sought "not to conquer and colonise, but to coexist, control, and exploit." They were able to manipulate the Spanish possessions of New Mexico and Texas for decades either side of the turn of the nineteenth century precisely because they refused to abide by Euro-American logics of cartographic fixity.[43] The Comanches developed an imperial formation based on movement and fluidity, avoiding maps at a time when neighbouring empires and nation-states were embracing them.[44] Having been assailed by these internal and external opponents, Mexico gained independence in 1821—a salient example of the variable power of imperial maps.

At this juncture, Iberian colonies in South America became sites in which substantially new forms of cartography developed. Neil Safier demonstrates that encounters with indigenes in the forests and mountains of the South American interior "unsettled" and "complicated" European efforts to know space and people.[45] Surveyors and explorers remained thoroughly dependent on local assistance and information, but simultaneously made concerted attempts to omit any signs of these people from geographical maps.[46] Safier argues, though, that this was not as one-dimensional a process of elision and removal as the likes of Brian Harley and Mary Louise Pratt suggested in work of the 1980s and early 1990s.[47] Rather than being silenced entirely, Indigenous people were symbolically displaced to other media, such as population maps and ethnographic prose and poetry. The work of the most influential cartographic innovator who journeyed to South America at this time, Alexander von Humboldt, typified these tensions between Indigenous absence and presence in European spatial representations. Humboldt's surveys, narratives, and biogeographical representations were products of his experiences in South America rather than mere exports from Europe. However, he overlooked both Indigenous and colonial testimonies when they worked against his own conjectures, and intermittently removed local inhabitants when presenting the region as one dominated by the immensity of nature.[48] Despite his ambivalent views of their knowledge structures and nation-building efforts, South American elites harnessed Humboldt to their aspirations for independence from Spain.[49] Once again, imperial cartography was open to renegotiation from diverse scientific and political agendas.

The elements that influenced mapping in colonial South America—simultaneous dependence on and elision of indigenes, creating separate representations for distinct fields of knowledge, and appropriating imperial cartographies for varied and often oppositional ends—also occurred in a host of other arenas of modern empire. Processes of gleaning and repackaging local knowledge were often most evident at the fringes of expanding empires. During the late eighteenth and early nineteenth centuries, British, French, Russian, and American voyages throughout the Pacific Ocean elicited geographical information and map images from islanders and coast dwellers.[50] Sakhalin Island, located between Japan and the Chinese and Russian Empires in the northwest Pacific, has proved a particularly productive site for competing recent theories of cartography and empire. Focusing on a French voyage to Sakhalin in 1787, Michael Bravo argues against Bruno Latour's notion of imperial maps as "immutable mobiles" that remained stable and intelligible across space and time.[51] Bravo instead demonstrates that incorporating islanders' knowledge into French maps was achievable only through a laborious process based on delicate encounters and mistranslations.[52] Sakhalin also reminds us that Westerners were not the only imperial players creating cartographic artefacts in and of these littoral spaces. In 1808 Japanese explorer Mamiya Rinzō was able to obtain geographical information from inhabitants of Sakhalin by employing translators unavailable to French voyagers to the island 20 years earlier. Mamiya's map followed European norms in symbolically emptying Sakhalin of the very Indigenous inhabitants whose cooperation enabled its construction, displacing them to an accompanying ethnographic tract. Like maps of southern Pacific islands from Captain James Cook's voyages of the 1760s and 1770s that combined local and British naval navigation expertise,[53] Mamiya's work "anticipated empire" by staking sovereign claims and providing a practical tool for Japanese expansion in the northern Pacific over the following century.[54]

Even as their cartographic institutions became increasingly grandiose during the nineteenth century, European empires continued to depend on exploiting the knowledge and labour of colonised people in making maps.[55] In British India, the monumental Survey of India overlaid the colony with a triangulated grid on an unprecedented scale (Figure 3.3).[56] Maps facilitated the extraction of taxes, imposition of colonial regimes of law and property, and spread of communications networks crucial to British rule.[57] Cartography in the imperial subcontinent also pursued prestigious scientific goals, such as accurately measuring the shape of the earth and

Figure 3.3 Index chart to the Great Trigonometrical Survey of India, 1870. Public domain, via Wikimedia Commons.

calculating the altitude of Himalayan peaks. Practices developed here were exported to other imperial realms, including the Middle East,[58] and became increasingly influential in the metropole.[59] Nonetheless, the Survey of India had limitations. Much of its work was, Matthew Edney contends, "haphazard" and "anarchic,"[60] and its conventional methods proved inadequate in the face of material and political challenges in mountains, deserts, and forests.[61] Here, British surveyors resorted to different techniques, including training and deputing so-called "native explorers" or "pundits," mostly from the fringes of the colonial subcontinent, to conduct route surveys. Racialised assumptions along with misgivings over the efficacy of their irregular instrumentation meant that these men were distrusted even as their work fed into the maps of an expanding colony.[62]

Similar paradoxes and inconsistencies marked French imperial cartography from Napoleon's expedition to Egypt between 1798 and 1801, to Southeast Asia and West Africa in the later nineteenth and early twentieth centuries.[63] Recent francophone scholarship highlights the conflicting aims of different cartographic circles within the French Empire.[64] For the British, too, maps were often sources and conduits of controversy. A prime example was competing representations of the River Niger in West Africa during the early nineteenth century. Drawn from heterogeneous sources, including slave testimony in the Caribbean, European explorers' accounts, and classical texts, these representations fuelled fierce debates over the river's course.[65] In this case as in many others, maps were not so much weapons of domination as expressions of dissensus between competing theories, interests, and institutions.

Given the varied circumstances of surveying and equally diverse representations that resulted, what power did cartography have in the era of high imperialism either side of the turn of the twentieth century? Many scholars now downplay the impact of imperial maps, going against the analyses of Harley's generation. Daniel Foliard's *Dislocating the Orient* typifies this turn, proclaiming that "most [British] maps of the Middle East were lacking in the characteristics which might have given them actual power over the territories they were supposed to represent … [and] nothing could be done effectively without local assistance."[66] Nonetheless, as Foliard and others suggest, many maps had significant communicative potential and enabled certain types of intervention and categorisation. Maps and globes were pedagogical devices closely linked to notions of modernity in imperial settings far beyond European metropoles.[67] Thematic maps allowed for more penetrating imperial interventions in realms from the military to the medical. Deborah Coen demonstrates that the Habsburg Empire underwent a "cartographic boom" during this period, as surveys and depictions of the distribution of elements such as language, architectural artefacts, and climate were harnessed to an imperial ideology of "unity in diversity."[68] In a similar vein, Timothy Mitchell contends that British maps of Egypt around the turn of the twentieth century were crucial to formulating "the economy" as a distinct object that could be acted upon by governments.[69] Following their advent in the first half of the nineteenth century, thematic maps became major tools of intervention and communication among empires as well as nation-states, enabling substantially new ways of exercising state control.[70]

Conclusion: imperial maps in the postcolonial era

Maps offered languages of anti-imperial protest as well as of imperial power. In East Central Europe during the late nineteenth century, advocates of Russian and Habsburg imperial consolidation shared with those pushing contrary agendas of national independence "the need for lavishly decorated cartouches expressing civilization, and 'scientific' maps and atlases providing information and symbols of national belonging and state power."[71] Similar cartographic borrowings marked the worldwide swell of anti-imperialism from the late nineteenth to the mid twentieth century. British India's spatial outline, reinterpreted in the gendered and divine form of "Mother India," circulated widely across the subcontinent. Such representations, Sumathi Ramaswamy comments, were "cheekily reliant on the [colonial] state's cartographic productions" even as they destabilised British rule by inspiring a vision of national territory as worthy of devotion and sacrifice.[72] Wresting the cartographic resources of modern empires to oppositional ends tended to be a delicate enterprise. Even as maps offered "means to challenge particular ways of 'thinking' the world," imperial modes of knowing and representing space maintained their allure and influence.[73]

Has cartography been decolonised? As Raymond Craib suggests, mapping can appear to tell us that "forms of domination have been adjusted rather than abolished."[74] New forms of knowing and manipulating terrestrial space developed through Cold War conflicts between superpowers whose detractors, at least, labelled them "empires."[75] As Daniel Immerwahr shows in the case of the omission of overseas territories from the United States' "logo map," elisions are essential means of "hiding" imperial possessions and ambitions in ostensibly postcolonial times.[76] Ever more portable, malleable, and ubiquitous cartographic technologies such as GPS have transitioned from military instruments to become integral to the forms of power exercised by global corporations and markets—today's postmodern "Empire" according to Michael Hardt and Antonio Negri.[77] At the same time, cartographies are major means of resisting power in the contemporary world, as the recent proliferation of counter-mapping projects drawing on these same technologies indicates.[78]

Cartography continues to make, break, and shape empires, as it has since the fifteenth century. From earliest early modernity to postmodernity, imperial formations have developed in conjunction with particular cartographies. Maps have been—and remain—essential instruments of imperial violence, extraction, and dispossession, facilitating vast inequalities and iniquities. They have also never been under the exclusive control of narrowly defined imperial elites, but reliant on the knowledge and labour of subject peoples and at least partially open to unexpected and resistant repurposing. Scholarship of recent decades has reminded us of the need to investigate spatial representations' power as a variable and relative quality rather than one to be taken for granted. Work during the coming years should pay attention to the many uses and receptions of maps to enable an even richer understanding of that most imperial of sciences, cartography.

Notes

1 J.L. Lagrange, quoted (approvingly) in Leo Bagrow, *The History of Cartography*, ed. R.A. Skelton (London: C.A. Watts, 1964), 22.

2 J.B. Harley and David Woodward, "Preface," in *The History of Cartography, Volume One: Cartography in Prehistoric, Ancient, and Medieval Europe and the Mediterranean*, ed. J.B. Harley and David Woodward (Chicago: University of Chicago Press, 1987), xv–xxi, here pp. xvi–xvii.

3 For example, Jane Burbank and Frederick Cooper, *Empires in World History: Power and the Politics of Difference* (Princeton, NJ: Princeton University Press, 2010).

4 For example, Pekka Hämäläinen, *The Comanche Empire* (New Haven, CT: Yale University Press, 2008).

5 For example, Dominic Lieven, *Empire: The Russian Empire and Its Rivals* (London: Pimlico, 2002), xiv–xv; Mark Harrison, "Science and the British Empire," *Isis* 96, no. 1 (2005): 56–63.

6 J.B. Harley, "Maps, Knowledge, and Power," in *The Iconography of Landscape: Essays on the Symbolic Representation, Design and Use of Past Environments*, ed. Denis Cosgrove and Stephen Daniels (Cambridge: Cambridge University Press, 1988), 277–312; reprinted in J.B. Harley, *The New Nature of Maps: Essays in the History of Cartography*, ed. Paul Laxton (Baltimore, MD: Johns Hopkins University Press, 2001), 51–81, here p. 57.

7 For critical reflections on Harley's work that deal substantially with his model of power, see "Deconstructing the Map: 25 Years On," ed. Reuben Rose Redwood, special issue, *Cartographica* 50, no. 1 (2015): 1–57.

8 On the Eurasian empires, see Stephen Frederic Dale, *The Muslim Empires of the Ottomans, Safavids, and Mughals* (Cambridge: Cambridge University Press, 2010). For a general overview of empires in world history during this period, see John Darwin, *After Tamerlane: The Rise and Fall of Global Empires, 1400–2000* (London: Penguin, 2007), 47–99.

9 On this point, see Jerry Brotton, *Trading Territories: Mapping the Early Modern World* (London: Reaktion, 1997), 26.

10 Jerry Brotton, *A History of the World in Twelve Maps* (London: Allen Lane, 2012), 114–45.

11 O.A.W. Dilke, "Cartography in the Byzantine Empire," in Harley and Woodward, *History of Cartography: Volume One*, 258–75, here pp. 267–72.

12 Kuei-Sheng Chang, "Africa and the Indian Ocean in Chinese Maps of the Fourteenth and Fifteenth Centuries," *Imago Mundi* 24 (1970): 21–30.

13 Brotton, *Trading*, 100–3.

14 Karen Pinto, "The Maps Are the Message: Mehmet II's Patronage of an 'Ottoman Cluster,'" *Imago Mundi* 63, no. 2 (2011): 155–79.

15 Brotton, *Trading*, 34–5.

16 Cordell D.K. Yee, "Reinterpreting Traditional Chinese Geographical Maps," in *The History of Cartography: Volume Two, Book Two: Cartography in the Traditional East and Southeast Asian Societies*, ed. J.B. Harley and David Woodward (Chicago: University of Chicago Press, 1994), 35–70, here pp. 52–3.

17 Brotton, *Trading*, 107–8. On Ottoman expansion overseas during the sixteenth century, see Giancarlo Casale, *The Ottoman Age of Exploration* (Oxford: Oxford University Press, 2010).

18 Denis Cosgrove, *Apollo's Eye: A Cartographic Genealogy of the Earth in the Western Imagination* (Baltimore, MD: Johns Hopkins University Press, 2001), 85; Ricardo Padrón, "Mapping Plus Ultra: Cartography, Space, and Hispanic Modernity," in *Empires of Vision: A Reader*, ed. Martin Jay and Sumathi Ramaswamy (Durham, NC: Duke University Press, 2014), 211–45, here pp. 231–5; Brotton, *Trading*, 46–60.

19 Brotton, *Trading*, 119–50; Barbara E. Mundy, *The Mapping of New Spain: Indigenous Cartography and the Maps of the Relaciones Geográficas* (Chicago: University of Chicago Press, 1996), 13–14.

20 Laura Hostetler, "Mapping, Registering, and Ordering: Time, Space, and Knowledge," in *Oxford World History of Empire*, ed. Peter Bang, Christopher Bayly, and Walter Scheidel (Oxford: Oxford University Press, forthcoming).

21 Kees Zandvliet, *Mapping for Money: Maps, Plans and Topographic Paintings and Their Role in Dutch Overseas Expansion during the 16th and 17th Centuries* (Amsterdam: Batavian Lion, 1998), 255–6. On "centres of calculation," see Bruno Latour, *Science in Action: How to Follow Scientists and Engineers through Society* (Cambridge, MA: Harvard University Press, 1987), 215–57.

22 Felipe Fernández-Armesto, "Maps and Exploration in the Sixteenth and Early Seventeenth Centuries,'" in *The History of Cartography: Volume Three: Cartography in the European Renaissance, Part One*, ed. David Woodward (Chicago: University of Chicago Press, 2007), 738–59, here pp. 749–55; Padrón, "Mapping," 223–4.

23 Yee, "Reinterpreting," 55–67; Susan Gole, *Indian Maps and Plans: From Earliest Times to the Advent of European Surveys* (New Delhi: Manohar, 1989); Giancarlo Casale, "Seeing the Past: Maps and Ottoman Historical Consciousness," in *Writing History at the Ottoman Court: Editing the Past, Fashioning the Future*, ed. H. Erdem Çıpa and Emine Fetvacı (Bloomington: Indiana University Press, 2013), 80–99.

24 On the cultural impact of new modes of cartography, see for example Neil Safier and Ilda Mendes dos Santos, "Mapping Maritime Triumph and the Enchantment of Empire: Portuguese Literature of the Renaissance," in Woodward, *History of Cartography: Volume Three, Part One*, 461–8; Brotton, *Trading*, 19–25; Robert C.D. Baldwin, "Colonial Cartography under the Tudor and Early Stuart Monarchies, ca. 1480–ca. 1640," in *The History of Cartography: Volume Three: Cartography in the European Renaissance, Part Two*, ed. David Woodward (Chicago: University of Chicago Press, 2007), 1754–80.

25 Brotton, *Trading*, 81–2.

26 Mundy, *Mapping*; David Buisseret, "Spanish Colonial Cartography, 1450–1700," in Woodward, *History of Cartography: Volume Three, Part One*, 1143–71, here p. 1156.

27 J.B. Harley, "New England Cartography and the Native Americans," in *New Nature*, 169–95; John Rennie Short, *Cartographic Encounters: Indigenous Peoples and the Exploration of the New World* (London: Reaktion, 2009); Benjamin Schmidt, "Mapping an Empire: Cartographic and Colonial

Rivalry in Seventeenth-Century Dutch and English North America," *The William and Mary Quarterly* 54, no. 3 (1997): 549–78.

28 Zandvliet, *Mapping*, 86–209.

29 Valerie A. Kivelson, "'Between All Parts of the Universe': Russian Cosmographies and Imperial Strategies in Early Modern Siberia and Ukraine," *Imago Mundi* 60, no. 2 (2008): 166–81, here pp. 169–73.

30 Benjamin Schmidt, "Mapping an Exotic World: The Global Project of Dutch Geography, circa 1700," in *Empires of Vision*, 246–66, here pp. 259–60.

31 Kivelson, "Between." See also Valerie Kivelson, *Cartographies of Tsardom: The Land and its Meanings in Seventeenth-Century Russia* (Ithaca, NY: Cornell University Press, 2006).

32 Peter C. Perdue, "Boundaries, Maps, and Movement: Chinese, Russian, and Mongolian Empires in Early Modern Central Eurasia," *The International History Review* 20, no. 2 (1998): 263–86.

33 On maps as weapons in this struggle, see Christine M. Petto, "'Notorious Abuse,' 'Traditious Ignorance,' and 'Ambitious Incroachment' in Seventeenth-Century Dutch, English, and French Maps of North America," *Cartographica* 53, no. 4 (2018): 241–61.

34 Laura Hostetler, "Imperial Competition in Eurasia: Russia and China," in *The Cambridge World History. Volume VI: The Constructions of a Global World, 1400–1800 CE. Part 1: Foundations*, ed. Jerry H. Bentley, Sanjay Subrahmanyam and Merry E. Wiesner-Hanks (Cambridge: Cambridge University Press, 2015), 297–322, here p. 307. Hostetler quotes the latter phrase from Charles H. Parker, *Global Interactions in the Early Modern Age, 1400–1800* (Cambridge: Cambridge University Press, 2010).

35 Perdue, "Boundaries," 279–81.

36 Laura Hostetler, *Qing Colonial Enterprise: Ethnography and Cartography in Early Modern China* (Chicago: University of Chicago Press, 2001), 24, 35–41

37 Mario Cams, *Companions in Geography: East–West Collaboration in the Mapping of Qing China (c. 1685–1735)* (Leiden: Brill, 2017).

38 Mario Cams, "Not Just a Jesuit Atlas of China: Qing Imperial Cartography and Its European Connections," *Imago Mundi* 69, no. 2 (2017): 188–201, here pp. 192–3.

39 Cams, *Companions*, 6; Hostetler, *Qing*, 17, 58.

40 On European publication of maps from the *Huangyu quanlan tu*, see Mario Cams, "The China Maps of Jean-Baptiste Bourguignon d'Anville: Origins and Supporting Networks," *Imago Mundi* 66, no. 1 (2014): 51–69.

41 For an overview of this era, see C.A. Bayly, *The Birth of the Modern World, 1780–1914: Global Connections and Comparisons* (Oxford: Blackwell, 2004), 27–169.

42 Magali Carrera, "Entangled Spaces: Mapping Multiple Identities in Eighteenth-Century New Spain," in *Decolonizing the Map: Cartography from Colony to Nation*, ed. James R. Akerman (Chicago: University of Chicago Press, 2017), 72–109.

43 Hämäläinen, *Comanche*, 4–5, 181–238.

44 Pekka Hämäläinen, "What's in a Concept? The Kinetic Empire of the Comanches," *History and Theory* 52 (2013), 81–90; Rachel St. John, "Imperial Spaces in Pekka Hämäläinen's *The Comanche Empire*," *History and Theory* 52 (2013), 75–80.

45 Neil Safier, *Measuring the New World: Enlightenment Science and South America* (Chicago: The University of Chicago Press, 2008), 272.

46 See, for example, D. Graham Burnett, *Masters of All They Surveyed: Exploration, Geography, and a British El Dorado* (Chicago: University of Chicago Press, 2000), 182–9.

47 Neil Safier, "The Confines of the Colony: Boundaries, Ethnographic Landscapes, and Imperial Cartography in Iberoamerica," in *The Imperial Map: Cartography and the Mastery of Empire*, ed. James R. Akerman (Chicago: University of Chicago Press, 2009), 133–83, here pp. 156–65. Harley's classic statement on this subject is "Silences and Secrecy: The Hidden Agenda of Cartography in Early Modern Europe," *Imago Mundi* 40 (1988): 57–76. Pratt discusses the erasure of indigenes in her concept of European "anti-conquest" narratives: see Mary Louise Pratt, *Imperial Eyes: Travel Writing and Transculturation* (London: Routledge, 1992), 38–68.

48 Gregory T. Cushman, "Humboldtian Science, Creole Meteorology, and the Discovery of Human-Caused Climate Change in South America," *Osiris* 26, no. 1 (2011): 19–44; Safier, "Confines," 183; Pratt, *Imperial Eyes*, 121–34.

49 Pratt, *Imperial Eyes*, 138–41, 182–8.

50 On British mapping in western Canada, see Daniel W. Clayton, *Islands of Truth: The Imperial Fashioning of Vancouver Island* (Vancouver: UBC Press, 1999). On Russian mapping in the north and eastern

Pacific, see Bronwen Douglas and Elena Govor, "Eponomy, Encounters, and Local Knowledge in Russian Place Naming in the Pacific Islands, 1804–1830," *The Historical Journal* 62, no. 3 (2019): 709–40. On American hydrographers in the Pacific, see D. Graham Burnett, "Hydrographic Discipline among the Navigators: Charting an 'Empire of Commerce and Science' in the Nineteenth-Century Pacific," in Akerman, *Imperial Map*, 185–260. On mapping in Australia, see Paul Carter, *The Road to Botany Bay: An Exploration of Landscape and History* (New York: Alfred A. Knopf, 1988).

51 Latour, *Science*, 215–29.

52 Michael T. Bravo, *The Accuracy of Ethnoscience: A Study of Inuit Cartography and Cross-Cultural Commensurability* (Manchester: Manchester Papers in Social Anthropology, 1996); Michael T. Bravo, "Ethnographic Navigation and the Geographical Gift," in *Geography and Enlightenment*, ed. David N. Livingstone and Charles W.J. Withers (Chicago: University of Chicago Press, 1996), 199–235.

53 Margaret Jolly, "Imagining Oceania: Indigenous and Foreign Representations of a Sea of Islands," *The Contemporary Pacific* 19, no. 2 (2007): 508–45, here pp. 508–11; Lars Eckstein and Anja Schwarz, "The Making of Tupaia's Map: A Story of the Extent and Mastery of Polynesian Navigation, Competing Systems of Wayfinding of James Cook's *Endeavour*, and the Invention of an Ingenious Cartographic System," *Journal of Pacific History* 54, no. 1 (2019): 1–95.

54 Brett L. Walker, "Mamiya Rinzō and the Japanese Exploration of Sakhalin Island: Cartography and Empire," *Journal of Historical Geography* 33 (2007): 283–313.

55 Raymond B. Craib, "Relocating Cartography," *Postcolonial Studies* 12, no. 4 (2009): 481–90.

56 Matthew Edney, *Mapping an Empire: The Geographical Construction of British India, 1765–1843* (Chicago: University of Chicago Press, 1997).

57 Bernard Cohn, *Colonialism and Its Forms of Knowledge: The British in India* (Princeton, NJ: Princeton University Press, 1996), 7. On taxation and mapping, see Bernardo A. Michael, "Making Territory Visible: The Revenue Surveys of Colonial South Asia," *Imago Mundi* 59, no. 1 (2007): 78–95.

58 Daniel Foliard, *Dislocating the Orient: British Maps and the Making of the Middle East, 1854–1921* (Chicago: University of Chicago Press, 2017).

59 Peter Collier and Rob Inkpen, "The Royal Geographical Society and the Development of Surveying 1870–1914," *Journal of Historical Geography* 29, no. 1 (2003): 93–108, here pp. 100–6.

60 Edney, *Mapping*, 138–9, 165–95.

61 Lachlan Fleetwood, "'No Former Travellers Having Attained such a Height on the Earth's Surface': Instruments, Inscriptions, and Bodies in the Himalaya, 1800–1830," *History of Science* 56, no. 1 (2018): 3–34; Thomas Simpson, "'Clean out of the Map': Knowing and Doubting Space at India's High Imperial Frontiers," *History of Science* 55, no. 1 (2017): 3–36.

62 Kapil Raj, *Relocating Modern Science: Circulation and the Construction of Knowledge in South Asia and Europe, 1650–1900* (Basingstoke: Palgrave Macmillan, 2007), 181–222; Felix Driver, "Intermediaries and the Archive of Exploration," in *Indigenous Intermediaries: New Perspectives on Exploration Archives*, ed. Shino Konishi, Maria Nugent, and Tiffany Shellam (Canberra: Australian National University Press, 2015), 11–29; Derek Waller, *The Pundits: British Exploration of Tibet and Central Asia* (Lexington: The University Press of Kentucky, 1990).

63 On the Egypt expedition's maps, see Simon Schaffer, "Oriental Metrology and the Politics of Antiquity in Nineteenth-Century Survey Sciences," *Science in Context* 30, no. 2 (2017): 173–212, here pp. 174–90. On French Indochina, see Marie de Rugy, *Aux confins des empires* (Paris: Éditions de la Sorbonne, 2018). On West Africa, see Camille Lefebvre, *Frontières de sable, frontières de papier: Histoire de territoires et de frontières, du jihad du Sokoto à la colonisation française du Niger, XIXᵉ-XXᵉ siècles* (Paris: Éditions de la Sorbonne, 2015).

64 For example, Hélène Blais, *Mirages de la carte: L'invention de l'Algérie coloniale* (Paris: Fayard, 2014). For an introduction to this scholarship in English, see "Feature: French Geography, Cartography and Colonialism," ed. Hélène Blais, Florence Deprest, and Pierre Singaravelou, special issue, *Journal of Historical Geography* 37, no. 2 (2011): 146–202.

65 David Lambert, *Mastering the Niger: James MacQueen's African Geography and the Struggle over Atlantic Slavery* (Chicago: University of Chicago Press, 2013). See also Charles W.J. Withers, "Mapping the Niger, 1798–1832: Trust, Testimony and 'Ocular Demonstration' in the Late Enlightenment," *Imago Mundi* 56, no. 2 (2004): 170–93.

66 Foliard, *Dislocating*, 239.

67 Benjamin C. Fortna, "Change in the School Maps of the Late Ottoman Empire," *Imago Mundi* 57, no. 1 (2005): 23–34; Sumathi Ramaswamy, *Terrestrial Lessons: The Conquest of the World as Globe* (Chicago: University of Chicago Press, 2017).

68 Deborah R. Coen, *Climate in Motion: Science, Empire, and the Problem of Scale* (Chicago: University of Chicago Press, 2018), 44–62.

69 Timothy Mitchell, *Rule of Experts: Egypt, Techno-Politics, Modernity* (Berkeley: University of California Press, 2002).

70 On thematic mapping, see Arthur H. Robinson, *Early Thematic Mapping in the History of Cartography* (Chicago: University of Chicago Press, 1982); Susan Schulten, *Mapping the Nation: History and Cartography in Nineteenth-Century America* (Chicago: University of Chicago Press, 2012), 1–7. On modern modes of framing governmental projects, see James C. Scott, *Seeing Like a State: How Certain Scheme to Improve the Human Condition Have Failed* (New Haven, CT: Yale University Press, 1998).

71 Steven Seegal, *Mapping Europe's Borderlands: Russian Cartography in the Age of Empire* (Chicago: University of Chicago Press, 2012), 14.

72 Sumathi Ramaswamy, "Maps, Mother/Goddesses, and Martyrdom in Modern India," *Journal of Asian Studies* 67, no. 3 (2008): 819–53, here p. 828. On reworkings of European spatial notions in South Asia, see also Sumathi Ramaswamy, *The Lost Land of Lemuria: Fabulous Geographies, Catastrophic Histories* (Berkeley: University of California Press, 2004).

73 Raymond B. Craib, "Cartography and Decolonization," in Akerman, *Decolonizing the Map*, 11–71, here p. 50. For a case study of these overlaps, see Jamie McGowan, "Uncovering the Roles of African Surveyors and Draftsmen in Mapping the Gold Coast, 1874–1957," in Akerman, *Decolonizing the Map*, 205–51.

74 Craib, "Cartography," 51.

75 William Rankin, *After the Map: Cartography, Navigation, and the Transformation of Territory in the Twentieth Century* (Chicago: University of Chicago Press, 2016). On the politics of labelling polities as "empires," see Lieven, *Empire*, 3–26.

76 Daniel Immerwahr, *How to Hide an Empire: A Short History of the Greater United States* (London: Penguin, 2019), 8–19.

77 Michael Hardt and Antonio Negri, *Empire* (Cambridge, MA: Harvard University Press, 2000). On the politics of GPS technologies, see Rankin, *After*, 287–99.

78 See for example Martin Dodge and Chris Perkins, "Reflecting on J.B. Harley's Influence and What He Missed in 'Deconstructing the Map,'" *Cartographica* 50, no. 1 (2015): 37–40.

4

RACIAL SCIENCE

James Poskett

Empires always seek to classify their subjects. From the early modern period onwards, colonial states deployed a range of sciences in order to identify and control different ethnic groups.[1] These sciences, particularly during the age of European imperial expansion, transformed into a distinct set of disciplines dedicated to racial classification. In this chapter, I situate the history of racial science within a global history of empire. I begin with the growth of Atlantic slavery in the eighteenth century, move through the formation of colonial and settler societies in the nineteenth century, and end with the period of decolonisation in the twentieth century. In doing so, I bring together a range of historiographies which are typically treated in isolation.

Drawing on recent work in the history of racial science, I make four broad arguments. First, I argue we need to move beyond the existing intellectual history of race. Traditionally, the history of racial science was treated as part of the history of ideas. This perspective was reinforced following the rise of the linguistic turn and the early cultural history of the 1980s. Race was understood as an "idea" or a "concept," something best analysed through the study of language. Nancy Stepan's *The Idea of Race in Science* (1982) is characteristic of this approach.[2] However, as more recent work has demonstrated, race was not just an idea. It was also something made through action. With this in mind, I follow the work of scholars such as Sadiah Qureshi and James Delbourgo in arguing that we need to study the history of racial science as part of the history of practice and performance, paying greater attention to sources grounded in material and popular culture.[3]

Second, I argue that we need to study a much broader set of racial sciences. Traditionally, historians of race focused on biological disciplines, such as physical anthropology and genetics. In this chapter, however, I uncover the role of a range of scientific disciplines, from philology to psychology, in the making of race. Rather than thinking of some sciences as racial, and others as not, we should therefore think of modern science as racial from the start. This approach also helps move away from an outdated narrative in which understandings of race are seen as becoming more "biological" over time. Such a narrative, whilst still featuring in many undergraduate history courses, has been thoroughly critiqued in recent years. As Qureshi argues, we in fact see a "proliferation" rather than a "homogenization" of racial ideas over time.[4]

Third, I argue that racial science was made in the colonial world. In developing this argument, I follow recent work by scholars such as Warwick Anderson and Bronwen Douglas. As

Anderson and Douglas argue, the history of racial science looks very different when we start from places like Latin America and the Pacific rather than Europe.[5] For this reason, I focus on colonial sites and colonial actors, rather than traditional metropolitan thinkers such as Johann Blumenbach and Charles Darwin.

Finally, I argue we need to pay greater attention to the agency of colonised and enslaved people in the history of racial science. Historians have increasingly emphasised the role played by people outside of Europe in the development of modern science. To date, much of this work has focused on natural history as well as physical sciences such as astronomy and surveying.[6] However, this approach has only recently been extended to the history of racial science. As I show in this chapter, colonised and enslaved people were not simply the objects of racial science. They were also important agents in this history. In many cases, colonised and enslaved people challenged the claims of European racial scientists. In other cases, they adapted racial theories to suit their own political goals.[7] Recovering these voices—and acknowledging the limits of such agency—remains an important challenge for historians of science, race, and empire.

Slavery and the African body

In December 1744, the Royal Society in London published an article claiming to identify "the Cause of the Colour of the Negroes." The author, a physician named John Mitchell, lived in Virginia, where he owned a number of slaves. In the article, Mitchell described how he had dissected the skin of an African man. Deploying the latest Newtonian theory of optics, Mitchell argued that African skin was thicker than European skin, and therefore led to "a Suffocation of the Rays of Light." Disturbingly, Mitchell boasted that his experiments had been conducted "on living Subjects." This was a marker of trustworthiness, as Mitchell complained that dead bodies might decompose prior to dissection, therefore rendering observations unreliable.[8] However, Mitchell's admission is also a reminder of the violence that went hand in hand with the scientific study of race in the eighteenth century. Mitchell was not the only person examining African bodies in an attempt to identify some underlying quality of "blackness."[9]

The exact relationship between race and slavery is widely debated.[10] For some historians, such as Joyce Chaplin, the "modern concept of race" was a product of the growth of Atlantic slavery in the eighteenth century.[11] Others, such as Seymour Drescher, place greater emphasis on the period of abolition during the nineteenth century.[12] Historians of science are well placed to intervene in this debate. Despite different accounts of the relationship between race and slavery, the traditional historiography is grounded almost exclusively in the history of ideas. However, as the experiments of John Mitchell remind us, race needs to be understood as part of a history of scientific practice.[13] Rather than simply an intellectual change, we should therefore see the growth of slavery as creating the conditions whereby a range of new sites, technologies, and practices came to play increasingly important roles in the making of race.

Over the course of the eighteenth century, the plantation emerged as a crucial site for the scientific study of race. Slaveowners employed physicians and surgeons to help manage health on the plantation, providing a minimal level of medical care which also functioned as a form of discipline.[14] The growth of plantation slavery therefore placed European medical men in an unprecedented position of power over African bodies. The difference between Africans and Europeans was then often conceived in medical terms, something Rana Hogarth refers to as the "medicalization of blackness."[15] Alongside debates over skin colour, physicians working on plantations developed racialised theories of disease. John Lining, a physician based in South Carolina, subscribed to the theory that Africans were particularly resistant to yellow fever. At the same time, a number of other diseases, such as the yaws, were understood to be peculiar to the African body.[16] As medical interest in racial difference increased, so too did demand for African bodies. By the middle of the nineteenth century, medical schools and museums across the United States and Europe housed collections of African anatomical specimens, often harvested from deceased slaves.[17]

New accounts of the difference between European and African bodies served a variety of political ends, ranging from supporting slavery through to the campaign for immediate abolition. In some cases, Africans were presented as a distinct "species," uniquely fit to undertake the backbreaking work of sugar farming in tropical climates. Writing in *The History of Jamaica* (1744), the slaveowner Edward Long argued that the African body was "peculiarly adapted to a hot climate." Long then concluded that "the White and the Negroe had not one common origin." The belief in different human origins—later known as polygenism—was controversial even at the time, particularly as it seemed to contradict the Biblical account of creation.[18] Others, such as Samuel Stanhope Smith, professor of moral philosophy at the College of New Jersey, argued that all human beings shared a common origin, and that physical variation was simply the result of climate.[19] Nonetheless, despite these differences in interpretation, the terms of the debate were set. By the beginning of the nineteenth century, both abolitionists and proslavery campaigners increasingly deployed scientific and medical evidence in making political arguments.[20]

It is important to recognise that Africans were not simply passive victims of racial science. Britt Rusert and others have recently uncovered the role that Africans themselves played in contesting racial science. This represents a move away from an older historiography of race focused solely on European dominance. From the late eighteenth century onwards, a number of African Americans and former slaves began to develop counternarratives, making use of the growth of cheap print and popular lecture circuits to reach new audiences.[21] Many were very critical of the use of anatomical and medical ideas to differentiate Europeans and Africans. James McCune Smith, the first African American to receive a medical degree, wrote a series of articles in the 1850s entitled "The Heads of the Colored People," ridiculing phrenological accounts of African mental inferiority. Similarly, the African American abolitionist Frederick Douglass delivered an influential lecture in 1854 entitled "The Claims of the Negro, Ethnologically Considered." In the lecture, Douglass noted that "if, for instance, a phrenologist, or a naturalist undertakes to represent in portraits, the difference between the two races... he will invariably present the *highest* type of the European, and the *lowest* type of the negro." Douglass then concluded with a statement that neatly sums up the relationship between racial science and slavery: "By making the enslaved a character fit only for slavery, they excuse themselves for refusing to make the slave a freeman."[22]

Racial science and the colonial state

Setting sail from Britain in 1783, William Jones composed a list of subjects he intended to study on arriving in India. These included "A Grammar of the Sanscrit Language," "A Translation of the Veda," and "A History of India before the Mahommedan conquest, from the Sanscrit-Cashmir Histories."[23] Jones's list—with its focus of translation and grammar—points to the importance of language for eighteenth-century studies of race. For Jones, as for many philologists at the time, the origins of different peoples were best traced through the study of language.[24] Over the next ten years, Jones worked alongside Indian pandits to translate many classical Indian texts. He concluded that ancient Indian languages, particularly Sanskrit, were closely related to classical European languages, such as Greek and Latin. There was, according to Jones, "a stronger affinity.... than could possibly have been produced by accident."[25] With this in mind, Jones argued that the people of Europe and India shared a common origin, constituting the "Aryan" race.[26]

Jones had been sent to India, not to study Sanskrit, but to work as a judge at the Supreme Court in Calcutta. The legal context here is important.[27] Once again, we need to pay close attention to particular sites and practices in the history of racial science, rather than ideas alone. The growth of European colonial states required the development of new legal institutions as well as forms of linguistic expertise. This was particularly the case from the middle of the eighteenth century, as European trading companies, such as the English East India Company, expanded into new territory and assumed the powers of administration and taxation. To understand the history of race and the colonial state, we therefore need to start with the sciences of philology and jurisprudence. Here, much of the earlier work of historians of the British Empire is instructive. Bernard Cohn long ago argued that British rule in India relied on the command of language. Similarly, Tony Ballantyne has identified the role played by "Aryanism" in shaping colonial law, both in British India and beyond.[28]

Jones's interest in Indian languages therefore stemmed not simply from curiosity, but a need to govern. Although European empires sometimes attempted to impose their own legal standards, most adopted a policy whereby "native" legal systems were applied to "native" peoples. All this required a definition of who the "native" was, and what kind of legal systems they would be subject to. Jones's linguistic studies were designed to aid in this—to uncover an "ancient Indian constitution"—and crucially separate the administration of Hindu and Muslim populations.[29] Similar policies were adopted by other European empires. In the early 1900s, the governor-general of German New Guinea, Albert Hahl, ordered a series of legal and linguistic surveys, supported by the Berlin Museum of Ethnology.[30] At around the same time, Alberto Pollera, an Italian colonial officer in Eritrea, produced a study of the Kunama language as part of an effort to reform the colonial legal system. (Pollera complained that the structure of the language was "very simple," indicating "a people so far behind in the civilisation ladder."[31]) As these examples suggest, the study of language—and the associated legal contexts—remained an important part of racial science well into the twentieth century.[32]

Linguistic studies of race operated alongside an interest in physical difference. The context here was that of colonial violence. As European empires expanded, they fought numerous bloody wars. Colonial states also enacted comprehensive regimes of physical and capital punishment. The battlefield and the execution block then provided the raw materials for studies of racial difference grounded in the body. Indeed, much recent work on the history of racial science, including by scholars such as Kim Wagner and Ricardo Roque, is characterised by a focus on these materials. In the 1820s, the Phrenological Society in Edinburgh received nine skulls from a British medical officer serving in Ceylon. One of the skulls, the phrenologists were told, had been taken from the body of an executed leader of a local rebellion.[33] Later in the century, the Coimbra University Museum in Portugal received 35 human skulls from East Timor. The heads had been severed from the enemy by a party of Timorese soldiers operating in the service of the Portuguese Empire. Conduct on the battlefield then fed into racial assessments back in the metropole. According to João Gualberto de Barros e Cunha, the anthropologist who analysed the skulls in the 1890s, the Timorese were a "cruel" race who "make war the savage way," taking the heads of their enemies as trophies.[34] This connection between race and colonial violence extended to the study of "martial races" and "criminal tribes." In India, particularly following the 1857 Rebellion, the British colonial government ordered a number of anthropometric and photographic surveys. The purported aim was to identify those groups that could be trusted, such as the "true Sikhs," as well as those that needed to be carefully monitored for signs of dissent.[35] By the end of the nineteenth century, most European capitals housed large collections of human skulls, often taken from the bodies of colonised people killed in war or executed as punishment.[36]

Sciences of settler colonialism

Addressing the Secretary of the Interior, John Wesley Powell warned of the impending fate of the Indigenous people of North America. "In a very few years it will be impossible to study our North American Indians in their primitive condition except from recorded history," explained Powell in 1879. Powell had recently been appointed director of the Bureau of Ethnology at the Smithsonian Institution in Washington, DC. Like many anthropologists in the late nineteenth century, he subscribed to the idea that Indigenous people living in settler colonies would be wiped out by the advance of civilisation. Powell suggested that the science of ethnology would help guide the US government in its policy towards Native Americans, explaining that "savagery … is a distinct status of society, with its own institutions, customs, philosophy, and religion."[37]

The discourse of "dying races" was not unique to the United States. In early twentieth-century South Africa, anthropologists identified the San people as what they called the "pure-breed" of the "Bushmen." Matthew Drennan, professor of anatomy at the University of Cape Town, took hundreds of photographs and plaster casts, claiming that the "Bushmen" represented "relics" of a distant evolutionary ancestor.[38] Similar claims were made concerning the fate of Aboriginal Australians, the Indigenous people of Canada, and even the Irish.[39] These people often formed the focus of human displays at colonial and international exhibitions. Popular from the middle of the nineteenth century onwards, these exhibitions presented Indigenous people, particularly in recently colonised lands, as living examples of the evolutionary past.[40]

The development of racial science in settler societies needs to be understood in the context of this specific form of colonialism. Over the course of the nineteenth and early twentieth centuries, a specific set of scientific institutions, technologies, and practices developed in order to explain and manage the introduction of European people into settler colonies. A range of sciences—including pathology, statistics, and eugenics—were used to identify so-called "dying races." Those same sciences were also used to isolate an underlying quality of "whiteness," something that distinguished the settler colonist from the Indigenous population. These two ideologies, of dying races and of whiteness, then operated in tandem in order to define and police settler colonies throughout the early twentieth century.[41]

In the eighteenth century, European people tended to fear that life in the tropics would lead to racial degeneration. Indeed, this was part of the justification for the use of enslaved labour in the Americas. However, during the late nineteenth century, this rhetoric underwent a dramatic reversal. By the start of the twentieth century, "whiteness" was increasingly understood as a necessary biological characteristic for adaptation to tropical environments. This change in understanding was closely tied to the development of new scientific practices and institutions, particularly those associated with tropical medicine. These racial sciences helped create what Warwick Anderson refers to as a "discourse of settlement."[42]

Much of the early scientific research on whiteness took place at the Australian Institute of Tropical Medicine, established in 1913 at Townsville, Queensland. Researchers at the institute collected data on blood pressure and red blood cell counts from white children living in northern Australia. In the end, the director of the institute, Anton Breinl, concluded that the tropical climate had no adverse effect on what he called "the white organism." By the 1920s, medical researchers were convinced that white settlers were uniquely suited to life in the tropics. Ronald Hamlyn-Harris, President of the Royal Society of Queensland, believed settlers represented "a type of human beings specially adapted to live in Tropical Queensland." This type was "based on British blood" but "amended by the sun and soil in appearance, physique, speech and temperament." There was a gendered element to this, as medical researchers tended to believe that the white male labourer was ideally suited to life in the tropics, whereas women were more liable to suffer.[43] Similar ideas were repeated across the settler colonial world. As early as the 1870s, French doctors in Algeria referred to the "robust, strong, and well-built" settlers, who apparently possessed "a great ability to adapt." By the end of the nineteenth century, French settlers in Algeria increasingly presented themselves as a "new white race," the next step along the evolutionary ladder.[44]

This specific notion of "whiteness" implied a system of segregation. Researchers in institutes of tropical medicine deployed the latest work in pathology and developmental biology in order to argue that the advantages of the white body could only be maintained in isolation from other races. In Australia, this manifest itself in the 1901 Immigration Restriction Act. Pathologists advised that immigrants, particularly those of the "black races," could still act as healthy carriers of disease. Through miscegenation immigrants might also dilute the "white stock." As John Cumpston, Director of the Commonwealth Quarantine Service, explained in 1921, "it is all very well to have a white Australia, but it must be kept white. There must be immaculate cleanliness."[45] Comparable claims, grounded in the language of biological science, were made to justify segregation policies in South Africa. Harold Fantham, a leading eugenicist and professor of zoology at the University of Witwatersrand, published a series of articles in the 1920s claiming that "racial admixture" led to physical abnormalities. In 1927, the South African government introduced strict laws regulating mixed-race pregnancies outside of marriage, a precursor to the 1949 apartheid legislation which outlawed all interracial sexual intercourse.[46]

Segregationist policies did not go unchallenged. Throughout the settler colonial world, a small number of scientifically trained individuals, often of Indigenous or mixed descent, developed counternarratives. In early twentieth-century New Zealand, the Maori anthropologist Te Rangi Hiroa—who had studied at the Otago Medical School—published an article based on anthropometric data arguing that racial mixing actually resulted in a superior physical type. "Miscegenation" would provide "the stepping-stone to the evolution of a future type of New-Zealander in which we hope the best features of the Maori race will be perpetuated forever," claimed Te Rangi Hiroa.[47] These views were echoed by the Brazilian anthropologist João Baptista de Lacerda, who promoted the evolutionary advantages of racial mixing at the Universal Races Congress in 1911.[48] In the early decades of the twentieth century, racial sciences were therefore put to radically different political uses, from the justification of segregation in South Africa to the promotion of multiracial states in Latin America.

Science, race, and anticolonialism

In early 1924, members of the Indian Psychoanalytic Society in Calcutta gathered to listen to a paper on the theme of "Hindu–Muslim Unity." The author, Owen Berkeley-Hill, worked as superintendent of the Ranchi Mental Hospital. He was also a keen psychoanalyst. Drawing on Sigmund Freud's studies of religion, Berkeley-Hill argued that "the feeling of hatred which most Hindus experience for Muslim is derived from two sources: (1) the mother-land complex. (2) The Cow-totem." At a time of growing religious tension, Berkeley-Hill provided a racialised explanation of Indian psychology. He also suggested a number of practical steps the British could take in order to contain religious and anticolonial fervour.[49]

Throughout the middle decades of the twentieth century, European empires increasingly turned to racial sciences—particularly sciences of the mind—in order to respond to the growth of anticolonial movements. This theme has been the subject of much recent research by scholars including Erik Linstrum, Sloan Mahone, and Shruti Kapila, amongst others. At the height of the Mau Mau Rebellion in Kenya, the British colonial government employed a psychiatrist named John Carothers to produce a report on the psychology of the insurgents. Published as *The Psychology of Mau Mau* (1954), Carothers rejected the idea of any significant biological difference between Europeans and Africans. Instead, he argued that the behaviour of the Mau Mau insurgents could be explained in psychological terms. Carothers claimed that Africans, because of a failure in child rearing, tended to feel a lack of security, and were liable to act rather like a "jilted lover." In response, Carothers suggested that the British adopt a system of "villagization," whereby Mau Mau insurgents would be forcibly relocated to small settlements, isolating them from the wider community and allegedly providing a sense of security.[50] The French in Algeria made similar use of racial psychology to explain the growth of anticolonialism. Psychiatrists at the Algiers Medical School argued that the violent tactics employed by the National Liberation Front in the 1950s could be attributed to Muslim "mental deficiency." One French psychologist went as far as to describe Islam as an "epidemic of religious madness."[51]

Racial sciences were not just a weapon of colonialism. From the early 1900s onwards, colonised elites began to adopt the languages and practices of racial science in order to promote their own political interests. Jomo Kenyatta, who went on to become the first president of independent Kenya, was just one of a number of anticolonial nationalists who studied anthropology under Bronisław Malinowski at the London School of Economics in the 1930s. Kenyatta completed his diploma in 1938, submitting a dissertation on the anthropology of his own people, the Kikuyu. This dissertation then formed the basis of Kenyatta's seminal work, *Facing Mount Kenya* (1938), in which he presented the Kikuyu people as possessing a coherent and stable culture. "It cannot be too strongly emphasised that the various sides of Gikuyu life here described are part of an integrated culture," concluded Kenyatta. By adopting the practices of European racial science, Kenyatta hoped to convince the British that the Kikuyu were already a nation, and therefore should be granted the right to independent government. As Bruce Berman argues, Kenyatta was "pursuing his politics as ethnography."[52]

A number of other anticolonial figures adopted the same strategy, employing racial sciences such as anthropology, psychology, and genetics in order to make the case for national or regional identity. Recent work by scholars including Omnia El Shakry and Projit Bihari Mukharji has explored this theme in the context of the Middle East and South Asia respectively. In Egypt, the anthropologist 'Abbas Mustafa 'Amma collected blood samples and anthropometric data throughout the 1940s, arguing for the unity of the Egyptian and Sudanese people, grounding his pan-Arab politics in racial science.[53] At the same time, the Bengali anthropologist Nirmal Kumar Bose—who went to prison for his involvement in the Quit India Movement—made the case for Indian national identity based on social and cultural characteristics. "In spite of the fact that [the] languages of India are many," Bose argued, "there is an overall unity of design … the sameness of traditions."[54] Others, such as the anthropologist Atul Krishna Sur, went beyond social and cultural analysis, making claims about Indian national and regional identity based on physical characteristics, such as the width of the forehead.[55] By the time that many colonial states gained independence, racial sciences often formed the basis of nationalist claims to identity.

Conclusion

Today, the sciences continue to play a significant role in making and enforcing racial boundaries. Biometric technologies are routinely used around the world, often with the effect—intended or otherwise—of profiling individuals according to race.[56] Anyone who has ever tried to enter the United States will be all too familiar with the assortment of fingerprint, body, and facial recognition scanners. Machine-learning technologies, which tend to reproduce and reinforce existing racial divisions, are also increasingly used in a wide variety of settings.[57] Recent studies have shown how algorithms used by American banks to assess mortgage applications are more likely to reject people of colour. The banks argue that this is not based on "race," by which they mean physical difference. But of course, these algorithms still enact a form of racial discrimination, making judgements based on long-standing geographic divisions, many of which are a product of the segregation era.[58] A study by computer scientist Joy Buolamwini also found that facial recognition software performs significantly worse when trying to identify people of African descent, particularly women. This has the potential to lead to an increase in "false positives," whereby people of colour are more likely to be wrongly identified as suspects by law enforcement agencies.[59]

Alongside computer technologies, the reduction in the cost of genetic testing has led to a resurgence in claims linking DNA to race. In some instances, genetic technologies have then been taken up by Indigenous people as well as the governments of former colonies. For some Native American groups, genetic testing provides a means to secure their identity and legal rights. However, as Kim TallBear warns, modern genetic testing often reproduces colonial categories, and erases Indigenous approaches to identity. For TallBear, as for many Indigenous people, there is more to being Native American than simply a genetic code.[60] Similarly, in an age of growing nationalism, genetic studies are being deployed by governments around the world to determine the racial character of the nation, as well as identify patterns of migration and miscegenation. The largest of these projects was launched by the Indian government in 2003, under the auspices of the Indian Genome Variation Consortium. After analysing 15,000 individuals, researchers at the Consortium produced a genetic map of India which seemed to perfectly reproduce many of the colonial categories described earlier in this chapter. The article, published in the journal *Human Genetics*, divided the subcontinent by language families and "morphological types," including the "Negrito" and the "Mongoloid."[61]

I hope that the history uncovered in this chapter points towards some of the ways in which we can start to grapple with the legacies of racial science in the postcolonial present. As I have argued, the history of racial science is best understood not just as a history of ideas, but rather as a history of particular sites, technologies, and practices. This emphasis on action rather than ideas should help focus our attention on the ways in which science and race come together in the present. In the first instance, we need to engage with the existence of large collections of human remains, and associated cultural objects, which are still housed in European and American museums.[62] Additionally, as the previous examples demonstrate, racial sciences today operate at new sites and deploy a range of new technologies. Some of these, such as genetic testing, are clearly biological. But many are not. As I suggested in this chapter, there is no meaningful sense in which understandings of race became more "biological" over time. Rather, we see biological, linguistic, psychological, and cultural approaches to race all operating side by side, even in the early decades of the twentieth century.

The myth that understandings of race became more biological over time is in fact a significant obstacle to the fight against scientific racism in the present.[63] Today, some of the most widespread technologies of racial discrimination are successful precisely because they ostensibly ignore biological characteristics. Machine learning in particular provides a new way for individuals, organisations, and governments to discriminate in practice whilst professing equality in principle. The power of racial science, then, does not lie in its homogeneity. Rather, racial science derives its power from its ability to constantly adapt.[64]

Acknowledgements

I would like to thank Andrew Goss, Thomas Simpson, and Charu Singh for their valuable comments on earlier drafts of this chapter.

Notes

1 For earlier sciences of human difference, see Surekha Davies, *Renaissance Ethnography and the Invention of the Human: New Worlds, Maps and Monsters* (Cambridge: Cambridge University Press, 2016); Joan-Pau Rubiés, "New Worlds and Renaissance Ethnology," *History and Anthropology* 6 (1993): 157–97; and Siep Stuurman, "François Bernier and the Invention of Racial Classification," *History Workshop Journal* 50 (2000): 1–21. For sciences of human difference in non-European empires, particularly the Qing, see Laura Hostetler, *Qing Colonial Enterprise: Ethnography and Cartography in Early Modern China* (Chicago: University of Chicago Press, 2005).

2 Nancy Stepan, *The Idea of Race in Science: Great Britain, 1800–1960* (London: Macmillan, 1982). See also Peter Robb, ed., *The Concept of Race in South Asia* (Delhi: Oxford University Press, 1995).

3 For this argument, see Sadiah Qureshi, *Peoples on Parade: Exhibitions, Empire, and Anthropology in Nineteenth-Century Britain* (Chicago: University of Chicago Press, 2011), 1–12; James Delbourgo, "The Newtonian Slave Body: Racial Enlightenment in the Atlantic World," *Atlantic Studies* 9 (2012): 186–7; James Poskett, *Materials of the Mind: Phrenology, Race, and the Global History of Science, 1815–1920* (Chicago: University of Chicago Press, 2019), 8, 256–61.

4 For a critique of this narrative, see Sadiah Qureshi, "Robert Gordon Latham, Displayed Peoples, and the Natural History of Race, 1854–1866," *The Historical Journal* 54 (2011): 163; Sujit Sivasundaram, "Race, Empire, and Biology before Darwin," in *Biology and Ideology from Descartes to Dawkins*, ed. Denis Alexander and Ronald Numbers (Chicago: Chicago University Press, 2010); Bruce Dain, *Hideous Monster of the Mind: American Race Theory in the Early Republic* (Cambridge, MA: Harvard University Press, 2003), vii; Poskett, *Materials of the Mind*, 256–61.

5 Warwick Anderson, "Racial Conceptions in the Global South," *Isis* 105 (2014): 782–92; Suman Seth, "Introduction: Relocating Race," *Isis* 105 (2014): 759–63; Chris Ballard, "The Cultivation of Difference in Oceania," in *Foreign Bodies: Oceania and the Science of Race, 1750–1940*, ed. Bronwen Douglas and Chris Ballard (Canberra: Australian National University Press, 2008), 339.

6 For example, Kapil Raj, *Relocating Modern Science: Circulation and the Construction of Knowledge in South Asia and Europe* (Basingstoke: Palgrave Macmillan, 2007); Benjamin Elman, *On Their Own Terms: Science in China, 1550–1900* (Cambridge, MA: Harvard University Press, 2009).

7 Poskett, *Materials of the Mind*, 141–7, 261–2; Thomas Simpson, "Historicizing Humans in Colonial India," in *Historicizing Humans: Deep Time, Evolution, and Nineteenth-Century British Sciences*, ed. Efram Sera-Shriar (Pittsburg, PA: University of Pittsburgh Press, 2018), 114–16; Britt Rusert, "The Science of Freedom: Counterarchives of Racial Science on the Antebellum Stage," *African American Review* 45 (2012): 291–308; Nancy Stepan and Sander Gilman, "Appropriating the Idioms of Science: The Rejection of Scientific Racism," in *The 'Racial' Economy of Science: Toward a Democratic Future*, ed. Sandra Harding (Bloomington: Indiana University Press, 1993).

8 Delbourgo, "Newtonian Slave Body," 185–202; John Mitchell, "An Essay upon the Causes of the Different Colours of People in Different Climates," *Philosophical Transactions* 43 (1744): 102. See also Todd Savitt, "The Use of Blacks for Medical Experimentation and Demonstration in the Old South," *The Journal of Southern History* 48 (1982): 331–48; Andrew Curran, *The Anatomy of Blackness: Science and Slavery in the Age of Enlightenment* (Baltimore, MD: Johns Hopkins University Press, 2011).

9 For earlier understandings of African bodies, see Anu Korhonen, "Washing the Ethiopian White: Conceptualising Black Skin in Renaissance England," in *Black Africans in Renaissance Europe*, ed. Thomas Earle and Kate Lowe (Cambridge: Cambridge University Press, 2005).

10 For an overview of this debate, see Timothy Lockley, "Race and Slavery," in *The Oxford Handbook of Slavery in the Americas*, ed. Mark Smith and Robert Paquette (Oxford: Oxford University Press, 2010).

11 Joyce Chaplin, "Race," in *The British Atlantic World, 1500–1800*, ed. David Armitage (Basingstoke: Palgrave Macmillan, 2002), 173.

12 Seymour Drescher, "The Ending of the Slave Trade and the Evolution of European Scientific Racism," *Social Science History* 14 (1990): 415–50.

13 Delbourgo, "Newtonian Slave Body," 186–7.

14 Classic studies include Todd Savitt, *Medicine and Slavery: The Diseases and Health Care of Blacks in Antebellum Virginia* (Urbana: University of Illinois Press, 1978); Richard Sheridan, *Doctors and Slaves: A Medical and Demographic History of Slavery in the British West Indies, 1680–1834* (Cambridge: Cambridge University Press, 1985).

15 Rana Hogarth, *The Medicalization of Blackness: Making Racial Differences in the Atlantic World, 1780–1840* (Chapel Hill: University of North Carolina Press, 2017), 2.

16 Suman Seth, *Difference and Disease: Medicine, Race, and the Eighteenth-Century British Empire* (Cambridge: Cambridge University Press, 2018), 196–8; Hogarth, *Medicalization of Blackness*, 10–25.

17 Stephen Kenney, "The Development of Medical Museums in the Antebellum American South: Slave Bodies in Networks of Anatomical Exchange," *Bulletin of the History of Medicine* 87 (2013): 32–62.

18 Seth, *Difference and Disease*, 208–37.

19 William Stanton, *The Leopard's Spots: Scientific Attitudes Toward Race in America, 1815–59* (Chicago: University of Chicago Press, 1960), 3–5; Dain, *Hideous Monster of the Mind*, 55–8.

20 Drescher, "Ending of the Slave Trade"; Poskett, *Materials of the Mind*, 56–76, 118–31; Curran, *Anatomy of Blackness*, 210–15; Adrian Desmond and James Moore, *Darwin's Sacred Cause: Race, Slavery and the Quest for Human Origins* (London: Allen Lane, 2009).

21 Britt Rusert, *Fugitive Science: Empiricism and Freedom in Early African American Culture* (New York: New York University Press, 2017).

22 Rusert, *Fugitive Science*, 30–1, 126–9; Frederick Douglass, *The Claims of the Negro, Ethnologically Considered* (Rochester, NY: Lee, Mann and Co, 1854), 15, 20. Some African Americans and former slaves did take up phrenology, turning racial science against claims of European superiority: see Poskett, *Materials of the Mind*, 129–31, 150–1; Rusert, "The Science of Freedom," 301–4.

23 John Shore, "A Discourse Delivered at a Meeting of the Asiatick Society", in *The Works of Sir William Jones* (London: John Stockdale, 1807), 3: xi.

24 Thomas Trautmann, *Aryans and British India* (Berkeley: University of California Press, 1997), 131–64.

25 Tony Ballantyne, *Orientalism and Race: Aryanism in the British Empire* (Basingstoke: Palgrave, 2002), 20–31; Simpson, "Historicizing Humans," 117–22.

26 See Ballantyne, *Orientalism and Race*, for the way in which Aryanism operated across the British Empire, beyond India.

27 Ballantyne, *Orientalism and Race*, 20; Bernard Cohn, *Colonialism and its Forms of Knowledge: The British in India* (Princeton, NJ: Princeton University Press, 1996), 16–75.

28 Cohn, *Colonialism,* and Ballantyne, *Orientalism and Race*.

29 Cohn, *Colonialism*, 75.

30 Rainer Buschmann, "Colonizing Anthropology: Albert Hahl and the Ethnographic Frontier in German New Guinea," in *Worldly Provincialism: German Anthropology in the Age of Empire*, ed. H. Glenn Penny and Matti Bunzl (Ann Arbor: University of Michigan Press, 2003).

31 Barbara Sòrgoni, "The Scripts of Alberto Pollera, an Italian Officer in Colonial Eritrea: Administration, Ethnography and Gender," in *Ordering Africa: Anthropology, European Imperialism, and the Politics of Knowledge*, ed. Helen Tilley and Robert Gordon (Manchester: Manchester University Press, 2007), 291.

32 This was also true in India: see Javed Majeed, *Colonialism and Knowledge in Grierson's Linguistic Survey of India* (London: Routledge, 2018).

33 Poskett, *Materials of the Mind*, 23–32; Nira Wickramasinghe, "The Return of Keppetipola's Cranium: Authenticity in a New Nation," *Economic and Political Weekly* 32 (1997): 85–92.

34 Ricardo Roque, *Headhunting and Colonialism: Anthropology and the Circulation of Human Skulls in the Portuguese Empire, 1870–1930* (Basingstoke: Palgrave Macmillan, 2010), 52–87, 119, 159–61.

35 Thomas Metcalf, *Ideologies of the Raj* (Cambridge: Cambridge University Press, 1994), 66–158; Clare Anderson, *Legible Bodies: Race, Criminality and Colonialism in South Asia* (Oxford: Berg, 2004); Crispin Bates, "Race, Caste and Tribe in Central India: The Early Origins of Indian Anthropometry," in *The Concept of Race in South Asia*, ed. Peter Robb (Delhi: Oxford University Press, 1995); Nicholas Dirks, "Castes of Mind," *Representations* 37 (1992): 56–78.

36 For skull collecting, see Kim Wagner, "Confessions of a Skull: Phrenology and Colonial Knowledge in Early Nineteenth-Century India," *History Workshop Journal* 69 (2010): 27–51; Simon Harrison, "Skulls and Scientific Collecting in the Victorian Military: Keeping the Enemy Dead in British Frontier Warfare," *Comparative Studies in Society and History* 50 (2008): 285–303; Ann Fabian, *The Skull Collectors: Race, Science, and America's Unburied Dead* (Chicago: University of Chicago Press, 2010); Poskett, *Materials of the Mind*, 19–50.

37 Curtis Hinsley, *Savages and Scientists: The Smithsonian Institution and the Development of American Anthropology, 1846–1910* (Washington, DC: Smithsonian Institution Press, 1981), 147–9; *Letter from the Acting President of the National Academy of Sciences, Transmitting a Report on the Surveys of the Territories* (Washington, DC: Government Printing Office, 1879), 20. See also Sadiah Qureshi, "Dying Americans: Race, Extinction and Conservation in the New World," in *From Plunder to Preservation: Britain and the Heritage of Empire, 1800–1950*, ed. Astrid Swenson and Peter Mandler (Oxford: Oxford University Press, 2013).

38 Saul Dubow, *Scientific Racism in Modern South Africa* (Cambridge: Cambridge University Press, 1995), 48–51.

39 Patrick Brantlinger, *Dark Vanishings: Discourse on the Extinction of Primitive Races, 1800–1930* (Ithaca, NY: Cornell University Press, 2013).

40 Qureshi, *Peoples on Parade*; Pascal Planchard, ed., *Human Zoos: Science and Spectacle in the Age of Colonial Empires* (Liverpool: Liverpool University Press, 2008); Sadiah Qureshi, "Peopling Natural History," in *Worlds of Natural History*, ed. Helen Curry, Nicholas Jardine, James Secord, and Emma Spary (Cambridge: Cambridge University Press, 2018).

41 Gregory Smithers, *Science, Sexuality, and Race in the United States and Australia, 1780s–1890s* (New York: Routledge, 2009), 44–69.

42 Warwick Anderson, *The Cultivation of Whiteness: Science, Health, and Racial Destiny in Australia* (Durham, NC: Duke University Press, 2006), 1–70.

43 Anderson, *Cultivation of Whiteness*, 100–38.

44 Ann Chopin, "Embodying 'The New White Race': Colonial Doctors and Settler Society in Algeria, 1878–1911," *Social History of Medicine* 29 (2016): 11.

45 Anderson, *Cultivation of Whiteness*, 98–147.

46 Dubow, *Scientific Racism*, 133–4, 181–2. Generally, eugenicists in settler colonies were more concerned with what they called the "poor white" than the Indigenous population. See Chloe Campbell, "Eugenics in Colonial Kenya," in *The Oxford Handbook of the History of Eugenics*, ed. Alison Bashford and Philippa Levine (Oxford: Oxford University Press, 2010); Saul Dubow, "South Africa: Paradoxes in the Place of Race," in Bashford and Levine, *Handbook of the History of Eugenics*.

47 Anderson, "Racial Conceptions in the Global South," 787; John Allen, "Te Rangi Hiroa's Physical Anthropology," *Journal of the Polynesian Society* 103 (1994): 20.

48 Anderson, "Racial Conceptions in the Global South," 32.

49 Shruti Kapila, "The 'Godless' Freud and his Indian Friends: An Indian Agenda for Psychoanalysis," in *Psychiatry and Empire*, ed. Meghan Vaughan and Sloan Mahone (Basingstoke: Palgrave Macmillan, 2007), 127.

50 Jock McCulloch, *Colonial Psychiatry and the "African Mind"* (Cambridge: Cambridge University Press, 1995), 64–71; Sloan Mahone, "East African Psychiatry and the Practical Problems of Empire," in Vaughan and Mahone, *Psychiatry and Empire*. The British and French made similar use of racial psychology during anticolonial struggles in Malaya and Indo-China: see Erik Linstrum, *Ruling Minds: Psychology in the British Empire* (Cambridge, MA: Harvard University Press, 2016), 155–7.

51 Richard Keller, *Colonial Madness: Psychiatry in French North Africa* (Chicago: University of Chicago Press, 2007), 166–70.

52 Bruce Berman and John Lonsdale, "Custom, Modernity, and the Search for Kihooto: Kenyatta, Malinowski, and the Making of *Facing Mount Kenya*," in Tilley and Gordon, *Ordering Africa: Anthropology*, 173–90.

53 Omnia El Shakry, *The Great Social Laboratory: Subjects of Knowledge in Colonial and Postcolonial Egypt* (Stanford, CA: Stanford University Press, 2007), 66–81.

54 Pradip Kumar Bose, "The Anthropologist as 'Scientist'? Nirmal Kumar Bose," in *Anthropology in the East: Founders of Indian Sociology and Anthropology*, ed. Patricia Uberoi and Satish Deshpande (Delhi: Permanent Black, 2007), 290–301.

55 Projit Bihari Mukharji, "The Bengali Pharaoh: Upper-Caste Aryanism, Pan-Egyptianism, and the Contested History of Biometric Nationalism in Twentieth-Century Bengal," *Comparative Studies in Society and History* 59 (2017): 454–55.

56 On the colonial origins of biometrics, see Keith Breckenridge, *Biometric State: The Global Politics of Identification and Surveillance in South Africa, 1850 to the Present* (Cambridge: Cambridge University Press, 2014).

57 Safiya Umoja Noble, *Algorithms of Oppression: How Search Engines Reinforce Racism* (New York: New York University Press, 2018).

58 Robert Bartlett, Adair Morse, Richard Stanton, and Nancy Wallace, "Consumer-Lending Discrimination in the FinTech Era," *NBER Working Paper* no. 25943 (2019).

59 Joy Buolamwini and Timnit Gebru, "Gender Shades: Intersectional Accuracy Disparities in Commercial Gender Classification," *Proceedings of Machine Learning Research* 81 (2018): 77–91.

60 Kim TallBear, *Native American DNA: Tribal Belonging and the False Promise of Genetic Science* (Minneapolis: University of Minnesota Press, 2013).

61 "The Indian Genome Variation Database (IGVdb): A Project Overview," *Human Genetics* 119 (2005): 1–11. For the colonial origins of this project, see Mukharji, "The Bengali Pharaoh," 452.

62 On the way in which institutions should respond, see Jeremiah Garsha, "Repatriating Histories: A Call for Global Policies on the Return of Human Remains," *History & Policy*, www.historyandpolicy. org/opinion-articles/articles/repatriating-histories-call-for-global-policies-on-return-of-human-remains.

63 This myth has its origins in the era of decolonisation, whereby European and American geneticists tried to distance themselves from earlier imperial projects: see Sebastián Gil-Riaño, "Relocating Anti-Racist Science: The 1950 UNESCO Statement on Race and Economic Development in the Global South," *The British Journal for the History of Science* 51 (2018): 281–303.

64 Qureshi, "Robert Gordon Latham," 163.

5

METEOROLOGY AND EMPIRE

Martin Mahony

Albert Walter was 20 years old when he took up residence in the observatory assistant's quarters where the previous two inhabitants had died of malaria. The Royal Alfred Observatory at Pamplemousses, Mauritius had been opened in 1875 on a site chosen for its clear view of approaching weather systems and its convenient transport links with the capital Port Louis, some eight miles to the southwest. Meteorology in colonial Mauritius had until that point primarily been a nautical science, concerned with the safety of maritime navigation, most particularly during the summer cyclone season. Yet a harbourside observatory had proven insufficient. While its location afforded ready access to the logbooks of docked ships, and in return navigation advice for anxious mariners, the building was cramped and regularly disturbed by soggy captains and by the noise, dust, and vibrations of the surrounding dockyards. Meteorology was a science that aspired to be at the heart of colonial life, but this was taking that integration too far. Pamplemousses could offer more seclusion, where the patient work of making and compiling observations, deriving laws, and issuing warnings could take place in peace. It was a site at which it was hoped that a new observatory could take its place among the roster of prominent scientific institutions then emerging across the European colonial world.

Soon after opening it was discovered that the site was riddled with malaria. The first Director, Charles Meldrum, suffered periodic bouts of fever, as did his family who lodged with him in the spacious new building. The fever was most abundant during the summer months, when the observatory's work of warning of approaching cyclones was most intense, and most consequential. Meldrum sought permission to live up-country during the summer where the threat from fever was reduced. Yet by the time Albert Walter arrived on the island to take up his role as observatory assistant, the authority of the observatory as a reliable foreteller of storms was under serious strain. Avoiding fever had occasionally meant missing approaching cyclones, and the Governor instituted a new rule that the observatory be staffed by a competent meteorologist at all hours of the day during the storm season. It largely fell to Walter during his early years to perform this lonely vigil by barometer and anemometer, and to weather the effects of the Pamplemousses climate on his own body.[1]

These shifting geographies of meteorology on Mauritius speak to wider relationships between the science and European imperial endeavour. Meteorology was a science which promised to make imperial trade robust to the vagaries of maritime weather, with captains fully equipped to make best use of seasonal winds, and ready to anticipate and avoid the worst consequences of an encounter with a tropical storm. Meteorology could also help make legible the strange, often threatening environments in which settler colonial societies sought to establish themselves. Yet it was a science which always struggled for authority as an interpreter of physical realities and a foreteller of futures, and which didn't always fulfil its practitioners' aim of being a key part of colonial governance.[2] It faced a number of practical obstacles, such as the coordination and standardisation of widespread data collection, as well as the climate itself—as Albert Walter would attest, weather and climate could often intervene in their own recording, making life exceedingly difficult for any ambitious colonial meteorologist.

This chapter surveys literature published in the last 10–15 years which has examined the relationship between empire and atmospheric science. It necessarily focuses largely on anglophone, particularly British Empire contexts, as this is where most of the existing research has focused. However, it also seeks to give visibility to work from other contexts which is helping to decentre British examples, and to allow the development of comparative perspectives. It argues that the most exciting work is seeking to unveil how meteorology, and the related field of climatology,[3] functioned in imperial settings not as a harbinger of a future science of the global atmosphere, but as a form of knowledge making and practice that directly contributed to colonial endeavours, and which was shaped by local and trans-local imperial politics.

Weather, climate, and early colonial expansion

The idea that climatic differences are behind apparent differences between groups of people—in dubious measures such as character, intelligence, or civilisational attainment—was a founding principle of European imperial thought. It offered a seemingly rational explanation of white European superiority, and in turn a source of justification for European expansion, invasion, settlement, and colonial rule. Climatic determinism has been a prominent feature of Western geographical thought since Classical times, and its spectres still haunt discussions of global climatic change.[4] Yet the period with which we are concerned here saw this mode of thought reaching its peak of influence, as well as undergoing a series of transformations through the encounter of European bodies, minds, and scientific instruments with the exotic climates which previously had largely been the subjects of hearsay, anecdote, and rumour.

As the British began to establish colonies in North America, natural philosophers grappled with the paradox that places on similar latitudes, on opposite sides of the Atlantic, could have such different climates. This realisation disrupted Classical ideas of "zonal" climates tied to latitudinal differences. As Sam White has shown, early European settlers grappled with the "puzzle" of the American climate and in so doing made a number of intellectual breakthroughs which helped lay the foundations of the science of climatology.[5] At the same time, debate raged about the influence that extensive human settlement would have on New World climates. Early-modern thinkers grappled with the question of whether deforestation in the Americas would "moderate" the climate, lessening its extremes and rendering it more "European" in character.[6] European states' growing array of small island colonies became particular points of concern over anthropogenic climate change, as settlers, governors, and savants grappled with the environmental impacts of tree felling and the spread of plantation agriculture.[7]

In the eighteenth and nineteenth century debates about human–climate interactions took another turn. Michael Osborne has shown how "acclimatisation" was, for many in this period, *the* paradigmatic colonial science, concerned not with the capacity of humans to alter climates, but to be altered by them. It was a body of knowledge and practice concerned with the relationships between plants, animals, and human bodies and the climates of new European colonial possessions, and whether and how the former could adapt to the latter. Its emergence speaks of the acceleration in the movement of people and things as networks of trade, exchange, and settlement grew and intensified. As an increasingly institutionalised science in the nineteenth century, it concerned itself with the geographies of settlement, agriculture, human health, and field sports, and for its promoters, it lay at the heart of the success of any colonial enterprise.[8]

But these new patterns of imperial mobility drew attention not just to climate, but to weather. Understanding the patterns of the winds was crucial to maritime exploration and expansion. Greg Bankoff argues that winds and currents were so crucial to the emergence of European empires that we should understand these political entities not as "maritime empires," but "aeolian empires."[9] Noting that if one superimposes "a map of European imperia on a chart of atmospheric circulation over the oceans of the world," one notices a "remarkable 'fit' between the prevailing winds and currents and the form and extent of the various European empires."[10] The practical knowledge of mariners helped improve the safety and efficiency of navigation and contributed to the development of early theories of global atmospheric circulation. But it wasn't until the nineteenth century that concerted efforts began at charting not just dominant wind patterns, but their variability and extremes.[11] Knowledge making and empire building proceeded in lockstep in this period, so that by the mid nineteenth century the production of weather data was happening with a new intensity and regularity.[12] Nonetheless, meteorology was not the internationalised science that we see today; in the British context, where Admiralty ships began to function as "itinerant observatories" from the 1840s, the geography of data collection "followed the paths of British ships and traced a geography of storms that conformed to the contours of Britain's imperial interests."[13]

Meteorology and the colonial observatories

Around the same time the infrastructure of land-based geophysical observation expanded greatly. This included meteorology, but largely as it clung on to the coat tails of the more fashionable pursuits of astronomy and terrestrial magnetism.[14] Physical observatories sprung up in Canada, South Africa, Saint Helena, India, Australia, and Singapore. Some of these began taking what would become largely continuous meteorological records, although the case of Singapore is illustrative of how the "Magnetic Crusade" quite quickly ran out of steam, to the cost of long-term record keeping.[15] Naylor and Goodman have documented how physical observatories nonetheless became culturally significant in the mid nineteenth century for broader ideas about imperial progress and civilisation. They were spaces apart from colonial societies, but also functioned as models for how such societies might be ordered on the basis of rational, disciplined action.[16] Metropolitan actors supported their establishment not just as means of expanding their own observational reach, but as a means of cultivating a scientific disposition at otherwise rough-and-ready colonial frontiers.

Nonetheless, it was a struggle—socially, politically, materially—to recreate the model observatories of London and Dublin in places like the Cape and Bombay. Instruments arrived broken, building works were beset with problems and delays, and once opened it might be discovered that the site had been ill chosen. The Colaba Observatory in Bombay had to contend with two neighbouring 18-pounder guns that would announce the arrival of ships in the harbour, causing untold problems for the building's delicate instruments. In Mauritius, where the establishment of the harbourside observatory was more a result of local initiative with limited funds, the Franco-Mauritian government observer clashed with the new British-dominated Meteorological Society (est. 1851) over recording techniques, to the extent that serious proposals were made to divide the observatory building in two, one half to accommodate the anglophone Society's meteorological practice, and the other the meteorology of the much-maligned creole observer.[17] Colonial meteorological observatories did not always live up to their billing as spaces defined by calm rationality and the peaceful pursuit of knowledge.

Some observatories, such as Colaba, became nodes in expanding regional meteorological networks that developed over the subsequent decades, and lost their centrality. Others, such as Toronto, became the hub of such networks. Toronto Observatory also became a hub of a project of rolling out meteorological observation to the province's schools, with the hope that engaging with weather watching would "foster habits of observation and an attention to nature in the schoolchildren," while helping to inculcate a "sense of Canadian community."[18] Here was a domesticated version of Edward Sabine's idea that geophysical observation, as practised through the Magnetic Crusade, could not only "elevate and refine the mind," but also form "one of those thousand links which form the chain of national feeling … tying together various classes of men in the Mother-country and in the Colonies."[19] To observe the meteorological elements was to join with a globally distributed community of weather watchers who were increasingly seeing—and feeling—empire as a culturally and politically coherent endeavour, and atmospheric phenomena as globally interconnected and patterned.[20]

It is nonetheless important to avoid the trappings of methodological nationalism in comprehending meteorology's imperial entanglements. Gregory Cushman has powerfully shown how, in the context of late-nineteenth-century hurricane prediction, knowledge networks were constituted as much by missionary science and transnational business interests as they were by imperial institutions.[21] Jesuit scientists established observatories in a range of colonial settings, from Havana to Shanghai, and made important contributions to the theory and practice of storm forecasting.[22] Knowledge, tools, and technique circulated transnationally, but Cushman concludes that the imperial drive to centralise and discipline, as well as to assert "white ethnic control"—illustrated by US actors in late-nineteenth-century Cuba for example—ultimately perverted the development of the science.[23]

Meteorology as colonial science

Historians have known for some time that meteorology benefitted greatly from imperial networks. Exchanges of data between places like India and Australia yielded, from the eighteenth century onwards, suggestive glimpses of global interconnections between geographically distant atmospheric phenomena.[24] However, such histories have tended towards the teleological, looking for the imperial origins of what is today taken to be reliable scientific truth. Building on this foundation though is work which attends more closely to the political contexts of such imperial knowledge making, approaching it not as a harbinger of future, global science but as part of contemporary, local systems of colonial rule. As Meredith McKittrick has argued, doing so often requires attending to intellectual "dead ends"—ideas and practices which we may with hindsight dismiss as inconsequential for the science of meteorology, but which in their own context were important for the exercise of colonial rule.[25]

In an important recent intervention Ruth Morgan demonstrates how, in Australia and India, "the development of colonial weather knowledge accompanied colonial efforts to consolidate territorial power."[26] Viewing meteorology as a *colonial* science—concerned not just with harnessing colonial systems for the benefit of the science, but with actively contributing to colonial rule—is a vital move. Unsurprisingly much work in this vein has come from historians located in places that were the forefront of settler colonial projects during meteorology's mid- to late-nineteenth-century professionalisation.[27] Modern-day Australia and New Zealand have provided rich case studies of meteorology's colonial role. In both contexts expectations of local climates being largely subject to the same rhythms as "home" quickly gave way to appreciations of volatility. As such, meteorologists acted "as important environmental interpreters for colonists and governments,"[28] helping to plan settlements, adapt agriculture to novel seasonalities, and make communities more resilient to extremes.[29]

Elsewhere, Robert Rouphail has demonstrated how the development of a novel form of "agricultural meteorology" in Mauritius around the turn of the nineteenth century sought to make the circulation of colonial capital robust to climatic extremes whose prediction remained stubbornly difficult.[30] Sandip Hazareesingh has examined how efforts to "modernise" cotton production in nineteenth-century India required the accentuated performance "of colonial authority over the lives of peasant cotton producers," which was achieved through the deployment of "new governmental technologies for monitoring populations and their environment, including censuses, crop surveys, and weather and rainfall reports."[31] Confident expert predictions of bounteous yields from the rational application of new agricultural techniques fell afoul of an ill-understood and ever-changing climate.[32] Conflicting climatic and meteorological knowledges were deployed by British agriculturalists and peasant cultivators, as well as by different imperial institutions—government foresters were far more attuned than their agricultural colleagues to the complexities and fragilities of local environments, and serve as a reminder that in considering meteorology as a colonial science, we need to approach the colonial enterprise as something marked by both political and epistemic struggles and disagreements.[33]

These struggles were often played out in boundary tussles between disciplines, but the cultural maps of disciplinary authority often looked different in the metropole and the colony.[34] Philipp Lehmann contends that a particular "colonial" approach to data gathering emerged in the German colonies, where the challenges of adhering to strictly scientific modes of observation, as defined in the metropole, saw the emergence of an approach to weather recording which emphasised sensory experience as much as quantified measurement.[35] In the early twentieth century the tools of agricultural meteorology seemed to again promise a new way of rationalising colonial agriculture, as well as being a field where the meteorologist could further demonstrate his colonial utility. Britain's Empire Marketing Board sponsored three days' worth of sessions on agricultural meteorology at London's 1929 Conference of Empire Meteorologists, with the hope of obtaining "a general view of the progress and prospects of agriculture in every part of the Empire," and of agreeing ways in which the science of meteorology could better those prospects.[36] The sessions were lively, with agriculturalists and meteorologists—largely the metropolitan representatives—tussling over the boundaries of their respective disciplines, and over the forms that data might take to enable cooperation. The colonial meteorologists in attendance were much more enthusiastic about cooperation and reported experiments underway from the Caribbean to East Africa in putting meteorological data to work in the service of settler farming. While metropolitan meteorologists were precious about the boundaries and scientificity of their discipline, colonial meteorologists embraced the liminal spaces between meteorology and neighbouring sciences, and between diverse modes of observation, as sites where local challenges of colonial practice could be collaboratively addressed.[37]

But meteorology served more than just practical purposes; the science also became part of the ideological machinery for justifying colonial rule itself. Late-nineteenth-century commentators would often equate the conduct of meteorology with "civilised" government.[38] Morgan likewise emphasises "the importance of predicting the future for the realisation of the colonial project." Liberal teleological ideas of how science would drive settler societies towards higher levels of civilisation percolated into meteorological circles. The science's own preoccupation with predicting the future meshed well with broader imperial ideas of "progress," whereby the future became something over which some measure of control could be exercised through the practices of enlightened government.[39] Claiming the space of the future became akin to the colonial claiming of territory, and likewise involved the exercise of new forms of power over colonial subjects. In India, Mike Davis suggests that efforts to develop seasonal forecasts of the monsoon, and thence anticipate potential drought, served to help "naturalise" the famines which devastated large swaths of the subcontinent during the late Victorian period, and locate responsibility for them not in colonial political economy, but in the vagaries of the weather.[40] The ability of colonial actors to predict the weather with some reliability also allowed them to distinguish themselves from the supposed irrationality of Indigenous peoples.

Although some evidence exists of two-way exchanges of weather knowledge between settler and Indigenous societies,[41] it seems that a newly professionalised science of meteorology was largely used as a means of performing the superiority of the European intellect, and of creating cultural distance between coloniser and colonised. This is perhaps best illustrated in cases where efforts to exercise control over the future became particularly literal—in parts of southern Africa for instance, appeals to meteorological authority were used to denigrate Indigenous rainmaking practices, and to emphasise the superiority of a scientific, mechanistic understanding of the atmosphere over more spiritual cosmologies, even as supposedly scientific rainmaking practices proved no more successful than their Indigenous rivals.[42] Through work such as this we can now confidently say that meteorology was more than just a "tool of empire," helping to enact "a project already imagined"; it was also a science which, through its contributions to the ideological justifications of empire, from the vestiges of climatic determinism to the cultural performance of European rationality, helped bring "into being the colonial project itself."[43]

This co-production of atmospheric science and empire, both ideational and practical, has been powerfully illustrated by Deborah Coen's work on Habsburg climatology. She has documented how the "Austrian idea" of seeking unity in diversity fed into, and was buttressed by, distinctive approaches to environmental knowledge making. A marrying of descriptive climatology and fluid dynamics was achieved within a scientific culture which placed particular emphasis on seeing across scales, and of articulating how local differences began to look more like unity at higher spatial levels. In so doing, Coen writes Habsburg climatology back into the history of the early-twentieth-century emergence of a dynamical science of the atmosphere, displacing the more traditional emphasis on later "Bergen School" methods. But at the same time, her work shows how such intellectual innovations were only made possible by an imperial context which fostered new ways of thinking across scales.[44] Empire helped birth modern atmospheric science not just by giving it a new spatial reach, but by providing the intellectual machinery to think in new ways about multi-scalar processes and phenomena.

Aerial imperialism

Other historians have likewise noted how meteorology became a model for wider imperial modes of thought and action. In British India it was held that the "atmosphere, like empire, required discipline on a monumental scale."[45] While earlier in the century the observatory was the model colonial space, late-nineteenth-century observation networks offered a model for a new kind of extended imperial endeavour; the kinds of standardisation and coordination required to effectively observe the weather at a continental scale inspired those who were campaigning for greater centralisation, harmonisation, and rationalisation of imperial governance in India. But across much European imperial space, it was the early-twentieth-century rise of aviation—both military and civilian—which prompted a concerted upscaling and coordination of meteorological activity. Meteorology had long since helped "to add a vertical dimension to the territorial impulses of colonisation," but aviation gave new meaning and prominence to the imperial conquest of the skies.[46]

Aviation was the chief motivation for London's calling of the 1929 Empire Meteorology conference—a moment in which hopes abounded that the geographically dispersed empire would soon be "welded together into a physical unity" by an emerging infrastructure of flight.[47] The rise to prominence of aviation during the Great War had seen meteorology cement its military role in European capitals of empire. In 1919 the UK Meteorological Office was moved into the new Air Ministry, to enable better coordination between meteorologists and military and civilian flyers. In Europe and North America short-haul flights quickly became established—the London–Paris route was inaugurated in 1920—and soon attention was turning to potential imperial routes. Adventurers like Alan Cobham were swiftly followed by official reconnaissance flights down prospective flyways such as Cairo to Cape Town, and "aviation meteorology" was institutionalised as a new branch of the science, with its own distinctive practices and norms of knowledge making.[48] The anticipation of aviation's needs for reliable weather observations and forecasts informed the rapid establishment of meteorological services in places such as British East Africa, Malaya, and much of the French colonial empire.[49]

For the most part the relationship between colonial territory and colonial airspace was conceived of in a relatively straightforward manner. The Versailles Treaty of 1919 gave national governments responsibility for securing and maintaining their airspace as "a knowable, navigable, and ultimately usable environment,"[50] and aviation boosters in settings like London were wedded to the idea that imperial air routes should be operated by national carriers, and reliant solely on national infrastructure, including meteorological services.[51] Yet in some parts of the world the roll-out of meteorological infrastructure in service of aviation could serve broader territorial aims. Zaiki and Tsukahara have shown how early-twentieth-century Japanese imperial expansion was accompanied by a rhetoric of scientific colonialism, with expansion to the west and south swiftly followed by meteorological infrastructure which could both demonstrate scientific superiority and open up the skies for aerial conquest.[52] Meanwhile, in the southwestern Pacific, an expansion of "British" meteorological networks was justified on the grounds of not only securing aerial hegemony in the region, but also of laying the foundations for a communications infrastructure which could prove vital to British interests in the case of military conflict.[53]

Empire-spanning aviation was a particularly metropolitan project. It meant that metropolitan meteorologists suddenly had a renewed interest in understanding the skies above colonial territories, and of seeing imperial atmospheres as interconnected wholes. As one British Air Minister put it retrospectively, "The need was ... to chart the air just as the sea had been charted by generations of hydrographers. Whilst it had taken centuries to chart the seas, we were attempting to chart the air in a few months."[54] But just as imperial imaginaries of aviation did not always chime with settler-colonial visions of the role of flight in furthering the imperial cause,[55] so did this metropolitan charting of the air, with its implied centralisation of data gathering as well as its new, hurried timescales of meteorological practice, not always sit easily with the priorities of colonial meteorologists. For Albert Walter, by this time running the British East African Meteorological Service, aviation meteorology regularly proved an annoying distraction from his own project of reconciling settler agriculture with the demands of East African climates.[56] With its focus on upper-air observations and the frenetic production of forecasts for pilots, aviation meteorology could detract from the slower work of making colonial territory on the ground.

Decolonising meteorology

Gregory Cushman has used the concept of neocolonialism to explore the relationship between aviation, empire and meteorology in early twentieth century South America.[57] In the absence (for the most part) of formal colonial rule, interwar South America provides a useful case study of how colonial relationships of dependency nonetheless persisted, or were newly established, between states, commercial entities, and scientists following formal decolonisation.[58] German scientists were prominent in South American meteorology in the early 1920s and helped establish a German presence in an emerging aviation system. The Deutsche Seewarte like-wise conducted extensive upper-air work over the Atlantic, in part to open up transoceanic air routes.[59] Yet by the 1930s Pan American Airways was the dominant force, and rather than relying on existing governmental meteorology, it organised its own private weather-forecasting system. Meanwhile, proselytisers of "Bergen School" methods of air mass analysis used the rise of aviation in the Americas to further the reach of their own scientific empire.[60] Cushman notes the contrasts between different national "styles" of neocolonial meteorology—in the German case, science was a leading edge, allowing in other forms of neocolonial domination; in the US case, science often followed other forms of political or economic dominance.[61] For Scandinavians, the geographical expansion of Bergen School methods allowed the paradigm to maintain its coherence even as cracks began to show in the underlying science, while the hearty embrace of such techniques by Canadian meteorologists helped them to assert their autonomy from British meteorological circles.[62]

This opens up wider questions about the impact of the loss of colonial territories on European meteorology. Philipp Lehmann has shown how interwar German meteorologists and climatologists dealt with the loss of colonial "field sites," and how this changed the emphasis of much German atmospheric science towards questions of regional particularity.[63] Deborah Coen has similarly argued that the collapse of the Habsburg Empire forced a re-scaling of Austrian atmospheric science, with mountain micro-climates taking the place of continental weather systems as a chief object of analysis.[64] It might be tempting then to imagine the shrivelling-up of European empires and their scientific infrastructures as we move into the Cold War period, and their replacement by a more benign, internationalist system of scientific cooperation between states on an even playing field.[65] We don't yet know enough about transformations in the science of meteorology wrought by decolonisation, although we do have glimpses of the per-sistence of highly unequal power relationships between "Western" scientific institutions and the makers and users of meteorological knowledge in postcolonial states. Transnational meteoro-logical cooperation continued to function as a means of exercising diplomatic power, of priv-ileging certain knowledge users over others, and of rendering the global atmosphere knowable to hegemonic military powers.[66]

There remains a great number of questions which historians might profitably address: for example, what was the impact of decolonisation on the meteorological institutions established by colonial powers? Historians of geography have recently had much to say about the fate of that discipline during decolonisation, and historians of meteorology might adopt similar methods.[67] Historians of the atmospheric sciences may also find much to say to ongoing efforts to decolonise the natural and physical sciences, in the sense of identifying the perpetuation of colonial modes of thought and practice. This might mean attending to what colonial meteor-ology and climatology helped erase or elide—Indigenous forms of environmental knowledge which sit at the heart of different ways of being in the world, and which have much to teach researchers and governments grappling with the challenges of mitigating and adapting to global climate change.

Attending to such processes of erasure, or perhaps co-option, will require new ways of approaching the long entanglement of meteorology and empire. It will require looking not just at formal state actors and prominent scientists, but at the people—both colonisers and colonised—who made up the socio-technical infrastructures of meteorology, the "go-betweens" who helped Europeans make sense of strange environments, and the contact zones where different forms of knowledge met and hybridised, and where transimperial dynamics played out.[68] Work in this vein is starting to emerge, alongside work which decentres European histories. Such work is challenging, as it requires reading across diverse and dispersed archival sources and recovering voices and stories that may be absent from official records. But a number of scholars are increasingly reading across the archival grain of imperial meteorology, with the promise of new ways of understanding the place of the science in the warp and weft of imperial endeavour.

Notes

1 See Joan M. Kenworthy, "Albert Walter, O.B.E (1877–1972) Meteorologist in the Colonial Service Part I: His Early Life and Work in Mauritius," *Occasional Papers on Meteorological History* (Reading: Royal Meteorological Society, 2013); Martin Mahony, "The 'Genie of the Storm': Cyclonic Reasoning and the Spaces of Weather Observation in the Southern Indian Ocean, 1851–1925," *British Journal for the History of Science* 51, no. 4 (2018): 607–33.

2 See for example Fiona Williamson, "Uncertain Skies: Forecasting Typhoons in Hong Kong, c. 1874–1906," *Quaderni Storici* 156, no. 3 (2017): 777–802.

3 A useful discussion of the unstable boundaries between meteorology and climatology features in Philipp N. Lehmann, "Whither Climatology? Brückner's Climate Oscillations, Data Debates, and Dynamic Climatology," *History of Meteorology* 7 (2015): 49–70.

4 David N. Livingstone, "Race, Space and Moral Climatology: Notes toward a Genealogy," *Journal of Historical Geography* 28, no. 2 (April 2002): 159–80.

5 Sam White, "Unpuzzling American Climate: New World Experience and the Foundations of a New Science," *Isis* 106, no. 3 (2015): 544–66; see also Vladimir Jankovic, "Climates as Commodities: Jean Pierre Purry and the Modelling of the Best Climate on Earth," *Studies in History and Philosophy of Science Part B – Studies in History and Philosophy of Modern Physics* 41, no. 3 (2010): 201–7.

6 James Rodger Fleming, *Historical Perspectives on Climate Change* (Oxford: Oxford University Press, 2005); Brant Vogel, "The Letter from Dublin: Climate Change, Colonialism, and the Royal Society in the Seventeenth Century," *Osiris* 26, no. 1 (2011): 111–28.

7 Richard H. Grove, *Green Imperialism: Colonial Expansion, Tropical Island Edens and the Origins of Environmentalism, 1600–1860*, vol. 3 (Cambridge: Cambridge University Press, 1995).

8 Michael A. Osborne, "Acclimatizing the World: A History of the Paradigmatic Colonial Science," *Osiris* 15 (2000): 135–51; see also Helen Cowie, "From the Andes to the Outback: Acclimatising Alpacas in the British Empire," *Journal of Imperial and Commonwealth History* 45, no. 4 (2017): 551–79.

9 Greg Bankoff, "Aeolian Empires: The Influence of Winds and Currents on European Maritime Expansion in the Days of Sail," *Environment and History* 23, no. 2 (2017): 163–96.

10 Bankoff, "Aeolian Empires," 170.

11 On the professionalisation and imperial role of ocean science in the nineteenth century, see Michael S. Reidy, *Tides of History: Ocean Science and Her Majesty's Navy* (Chicago: University of Chicago Press, 2008); Michael S. Reidy and Helen M. Rozwadowski, "The Spaces In Between: Science, Ocean, Empire," *Isis* 105, no. 2 (2014): 338–51; Fabien Locher, "L'émergence d'une techno-science de l'environnement océanique: le cas du programme français de cartographie des vents en mer, 1860–1914," *Nuncius: Annali Di Storia Della Scienza* 22, no. 2 (2007): 49–68.

12 Azadeh Achbari, "Building Networks for Science: Conflict and Cooperation in Nineteenth-Century Global Marine Studies," *Isis* 106, no. 2 (2015): 257–82.

13 Simon Naylor, "Log Books and the Law of Storms: Maritime Meteorology and the British Admiralty in the Nineteenth Century," *Isis* 106, no. 4 (2015): 782.

14 Fabien Locher, "The Observatory, the Land-Based Ship and the Crusades: Earth Sciences in European Context, 1830–50," *British Journal for the History of Science* 40, no. 147 (2007): 491–504.

15 Fiona Williamson, "Weathering the Empire: Meteorological Research in the Early British Straits Settlements," *The British Journal for the History of Science* 48, no. 3 (2015): 475–92.

16 Simon Naylor and Matthew Goodman, "Atmospheric Empire: Historical Geographies of Meteorology at the Colonial Observatories," in *Weather, Climate, and the Geographical Imagination*, ed. Martin Mahony and Samuel Randalls (Pittsburgh, PA: University of Pittsburgh Press, 2020), 25–42; see also Pedro M.P. Raposo, "Time, Weather and Empires: The Campos Rodrigues Observatory in Lourenço Marques, Mozambique (1905–1930)," *Annals of Science* 72 (2015): 279–305; Frédéric Soulu, "Observatoires français dans l'Algérie Coloniale: Forme et spatialité." Actualité des recherches du Centre François Viète, Série III (Nantes: Centre François Viète, 2018).

17 Mahony, "The 'Genie of the Storm.'"

18 Naylor and Goodman, "Atmospheric Empire," 41.

19 Quoted in Williamson, "Weathering the Empire," 488.

20 Naylor and Goodman, "Atmospheric Empire," 42.

21 Gregory T. Cushman, "The Imperial Politics of Hurricane Prediction: From Calcutta and Havana to Manila and Galveston, 1839–1900," in *Nation-States and the Global Environment*, ed. Mark Lawrence, Erika Bsumek, and David Kinkela (Oxford: Oxford University Press, 2013), 137–62.

22 Kerby C. Alvarez, "Instrumentation and Institutionalization: Colonial Science and the Observatorio Meteorológico de Manila, 1865–1899," *Philippine Studies: Historical and Ethnographic Viewpoints* 64, no. 3–4 (2016): 385–416; Aitor Anduaga, "Spanish Jesuits in the Philippines: Geophysical Research and Synergies between Science, Education and Trade, 1865–1898," *Annals of Science* 71, no. 4 (2014): 497–521.

23 Cushman, "Imperial Politics of Hurricane Prediction." On transimperial networks, see also R. Wille, "Colonizing the Free Atmosphere: Wladimir Köppen's 'Aerology,' the German Maritime Observatory, and the Emergence of a Trans-Imperial Network of Weather Balloons and Kites, 1873–1906," *History of Meteorology* 8 (2017): 95–123.

24 See for example Richard H. Grove, "The East India Company, the Raj and the El Niño: The Critical Role Played by Colonial Scientists in Establishing the Mechanisms of Global Climate Teleconnections 1770–1930," in *Nature & The Orient*, ed. Richard H. Grove, Vinita Damodaran, and Satpal Sangwan (Oxford: Oxford University Press, 1998), 301–23. Richard H. Grove and George Adamson, *El Niño in World History* (Basingstoke: Palgrave, 2018).

25 Meredith McKittrick, "Theories of 'Reprecipitation' and Climate Change in the Settler Colonial World," *History of Meteorology* 8 (2017): 75.

26 Ruth Morgan, "Prophecy and Prediction: Forecasting Drought and Famine in British India and the Australian Colonies," *Global Environment* 13, no. 1 (2020): 99.

27 See for instance Zeke Baker, "Meteorological Frontiers: Climate Knowledge, the West, and US Statecraft, 1800–50," *Social Science History* 42, no. 4 (March 7, 2018): 731–61.

28 Emily O'Gorman, "'Soothsaying' or 'Science'? H.C. Russell, Meteorology, and Environmental Knowledge of Rivers in Colonial Australia," in *Climate, Science, and Colonization: Histories from Australia and New Zealand*, ed. James Beattie, Emily O'Gorman, and Matthew Henry (Basingstoke: Palgrave Macmillan, 2014), 180.

29 See for example D.W. Meinig, "Goyder's Line of Rainfall: The Role of a Geographical Concept in South Australian Land Policy and Agricultural Settlement," *Agricultural History* 35, no. 4 (1961): 207–14; Kirsty Douglas, "'For the Sake of a Little Grass': A Comparative History of Settler Science and Environmental Limits in South Australia," in Beattie, O'Gorman, and Henry, *Climate, Science, and Colonization*, 99–118; Chris O'Brien, "Imported Understandings: Calendars, Weather, and Climate in Tropical Australia, 1870s-1940s," in Beattie, O'Gorman, and Henry, *Climate, Science, and Colonization*, 195–212.

30 Robert M. Rouphail, "Cyclonic Ecology: Sugar, Cyclone Science, and the Limits of Empire in Mauritius and the Indian Ocean World, 1870s–1930s," *Isis* 110, no. 1 (March 2019): 48–67.

31 Sandip Hazareesingh, "Cotton, Climate and Colonialism in Dharwar, Western India, 1840–1880," *Journal of Historical Geography* 38, no. 1 (2012): 11.

32 For a comparable mid-twentieth-century example, see Martin Mahony, "Weather, Climate, and the Colonial Imagination: Meteorology and the End of Empire," in Mahony and Randalls, *Weather, Climate, and the Geographical Imagination*, 168–89.

33 Similar points are articulated in Richard H. Grove, *Ecology, Climate and Empire: Colonialism and Global Environmental History, 1400–1940* (Cambridge: White Horse Press, 1997); see also Helen Tilley, *Africa as a Living Laboratory: Empire, Development, and the Problem of Scientific Knowledge, 1870–1950*

(Chicago: University of Chicago Press, 2011). On the links between imperial forestry and a nascent environmentalism, see Gregory Allen Barton, *Empire Forestry and the Origins of Environmentalism* (Cambridge: Cambridge University Press, 2002). McKittrick has also documented interesting patterns of epistemic authority in South African settler community debates over climate change. See Meredith McKittrick, "Talking about the Weather: Settler Vernaculars and Climate Anxieties in Early Twentieth-Century South Africa," *Environmental History* 23 (2018): 3–27.

34 On boundary struggles in meteorology, see e.g. Matthew Henry, " 'Inspired Divination': Mapping the Boundaries of Meteorological Credibility in New Zealand, 1920–1939," *Journal of Historical Geography* 50, no. October 2015 (2015): 66–75.

35 Philipp Lehmann, "Average Rainfall and the Play of Colors: Colonial Experience and Global Climate Data," *Studies in History and Philosophy of Science Part A* 70 (May 2018): 38–48.

36 Napier Shaw, quoted in the *Report of the Agricultural Section of the Conference of Empire Meteorologists* (London: His Majesty's Stationery Office, 1929).

37 Martin Mahony, "For an Empire of 'All Types of Climate': Meteorology as an Imperial Science," *Journal of Historical Geography* 51 (2016): 29–39. On the social variability of the boundaries of scientific disciplines, see Thomas F. Gieryn, *Cultural Boundaries of Science: Credibility on the Line* (Chicago: University of Chicago Press, 1999).

38 Katharine Anderson, *Predicting the Weather: Victorians and the Science of Meteorology* (Chicago: University of Chicago Press, 2005).

39 Morgan, "Prophecy and Prediction," 130.

40 Mike Davis, *Late Victorian Holocausts: El Niño Famines and the Making of the Third World* (London: Verso, 2001).

41 Peter Holland and Jim Williams, "Pioneer Settlers Recognizing and Responding to the Climatic Challenges of Southern New Zealand," in Beattie, O'Gorman, and Henry, *Climate, Science, and Colonization*, 81–96.

42 Meredith McKittrick, "Race and Rainmaking in Twentieth Century Southern Africa," in Mahony and Randalls, *Weather, Climate, and the Geographical Imagination*, 152–67.

43 Suman Seth, "Putting Knowledge in Its Place: Science, Colonialism, and the Postcolonial," *Postcolonial Studies* 12, no. 4 (2009): 375.

44 Deborah R. Coen, *Climate in Motion: Science, Empire, and the Problem of Scale* (Chicago: University of Chicago Press, 2018).

45 Anderson, "Predicting the Weather," 284.

46 Morgan, "Prophecy and Prediction," 99–100.

47 C. Dennistoun Burney, *The World, the Air and the Future* (London: A.A. Knopf, 1929), 21.

48 Roger Turner, "Weathering Heights: The Emergence of Aeronautical Meteorology as an Infrastructural Science" (PhD thesis, University of Pennsylvania, 2010); Martin Mahony, "Historical Geographies of the Future: Airships and the Making of Imperial Atmospheres," *Annals of the American Association of Geographers* 109, no. 4 (2019): 1279–99.

49 Mahony, "For an Empire of 'All Types of Climate'"; Pierre Duvergé, "Le Service Météorologique Colonial," *La Météorologie* (April 1995): 46–51.

50 Matthew Henry, "Australasian Airspace: Meteorology, and the Practical Geopolitics of Australasian Airspace, 1935–1940," in Beattie, O'Gorman, and Henry, *Climate, Science, and Colonization*, 235.

51 See for example Gordon Pirie, *Air Empire: British Imperial Civil Aviation, 1919–39* (Manchester: Manchester University Press, 2009).

52 M. Zaiki and T. Tsukahara, "Meteorology on the Southern Frontier of Japan's Empire: Ogasawara Kazuo at Taihoku Imperial University," *East Asian Science, Technology and Society* 1, no. 2 (2007): 183–203.

53 Henry, "Australasian Airspace"; see also Matthew Henry, "Assembling Meteorology: Balloons, Leaking Gas, and Colonial Relations in the Making of New Atmospheres," *Journal of the Royal Society of New Zealand* 47, no. 2 (April 2017): 162–8.

54 Sir Samuel Hoare, *Empire of the Air: The Advent of the Air Age, 1922–1929* (London: Collins, 1957), 225.

55 See for example Robert L. McCormack, "Imperialism, Air Transport and Colonial Development: Kenya, 1920–46," *The Journal of Imperial and Commonwealth History* 17, no. 3 (May 1989): 374–95.

56 Mahony, "Weather, Climate, and the Colonial Imagination." See also Henry, "Making Weather Vertical: Meteorology and the Anxious Temporalities of Infrastructural Atmospheres in New Zealand, c. 1920," *Centaurus* 62, no. 4 (2020): 744–762.

57 Gregory T. Cushman, "The Struggle over Airways in the Americas, 1919–1945: Atmospheric Science, Aviation Technology, and Neocolonialism," in *Intimate Universality: Local and Global Themes in the History of Weather and Climate*, ed. James Rodger Fleming, Vladimir Janković, and Deborah R. Coen (Sagamore Beach, MA: Science History Publications, 2006), 175–222.

58 See also Gregory T. Cushman, "Enclave Vision: Foreign Networks in Peru and the Internationalization of El Niño Research during the 1920s," *Proceedings of the International Commission on History of Meteorology* 1 (2004): 65–74.

59 Penelope K. Hardy, "Meteorology as Nationalism on the German Atlantic Expedition, 1925–1927," *History of Meteorology* 8 (2017): 124–44.

60 See also Robert Marc Friedman, *Appropriating the Weather: Vilhelm Bjerknes and the Construction of a Modern Meteorology* (Ithaca, NY: Cornell University Press, 1993).

61 Jamie L. Pietruska, "Hurricanes, Crops, and Capital: The Meteorological Infrastructure of American Empire in the West Indies," *The Journal of the Gilded Age and Progressive Era* 15, no. 4 (2016): 418–45.

62 Cushman, "The Struggle over Airways."

63 Philipp Lehmann, "Losing the Field: Franz Thorbecke and (Post-)Colonial Climatology in Germany," *History of Meteorology* 8 (2017): 145–58.

64 Deborah R. Coen, "Scaling Down: The 'Austrian' Climate between Empire and Republic," in Fleming, Janković, and Coen, *Intimate Universality*, 115–40; Deborah R. Coen, "The Storm Lab: Meteorology in the Austrian Alps," *Science in Context* 22, no. 3 (2009): 463–86.

65 On the shift from "voluntary internationalism" to "quasi-obligatory globalism" in post-World War II meteorology, see Paul N. Edwards, "Meteorology as Infrastructural Globalism," *Osiris* 21, no. 1 (2006): 229–50.

66 See Clark A. Miller, "Scientific Internationalism in American Foreign Policy: The Case of Meteorology, 1947–1958," in *Changing the Atmosphere: Expert Knowledge and Environmental Governance*, ed. Clark A. Miller and Paul N. Edwards (Cambridge, MA: MIT Press, 2001), 167–217; Clark A. Miller, "Resisting Empire: Globalism, Relocalization, and the Politics of Knowledge," in *Earthly Politics: Local and Global in Environmental Governance*, ed. Sheila Jasanoff and Marybeth Long Martello (Cambridge, MA: MIT Press, 2004), 81–102; Paul N. Edwards, "Entangled Histories: Climate Science and Nuclear Weapons Research," *Bulletin of the Atomic Scientists* 68, no. 4 (2012): 28–40.

67 Ruth Craggs and Hannah Neate, "What Happens If We Start from Nigeria? Diversifying Histories of Geography," *Annals of the American Association of Geographers* 110, no. 3 (2020): 899–916.

68 Kapil Raj, "Go-Betweens, Travelers, and Cultural Translators," in *A Companion to the History of Science*, ed. Bernard Lightman (Chichester: Wiley, 2016), 39–57; Martin Mahony and Angelo Matteo Caglioti, "Relocating Meteorology," *History of Meteorology* 8 (2017): 1–14.

6

COLONIAL PSYCHIATRY

Matthew M. Heaton

Psychiatry is generally defined as the science of the treatment of mental illness. However, defining mental illness requires the ability to differentiate "abnormal" psychological function from "normal" thought and behaviour. Because the medical science of psychiatry developed in Western Europe, its definitions of normal psychological functioning were originally (many would say remain) rooted deeply in European cultures, traditions, and worldviews. As psychiatry spread around the world alongside European imperialism in the nineteenth and twentieth centuries, complex tensions emerged as colonial governments and institutions attempted to use European psychiatry to define, diagnose, and treat mental illness in non-European populations. On the one hand, colonial psychiatrists believed in the universal nature of their science: they believed they understood the basic workings of the human mind better than practitioners of any other health system and that European medicine should be used to define, diagnose, and treat severe mental illnesses regardless of cultural context. On the other hand, European imperialism rested on the notion that non-European peoples were different from Europeans, constitutionally and/or culturally inferior, and therefore in possession of "primitive" psyches both in need of and threatened by European "civilisation."

This tension between the universalist, egalitarian ethos of European scientific process and the particularist, hierarchical nature of European social constructs has shaped the academic discourse on colonial and postcolonial psychiatry since at least the early twentieth century. Since the 1980s, historians have made important contributions to understanding the political, social, and cultural forces that shaped developments in colonial psychiatry. In order to do this, however, psychiatry has to be put into a broader conversation with the beliefs and practices of colonial ideology and society. In order for colonial psychiatrists to study mental illness, they had to understand what constituted "normal" behaviour in colonial subjects. How they defined normal and abnormal behaviour was deeply embedded in the imperial mindset of European politics, culture, and science at the time. If European cultures were considered superior to non-European ones, and Europeans as more rational and of higher intelligence, then non-European cultures were inherently less capable of producing psychologically stable individuals, and non-European individuals were innately less intelligent and more irrational by nature. Colonial psychiatry therefore built from an original discourse in which "normal" psychology of colonised peoples was pathological compared to Europeans, and mentally ill colonial subjects were pathological compared to "normal" colonial subjects. Colonial psychiatry used the supposedly universal processes of scientific discovery to engage in what was ultimately a highly subjective pursuit of othering by delineating the psychologically abnormal from what they already considered to be exotic, strange, and pathological cultures.[1]

This chapter examines this key tension through several different lenses. It starts with an examination of colonial lunatic asylums, which existed to enforce some kind of physical demarcation between the sane and insane. It will then examine psychiatric research on the nature of the "native mind" and its pathologies, including colonial discourse on psychoanalytic theory. The chapter concludes with a discussion of psychiatric and psychoanalytic critiques of colonial psychiatry and the effects of decolonisation on the transformation of psychiatric theory and practice in what was soon to become the "Third World." While colonial psychiatry played a role in reinforcing the boundaries and ideology of colonial societies, it did so in very limited, contingent, and always contested ways.

Institutions

The asylum looms large in the historiography of colonial psychiatry. Most colonies offered little to nothing in the realm of outpatient treatment or therapeutic intervention, making the custodial asylum the primary locus of patient experience and knowledge production that has formed the basis of so much historical study. British India stands out as having a particularly long history of colonial lunatic asylums, with the earliest established in the late eighteenth century.[2] Many colonies in the Caribbean established asylums in the 1840 and 1850s.[3] The first asylum in sub-Saharan Africa was established at Kissy, in Sierra Leone, in 1820.[4] The late nineteenth and early twentieth centuries saw a major increase in the growth of colonial asylums, as European power spread out to occupy much of Africa and Asia. South Africa's Cape Colony used the infamous Robben Island to confine the insane from 1846, but the first official asylum opened in Natal in 1880.[5] Other newly acquired British colonies followed suit, with asylums in Accra (Gold Coast) in 1888,[6] Calabar and Yaba (Nigeria) in 1906, and Nairobi (Kenya) in 1910. French colonies also established asylums around the same time, in Algeria in 1911,[7] and Indochina in 1919.[8] The Dutch East Indies colony opened its first asylum in 1881 and claimed four by 1930.[9] Portuguese colonies established asylums a bit later, with the first asylum in Mozambique opening in 1930 and in Angola in 1946.[10]

Colonial asylums have often been conceptualised as institutions of social control in the sense that they existed almost exclusively as a space to sequester violent or otherwise uncontrollable individuals from the general population. However, as Megan Vaughan has astutely noted, the Foucauldian notion of a "Great Confinement" of the insane applies even less meaningfully in colonial contexts than it ultimately did in European metropolitan settings. Most colonies provided such limited resources to their asylums that the vast majority of mentally ill individuals in colonial contexts never experienced any colonial medical intervention whatsoever, relying instead on far more culturally legitimate and locally available Indigenous healers to understand and treat psychiatric disorders.[11] The asylums that did exist were chronically overcrowded and ineffective in aiding the plight of their inmates.

Historians have recounted how the day-to-day organisation of colonial asylums was deeply racialised and gendered. Most asylums were segregated with European and "native" wings, with Europeans receiving better treatment, less overcrowding, better food, etc.[12] Women's wings were kept separate from men's accommodations as well, and women's diagnoses often differed from men's in gender-specific ways. In Fiji, for example, most women's diagnoses were related to their sexual and reproductive functions, with many exhibiting symptoms of postpartum depression.[13] Europeans were often only temporarily housed in colonial asylums awaiting repatriation to their home countries. Non-white colonial subjects who happened to find their way into asylums were often repatriated to their homelands as well, particularly in the UK, on the notion that it would be in their interests to be returned to a more familiar, "primitive" environment.[14]

Diagnosis of mental disorders was far from uniform across or within colonies and empires in the nineteenth and twentieth centuries. Sometimes trends in diagnosis reflected European stereotypes about different colonial peoples. Malays were considered prone to group psychosis; North and West African Muslims to psychosis brought on by religious frenzy; sub-Saharan Africans were considered unlikely to develop depression because of perceptions that they were incapable of internalising guilt. Diagnoses also varied because of shifting definitions and criteria in psychiatry in the early twentieth century, as well as the fact that many of the colonial officials running asylums and making decisions about the condition of individual cases were not actually trained psychiatrists but rather general medical officers, police, or officers of the court. It therefore makes sense to see colonial asylums less as places where orthodox psychiatry was practised than as places where psychiatric science was experimented with and shaped to the political and social needs of colonial regimes.

Because colonial asylums generally lacked the knowledge, resources, or capacity to treat mental illness in Indigenous patients, they also lacked legitimacy with local populations, who generally saw them as places to be avoided.[15] Far from establishing hard boundaries between sanity and insanity, confinement and freedom, colonial asylums often had quite porous walls, with patients going in and out of custody for a variety of reasons over the course of their lives.[16] Patient perspectives have often been difficult to recover because of the limited availability of first-hand accounts of life in colonial asylums. Colonial accounts of individual patients have been recovered and analysed by historians reading case files or court records "against the grain" to reveal aspects of patient agency that might otherwise be impossible to trace. Sometimes these works have sought to deconstruct patients' symptom presentation for its political and social messages.[17] Others have explored the cultural content of individual cases to attempt to recover the world of the colonised subject as they might have understood it.[18] Such historical case studies have revealed that how Indigenous communities and patients understood mental illness and its treatment did not always align with the ways colonial officials understood things. Families regularly petitioned asylums for the release of their loved ones on the basis of their own experiences with them or the opinions of traditional healers more than the diagnosis of a colonial medical officer. In some contexts, communities sought assistance in dealing with mentally ill members for behaviours or problems that colonial officials found insufficient to act upon. Overall, studies of colonial asylums have demonstrated the contingencies and limitations of colonial power as well as Indigenous adaptations and resistance to colonial social, legal, and cultural impositions.

Colonial ideology and psychiatric knowledge

Although colonial psychiatric institutions remained significantly underdeveloped, they were nevertheless key sites for the creation of knowledge about the nature and function of the "native mind." Much of the research that took place in African and Asian asylums and hospitals centred on questions of what the "normal" psychological processes of colonised peoples were, what the causes of mental illness in colonised peoples were, and how Indigenous psychology and psychopathology compared to that of Europeans. Richard Keller has noted that in French North African colonies psychiatric research actually enjoyed a greater degree of freedom and experimentation than in the French metropole because of how far removed it was from the traditional power centre, both geographically and in terms of the socio-political environment.[19] While not all colonies produced on the scale of this "Algiers School," many were involved in efforts to understand the psychology of colonised populations for the twin sakes of science and to contribute to their idea of effective colonial governance.

Accounting for racial and cultural difference was the preoccupying concern of much of the psychological science that took place in European colonies in the nineteenth and twentieth centuries. On the one hand, a belief in the underlying basic function of the human mind drove psychiatric science in general. For example, Emil Kraepelin famously went to the Dutch East Indies in 1904 to examine mental illness among Indigenous peoples and concluded that both depressive illness and schizophrenia (dementia praecox) existed in these populations, although with different prognoses and cultural expressions of symptoms. On the other hand, however, colonial rule and the ideology of European power required the imposition of a racialised hierarchy in which Europeans represented the best, smartest, most "civilised" human societies, while non-white, colonised peoples were by definition inferior to Europeans, defined by Europeans in terms of what they were perceived to lack in relation to Europeans. Researchers were of different opinions about whether fundamental biological differences or cultural difference accounted for the inferiority of colonised peoples, and much research in the mental sciences was framed implicitly or explicitly around this question. Kraepelin, for example, explained the differences in symptomatology that he saw in Indonesia largely in terms of what he assumed to be differences between "Eastern" and "Western" brains, as well as the "primitive" simplicity of Javanese culture. The East Africa School that emerged out of Mathari Hospital in Nairobi in the 1920s was particularly focused on establishing a biological basis of psychological difference, with psychiatrists such as H.L. Gordon, F.W. Vint, and ultimately J.C. Carothers making the case for weak frontal-lobe function in Africans.[20] As late as 1951, Carothers infamously declared that the mental function of the "normal" African resembled that of the leucotomised European.[21]

Although biological explanations for psychological and psychopathological differences were common in the early twentieth century, cultural explanations interacted with them and ultimately became more prominent over time. "Othering" of colonised subjects remained central to these explanations, with colonised peoples' cultures taking on a pathological quality that warped non-Western minds. "Traditional" cultures were viewed by Europeans as simpler, dominated by religious superstitions, and designed to create dependent, rather than independent, adults. The result, Europeans argued, were adults who in their traditional environments were probably much less likely to develop severe mental illnesses, as their worldviews were rarely challenged and required them to take very little personal responsibility, thereby limiting psychological turmoil.[22] However, when these traditional worldviews came into contact with European modernity through the colonial encounter, the likelihood of developing mental illness grew because "primitive" non-Westerners were not psychologically equipped to handle the transition. Those "deculturated" or "detribalised" colonial subjects who attempted to assimilate European ideas through Western education, employment, or lifestyle were considered to be the most vulnerable to breakdown. This "clash of cultures" theory predominated in many different colonial environments, and led to tensions regarding the proper response. On the one hand, it clearly pathologised Indigenous cultures—usually defined in racial terms—as less sophisticated than European culture. On the other, it also contained an implicit colonial critique, arguing that the colonial encounter was itself pathological, endangering the psychological well-being of those it sought to transform.

Historians have since debunked many of the scientific studies that supposedly supported notions of non-European psychological inferiority. Most of the studies on mental illness in colonised populations were based strictly on examination of asylum inmates, who generally represented a fraction of the most severely afflicted individuals in the larger society. For example, studies of psychiatric patients in British African territories regularly found institutionalisation rates more than ten times lower than in the UK. As of 1939, the rate of institutionalisation for mental derangement in England and Wales stood at four per 1,000 of the total population. In contrast, colonial statistics indicated that institutionalisation rates in African colonies were much lower. Geoffrey Tooth estimated the rate of institutionalisation in the Gold Coast to be 0.3 per 1,000 in 1950.[23] In Nyasaland (Malawi), Shelley and Watson estimated the rate to be 0.06 per 1,000 as of 1936.[24] Carothers estimated an institutionalisation rate of 0.1 per 1,000 in Kenya in 1953,[25] Despite recognised problems with the data, colonial psychiatrists regularly extrapolated from observations of those diagnosed with mental illness to make claims about the nature of rates of mental illness in the society at large and the "normal" function of general populations in ways that are not scientifically supported by the nature of the evidence. Perhaps most famously, Carothers argued, based on his experience at Mathari hospital, that the Mau Mau uprising in Kenya in the 1950s was a form of group psychosis brought on by the "deculturating" effects of urbanisation and proximity to European "civilisation."[26]

Not all colonial mental science supported such racist conclusions. A number of psychological studies done in general populations in the early twentieth century actually supported the idea that psychological functioning of individuals who had not been diagnosed with severe mental illness was actually much more similar to that of Europeans than was generally assumed.[27] Concerns about cultural bias in intelligence testing were already well documented in the 1920s. By the late 1950s, some anthropologists and psychiatrists were also arguing that rates of mental illness were much higher in general populations than previous estimates based on asylum admissions suggested.[28] Margaret Field's seminal 1960 work, *Search for Security*, documented a large number of cases of what she believed to be depressed women attending Indigenous shrines in northern Ghana, challenging prevailing notions that Africans rarely suffered from depression.[29] But studies that implied colonial subjects should receive similar psychiatric care and attention to what Europeans received in their home countries tended to be less politically useful to colonial regimes, and generally failed to influence government action significantly.[30]

The discourse surrounding the importance of cultural difference in understanding non-European psychology and psychopathology also produced debates about the appropriate treatment of mental illness in colonised populations. If culture determined psychological function and symptomatology of mental illness, might not culture also be important in shaping effective treatment of individuals of different backgrounds? While colonial psychiatrists regularly argued that their knowledge about the nature and diagnosis of mental illness was superior to that of Indigenous cultures, they nevertheless often begrudgingly noted that Indigenous healers might be in a better position to treat mental illness due to their deep cultural knowledge and legitimacy amongst local communities, which colonial psychiatry certainly did not enjoy. It was not at all uncommon for colonial psychiatrists to argue that Indigenous healers were better equipped to handle less severe forms of mental illness, such as depression, and "neuroses," while Western psychiatry was more effective in treating psychosis or organic brain diseases like epilepsy and neurosyphilis.[31]

Psychoanalysis

In addition to the scientific discourse of colonial psychiatry, Sigmund Freud's theory of the unconscious as the primary underlying force shaping individual behaviour had a profound global impact in the twentieth century. Freudian psychoanalysis and its offshoots (Jungian,

Lacanian, Kleinian, etc.) were powerful models not simply for their psychotherapeutic interventions but also for theorising about the nature of the human mind and the relationship between nature and nurture. Is the unconscious universal? Does it work the same way in all people? How do culture and environment shape psychoanalytic processes, and are differences fundamental or simply in need of translation? As with the psychiatric research described above, European colonies became particularly salient spaces for the examination of questions about the universal principles of psychoanalysis, with colonised populations often serving as test cases for comparisons between "primitive" and "modern" societies. Psychoanalysis both shaped and was shaped by colonial ideologies in profound ways that historians have begun to examine since the early twenty-first century, mostly in terms of postcolonial and globalisation critiques that emphasise local adaptation and implementation of knowledge rather than pure diffusion. As a result psychoanalysis serves different cultural and political purposes in different places, even as it draws on a notion of the universality of the self.

Much of the historical study of the relationship between psychoanalysis and colonialism has applied postcolonial critique to the writings of Freud and his acolytes. Freud's theories have long been recognised as being deeply influenced by *fin de siècle bourgeois* European political, social, and cultural context. However, historians have also been paying more attention to the ways that European colonial expansion shaped the worldview of those *bourgeois* Europeans, Freud included. Racialised conceptions of the non-European "other" are operational in Freudian psychoanalytic theory just as they were in colonial psychiatric science. The notion of human cultural evolution as a progression from the "primitive" to the "modern" (both of which are pathological, of course) was grafted on to a colonised world in which the coloniser represented "modernity" and the colonised represented the "primitive" cultures out of which the "modern" European had evolved (although elements of primitivity continue to pervade the modern mind in Freud's view). The result was a largely racialised construction in which much psychoanalytic interpretation of colonial peoples equated European children and neurotic adults to "normal" non-European adults. On a basic level, the universal aspects of the unconscious applied to all peoples, but non-Western cultures were innately more primitive, pathological, and in need of guidance by enlightened, modern, Europeans.[32]

However, recognising the colonial influence on psychoanalytic knowledge production says more about psychoanalysis than about colonialism. Indeed, as Ranjana Khanna puts it, "to say so is to say very little."[33] It still places Europeans at the centre of the story and tells us little about how colonial subjects have engaged with or otherwise been affected by psychoanalytic ideas. Historians trained in Area Studies in different parts of the world have examined the impact of psychoanalysis in a variety of colonial and postcolonial spaces. These works engage partly in postcolonial critique, recognising psychoanalytic discourse as a means through which to examine social constructions of race, gender, class, ethnicity, and sexuality in particular colonial contexts. However, they also draw on methodologies in science studies and globalisation studies to examine the ways that psychoanalytic knowledge produced and adapted in colonial contexts circulated between colonies, imperial motherlands, and international intellectual circuits in the twentieth century.

Historical examination of efforts to translate psychoanalytic texts and apply them on non-Western patient populations reveals that such experiments took place in limited, specific settings, but deep analysis of individual patients did not spread very widely. Nevertheless, study of applications of psychoanalysis in non-European contexts offered opportunities to theorise about the psychological relationship between colonised and coloniser, as well as the psycho-logical implications of colonial rule itself. For example Wulf Sachs's *Black Hamlet* famously psychoanalysed a single South African patient named John Chavafambira, making a case for the existence of an inner psychoanalytic self in black Africans that was a fairly radical concept at the time of its publication in 1937.[34] The most famous example of an effort to apply psychoanalysis in a colonial context is that of Girindrasekhar Bose, sometimes called the "Freud of India," who established a psychoanalytic practice in Calcutta and founded the Indian Psychoanalytic Society in 1922. As Christine Hartnack has shown, Bose found that to practise psychoanalysis amongst Bengalis, he had to make changes to the clinical environment. He also engaged in efforts to apply psychoanalysis to ancient Hindu texts, on the assumption that both fundamentally revealed universal truths about the self in different cultural landscapes. While Freud was apparently flattered by Bose's work, he did not approve of the derivation of his theories.[35] Similarly, Omnia El Shakry's elucidation of the work of Egyptian scholar Yusuf Murad reveals the complexity, but also the intellectual sophistication, of colonial subjects' efforts to translate psychoanalysis cross-culturally. Murad spent much of the 1940s and 1950s working to apply modern, secular psycho-analytic theory to Islamic religious thought in ways similar to Bose's work with Hindu scripts.[36]

Psychoanalytic concepts were as much political as scientific terrain, particularly in the late colonial era. Whereas many Europeans had found in psychoanalysis justifications for racialised worldviews and colonial governance, by the 1940s some thinkers were using psychoanalysis to critique colonial society and the effect that it had on both colonised and coloniser. For example, the French Lacanian psychoanalyst Octave Mannoni published *Prospero and Caliban,* a scathing critique of colonialism in Madagascar, in 1950. Using the relationship of the master and servant from Shakespeare's *The Tempest* as the central metaphor, Mannoni argued that the coloniser projected his guilt on to the colonised "other" in violent and controlling ways, while the colonised became dependent and suffered from an inferiority complex as a result. The relationship was pathological on both sides.[37]

The most famous critic of European colonialism's psychological impacts was Frantz Fanon. Born in Martinique in 1925, Fanon trained in psychiatry in France after serving in the French military during the Second World War. Beginning in 1953, he practised psychiatry at the Blida-Joinville Psychiatric Hospital in Algeria, where he became involved in the National Liberation Front's (FLN's) national war of liberation from 1954 until his death in 1961. Fanon wrote extensively about the role of violence both in maintaining the colonial order and in overturning it. In so doing, he made appeals to psychoanalytic concepts to criticise European racism generally, and colonial violence specifically. In *Black Skin, White Masks,* Fanon argued that colonialism and racism created a double consciousness in black people that bred an inferiority complex in which black subjects saw themselves as neither authentically connected to their ancestral cultures nor fully able or allowed to assimilate into the society of the colonial mother country.[38] Much of Fanon's psychoanalytic critique of colonialism was explicitly sexual and gendered, arguing that structural racism did not allow black males to become full men (relative to the white master), and that French efforts to unveil Muslim Algerian women were a symbolic act of sexual violence reflecting French men's sexual desires to possess Algerian women on the one hand, and colonialist desires to penetrate and overpower the Algerian nation on the other.[39]

Psychoanalytic concepts could also be used for counter-revolutionary purposes, however. In the Dutch East Indies, the psychiatrist Pieter Mattheus van Wulfften Palthe explained away the post-war nationalist movement as a confused displacement of aggression on to the Dutch resulting from the surrender of the Japanese, the absent "all-dominant father figure" that Indonesians really wanted to fight.[40] And Carothers's opinion, mentioned above, that the Mau Mau rebellion was caused by the psychological conflict generated by a "primitive" people's inability to effectively assimilate to an abruptly encountered "modernity" and not by the material grievances of Kikuyu peasants, served the repressive politics of the British colonial state in Kenya.[41]

Decolonisation

Fanon's psychiatric and psychoanalytic critiques of colonialism have been highly influential in discourses on decolonisation and postcolonialism. In recent decades, so much analysis of Fanon's contributions has been penned that some scholars reference a burgeoning field of "Fanon studies."[42] To the extent that colonial subjects were "mad," he argued, colonial rule had made them that way by creating conditions in which fear, anger, violence, and psychological disorder were natural responses. In this way, Fanon's work played into notions of the psychopathology of colonised peoples that colonial psychiatry had promulgated, but used them to condemn the nature of European power rather than Indigenous cultures. This ability to turn colonial knowledge against the coloniser, to use knowledge created by the coloniser about the colonised as a mirror reflecting back on the coloniser, ultimately became one of the foundational methodological principles of postcolonialism, an interdisciplinary theoretical framework that has been one of the most significant transformers of humanistic scholarship about "non-Western" societies since the 1980s.

However, decolonisation produced a variety of other kinds of critiques and transformations of colonial-era psychiatry that historians have only recently begun to examine in detail. In the decades after the Second World War, most European colonies undertook the political transition to independent nation-states. Processes of decolonisation took different forms, ranging from the gradual transfer of power through peaceful negotiation in places like Nigeria, French West Africa, and India to forced European withdrawal as the outcome of wars of liberation as in Algeria, Indochina, Indonesia, and Portuguese colonies in Africa. Decolonisation brought about changes both to the structures and practices of psychiatry in these transitioning spaces. In cases of gradual, negotiated decolonisation efforts were often made to increase the capacity of psychiatric institutions from the 1940s to the 1960s. Expansion of educational opportunities and the indigenisation of the civil service and professions meant that many colonies saw their first psychiatrists of Indigenous background during this period. E.F.B. Forester of Ghana (a Gambian by birth), Tigani El-Mahi of Sudan, and Thomas Adeoye Lambo of Nigeria became the first three African psychiatrists practising in sub-Saharan Africa in the early 1950s. Indian psychiatrists had been practising since the early twentieth century, but their total numbers grew rapidly in the decade after independence and partition in 1947.[43]

The indigenisation of psychiatry in the colonies corresponded with transformations in psychiatric emphasis in the West after the Second World War as well. The massive psychological trauma of the war, coupled with the enormous refugee crisis after the war, led to a growing concern for psychiatry as both a transcultural and a global necessity. Efforts to apply psychiatric principles across cultural boundaries and to research specific cultural beliefs and practices related to mental illness grew significantly from the 1950s on, with many psychiatrists in the colonies leading the way in cutting-edge research. Lambo's tenure at Aro Mental Hospital in Nigeria established a community therapy model in which patients lived in local villages and received treatment from both trained medical staff and Indigenous healers. He also produced research into the prevalence of mental illness in rural communities, concluding that depression was actually much more common than earlier colonial psychiatrists would have thought possible.[44] The Fann School in Dakar, Senegal, became famous for its study of Indigenous perspectives on mental illness and psychotherapy under the leadership of the French ethnopsychiatrist Henri Collomb.[45] Indian psychiatrists at the All Indian Institute for Mental Health in Bangalore studied ayurvedic approaches to diagnosing and treating mental illness at the same time that they undertook clinical trials into psychopharmaceuticals.[46] The widely prescribed antipsychotic drug reserpine was famously isolated from the plant *Rauwolfia serpentina,* which had been used to treat mental illness in India for hundreds of years.[47] Overall availability of treatment at psychiatric hospitals increased dramatically from the 1950s in many (former) European colonies, although supply has remained far below need in most places ever since colonial times.

At the same time, much of the new research of the decolonisation and post-independence eras tended to focus more on the cross-cultural psychological and psychopatholotical similarities between peoples, challenging the conclusions of previous colonial-era research that had emphasised the role that the primitive nature of Indigenous cultures played in establishing fundamental psychological and psychopathological differences between Europeans and non-Europeans. This shift toward including non-Europeans as psychological equals in psychiatric discourse allowed psychiatrists practising in postcolonial contexts to further exclaim the universality of their knowledge and craft, while simultaneously demonstrating the contributions that both psychiatric science and Indigenous intellectual labour could make toward the modernisation of independent nation-states. Still, critics have argued that the "globalisation" of Western psychiatric categories and treatments that accompanied this modernisation has led to the marginalisation of cultural competency as a meaningful tool of psychiatric care. As Arthur Kleinman famously declared, the spread of Western psychiatry created a "category fallacy" in which psychiatrists always "*find* what is 'universal' and systematically *miss* what does not fit its tight parameters."[48] The critique has carried forward into twenty-first-century analysis of the movement for global mental health, which aims to fill what proponents call the "treatment gap" by implementing basic psychotherapeutic and psychopharmaceutical interventions in ways that detractors say is not culturally attuned. Some see the predominance of biological psychiatry and its emphasis on pharmaceuticalisation as reflective of colonial psychiatry's top-down approach to understanding and treating mental illness.[49] Others see global mental health as being innately dependent upon local cooperation and the training of more mental health practitioners, similar to the transformations begun during the times of decolonisation.[50] Yet the fundamental questions that historians have identified as confounding colonial psychiatry persist today: is there universal knowledge about the nature of mental illness in humans and how can it be obtained? What kind of mental health care is appropriate for people of different cultures? And how should resources be allocated to achieve desired outcomes? The legacies of colonialism and decolonisation live in contemporary psychiatric debates and discourses.

Conclusion

The history of colonial psychiatry, while taking different shapes in different places and times, nevertheless is an object lesson in the fundamental principles of social and cultural histories of science and medicine. Scientific knowledge reflects the social and cultural beliefs and priorities of the powers that create it, and medical practice serves bodies and minds in ways that reflect the perceived roles they are expected to play in society. Colonial psychiatry was always significantly constrained both in terms of resources and impact, but its discourses both reinforced and critiqued the colonial project at large. Whether psychiatric science possessed the tools to understand and treat mental illness in all humans universally, or whether it could hope to acquire those tools, remains a central debate in efforts to achieve global mental health today. The question of how to account for colonial legacies in the development of international and postcolonial mental health initiatives remains an open one. If there is one thing that we have learned from the last four decades of historical scholarship, it is that the study of colonial psychiatry can teach us as much about the nature and practices of colonialism generally as it does those of psychiatry specifically.

Notes

1 Sloan Mahone and Megan Vaughan, eds., *Psychiatry and Empire* (New York: Palgrave/Macmillan, 2007).

2 Waltraud Ernst, "Asylum Provision and the East India Company in the Nineteenth Century," *Medical History* 42 (1998): 476–502; Waltraud Ernst, "Medical/Colonial Power – Lunatic Asylums in Bengal, c. 1800–1900," *Journal of Asian History* 40, no. 1 (2006): 49–79; James Mills, "The History of Modern Psychiatry in India, 1858–1947," *History of Psychiatry* 12 (2001): 431–58; Dinesh Bhugra, "The Colonized Psyche: British Influence on Indian Psychiatry," in *Colonialism and Psychiatry*, ed. Dinesh Bhugra and Roland Littlewood (Oxford: OUP, 2001), 46–76.

3 Leonard Smith, *Insanity, Race and Colonialism: Managing Mental Illness in the Post-Emancipation British Carribean, 1838–1914* (New York: Palgrave, 2014), 3.

4 Leland V. Bell, *Mental and Social Disorder in Sub-Saharan Africa: The Case of Sierra Leone, 1787–1990* (Westport, CT: Greenwood Press, 1991).

5 Julie Parle, *States of Mind: Searching for Mental Health in Natal and Zululand, 1868–1918* (Scottsville, South Africa: University of KwaZulu-Natal Press, 2007), 4.

6 K. David Patterson, *Health in Colonial Ghana: Disease, Medicine and Socioeconomic Change, 1900–1955* (Waltham, MA: Crossroads Press, 1981), 7.

7 Richard C. Keller, *Colonial Madness: Psychiatry in French North Africa* (Chicago: University of Chicago Press, 2007).

8 Claire E. Edington, *Beyond the Asylum: Mental Illness in French Colonial Vietnam* (Ithaca, NY: Cornell University Press, 2019).

9 Hans Pols, "The Psychiatrist as Administrator: The Career of W.F. Theunissen in the Dutch East Indies," *Health and History* 14, no. 1 (2012): 145.

10 Emmanuel Akyeampong, "A Historical Overview of Psychiatry in Africa," in *The Culture of Mental Illness and Psychiatric Practice in Africa*, ed. Emmanuel Akyeampong, Allan G. Hill, and Arthur Kleinman (Bloomington: Indiana University Press, 2015), 29.

11 Megan Vaughan, "Idioms of Madness: Zomba Lunatic Asylum, Nyasaland, in the Colonial Period," *Journal of Southern African Studies* 9, no. 2 (1983): 218–38.

12 Edington, *Beyond the Asylum*.

13 Jacqueline Leckie, *Colonizing Madness: Asylum and Community in Fiji* (Honolulu: University of Hawaii Press, 2019).

14 Matthew M. Heaton, "Contingencies of Colonial Psychiatry: Migration, Mental Illness, and the Repatriation of Nigerian 'Lunatics,'" *Social History of Medicine* 27, no. 1 (2014): 41–63.

15 Jonathan Sadowsky, *Imperial Bedlam: Institutions of Madness in Colonial Southwest Nigeria* (Berkeley: University of California Press, 1999).

16 Waltraud Ernst, *Mad Tales from the Raj: The European Insane in British India, 1800–1858* (London: Routledge, 1991). Lynette Jackson, *Surfacing Up: Psychiatry and Social Order in Colonial Zimbabwe, 1908–1968* (Ithaca, NY: Cornell University Press, 2005); Matthew M. Heaton, *Black Skin, White Coats: Nigerian Psychiatrists, Decolonization, and the Globalization of Psychiatry* (Athens: Ohio University Press, 2013); Edington, *Beyond the Asylum*.

17 Jean Comaroff and John Comaroff, "The Madman and the Migrant: Work and Labor in the Consciousness of a South African People," *American Ethnologist* 14 (1987): 191–209.

18 Jonathan Sadowsky, "The Confinements of Isaac O.: A Case of 'Acute Mania' in Colonial Nigeria," *History of Psychiatry* 7 (1996): 91–112.

19 Keller, *Colonial Madness*.

20 Jock McCulloch, *Colonial Psychiatry and the African Mind* (Cambridge: Cambridge University Press, 1995).

21 J.C. Carothers, "Frontal Lobe Function and the African," *Journal of Mental Science* 97, no. 406 (1951): 12–48.

22 For a synthesis of colonial-era scholarship on this point, see J.C. Carothers, *The African Mind in Health and Disease: A Study in Ethnopsychiatry* (Geneva: World Health Organization, 1953).

23 Geoffrey C. Tooth, *Studies in Mental Illness in the Gold Coast* (London: His Majesty's Stationery Office, 1950), 63.

24 H.M. Shelley and W.H. Watson, "An Investigation Concerning Mental Disorder in the Nyasaland Natives," *Journal of Mental Science* 82 (1936): 704.

25 Carothers, *The African Mind in Health and Disease*, 125.

26 J.C. Carothers, *The Psychology of Mau Mau* (Nairobi: Government Printer, 1954).

27 Erik Linstrum, *Ruling Minds: Psychology in the British Empire* (Cambridge, MA: Harvard University Press, 2016).

28 T. Adeoye Lambo, "Further Neuropsychiatric Observations in Nigeria with Comments on the Need for Epidemiological Study in Africa," *British Medical Journal* 2 (1960): 1696–1704.

29 M.J. Field, *Search for Security: An Ethno-Psychiatric Study of Rural Ghana* (Evanston, IL: Northwestern University Press, 1960).

30 Heaton, *Black Skin, White Coats*, chap. 1.

31 Heaton, *Black Skin, White Coats*, chap. 5.

32 Warwick Anderson, Deborah Jenson, and Richard C. Keller, eds., *Unconscious Dominions: Psychoanalysis, Colonial Trauma, and Global Sovereignties* (Durham, NC: Duke University Press, 2011); Ranjana Khanna, *Dark Continents: Psychoanalysis and Colonialism* (Durham, NC: Duke University Press, 2003).

33 Ranjana Khanna, "Concluding Remarks: Hope, Demand, and the Perpetual," in Anderson, Jenson, and Keller, *Unconscious Dominions*, 247.

34 Wulf Sachs, *Black Hamlet* (London: Geoffery Bles, 1937). Discussed at length in Jock McCulloch, *Colonial Psychiatry and the African Mind* (Cambridge: Cambridge University Press, 1995), chap. 6.

35 Christine Hartnack, *Psychoanalysis in Colonial India* (New Delhi: Oxford University Press, 2001); Ashis Nandy, "The Savage Freud: The First Non-Western Psychoanalyst and the Politics of Secret Selves in Colonial India," in *The Savage Freud and Other Essays on Possible and Retrievable Selves* (Princeton, NJ: Princeton University Press, 1995), 81–144.

36 Omnia El Shakry, *The Arabic Freud: Psychonalaysis and Islam in Modern Egypt* (Princeton, NJ: Princeton University Press, 2017).

37 Octave Mannoni, *Prospero and Caliban: The Psychology of Colonization* (Ann Arbor: University of Michigan Press, 1990). First published 1960.

38 Frantz Fanon, *Black Skin, White Masks*, trans. Charles Lam Markmann (New York: Grove Press, 1967), first published 1952; Frantz Fanon, *The Wretched of the Earth*, trans. Constance Farrington (New York: Monthly Review Press, 1978), first published 1961.

39 Frantz Fanon, *A Dying Colonialism*, trans. Haakon Chevalier (New York: Grove Press, 1965), first published 1959.

40 Hans Pols, "The Totem Vanishes, the Hordes Revolt: A Psychoanalytic Interpretation of the Indonesian Struggle for Independence," in Anderson, Jenson, and Keller, *Unconscious Dominions*, 141–65.

41 J.C. Carothers, *The Psychology of Mau Mau* (Nairobi: Government Printer, 1955).

42 Lewis R. Gordon, *What Fanon Said: A Philosophical Introduction to His Life and Thought* (London: Hurst, 2015), 3; Lewis Gordon, T. Denean Sharpley-Whiting, and Rnee T. White, *Fanon: A Critical Reader* (New York: Wiley-Blackwell, 1996).

43 Sanjeev Jain, Alok Sarin, Nadja van Ginneken, Pratima Murthy, Christopher Harding, and Sudipto Chatterjee, "Psychiatry in India: Historical Roots, Development as a Discipline, and Contemporary Context," in *Mental Health in Asia and the Pacific*, ed. H. Minas and M. Lewis (Boston: Springer, 2017), 39–57.

44 Heaton, *Black Skin, White Coats*.

45 Katie Kilroy-Marac, *An Impossible Inheritance: Postcolonial Psychiatry and the Work of Memory in a West African Clinic* (Berkeley: University of California Press, 2019); Alice Bullard, "Imperial Networks and Postcolonial Independence: The Transition from Colonial to Transcultural Psychiatry," in Mahone and Vaughan, *Psychiatry and Empire*, 197–219.

46 For example, N.C. Surya, K.P. Unnikrishnan, R. Shivathanuvan Thampi, K. Sathyavathi, and N. Sundararaj, "Ayurvedic Treatment in Mental Illness – A Report," *Transactions of the All-India Institute of Mental Health* 5 (1965): 28–39; K.P. Unnikrishnan, "Research in Ayurvedic Psychiatry," *Indian Journal of Psychiatry* 8, no. 1 (1966): 56–9.

47 P.K. Ray, "The Use of *Rauwolfia serpentina* in Psychiatry," *Indian Journal of Neurology and Psychiatry* 3, no. 4 (1952): 380–98.

48 Arthur M. Kleinman, "Depression, Somatization and the 'New Cross-Cultural Psychiatry," *Social Science & Medicine* 11 (1977): 4.

49 Derek Summerfield, "Global Mental Health is an Oxymoron and Medical Imperialism," *British Medical Journal* (2013): 346.

50 V. Patel, A. Cohen, H. Minas, and M.J. Prince, eds., *Global Mental Health: Principles and Practice* (Oxford: Oxford University Press, 2014), 27–40.

7

ANTHROPOLOGY AND EMPIRE

Fenneke Sysling

The history of anthropology

There are many places and times to locate the start of anthropology, the study of mankind. In the fifth century BCE, Herodotus wrote about the customs of the nomadic people north of Greece, and four centuries later the historian Sima Qian in China wrote about the neighbouring Xiongnu tribes. Both authors lived in societies that could spare manpower for writing histories (classical Greek city states and the Han dynasty), and they both travelled extensively and grappled with questions about how to explain differences between one's own people and those outside the boundaries of one's own society.[1]

It is no surprise, then, that the history of anthropology is entangled with the history of the explorations of Europeans from the fifteenth century onward, and subsequently became fully intertwined with their fully-fledged empires. The growing importance of systematic study in Enlightenment Europe and the increasing number of encounters with non-Europeans were stimuli for descriptions of what was different (and what was similar) about these people: their bodies, their dress, their customs, their languages, their behaviour, their kinship systems and their norms, values, and beliefs. These descriptions could be pejorative or positive, universalistic or relativistic, and they reflect on Europeans themselves as much as those they described. Together with these descriptions, Europeans also developed ideas about how societies work (or should work), and proposed models of development and progress that they also used to classify those who were outside Europe's fast developments.[2]

It was only in the nineteenth century that the missionary or administrative ethnographies of the earlier period were transformed into the more systematic, modern, and scientific discipline of anthropology, while the balance of power tilted decisively in favour of Europeans. But the fact that anthropology became recognizable as a discipline does not mean that it was uniform. One important difference, for example, was between medically educated anthropologists who focused on biological (i.e. racial) differences and similarities, and others who specialised in cultural traits. Different traditions and theoretical frameworks also developed across Europe—with distinct national traditions of anthropology in Germany, France, and Britain. The four-field approach—integrating physical (biological) anthropology, archaeology, linguistics, and cultural anthropology—propagated by Franz Boas at the start of the twentieth century came to dominate academic anthropology in the United States. Over time, new conceptions of culture succeeded one another: evolutionary frameworks were replaced by structuralist theories and later by postmodern theories.

The developments outlined above mean that some heavily "anthropologised" regions of the world have seen different generations of anthropologists come and go. The island of Bali, in today's Indonesia, in particular became a "test case for new methods of research and theoretical approaches in anthropology."[3] The Balinese first encountered Europeans when Portuguese and Dutch traders visited their island in the sixteenth century. When Cornelis de Houtman's ships called at the island in 1597, the shore-going party included Aernoudt Lintgensz, who wrote a short note about his visit and his encounter with the local king and the latter's entourage of nobles and deformed individuals.[4] In 1847, in the course of another Dutch expedition to Java, Madura, and Bali, W.R. van Hoëvell, a Dutch protestant minister, scholar, and later politician, described the Balinese as having "nobler and more symmetrical features" than the neighbouring Javanese.[5]

From the late nineteenth century onwards, once the whole of the Dutch East Indies had come more firmly under Dutch imperial control, many other European scholars followed. Between 1896 and 1898, the islands of Java, Bali, and Lombok were visited, for example, by the German anthropologist Adolf Bastian (1826–1905), a critic of the evolutionism then current in Britain. Bastian thought all cultures had a common origin. He was an avid traveller and brought major collections of ethnographic artefacts back to Germany. In the early twentieth century, "baliology" came of age, partly through work done by Dutch ethnologists who were also part of the colonial government. Due to the work of these Dutch ethnographers, such as Julius Jacobs and J.E. Liefrinck, the island also featured in works with a broader scope, such as *The Golden Bough* (1906) by the evolutionist anthropologist James George Frazer, who mentions the Balinese in writing about beliefs in demons and rituals surrounding the harvest of rice.

Biological anthropologists too found the island interesting. In the late 1930s, Ernst Rodenwaldt, a German specialist of tropical medicine, anthropologist, and later professor at the pro-Nazi medical school of University of Heidelberg, investigated racial differences between the different Hindu castes in Bali.[6] In 1938, the Dutch biological anthropologist J.P. Kleiweg de Zwaan also did fieldwork on Bali, measuring the bodily dimensions of the Balinese. He was interested in Bali because it was a place where different racial influences intersected. The fact that the inhabitants of Bali included the Bali-Aga, a non-Hindu people who were believed to have been there for longer than the other Balinese peoples, made the island, in the words of Frie Kleiweg de Zwaan-Vellema, Kleiweg's wife and research assistant, "an exquisite restaurant for anthropologists and ethnologists."[7]

Kleiweg de Zwaan may have encountered Margaret Mead on the island in 1938, where Mead carried out research together with her then husband, the fellow anthropologist Gregory Bateson. Mead represented a different anthropological tradition again. Racial science had been criticised right from the start, and Margaret Mead was one of its most outspoken American critics, exposing the flaws of racial research and thinking.[8] She was a student of Franz Boas and emphasised culture and personality (and how the former influenced the latter), and the method of observation and visual anthropology. Bali was one of the field sites where this new direction was practised.[9]

Anthropological interest in Bali of course continued after the colonial era ended with Indonesia's independence in 1945. But these political changes also triggered new scholarly directions. From the 1960s, scholars of the postcolonial generation started to engage critically with their discipline's colonial past, perhaps most famously in Talal Asad's 1973 book *Anthropology and the Colonial Encounter*.[10] The anthropologists' complicity in the colonial order was severely criticised, and this was tied to criticisms of objectivity and positivism. South

African sociologist Bernard Magubane, for example, added fuel to the debate in an analysis of studies on social change in Africa, published in 1971, which held that the only contribution of these anthropological works was "their implicit consecration of the hegemony of white colonialists."[11] This debate took off not least thanks to the protests and academic criticism from the newly independent nations and from academics of colour based at universities in the West.

The criticism was aimed especially at British social anthropology, the branch of anthropology that made a lasting impact on the discipline in the first half of the twentieth century with its emphasis on fieldwork, and on the idea that every society was "historically rooted and environmentally conditioned" instead of following a general evolutionary law. Critical engagement received a new impetus in the 1990s, when the criticism was broadened to include more kinds of anthropology, such as evolutionary anthropology or biological anthropology; more kinds of anthropologists, such as missionary ethnographers; and a wider variety of locations.

These generations of postcolonial critics have laid bare the various entanglements between anthropology and empire, highlighting several aspects or levels of the relation between the two, which we will trace in the following pages. First, anthropologists found practical support in the empires and their infrastructure. Second, they produced knowledge that was then applied by administrators, thus providing fuel for colonial governance. Third, even if the knowledge anthropologists produced was not directly useful, anthropological work still supported the empire as an ideology, and the empire was equally an ideological driving force for anthropologists. Fourth, this also meant that the anthropologists' work was flawed, because they did not do justice to the people they described and failed to see the impact of colonialism on their daily lives.

Practical entanglements

Colonialism afforded opportunities for anthropologists to travel to and in colonial territories, to network with missionaries and administrators already there, and to live among communities for shorter or longer periods. They were also dependent on official permission, patronage, or material support from colonial institutions. When Thomas Huxley wanted to collect anthropological photographs of as many "tribes" of colonised people as possible for anthropologists to study, he contacted the British Colonial Office in 1869, who sent a notice to their governors asking them to support the project and to send in photographs according to the instructions Huxley had given. Forty sets of images returned to Britain. The French Institut d'Ethnologie at the Sorbonne in Paris, founded in 1925, was funded by taxes collected in the French Empire, which the institute was supposed to serve.[12]

For anthropologists, there was usually some colonial paperwork involved in travelling to and through colonies. They, like other scientists, needed letters of support from colonial governors before they were allowed in. Though French anthropologists often studied French colonial subjects, and so on, there was not always such a neat national division. German and Swiss anthropologists, for example, were also welcome in the Dutch Indies or British India. When the Dutch anthropologist J.P. Kleiweg de Zwaan paid a visit to the governor-general of the Dutch East Indies, he was given permission to do fieldwork on the island of Nias but was warned not to collect skulls, because the government feared the same unrest as was caused by his Italian counterpart Elio Modigliani when he visited the island.

Long periods of fieldwork were a new and defining feature of social anthropology in the early twentieth century. This brought interaction and increasing closeness with research subjects, with the anthropologists the only ones who could move out whenever they wanted to. Paradoxically, although these fieldwork specialists needed the infrastructure of Empire, anthropologists preferred to study people they thought were not yet corrupted by modernity and colonialism. George Stocking, a historian of anthropology, refers to the view of the nineteenth-century anthropologist William Rivers that "the most favourable moment for anthropological work" was about 10–30 years after a people had been brought under the "mollifying influences of the official and the missionary": long enough to be able to count on a "friendly reception and peaceful surroundings" but "not long enough to have allowed any serious impairment of the native culture."[13]

Biological anthropologists had different needs than cultural or social anthropologists. Biological anthropology, the branch of anthropology concerned with the biological or racial characteristics of humans, was based on measurements of bodily dimensions: either of skeletons (usually carried out in Europe), or of living human beings. This was seen as a good "first-contact science," as the only communication it required was to persuade people to be measured, and these measurements could be done in days or weeks. When empire came to Europe, in the form of visits by non-Europeans—as part of expositions known as "human zoos," for example—biological anthropologists would take the opportunity to do measurements there, whereas for social anthropologists there was little to gain from these events.

Many anthropologists criticised the colonial project because they saw the impact of colonialism on the subjects they studied, even though at the same time they themselves were pushing the boundaries of empire. Their fear of cultural and actual decline could be motivated by humanitarian concerns, but also by selfish ones. However, as Wendy James writes,

> [a]s an individual, the anthropologist can often appear as a critic of colonial policy … and he was usually at odds with the various administrators, missionaries and other Europeans he had dealings with … [but] in the inter-war period at least, open political dissent was scarcely possible within colonial society.[14]

Useful knowledge for the empire

Early anthropologists, not unlike many of us today, often justified their projects by insisting (or exaggerating) that their knowledge was useful for, and encouraged by, the state. One of the ways in which empire and anthropology were intertwined was a result of anthropologists arguing that their expertise could be directly applied to colonial governance. And in turn, colonial officials accepted that anthropology, social anthropology especially, was producing useful knowledge. As Helen Lackner noted about Nigeria:

> Functionalist theory in social anthropology is the type of analysis of colonial communities that answers best the questions asked by the ideologues of Indirect Rule. The studies of Ibo social structure helped the administration by providing it with valuable information about institutions that it needed to understand: what chiefs were there and how much and what kind of power did they have? How institutionalized and how close-knit were the women's organizations?[15]

Another of those questions was "who owns the land"? In Fiji, for example, as critics of anthropology have shown, anthropologists influenced colonial (and postcolonial) structures of land tenure. They singled out one of Fiji's social structures, *mataqali*, as the group that could register land, while in fact there was not just one system, and other groups such as families could own land too. This colonial production of knowledge was appropriated by Fijians themselves, who used it in postcolonial times to exclude Indian migrants from buying land.[16]

On Bali in the Dutch East Indies, as Henk Schulte Nordholt has shown, ethnographers provided colonial bureaucrats with a model of Balinese society and Hinduism that then became rooted and cemented in the bureaucratic regime. The regime thus imposed caste structures that had been more flexible and variable in the past, and used them, for example, to define who had to do forced labour. The lower castes had to do forced labour as colonial "corvée," while the higher casts were exempt. Corvée labour then became the defining marker of difference between casts. One result was that there was also a run on noble titles. This was an invented tradition of a rigid caste system.[17]

In the Italian colony of Eritrea, Alberto Pollera was a colonial judge and one of the few Italian colonial ethnographers; his ethnographical knowledge, as Barbara Sorgoni writes, clearly influenced his administrative practice. Pollera's ethnographic work persuaded him that the "straight adoption of Italian justice and penal sanctions could not fit with radically different customs." When he was confronted with the case of the ritual killing of an enemy by young Kunama men, Pollera manipulated what he perceived to be an Indigenous tradition to serve his colonial needs. In 1909 he developed and introduced a traditional oath that the local Kunama chiefs had to swear at a sacred stone, by which they promised not to kill any more.[18]

Knowledge as ideology

A lot of knowledge that was produced however—by social anthropologists of the interwar period, or by others such as evolutionary anthropologists or biological anthropologists—was not directly useful in practice. Although anthropologists usually suggested that what they did was useful for colonial governance, officials often felt that much of anthropologists' work did not relate much to their daily administrative problems. Talal Asad for example agreed that most anthropological knowledge was "trivial" or "too esoteric for government use."[19]

That does not mean, however, that this knowledge production was entirely divorced from the larger imperial project. At the core of anthropology as a science lay the endeavour to define and explain the differences between European people and those outside Europe; and by defining who was civilised and who was not, it provided a mental framework to justify the larger project of empire, and sketched possible pathways of extinction or slow progress under European guidance. As Archie Mafeje has pointed out: "The intellectual effort was a service to colonialism not because of crude suppositions about direct conspiracy or collusion but because of the ontology of its thought categories."[20]

Critics of anthropology have argued that even the gathering of data already legitimised the subordination of the subject. As Diane Lewis stressed in 1973, the relation between the anthropologist and his or her subjects was dangerously close to that of the coloniser and colonised: "The anthropologist, like the other Europeans in a colony, occupied a position of economic, political, and psychological superiority vis-à-vis the subject people."[21] Even if administrators found the anthropologists' work irrelevant, and vice versa, both groups still worked within a framework in which they were on the same side; in working to understand and improve those on the other side, they shaped these roles and strengthened each other's legitimation in the process.

If this was true for social anthropology, it was perhaps even truer for the more sweeping evolutionary approach in which anthropologists attempted to identify stages and laws of human (or cultural) evolution. It was also true for biological anthropology, which had "race" as its central object of study. Emmanuelle Sibeud argued in 2012 that French physical anthropology was a "useless" colonial science. Especially considering the investments it took to do the "tedious, seemingly endless, and hardly rewarding task of collecting bones or measuring reluctant Natives" and to define racial differences, French anthropologists were at a loss to produce useful theories and support for the French Empire.[22] Their research, according to Sibeud, blurred colonial categories rather than offering practical support for colonial policies. But even if we accept Sibeud's arguments about anthropology's lack of usefulness to the colonial state, physical anthropologists propagated the idea that race was a valid category and thus provided an intellectual foundation for colonial, segregationist policies and practices.

Anthropological work and research subjects' lives

Even though anthropological work did not have to be relevant or even accurate to support the colonial power structures, this was still a mutually constitutive process, according to critics such as Asad. In his view, "the fact of European power, as discourse and practice, was always part of the reality anthropologists sought to understand, and of the way they sought to understand it."[23] This entanglement of anthropologists with the empire and with the larger imperialist mindset also influenced the theoretical choices anthropologists made, and meant, according to critics, that anthropologists did not do justice to their research subjects' lives.

The elements of anthropological theorising that received the most criticism were the ways in which anthropologists associated their research subjects with the past and with earlier evolutionary stages, thus defining them as primitive people or as noble savages.[24] Anthropologists also constructed and cemented categories such as "race" and "tribe," and ignored the role the changes of European colonialism played in people's lives.[25] As Catherine Gough argued in a provocative piece in 1968: "We have virtually failed to study Western imperialism as a social system, or even adequately to explore the effects of imperialism on the societies we studied."[26]

For the history of Africa such failures meant that African structures were often described without taking into account the fact that local chiefs fell under European power and ultimately depended on that. Talal Asad describes the work of the anthropologist Meyer Fortes, for example, who worked among the Tallensi in modern Ghana: In his book about the Tallensi, Fortes hardly mentioned British rule, despite having noted earlier that "the political and legal behaviour of the Tallensi … is as strongly conditioned by the ever-felt presence of the District Commissioner as by their own traditions."[27]

This approach disguised the impact of the colonial presence in the life of locals, and continued the portrayal of non-Europeans as if they were living in a timeless past. And if anthropologists did write about change, about urbanisation for example, they found it hard to explain why people still held on to ideas and practices that did not seem to fit with their place in the modern world and which were then depicted as "'myths' that help people cope with disorientation or resist oppression."[28] The anthropologists' method and analysis, Magubane's verdict ran, "has provided ideological blinders as to the true condition of urban Africans."[29]

Recent generations of scholars

Since the turn of the millennium, a new generation of historians of anthropology have presented themselves as moving away from the earlier critics. They position themselves as less inclined to portray colonialism as a "totalising project" and anthropology as its ally in that project. Instead, they stress the agency of non-Europeans, the limitations and ambiguities of colonial power, and the anxieties of the colonial officers. Hesitant, vulnerable, messy, and contradictory are the keywords here (although it must be admitted that many of the earlier generation were also quite open to nuance and ambiguity). The new generation have also started to look at more localities, more forms of colonial power, and more kinds of producers, users, and brokers of knowledge, from professional anthropologists to missionaries, and from local teachers to traditional healers.

Lyn Schumaker, for example, writes that "looking at anthropologists' relationships with particular colonial projects rather than evoking the dominant influence of a hegemonic and homogeneous colonialism promises a more productive approach."[30] Helen Tilley suggests that "it seems fair to say that anthropology's significance to colonial regimes was usually much less than that of other fields, such as the environmental or medical sciences." But anthropology was still "deeply imbricated in imperial affairs."[31] For Germany, H. Glenn Penny and Matti Bunzl argued that German anthropology was "tightly bound up in a range of intellectual traditions that were much richer and more multifarious than a simple colonialist drive." These traditions did not easily change once Germany became a colonial nation, or again when it lost its colonies, and German anthropologists maintained a very wide geographical interest, with many studies of Brazil for example.[32]

This new generation have also presented case studies that show that anthropologists were not always in line with the colonial discourse, and sometimes used their careers and anthropological tools to criticise the system from within. Alice Conklin writes that the experience of intensive field work in the French Empire in the early twentieth century "relativized younger ethnologists' understanding of race and culture." As a result they analysed the "ravages of Empire" in their works and challenged older ideas about race.[33] Similarly, American biological anthropologists such as Harry L. Shapiro started to doubt the value of race when they arrived in Hawai'i for fieldwork in the 1920s and 1930s.[34] Germany in some sense showed the opposite of English and French development, with its liberal stance in the nineteenth century but increasingly nationalist and racist tradition in the twentieth.

The attempt to avoid a binary anthropologist-versus-object relation also meant that this new generation paid more attention to the different roles of colonised people in the shaping of knowledge. Both coloniser and colonised were affected by the encounter, writes John MacKenzie, and they were "always interactive in some shape or form."[35] Colonised people gathered and supplied data, for example, or acted as Indigenous translators, or themselves became anthropologists who incorporated some aspects of Western anthropological thinking and dismissed others. Some of the Indigenous authors were very critical of Western categories; others used an idiom that "resonated with their patrons' existing assumptions."[36] Omnia El Shakry, for example, showed that the anthropological tradition in Egypt, already vibrant under colonialism, sought to demonstrate "the uniqueness of the Egyptian culture, race, and history." The Egyptian nationalist intelligentsia resisted the totalising European claims and developed an alternative, but also incorporated Western ideas such as an emphasis on the moral uplifting of the masses, and an Indigenous discourse on race.[37]

In French West Africa, West Africans have been writing ethnographical texts since the 1920s. These men were often schoolmasters who had received Western education and worked for a certain period as assistants to European anthropologists, who did not like them to go beyond simple anthropological reporting. As J-H. Jezequel describes, Mamby Sidibé, a schoolmaster and ethnographer, wrote a study in 1923 about the Banfora region (Upper Volta); this study "protested against colonial clichés" while also criticising Indigenous societies.[38] While some of these texts found their way into European publications, others had a local impact of their own, to strengthen local dynasties for example and to reinvent local history.

This new generation of scholars have also turned the focus back on Europe to show how, through museums, anthropology was connected to a mass culture in Europe. This work takes the museum, and the histories and trajectories of its objects, seriously as research subjects. Through a focus on the collecting, ordering, and display of objects such as boomerangs, skulls, or what used to be known as "primitive" or "tribal" art, it shows how museums were important places of anthropological knowledge making and colonial governance.[39]

Decolonising anthropology

The critique on the historical relationship between anthropology and empire has produced shelves full of books and articles, but for many the process has not gone far enough. The independence of former colonies did not make an end to unequal power relations, and many of the categories once produced were not discarded with independence. Museum categories, vocabulary, and shelving practices, for example, are hard to change, so older anthropological ideas have lingered in museums, still shaping ideas about non-Europeans. On the other hand, non-Europeans have used anthropological works to insist on their indigeneity, and to decide which cultural forms are authentic and which ones are not. Melanesians, for example, have used the concept of "Melanesia"—once a racial ethnographer's term that had no local equivalent—as an empowering concept.[40] On the island of Bali too, according to Schulte Nordholt, colonial bureaucratic knowledge was reproduced by Balinese intellectuals.[41]

It is not for academics to decide which cultural forms and beliefs people should keep and which they should discard, but even within academia it is not as easy as it seems to fully expunge the colonial mindset. Today's anthropology is still shaped by these relations, and by the work of earlier giants. So how can we get the empire out of anthropology? Every generation of students will encounter new versions of this discussion and will need to be taught how not to copy colonial discourse, how to critically address the categories anthropologists use, and how to assess their own position in the system and their responsibility towards others. Awareness of the past's influence on the present and this continuing discussion has made anthropology one of the most self-reflexive disciplines in academia, from which other disciplines once involved in colonial projects have much to learn.

Notes

1 Siep Stuurman, "Herodotus and Sima Qian: History and the Anthropological Turn in Ancient Greece and Han China," *Journal of World History* 19, no. 1 (2008): 1–40.
2 Thomas Hylland Eriksen and Finn Sivert Nielsen, *A History of Anthropology* (London: Pluto Press, 2001).
3 T.A. Reuter, *Custodians of the Sacred Mountains. Culture and Society in the Highlands of Bali* (Honolulu: University of Hawai'i Press, 2002), 1.
4 A. Lintgensz and P.A. Leupe, "Bali 1597," *Bijdragen tot de Taal-, Land- en Volkenkunde van Nederlandsch-Indië* 5 (1856): 203–34.
5 W.R. van Hoëvell, *Reis over Java, Madura en Bali, in het midden van 1847*, part 3 (Amsterdam: Van Kampen, 1854), 22.
6 E. Rodenwaldt, "Die nicht gemeinsamen Rasseelemente der Balischen Kasten," *Archiv für Rassen- und Gesellschaftsbiologie* 32 (1938): 111–42.
7 Notes by Mrs. Kleiweg de Zwaan-Vellema, n.d. [after 1945], 3, Kleiweg de Zwaan Family private archive, The Netherlands.
8 Margaret Mead, "The Methodology of Racial Testing: Its Significance for Sociology," *American Journal of Sociology* 31, no. 5 (1926): 465–8.
9 Ira Jacknis, "Margaret Mead and Gregory Bateson in Bali: Their Use of Photography and Film," *Cultural Anthropology* 3, no. 2 (1988): 160–77.
10 Talal Asad, ed., *Anthropology and the Colonial Encounter* (Atlantic Highlands, NJ: Humanities Press, 1973).
11 Bernard Magubane, "A Critical Look at Indices Used in the Study of Social Change in Colonial Africa," *Current Anthropology* 12, no. 4/5 (1971): 419–45, here p. 427. See also Archie Mafeje, "The Problem of Anthropology in Historical Perspective: An Inquiry into the Growth of the Social Sciences," *Canadian Journal of African Studies/La Revue Canadienne des études Africaines* 10, no. 2 (1976): 307–33.
12 Alice Conklin, *In the Museum of Man: Race, Anthropology, and Empire in France, 1850–1950* (Ithaca, NY: Cornell University Press, 2013).

13 W.H.R. Rivers, "Report on Anthropological Research outside America," in *Reports upon the Present Condition and Future Needs of the Science of Anthropology*, ed. W.H.R. Rivers, A.E. Jenks, and S.G. Morley (Carnegie Institution of Washington, 1913), 5–28, here p. 7, quoted by George W. Stocking, Jr., "Maclay, Kubary, Malinowski: Archetypes from the Dreamtime of Anthropology," in *Colonial Situations: Essays on the Contextualization of Ethnographic Knowledge*, ed. G.W. Stocking (Madison: University of Wisconsin Press, 1991), 9–74, here p. 10.

14 Wendy James, "The Anthropologist as Reluctant Imperialist," in Asad, *Anthropology and the Colonial Encounter*, 41–69, here p. 42.

15 Helen Lackner, "Social Anthropology and Indirect Rule. The Colonial Administration and Anthropology in Eastern Nigeria: 1920–1940," in Asad, *Anthropology and the Colonial Encounter*, 123–51.

16 John Clammer, "Colonialism and the Perception of Tradition," in Asad, *Anthropology and the Colonial Encounter*, 199–220; Adrian Tanner, "On Understanding Too Quickly: Colonial and Postcolonial Misrepresentation of Indigenous Fijian Land Tenure," *Human Organization* 66, 1 (2007): 69–77.

17 Henk Schulte Nordholt, "The Making of Traditional Bali: Colonial Ethnography and Bureaucratic Reproduction," *History and Anthropology* 8, no. 1–4 (1994): 89–127.

18 Barbara Sorgoni, "The Scripts of Alberto Pollera, an Italian Officer in Colonial Eritrea: Administration, Ethnography and Gender," in *Ordering Africa: Anthropology, European Imperialism and the Politics of Knowledge*, ed. Helen Tilley with Robert Gordon (Manchester: Manchester University Press, 2010), 285–308, here p. 290.

19 Talal Asad, "Afterword: From the History of Colonial Anthropology to the Anthropology of Western Hegemony," in Stocking, *Colonial Situations*, 314–324, here p. 315.

20 Mafeje, "The Problem of Anthropology," 318.

21 Diane Lewis, "Anthropology and Colonialism," *Current Anthropology* 14, no. 5 (1973): 581–602, here p. 582.

22 Emmanuelle Sibeud, "A Useless Colonial Science? Practicing Anthropology in the French Colonial Empire, circa 1880–1960," *Current Anthropology* 53, no. S5 (April 2012): S83–94, here p. S84.

23 Asad, "Afterword," 315.

24 Johannes Fabian, *Time and the Other: How Anthropology Makes its Object* (New York: Columbia University Press, 1983).

25 Archie Mafeje, "The Ideology of 'Tribalism,'" *The Journal of Modern African Studies* 9, no. 2 (1971): 253–61; Terence Ranger, "The Invention of Tradition in Colonial Africa", in *The Invention of Tradition*, ed. Eric Hobsbawm and Terence Ranger (Cambridge: Cambridge University Press, 1983).

26 Catherine Gough, "Anthropology and Imperialism," *Monthly Review* 19, no. 11 (1968): 12–27, here p. 19.

27 Talal Asad, "Two European Images of Non-European Rule Compared," in Asad, *Anthropology and the Colonial Encounter*, 103–118, here p. 108.

28 Asad, "Afterword," 317.

29 Magubane, "A Critical Look," 427.

30 Lyn Schumaker, *Africanizing Anthropology: Fieldwork, Networks, and the Making of Cultural Knowledge in Central Africa* (Durham, NC: Duke University Press, 2001), 7.

31 Helen Tilley, "Introduction: Africa, Imperialism, and Anthropology," in Tilley and Gordon, *Ordering Africa*, 1–45, here p. 6.

32 H. Glenn Penny and Matti Bunzl, "Introduction," in *Worldly Provincialism: German Anthropology in the Age of Empire*, ed. H. Glenn Penny and Matti Bunzl (Ann Arbor: University of Michigan Press, 2003), 1–29, here p. 7.

33 Conklin, *In the Museum of Man*, 4.

34 Warwick Anderson, "Racial Hybridity, Physical Anthropology, and Human Biology in the Colonial Laboratories of the United States," *Current Anthropology* 53, no. S5 (2012): 95–S107.

35 John MacKenzie, "General Editor's Introduction," in Tilley and Gordon, *Ordering Africa*, xiii.

36 Tilley, "Introduction: Africa, Imperialism, and Anthropology," 18.

37 Omnia El Shakry, *The Great Social Laboratory: Subjects of Knowledge in Colonial and Postcolonial Egypt* (Stanford, CA: Stanford University Press, 2007).

38 J-H. Jezequel, "Voices of their Own? African Participation in the Production of Colonial Knowledge in French West Africa, 1910–1950," in Tilley and Gordon, *Ordering Africa*, 145–72, here p. 154.

39 Andrew Zimmerman, *Anthropology and Antihumanism in Imperial Germany* (Chicago: University of Chicago Press, 2001), esp. 240–2; Ricardo Roque, *Headhunting and Colonialism: Anthropology and the Circulation of Human Skulls in the Portuguese Empire, 1870–1930* (Basingstoke: Palgrave Macmillan, 2010); Fenneke Sysling, *Racial Science and Human Diversity in Colonial Indonesia* (Singapore: NUS Press, 2016); Tony Bennett, Fiona Cameron et al., ed., *Collecting, Ordering, Governing. Anthropology, Museums and Liberal Government* (Durham, NC: Duke University Press, 2017).

40 Melanie Lawson, " 'Melanesia,' The History and Politics of an Idea," *The Journal of Pacific History* 48, no. 1 (2013): 1–22.

41 Schulte Nordholt, "The Making of Traditional Bali," 90.

8

NATURAL HISTORY COLLECTIONS AND EMPIRE

Andreas Weber

Natural history museums and herbaria in the Global North owe much of their authority to plants, animals, and minerals collected in the Global South. Naturalis Biodiversity Centre in Leiden, which is one of the largest natural history museums in Europe, houses for instances one of the world's largest collection of plants and animals from Indonesia, a former Dutch colony in insular Southeast Asia.[1] Recent estimates assume that a large majority of natural objects in repositories in Europe and the United States stem from former colonial areas in South America, Asia, and Africa.[2] This unequal distribution of the planet's natural heritage is the historical result of intimate and often invisible linkages between natural history repositories and evolving schemes of colonial exploitation, violence, and commerce.[3] However, in institutional discussions about future research on natural historical collections the colonial provenance of such collections is usually not acknowledged.[4] This is truly astonishing, since in particular the large-scale digitisation of specimen collections and accompanying archival holdings offers researchers access to a wealth of new source material which has the potential to shed fresh light on how natural history and empire co-produces each other from the sixteenth century to the present.[5] This new type of source material allows historians to deepen our understanding of the daily practices and polycentric networks of collecting and natural historical knowledge production in former colonial areas.[6] Moreover, it will allow for studies on how *and* with what implications natural objects have travelled to affluent individuals and natural historical institutions across the globe.[7] By acknowledging the geographical imbalance of natural historical collections in the Global North, this essay argues that natural historical archives and specimen collections are an underused starting point for historical inquiries into the colonial roots of our present-day understanding of nature and its diversity.

Since antiquity, the term natural history has been used to denote attempts to collect, classify, name, and systematically understand plants, animals, and minerals. Although most historians of natural history have taken secluded spaces of cabinets, laboratories, and museums as the starting point for their historical analyses, the impact of the transfer of a myriad of natural objects and related knowledges within colonial areas and to natural historical institutions in the West has not remained unnoticed.[8] Since the inception of the wider research field of Science and Empire by George Basalla in the 1960s, historians of natural history have studied local conditions of natural historical knowledge production in South America, the Caribbean, the Cape in South Africa, India, China, Japan, Sri Lanka, Southeast Asia, and the Pacific Islands.[9] What emerges from these studies is a rich picture of complex local encounters which make it often difficult to draw clear boundaries between local *and* European agency. Instead of considering European naturalists and their institutional patrons in the West as the main motor of scientific development, authors in this field have carefully examined how and under which political and socioeconomic circumstances natural historical knowledge and collections were produced. Often concepts such as "contact zones," "middle ground," or "borderlands" have helped to render hybrid forms of local agency visible.[10] What has become clear from many of these studies is that, until a natural object and related visual or handwritten material reached museums and repositories in the West, it had often travelled within extensive regional networks of exchange. Military men, specimen traders, merchants, enslaved people, naval personal, and medical practitioners with highly heterogeneous socioeconomic backgrounds played a pivotal role in these networks.[11] A detailed reconstruction and analysis of such regional and global itineraries of natural historical objects allows historians of natural history and empire to enrich available studies on centres and peripheries with a new analytical framework.

Regional networks of exchange existed long before Europeans set out to inventory nature and natural resources for learned and economic purposes in the sixteenth century. A good example in this respect is the trade of birds of paradise from the islands of Eastern Indonesia and New Guinea in Southeast Asia. Prior to the colonisation of the islands by Europeans in the sixteenth century, the birds were traded within insular Southeast Asia for centuries.[12] Next to a decorative function, in particular the birds' plumes were locally used as bride price. Owing to such a rich history, it is therefore not surprising that the Portuguese were offered birds of paradise skins as valuable diplomatic gifts when they arrived in the area. In the centuries to come, not only the skins and feathers, but also living specimens and knowledge in the form of stories and myths about birds of paradise circulated to courts and empires across the globe. In 1912 and 1913 alone, more than 100,000 skins of birds of paradise left New Guinea to be auctioned, among other places, in London.[13] British hat makers depended on the bird's feathers in order to be able to satisfy their wealthy customers.[14] Since birds of paradise could be relatively easily transported on ships, living specimens found their ways into Western public zoos and private aviaries.[15] The increased interest in exotic living and dead animals from New Guinea and other biodiverse areas of the world also triggered resistance, which sought to ban illegal trade and consumption. Since the mid-1920s, the trade of birds of paradise and other exotic birds from insular Southeast Asia for commercial purposes has been officially prohibited.[16] However, until the present day, scientific institutions and museums are exempt from these rules, and all over the world, natural history museums house thousands of bird of paradise skins, feathers, and specimens. The largest scientific collection of birds of paradise can today be found at the American Museum of Natural History in New York.

This brief excursion into the history of birds of paradise serves as an important reminder that the historical itinerary of a group of specimens can be a fruitful starting point for historians of natural history and empire. Instead of separating the world in European centres and colonial peripheries, the historical journeys of birds of paradise and other natural objects rather require us to work towards an "entangled" history of natural history *and* empire.[17] Guiding in this respect can be a study by Nicholas Thomas who in the early 1990s used the term "entanglement" to analyse different modalities of cultural interactions in the context of museum objects. Seen from his perspective the collection as well as the movement of natural objects can be best studied as historically open-ended, networked, and polycentric process.[18] While an entangled history of birds of paradise is well under way, new studies on, for instance, plants (e.g., orchids) and animals (e.g., amphibians, insects, monkeys, rhinoceroses) have the potential to deepen

Figure 8.1 Hat feather probably composed of a prepared head of a bird of paradise, feathers of a *Colibri* and other tropical birds, anonymous, c. 1910–c. 1915. Rijksmuseum Amsterdam.

our understanding and long-term impact of encounters and exchanges at mutually dependent colonial localities all over the world.[19] Next to already available digital resources such as the Biodiversity Heritage Library (BHL), new digital portals of natural history museums, and new large-scale infrastructures aimed at interconnecting available geographical information on specimens can be of invaluable help in carrying out such research.[20]

The birds of paradise episode also reminds us that every transfer of natural objects was a complex procedure in which asymmetries of power played a pivotal role. First objects needed to be collected. This often required a close collaboration with colonial administrations, collectors, adventurers, private landowners, and local rulers. Like birds of paradise, many natural objects were collected in areas in which military violence was common. Not only in New Guinea and other parts of Asia, but also in South America practising natural history remained closely linked to violent forms of colonialism and exploitation. An interesting example in this respect are probably the collecting activities of the Prussian traveller and naturalist Alexander von Humboldt (1763–1859) and Aimé Bonpland (1773–1858) in Spanish South America. During their five years' stay in South America, both travellers visited various parts of the Spanish Empire and gathered a large number of natural objects, took extensive measurements, and recorded numerous observations. Today hailed by some scholars as the most important naturalist of the nineteenth century, an analysis of his fieldwork and natural historical collections in the area cannot be disconnected from the economic and political interests of the Spanish crown in the area.[21] As Jorge Cañizeras-Esguerra has shown, Humboldt and his companion Aimé Bonpland drew extensively upon colonial infrastructure and local expertise during their travels in South America.[22] Following this line of analysis, historians have started to probe why Humboldt's travel narrative remained silent about the often violent political context in which his collections—now stored in natural history museums in France and Germany—were gathered.[23]

Figure 8.2 An unopened package of botanical field notes by Pieter Willem Korthals (1807–1892).
Andreas Weber, photograph taken in the archives of the former National Herbarium of the Netherlands
(now Naturalis Biodiversity Center), 2010.

Next to collecting specimens in the field, naturalists were also challenged to prepare specimens for shipping to Europe or other parts of the world. In the context of natural historical inquiry in the Dutch colonies in insular Southeast Asia, we know that colonial institutions such as the botanical garden in Buitenzorg (now Bogor) played an important role.[24] Situated in the hinterland of Batavia, the administrative seat and main harbour of the Dutch in the area, the garden and its facilities served naturalists as a site at which natural objects could be gathered, dissected, prepared, described, compared, and prepared for shipping to gardens and museums in, among others, Vienna, Leiden, Berlin, London, Philadelphia, Kolkata, Sri Lanka, and Guangzhou.[25] In the early 1820s, the garden in Bogor was, for instance, used to dissect and describe the anatomy of a female elephant (*Elephas indicus*). The notes and drawings which were made during the dissection were later shipped to Europe. In the decades before and after 1900, the garden and its botanical laboratories witnessed the rise of plant embryology.[26] In the 1940s, Kees van Steenis capitalised on the botanical garden and its infrastructure when he was preparing his Flora Malesiana, a large-scale and still ongoing attempt to map the flora of Indonesia, Malaysia, Singapore, Brunei, the Philippines, and Papua New Guinea in the form of a series of books and conferences.[27] Although originally designed as a repository of economically viable plants, botanical gardens such as the one in Bogor also played an important role in the study of animals in the nineteenth and twentieth centuries.

Writing and publishing in the field of natural history heavily depended on the availability of a large number of natural objects. However, in particular in times of long-distance travel on ships, many plants and animals reached natural historical institutions in Europe and North America in a poor shape.[28] In order to compensate for humidity on ships, shipwreck, vermin, theft, and other threats of long-distance transfer, naturalists made sure that specimens were accompanied with detailed handwritten field notes, lists, and drawings.[29] Owing to the often intricate recording practices which naturalists used to note their observations, historians of science have started to examine such natural historical paper heritage as "paper technologies."[30] Such "paper technologies," which were often inextricably entangled with daily practices of global commerce and colonial governance, helped naturalists in Europe and other parts of the world to understand and reconstruct observations made en route.[31] In particular for a new generation of historians of natural history and empire, the growing digital availability of such unpublished handwritten and hand-drawn material can be a real treasure trove. Initiatives such as the Smithsonian Digital Volunteers: Transcription Center as well as the rise of automated handwriting recognition and semantic in the context of handwritten natural historical archives will likely in the future enable historians to intensify and deepen their study of natural objects on the move.[32] However, this will only be successful if interdisciplinary teams of historians, computer scientists, *and* biologists find ways to link specimen labels, handwritten field notes, travel diaries, illustrations, and publications in new ways, laying the valuable groundwork for a new historical contextualisation of natural historical collections and their mobile imperial past.[33]

Taken together this chapter has made three interrelated points: first, it argues that historians of natural history and empire are looking into a bright future. In particular the fast-growing digitisation of plants and animals, as well as related handwritten field notes and illustrations, offers historians and other scholars a fascinating range of new textual and visual source material. Such new sources allow historians of natural history to intensify their reconstructions of the provenance and global mobility of natural objects. Second, over the last 400 years, millions of plants and animals have received a new home in natural history institutions in the Western world. While they serve today mainly as an archive of nature, used by biologists to map and understand shifts in global biodiversity, they are also a shared heritage of a mobile imperial past. And third, as historians of natural history and empire we can help in reading natural historical collections and related textual and visual material as the product of a process of entanglement in which local expertise about flora and fauna, natural history, and often violent forms of colonialism played a pivotal role. This should prod natural museums, which are investing heavily in the development of digital infrastructures to enrich and link collections and archives across institutions, to realise that the authority of "biodiversity heritage" in the Western world is rooted in histories of empires which from the eighteenth to the twentieth century shaped the political reality in large parts the world.

Notes

1 According to the database BioPortal, almost 900,000 specimens of the museum's specimens stem from Indonesia. This excludes specimens which are not registered in the database. See https://bioportal. naturalis.nl/.

2 Joshua Drew, Corrie Mureau, and Melanie Stiassny, "Digitization of Museum Collections Holds the Potential to Enhance Researcher Diversity," *Nature Ecology & Evolution* (2017): 1789–90, doi.org/ 10.1038/s41559-017-0401-6.

3 John M. MacKenzie, *Museums and Empire: Natural History, Human Culture and Colonial Identities* (Manchester: Manchester University Press, 2009).

4 Christopher A. Norris, "The Future of Natural History Collections," in *The Future of Natural History Museums,* ed. Eric Dorfman (London: Routledge, 2017), 13–28; Freek T. Bakker, et al. "The Global

Museum: Natural History Collections and the Future of Evolutionary Biology and Public Education" (2020), doi.org/10.7717/peerj.8225.

5 B.P. Hedrick, et al. "Digitization and the Future of Natural History Collections," *BioScience* 70, no. 3 (2020): 243–51, doi.org/10.1093/biosci/biz163; Maarten Heerlien, J. van Leusen, S. Schnörr, S. De Jong-Kole, N. Raes, and K. Van Hulsen, "The Natural History Production Line: An Industrial Approach to the Digitization of Scientific Collections," *Journal on Computing and Cultural Heritage* 8, no. 3 (2015).

6 Paula Findlen and Anna Toledano, "The Materials of Natural History," in *Worlds of Natural History*, ed. Helen A. Curry, Nicholas Jardine, Jim Secord, and Emma Spary (Cambridge: Cambridge University Press, 2018), 151–69.

7 James Secord, "Knowledge in Transit," *Isis* 95, no. 4 (2004): 654–72; Neil Safier, "Global Knowledge on the Move: Itineraries, Amerindian Narratives, and Deep Histories of Science," *Isis* 101, no. 1 (2010): 133–45.

8 For three excellent more general introductions to the field see: Arthur MacGregor, *Naturalists in the Field* (Leiden: Brill, 2018); Curry, Jardine, Secord, and Spary, *Worlds of Natural History*; N. Jardine, J. Secord, and E.C. Spary, eds., *Cultures of Natural History* (Cambridge: Cambridge University Press, 1996).

9 Fa-Ti Fan, *British Naturalists in Qing China: Science, Empire and Cultural Encounter* (Cambridge, MA: Harvard University Press, 2004); Harold Cook, *Matters of Exchange: Commerce, Medicine and Science in the Dutch Golden Age* (New Haven, CT: Yale University Press, 2007); Juan Pimentel, "Green Treasures and Paper Floras: the Business of Mutis in New Granada (1783–1808), *History of Science* 52, no. 3 (2014): 277–96; Bernhard Schär, *Tropenliebe. Schweizer Naturforscher und niederländischer Imperialismus in Südostasien um 1900* (Frankfurt am Main: Campus, 2015); Londa Schiebinger, *Plants and Bioprospecting in the Atlantic World* (Cambridge, MA: Harvard University Press, 2004); Andrew Goss, *The Floracrats: State Sponsored Science and the Failure of Enlightenment in Indonesia* (Madison: University of Wisconsin Press, 2011).

10 Lissa Roberts, "Situating Science in Global History: Local Exchanges and Networks of Circulation," *Itinerario* 33 (2009): 9–30; Fa-ti Fan, "Science in Cultural Borderlands: Methodological Reflections on the Study of Science, European Imperialism, and Cultural Encounter," *East Asian Science, Technology and Society* 1 (2007): 213–31.

11 Genie Hoo, "Wars and Wonders: The Inter-Island Information Networks of Georg Everhard Rumphius," *The British Journal for the History of Science* 51, no. 4 (2018): 559–84; Dorit Brixius, "From Ethnobotany to Emancipation: Slaves, Plant Knowledge, and Gardens on Eighteenth-Century Isle de France," *History of Science* 58, no. 1 (2019): 51–75; Simon Schaffer, Lissa Roberts, Kapil Raj, and James Delbourgo, *The Brokered World: Go-Betweens and Global Intelligence, 1770–1820* (Sagamore Beach, MA: Science History Publications, 2009).

12 Leonard Y. Andaya, "Flights of Fancy: The Bird of Paradise and its Cultural Impact," *Journal of Southeast Asian Studies* 4, no. 3 (2017): 372–89; Pamela Swadling, *Plumes from Paradise: Trade Cycles in Outer Southeast Asia and Their Impact on New Guinea and Nearby Islands until 1920* (Boroko: Papua New Guinea National Museum, 1996).

13 In the early twentieth century, the Missiemuseum in Steijl, the Netherlands, was an important nodal point for the import of plumes from birds of paradise to Europe. Justin J.F.J. Jansen, "Missiemuseum, Steijl, the Netherlands – The History of a Little-Known Collection," *Bulletin British Ornithologists' Club* 132, no. 3 (2012): 212–14.

14 For a more general history of the feather trade in this period, see Dirk H.R. Spennemann, "Exploitation of Bird Plumages in the German Mariana Islands," *Micronesica* 31, no. 2 (1999): 309–18. I thank Marc Argeloo for this reference.

15 Anne Coote, Alison Heynes, Jude Philp, and Simon Ville, "When Commerce, Science, and Leisure Collaborate: The Nineteenth-Century Global Trade Boom in Natural History Collections," *Journal of Global History* 12 (2017): 319–39.

16 Clifford B. Frith and Bruce M. Beehler, *The Birds of Paradise* (Oxford: Oxford University Press, 1998), chaps. 1 and 2.

17 Nicholas Thomas, *Entangled Objects: Exchange, Material Culture, and Colonialism in the Pacific* (Cambridge, MA: Harvard University Press, 2009).

18 Ralph Bauer and Marcy Norton, "Introduction: Entangled Trajectories: Indigenous and European Histories," *Colonial Latin American Review* 26, no. 1 (2017): 1–17; Sebastian Kroupa, Stephanie J. Mawson, and Dorit Brixius, "Science and Islands in the Indo-Pacific Worlds," *The British Journal for the History of Science* 51, no. 4 (2018): 541–58.

19 See for instance Marco Masseti, "New World and Other Exotic Animals in the Italian Renaissance: The Menageries of Lorenzo Il Magnifico and his Son, Pope Leo X," in *Naturalists in the Field: Collecting, Recording and Preserving the Natural World from the Fifteenth to the Twentieth-First Century*, ed. Arthur MacGregor (Leiden: Brill Publishers, 2018), 40–75; Jim Endersby, *Orchid: A Cultural History* (Chicago: Chicago University Press, 2016).

20 See for instance https://data.nhm.ac.uk/ and the project website of the project "Distributed System of Scientific Collections (DISCCo)," www.dissco.eu/.

21 For a very positive view on Humboldt, see Sandra Rebok, "Humboldt's Exploration at a Distance," in Curry, Jardine, Secord, and Spary, *Worlds of Natural History*, 319–34.

22 Jorge Cañizares-Esguerra, "How Derivate Was Humboldt? Microcosmic Nature Narratives in Early Modern Spanish America and the (Other) Origins of Humboldt's Ecological Sensitivities," in *Colonial Botany: Science, Commerce, and Politics in the Early Modern World*, ed. Londa Schiebinger and Claudia Swan (Philadelphia: University of Pennsylvania Press, 2007), 148–65.

23 Peter Giere, Peter Bartsch, and Christiane Quaisser, "BERLIN: From Humboldt to HVac—The Zoological Collections of the Museum für Naturkunde Leibniz Institute for Evolution and Biodiversity Science in Berlin," in *Zoological Collections of Germany*, ed. Lothar Beck (Cham: Springer, 2018), 89–122.

24 Andreas Weber, "Collecting Colonial Nature: European Naturalists and the Netherlands Indies in the Early Nineteenth Century," *BMGN—Low Countries Historical Review* 134, no. 3 (2019): 72–95.

25 Andreas Weber, "A Garden as Niche: Botany and Imperial Politics in the Early Nineteenth Century Dutch Empire," *Studium* 11, no. 3 (2018): 178–90; Emma Spary, *Utopia's Garden: French Natural History from Old Regime to Revolution* (Chicago: University of Chicago Press, 2000); Richard Drayton, *Nature's Government. Science, Imperial Britain and the "Improvement" of the World* (New Haven, CT: Yale University Press, 2000). For a more general history of the emergence and function of British colonial botanical gardens see Richard Grove, *Green Imperialism: Colonial Expansion, Tropical Island Edens and the Origins of Environmentalism* (Cambridge: Cambridge University Press, 1995), 309–48; Miles Ogborn, "Vegetable Empire," in Curry, Jardine, Secord, and Spary, *Worlds of Natural History*, 271–86.

26 Robert-Jan Wille, "From Laboratory Lichens to Colonial Symbiosis. Melchior Treub Bringing German Evolutionary Plant Embryology to Dutch Indonesia, 1880–1909," *Studium* 11, no. 3 (2018): 191–205.

27 Andrew Goss, "Reinventing the Kebun Raya in the New Republic: Scientific Research at the Bogor Botanical Gardens in the Age of Decolonization," *Studium* 11, no. 3 (2018): 206–19.

28 Sarah Easterby-Smith, "Recalcitrant Seeds: Material Culture and the Global History of Science," *Past & Present* 242, Supplement 14 (2019): 215–42.

29 Anne Coote, Alison Haynes, Jude Philp, and Simon Ville, "When Commerce, Science, and Leisure Collaborated: The Nineteenth-Century Global Trade Boom in Natural History Collections," *Journal of Global History* 12, no. 3 (2017): 319–39.

30 A. Cooper, "Placing Plants on Paper: Lists, Herbaria, and Tables as Experiments with Territorial Inventory at the Mid-Seventeenth Century Gotha Court," *History of Science* 56 (2018): 257–77; Staffan Müller-Wille and Isabelle Charmantier, "Natural History and Information Overload: The Case of Linnaeus," *Studies in History and Philosophy of Biological and Biomedical Sciences* 43 (2012): 4–15; A. te Heesen, "The Notebook: A Paper Technology," in *Making Things Public: Atmospheres of Democracy*, ed. Bruno Latour (Karlsruhe, 2005), 582–9.

31 The entangled character of "paper technologies" is highlighted in Volker Hess and J. Andrew Mendelsohn, "Case and Series: Medical Knowledge and Paper Technology, 1600–1900," *History of Science* 48 (2010): 287–314.

32 "Smithsonian Digital Volunteers: Transcription Center," https://transcription.si.edu/; L. Stork, et al., "Semantic Annotation of Natural History Collections," *Journal of Web Semantics* 59 (2019), 100462; L. Schomaker, "Lifelong Learning for Text Retrieval and Recognition in Historical Handwritten Document Collections," in *Handwritten Historical Document Analysis, Recognition, and Retrieval—State of the Art and Future Trends*, ed. Andreas Fischer, Marcus Liwicki, Rolf Ingold (World Scientific Publisher, 2020), https://arxiv.org/abs/1912.05156.

33 Andreas Weber, Mahya Ameryan, Katherine Wolstencroft, Lise Stork, Maarten Heerlien, and Lambert Schomaker, "Towards a Digital Infrastructure for Illustrated Handwritten Archives," in *Digital Cultural Heritage: Lecture Notes in Computer Science*, vol. 10605, ed. M. Ioannides (Cham: Springer, 2018), 155–66.

9

NON-WESTERN COLLECTORS AND THEIR CONTRIBUTIONS TO NATURAL HISTORY, C. 1750–1940

Jennifer R. Morris

On the penultimate page of George Basalla's renowned 1967 essay, "The Spread of Western Science," an illustration catches the eye. It depicts a natural history specimen collector, probably of African origin, walking through a tropical landscape. He carries a butterfly net, a large case, and a forked stick from which a snake dangles. Butterflies and other sizeable insects appear to be pinned to the rim of his hat. The caption explains that the image is taken from an 1845 travel memoir by Methodist missionary Daniel Parish Kidder, and that it shows a man collecting and preparing specimens on behalf of a Brazilian naturalist. Who was this man and under what circumstances did he take on the role of natural history collector in Brazil? He is depicted as replete with specimens, but what became of the collections he accumulated and what knowledge was gained from them? Basalla did not touch upon the answers to any of these questions, but instead used the image to illustrate his discussion of the process by which a colonial territory might develop its own independent scientific tradition. He uncritically accepted the opinions of Louis Agassiz on the elite naturalists of Brazil, whose reliance on "the socially inferior" to carry out the manual tasks of collecting apparently prevented their full engagement with local flora and fauna and thus "retarded" Brazil's progress towards scientific independence.[1]

Of course, much of Basalla's framework for the dissemination of "Western science" via European colonialism has been challenged since the 1960s. While his assertion regarding the deficiencies of the nineteenth-century Brazilian naturalists seems somewhat unfair—given that countless natural scientists in all localities built their collections with the assistance of their servants or slaves—his failure to acknowledge the agency of the collector depicted in the illustration is symptomatic of a wider problem that still faces the history of science today. We have ample evidence of the reliance of professional naturalists, metropolitan specimen traders, amateur "scholar-administrators," and museum curators alike on expansive networks of collectors, traders, and specimen preparators. The contributions to science of Europeans from lower social strata via these collecting networks have been widely recognised, particularly the numerous individuals who travelled the world collecting for famous names such as Joseph Banks and Carl Linnaeus.[2] But natural history collecting in colonial contexts inevitably crossed boundaries of ethnicity as well as class. The non-Western actors who became involved in this process, whether through employment or coercion, remain largely absent from most narratives of scientific endeavour during the age of empire, despite the ongoing efforts of postcolonial historians. This chapter explores the nature of this historiographical lacuna and some recent efforts to rectify it via new approaches and alternative source material, and through initiatives to "decolonise" the natural history museum. In reviewing the existing scholarship on this theme, I suggest that, in order to adequately recognise and understand the transformational role of non-Western actors in the practice of natural history, further reconsideration of our research methods is urgently required. The change in perspective that is currently in progress, placing the colony and the natural history collection itself at the heart of investigations, has the potential to reap much-needed rewards.

The term "non-Western collector" is far from ideal, but, for reasons of space, it is used here to refer to all those engaged in the hands-on work associated with natural history collecting whose cultural backgrounds differed from the empirical, Enlightenment-inspired worldviews of European origin. Many of the collectors considered here were indigenous to the regions in which they worked, but many also travelled widely, either through the demands of the job or as a result of migration. Much of this migration was forced: the transatlantic slave trade provided a great deal of the infrastructure and manpower that supported travelling naturalists on all sides of the Atlantic between the seventeenth and nineteenth centuries.[3]

Local knowledge and the field sciences

In the decades since Basalla penned his seminal paper, significant shifts have occurred in the historiography of science, particularly with regard to natural history. While earlier paradigms emphasised the ideas of great men, usually formulated in metropolitan laboratories and lecture halls, the social and cultural turns have prompted greater interest in the material and social practices of scientific endeavour, and the global networks these practices sustained. As a result, the field sciences, previously neglected by historians, have enjoyed increasing exposure. Henrika Kuklick and Robert E. Kohler's 1996 edition of *Osiris*, for example, foregrounded the field sciences as a prime subject for "history from below," noting the large and socially diverse communities that contributed to their practice.[4] These communities spanned divides of class, gender, and ethnicity.

That non-Western actors contributed a great deal to scientific endeavour across the globe during the colonial period is well known. The central role of local intermediaries in the production of colonial knowledge, both scientific and otherwise, has been a key feature of discourse on cultural tools of power. Many historians of empire argue these were fundamental to the maintenance of control in colonised territories.[5] The research of scholars including Nicholas Dirks, Simon Schaffer, Kapil Raj, and Sandra Manickam highlights the ways in which local elites negotiated the form and depth of the scientific knowledge gathered by Europeans, acting as intermediaries, translators, and researchers on the ground in areas as diverse as ethnography and surveying.[6] In the history of medicine, too, recent scholarship reveals the extent of European reliance on local botanical knowledge in efforts to eradicate diseases that threatened the progress of colonial expansion in unfamiliar tropical environments.[7] Opinions are divided on how to interpret the nature of these contributions in light of the multifaceted power structures in colonial territories. Deepak Kumar contends that scholars such as Raj have overstated the case for the impact of Indigenous interlocutors on the history of science in India, and that depicting local assistants as part of a collaborative cross-cultural network of knowledge production serves only to "blunt … the violent edge of colonial modernity."[8]

The gathering and preparation of natural history specimens have seldom been examined from these perspectives. This is despite the profusion of zoological, botanical, and geological collections formed during the colonial era which fill museum stores the world over. Both the histories of science and of museums have, until very recently, focused on the stages after collections were accessioned into museum catalogues or received by eager European naturalists.[9] Calls from historians such as Brett Bennett and Joseph Hodge for the adoption of a networked and global perspective on colonial science, combined with a reassessment of the colony as a scientific centre in itself (rather than a conduit by which knowledge and objects were siphoned to the metropole) have inspired new research into both collecting expeditions and the scientific institutions established in colonial territories.[10]

Collectors in the archives: the cases of John Edmonston and Ali

The contributions of Indigenous or other non-Western actors to these processes, though beginning to receive acknowledgement, remain mostly unexplored by scholars. This is largely, and predictably, due to the absence of source material. Travelling European naturalists occasionally mentioned their local collectors in expedition reports, collection labels, and other writings, but rarely by name or with any biographical details, leaving Indigenous assistants and informants all "mere 'shadows'" in the textual record.[11] Crediting non-Western assistants became even less acceptable among scientific writers as the pioneering era of colonisation gave way to the High Imperial period.[12] In the later nineteenth century, the conventions of different genres of writing also governed the frequency with which such names were recorded. Itinerant naturalists who named regular employees and described their contributions in travelogues were quick to omit these details from formal descriptions of their specimens.[13] As Steven Shapin has noted, trust in scientific networks was very much dependent on deeply rooted social mores.[14] While interactions with Indigenous people might have made for a thrilling anecdote for the metropolitan public, scientific information obtained in the field was already considered less "rational" and rigorous than that sought in the controlled environment of the laboratory.[15] Packaging this knowledge as produced by a suitably educated man of science was thus essential for its acceptance into intellectual discourse.

Some information about non-Western collectors can, however, be gleaned from naturalists' travelogues, diaries, and correspondence. In-depth research into these individuals has usually been prompted by their personal connections with particularly prominent imperial scientists. As early as 1978, for example, Richard Freeman used municipal and museum archives to trace the life of John Edmonston, a skilled taxidermist who was trained while enslaved in British Guiana and later worked for various Scottish museums. Freeman's investigations were prompted by the desire to identify the unnamed taxidermist who taught the young Charles Darwin in the 1820s, and he thus characterised Edmonston as "Darwin's negro bird-stuffer."[16]

Darwin's counterpart and collaborator, Alfred Russel Wallace, also provided an unusually rich trove of evidence about the life and work of an Indigenous collector. While Wallace employed numerous men to collect for him during his travels through the Malay Archipelago, to say nothing of the hundreds who worked for him in other capacities, his most trusted assistant in the region was a Malay man named Ali. Not only did Wallace's *The Malay Archipelago* include plentiful information about Ali's contributions to Wallace's collections and, indeed, to the naturalist's survival in Southeast Asia, but his 1905 memoir, *My Life: A Record of Events and Opinions,* also featured a portrait photograph of Ali.[17] The combination of this comparatively rich source material and Wallace's own elevation in the public imagination over the last two decades has seen Ali become probably the most thoroughly studied non-Western collector from this period.[18] Numerous published papers reveal a range of perspectives on Ali's role in Wallace's work in Asia. Scholars have been particularly preoccupied by Ali's status in relation to his employer and to the knowledge they produced. All acknowledge that Ali was crucial to Wallace's success in the field but differ in their interpretation of the pair's relationship on both professional and personal levels. Ali has been characterised as everything from an unskilled assistant to a "virtual collaborator" and a dedicated naturalist.[19] First hired by Wallace as a teenager, Ali initially worked as a cook, but learnt to hunt, collect, and practise taxidermy during their years travelling together. Most commentators paint a congenial picture of these expeditions. John van Wyhe and Gerrell Drawhorn emphasised Ali's mastery of the aforementioned zoological skills. They identified Wallace's descriptions of Ali's enthusiastic bird collecting in the Moluccas as evidence that "clearly he did not just work for money" but had "entered actively into the spirit" of a scientific expedition.[20] Lord Cranbrook and Adrian Marshall maintained a more traditional perspective on colonial knowledge production, asserting that, although the local collectors deserved recognition for their work, they were ultimately facilitators of Wallace's vision: "[Wallace's] claim to ownership of this large and varied lot of animal specimens is unchallengeable. His was the drive and initiative that directed their collection; he was responsible for the logistics, chose the locations, curated and catalogued the specimens."[21] Carey McCormack, on the other hand, viewed this balance of power through a more postcolonial lens. She charged Wallace with "appropriating" his assistants' work and dismissing their ability to produce knowledge due to their lack of education "in the Linnaean tradition."[22]

A somewhat different angle is found in the work of Jane Camerini, who described Ali primarily as "a loyal servant who skilfully adapted himself to his master's needs" but emphasised the personal nature of his relationship with Wallace, who depended on him not just as collector but as teacher and companion.[23] Without Ali's skilled mediation and translation, she argued, Wallace could not have successfully interacted with the local communities who assisted in his search for birds of paradise and other valuable specimens. The feelings of trust and respect established between the pair during their shared trials and tribulations in the field were just as vital to the practice of natural history as was educated reason.

An oft-repeated anecdote from Harvard zoologist Thomas Barbour illustrates the enduring and personal nature of Ali's relationship with Wallace. Barbour claimed to have encountered an elderly man in Ternate in 1907 who introduced himself as "Ali Wallace." That Ali had apparently chosen to take Wallace's name, and continued to use it nearly five decades after their parting, has been cited as evidence that Ali's work with the Briton formed "the central and controlling incident of his life."[24] While Cranbrook and Marshall implied that this was due to the profundity of his experiences with the great naturalist, Camerini suggested that, although the use of the name does reflect an enduring and familial relationship, it also recalls "a history of slavery and servitude in the colonial East Indies in which the line between the two was fuzzy."[25]

In the 2010s, these debates were complicated further by controversy over Ali's origins. Researchers scoured archives and oral histories in attempts to pinpoint the collector's birth in the state of Sarawak, Malaysian Borneo, while others maintained that he was most likely to have originated in Ternate, now an island of Indonesia.[26] The compelling yet still enigmatic nature of Ali's biography continues to inspire research serving a wide variety of agendas and products, from identity politics to art installations.[27] These diverse interpretations of his life nonetheless remain inextricably bound up with Wallace's shifting position in the historiography of science and ideas.

Alternative approaches: collections and the case for decolonisation

While "Ali Wallace" is unusual in having been immortalised in a portrait photograph, glimpses of the contributions of non-Western natural history collectors can sometimes be clearer in images than in texts. The biography of another, earlier Ali, Swedish naturalist Claës Fredrik Hornstedt's Javanese slave, for example, remains largely mysterious. Indeed, we do not even know his real name, as Hornstedt replaced this "long Javanese name" with "Ali" at the point of his purchase.[28] A striking 1788 watercolour, however, depicts Hornstedt and Ali side by side, working at a table strewn with zoological paraphernalia. Ali is pictured handing a scorpion specimen to Hornstedt, suggesting active involvement in the latter's collecting activities. Two recent exhibitions about colonial collecting mounted by Singapore's National Heritage Board have chosen to display this image to represent the local networks European collectors relied upon: a small but visually arresting acknowledgement of Indigenous contributions to the practice of science.[29] Visual sources like this one are an additional way in which we can access the stories of non-Western collectors, just as Basalla's choice of illustration hinted at narratives more complex than his Eurocentric framework recognised. In order to more fully circumvent the sparsity of the archival record, however, it is necessary to consider both alternative source material and new, interdisciplinary approaches encompassing perspectives from social history, anthropology, and museology.

The Royal Society of London's gift shop offers a poster print for sale entitled "Landscape with termite hills," taken from Henry Smeathman's entomologically significant 1781 paper on termites. In the illustration, an unnamed African "worker" stands alongside a termite mound that he has clearly just broken open to reveal its internal structure. Smeathman's writings from his time in Sierra Leone make no mention of the identities of any of the countless Africans who collected, informed, and carried out manual labour on his behalf during his quest to understand the complicated workings of the termite colony. While it is difficult to draw out information about Smeathman's local collectors from his entomological writings, a 2018 book by Deirdre Coleman examines Smeathman's archive from an alternative perspective.[30] Coleman looks beyond his zoological work and considers his experiences in Sierra Leone holistically, including his shifting opinions about the slave trade in which he found himself inescapably embroiled. The result is a ground-breaking examination of the deep and inextricable links between the world of natural history and the slave trade. Coleman charts the simultaneous development of Smeathman's increasing empirical understanding of the termites and his feelings towards and perceptions of the slaves and free Africans he lived and worked alongside. It was a combination of the two which ultimately inspired his plans for a free plantation settlement in Sierra Leone, as an alternative for the slave-driven plantation systems he had witnessed during later travels in the Caribbean. Coleman extends this analogy to Smeathman's collections and collectors. In contemplating the process by which Smeathman's insect specimens were collected and shipped to Europe, she writes: "it is hard not to think of the parallel scene animating this book's core narrative—the many men, women, and children who, captured and commodified in slavery's net, were also consigned to the world of things."[31]

Coleman's book is timely, indicative of increasing awareness of unsavoury elements in the story of natural history. While ethnographic museums founded during the colonial period have been grappling with the question of decolonisation for some time, it has taken longer for natural history collections to receive the same scrutiny. With the growing prominence of decolonising narratives and movements such as the international Rhodes Must Fall campaign, however, historians of natural science and museums with historical natural history collections can no longer avoid engaging with these issues. In 2018, Subhadra Das and Miranda Lowe, both curators of scientific collections in London, charted the perpetuation of colonial worldviews by contemporary natural history museums that consistently fail to acknowledge the contexts in which their collections were formed.[32] Their call for museum professionals to address this structural racism and decolonise museum spaces is equally relevant for historians of science and museums.

It is in the museum that we may find one possible answer to our quandary regarding the absence of non-Western contributors from the archives of scientific knowledge. Das and Lowe contended that "natural history museums are well placed to relate decolonial narratives because the stories, work, and knowledge of non-white peoples remain manifest in natural history collections and museum spaces."[33] Rather than working from a top-down perspective via the work of great men of science, switching our standpoint to foreground museum collections and their accompanying archives can draw out myriad stories of the ordinary people around the globe who built them.

The London Natural History Museum's 2007 project entitled *Slavery and the Natural World* provides a case study of this approach in museological practice. Coinciding with the bicentenary of abolition of the slave trade in the British Empire, the project brought scientists, historians, grass-roots organisations, and members of the public together to explore the impact of the transatlantic slave trade on the development of the Museum's collections, and to discuss how this history might be better acknowledged in its galleries. The project delved into the Museum's extensive archives, revealing previously unrecognised examples of non-Western contributions to the Museum's collections and to British knowledge about the natural world. It brought little-known narratives into the public arena, such as that of enslaved eighteenth-century Ghanaian healer Graman Kwasimukambe, also known as Quacy or Kwasi, who traded knowledge of medicinal plants with Europeans in Dutch Guiana.[34] Linnaeus gave Kwasi's name to the *Quassia amara,* a plant which the Ghanaian had identified as having medicinal properties. The *Quassia amara* appears in the botanical artwork decorating the "Gilded Canopy" of the Natural History Museum's central Hintze Hall, under which all visitors to the museum pass. Since the project, this image and the story behind the plant's name are now highlighted in museum tours and literature.[35] Many such legacies of slavery and non-Western contributions to science can be found within historic museum halls, visible yet often still unseen.

Inverting perspectives and new narratives

My own doctoral research utilised similarly interdisciplinary methods to explore the development of natural history in nineteenth-to-twentieth-century Borneo, with exciting results. A study of the Sarawak Museum, my thesis considered the state of Sarawak under the Brooke Rajahs (the British family who governed north-west Borneo as an independent state and hereditary monarchy from 1843 to 1946) as a scientific hub in its own right.[36] Rather than interpreting the museum, founded in Kuching in 1886, as an acolyte of the British Museum, I conceptualised the institution as a centre for scientific knowledge production through cross-cultural negotiation, and a node in an intercolonial, transimperial network through which knowledge flowed in all directions. From this Asia-centric starting point, I used a combination of collective biography and object biography to examine the museum's colonial-era collections. Foregrounding objects and donation records allowed me to identify several Sarawakian individuals who interacted with the museum during the Brooke period. Further research in government archives, local newspapers, museum annual reports and collections catalogues revealed that the museum was the centre of a network of Indigenous natural history collectors, most of whom were Iban—the majority ethnic group in Sarawak—from the Saribas River region.

Figure 9.1   The Sarawak Museum in Kuching, a focal point for a region-wide network of Iban natural history collectors, c. 1900. KITLV collection, Leiden University Libraries. CC BY 4.0.

The Iban were perceived by the British government officers in the state as particularly well suited to natural history work, due to a combination of cultural and political factors.[37] The first documented Iban collector was hired by the Sarawak Museum in 1891, and over the next few decades, British perceptions combined with kinship ties led to more and more Saribas Iban taking up employment as collectors for both the Sarawak Museum and for amateur scholar-administrators amassing private collections.[38] The names of these collectors appear throughout the original zoology catalogues that survive in the museum archive, confirming that they contributed a vast number of specimens to the historic collections.

By building up object biographies for specimens and objects collected in Sarawak, I was able to ascertain that the Sarawakian collectors not only contributed to collections in Washington DC, London, Oxford, Cambridge, Paris, and Stockholm, but also travelled overseas themselves in pursuit of this career path. Word of the apparent innate skill of the Saribas Iban in the arts of natural history spread through colonial networks, and museums in Singapore, British Malaya, and further afield were soon seeking to employ Sarawakians. The "careers" of specimens, as well as ethnographic objects, in museum collections around the globe revealed a transboundary network of highly skilled Iban hunters, collectors, and taxidermists, who maintained ties with each other and with the Sarawak Museum while filling the collections stores of museums across Southeast Asia and beyond.[39] Close scrutiny of museum catalogues reveals that the most successful of these men also developed skills in taxonomic identification and contributed knowledge of flora and fauna based on their own observations in the field.[40] The Iban collectors were not just assistants or intermediaries, but formed a professional scientific network which operated in parallel to Western webs of knowledge circulation, and whose legacy for natural history collections has global significance. While the existence of hybrid cultures of knowledge production remains controversial, the evidence suggests that the Iban can justifiably be understood as collaborators in the European scientific enterprise in Southeast Asia.[41] Research into the Iban collecting network is ongoing, but strongly suggests that it may indeed be possible to reconstruct the biographies of more non-Western natural history collectors across the European empires.

Conclusions

The extent to which Indigenous and non-Western collectors can be considered collaborators, as opposed to assistants or intermediaries in the production of scientific knowledge globally, is, as we have seen, contentious. Relationships between naturalists, collectors, specimens, and ideas were, of course, partly shaped by the local cultural and political contexts in which they interacted. The historiographical shifts outlined above are the first steps towards a more thorough understanding of these processes on local and global scales.

These developments should continue apace, as it is only by inverting Eurocentric perspectives that more balanced narratives may begin to emerge from the shadows cast by the "great men" of "Western science." While conventional scientific archives remain frustratingly silent on the roles of non-Western collectors, moving away from top-down intellectual history and placing scientific networks and museum collections themselves in the spotlight can, as we have seen, reap rewards. The histories of museums and collections founded in extra-European territories, in particular, cry out for further attention. The growing utility of network studies in the history of science also proves beneficial here. Conceptualising museums and collectors in colonial territories as nodes in many-layered, transboundary webs, which disseminated knowledge in multiple directions, allows us to trace the movement of people, specimens, and ideas unencumbered by the persistent but often misleading metropole–periphery dichotomy.[42]

Further investigation of non-Western actors and their contributions to natural history promises to enhance our understanding of wider issues in the history of imperial scientific knowledge-production. Not least, the intertwined relationships between scientific networks and the slave trade are revealed by the important work of Coleman and others. Such work also sheds light on the complex interactions between colonial power and scientific knowledge, exploring issues of race, class, and trust in the cross-cultural encounters that formed the foundations of much colonial scientific endeavour.

Current developments in the discourse surrounding museums, decolonisation, and repatriation make this work even more pressing. The inherent racism in natural history museum collections and displays must be addressed, by museum professionals but also by historians of science through in-depth research into the diverse forces that shaped imperial-era natural history at the most fundamental level. The story of "colonial natural science" is not, as Basalla implied, that of the elite naturalists alone, but of all those who made up the vast community of hunters, collectors, specimen preparators, and taxidermists without whom the collections contained within our museums, still invaluable to researchers today, would be unrecognisable. To fully understand their role, we must consider them not simply as the acolytes of European scientists but as agents (sometimes highly skilled) of global scientific endeavour in their own right—a perspective which may seem self-evident, but remains a historiographical work-in-progress.

Notes

1 George Basalla, "The Spread of Western Science," *Science* 156, no. 3775 (May 1967): 617–19.

2 David MacKay, "Agents of Empire: The Banksian Collectors and Evaluation of a New Land," in *Visions of Empire: Voyages, Botany and Representations of Nature,* ed. David Miller and Peter Reill (Cambridge: Cambridge University Press, 1996); Londa Schiebinger, *Plants and Empire: Colonial Bioprospecting in the Atlantic World* (Cambridge, MA: Harvard University Press, 2007), 57–9.

3 Deirdre Coleman, *Henry Smeathman, the Flycatcher: Natural History, Slavery and Empire in the Late Eighteenth Century* (Liverpool: Liverpool University Press, 2018); Natural History Museum London, *Slavery and the Natural World* (London: Natural History Museum, 2018).

4 Henrika Kuklick and Robert E. Kohler, "Introduction," *Osiris* 11 (1996): 2.

5 Bernard Cohn, *Colonialism and its Forms of Knowledge* (Princeton, NJ: Princeton University Press, 1996); Christopher A. Bayly, *Empire and Information: Intelligence Gathering and Social Communication in India: 1780–1870* (Cambridge: Cambridge University Press, 1996).

6 Nicholas Dirks, "Colonial Histories and Native Informants: Biography of an Archive," in *Orientalism and the Postcolonial Predicament: Perspectives on South Asia,* ed. C. Breckenridge and P. van der Veer (Philadelphia: University of Pennsylvania Press, 1993); Simon Schaffer, Lissa Roberts et al., *The Brokered World: Go-betweens and Global Intelligence, 1770–1820* (Sagamore Beach, MA: Science History Publications, 2009); Kapil Raj, *Relocating Modern Science: Circulation and the Construction of Knowledge in South Asia and Europe, 1650–1900* (Basingstoke: Palgrave Macmillan, 2007); Sandra Khor Manickam, *Taming the Wild: Aborigines and Racial Knowledge in Colonial Malaya* (Singapore: NUS Press, 2015).

7 Schiebinger, *Plants and Empire,* 75–82.

8 Deepak Kumar, "Botanical Explorations and the East India Company: Revisiting 'Plant Colonialism,'" in *The East India Company and the Natural World,* ed. Vinita Damodaran, Anna Winterbottom and Alan Lester (London: Palgrave Macmillan, 2015), 30.

9 The foundational works of historical scholarship on colonial natural history museums make little mention of the practicalities of the collecting process: Susan Sheets-Pyenson, *Cathedrals of Science: the Development of Colonial Natural History Museums During the Late Nineteenth Century* (Montreal: McGill-Queen's University Press, 1988); John M. MacKenzie, *Museums and Empire: Natural History, Human Cultures and Colonial Identities* (Manchester: Manchester University Press, 2009).

10 Brett M. Bennett and Joseph M. Hodge, *Science and Empire: Knowledge and Networks of Science Across the British Empire, 1800–1970* (London: Palgrave Macmillan, 2011); Timothy P. Barnard, *Nature's Colony: Empire, Nation and Environment in the Singapore Botanic Gardens* (Singapore: NUS Press, 2016).

11 Maria-Theresia Leuker, "Knowledge Transfer and Cultural Appropriation: Georg Everhard Rumphius's 'D'Amboinsche Rariteitkamer' (1705)," in *The Dutch Trading Companies as Knowledge Networks,* ed. Siegfried Huigen, Jan L. de Jong, and Elmer Kolfin, (Leiden: Brill, 2010), 168.

12 Richard Drayton, *Nature's Government: Science, Imperial Britain and the 'Improvement' of the World* (New Haven, CT: Yale University Press, 2000), 92.

13 Jane R. Camerini, "Wallace in the Field," *Osiris* 11 (1996): 61. It has also been suggested that nineteenth-century zoologists were less inclined to acknowledge their local assistants than botanists, due to popular perceptions of hunting as a pastime of the European elite in colonial territories: Jeyamalar Kathirithamby-Wells, "Peninsular Malaysia in the Context of Natural History and Colonial Science," *New Zealand Journal of Asian Studies* 11, no. 1 (June 2009): 361.

14 Steven Shapin, *A Social History of Truth: Civility and Science in Seventeenth-century England* (Chicago: University of Chicago Press, 1994).

15 Kuklick and Kohler, "Introduction."

16 Richard Freeman, "Darwin's Negro Bird-stuffer," *Notes and Records of the Royal Society of London* 33, no. 1 (August 1978): 83–6.

17 Alfred Russel Wallace, *The Malay Archipelago* (New York: Dover, 1962); Alfred Russel Wallace, *My Life: A Record of Events and Opinions* (New York: Dodd, Mead and Co., 1905).

18 A campaign to obtain greater recognition for Wallace, placing him on a par with Darwin, has gone from strength to strength since the early 2000s. This has resulted in portraits and statues being erected in natural history museums in London and Singapore, the making of television documentaries, and the Wallace Correspondence Project, among other initiatives to bring Wallace's work to public attention: "The Alfred Russel Wallace Correspondence Project," http://wallaceletters.info; "The Alfred Russel Wallace Website," http://wallacefund.info.

19 Earl of Cranbrook and Adrian G. Marshall, "Alfred Russel Wallace's Assistants, and Other Helpers, in the Malay Archipelago 1854–62," *Sarawak Museum Journal* 94 (December 2014); Kuklick and Kohler, "Introduction," 10; Kathirithamby-Wells, "Peninsular Malaysia in the Context of Natural History," 361.

20 John van Wyhe and Gerrell M. Drawhorn, "'I Am Ali Wallace': The Malay Assistant of Alfred Russel Wallace," *Journal of the Malaysian Branch of the Royal Asiatic Society* 88, no. 1 (June 2015): 19.

21 Cranbrook and Marshall, "Alfred Russel Wallace's Assistants," 75.

22 Carey McCormack, "Collection and Discovery: Indigenous Guides and Alfred Russel Wallace in Southeast Asia, 1854–1862," *Journal of Indian Ocean World Studies* 1 (2017): 127.

23 Camerini, "Wallace in the Field," 54.

24 Cranbrook and Marshall, "Alfred Russel Wallace's Assistants," 109.

25 Camerini, "Wallace in the Field," 55.

26 Gerrell M. Drawhorn, "The Alienation of Ali: Was A.R. Wallace's Assistant from Sarawak or Indonesia?" (paper presentation, Rainforest Fringe Festival, Kuching, Malaysia, 14 July 2018).

27 See, for example, the 2016 docu-drama by Lord Cranbrook and Jamie Curtis-Hayward, *Searching for Ali Wallace* and Singapore-based artist Isabelle Desjeux's project entitled "Buang, the Lost Malay Scientist." "*Searching for Ali Wallace,*" 2016, Internet Movie Database, www.imdb.com/title/tt6437390/; "Buang, the Lost Malay Scientist", Isabelle Desjeux, https://isabellecreates.wordpress.com/projects/buang-the-lost-malay-scientist/.

28 Ann Kumar, "A Swedish View of Batavia in 1783–4: Hornstedt's Letters," *Archipel* 37, no. 1 (1989): 248.

29 The watercolour, part of the Singapore national collection, appeared in the Asian Civilizations Museum's *Raffles in Southeast Asia* exhibition in February–April 2019 and the National Museum of Singapore's *An Old New World: From the East Indies to the Founding of Singapore 1600s-1819* in September 2019–March 2020.

30 Coleman, *Henry Smeathman, the Flycatcher.*

31 Coleman, *Henry Smeathman, the Flycatcher,* 242.

32 Subhadra Das and Miranda Lowe, "Nature Read in Black and White: Decolonial Approaches to Interpreting Natural History Collections," *Journal of Natural Science Collections* 6 (2018), http://natsca.org/article/2509.

33 Das and Lowe, "Nature Read in Black and White," 8.

34 Simon Schaffer, Lisa Roberts et al., "Introduction," in Schaffer, Roberts et al., *The Brokered World,* xxix.

35 Kerry Lotzof, "Who was Graman Kwasi?," National History Museum, London, posted 4 October 2018, www.nhm.ac.uk/discover/who-was-graman-kwasi.html; Sabrina Imbler, "In London, Natural History Museums Confront their Colonial Histories," *Atlas Obscura,* 14 October 2019, www.atlasobscura.com/articles/decolonizing-natural-history-museum.

36 Jennifer R. Morris, "Museumising Sarawak: Objects, Collectors and Scientific Knowledge-production Under the Brooke State, c.1840–1940" (PhD diss., National University of Singapore, 2019).

37 Morris, "Museumising Sarawak," 202.

38 Insights into the Saribas Iban communities and culture from anthropologists, particularly Peter Kedit, were of great help in piecing together the development of these networks: Peter M. Kedit, *Iban Bejalai* (Kuala Lumpur: Sarawak Literary Society, 1993).

39 Morris, "Museumising Sarawak," 213–27.

40 Morris, "Museumising Sarawak," 213, 219.

41 Kumar, "Botanical Explorations."

42 For the conceptualisation of scientific networks as webs, see Joseph M. Hodge, "Science and Empire: An Overview of the Historical Scholarship," in Bennett and Hodge, *Science and Empire,* 16.

10

ENERGY AND EMPIRE

Nathan Kapoor

And we ought not at least to delay dispersing a set of plausible fallacies about the economy of fuel, and the discovery of substitutes for coal, which at present obscure the critical nature of the question, and are eagerly passed about among those who like to believe that we have an indefinite period of prosperity before us.

William Stanley Jevons (1866)[1]

As illustrated by this passage from William Stanley Jevons's *The Coal Question*, the search for new sources of energy to power Great Britain and its expanding empire generated anxiety about the sustainability of their energy demands. Furthermore, he recognised that coal represented more than an energy resource that powered the machines of the British Empire. Coal enabled and defined the prosperity of that empire. Indeed, during the late-nineteenth and early-twentieth century, the drive to alter existing energy systems or transition to new ones featured heavily in the imperial agenda of countries like Great Britain, China, and the United States.[2] In the last few decades, historians and sociologists studying the environment, energy, technology, and empire have demonstrated that interconnections between energy systems and empire are an increasingly important topic of analysis in colonial scholarship.[3] "Energy Systems" include the resources humans extract and harness, the technological systems which facilitate that action, and the social structures which participate in those processes. By analysing efforts to extract energy resources and construct technological systems to distribute it, historians of colonialism are poised to track how energy shaped colonial and imperial aims and locate the legacies of empire in current energy infrastructures.

Approaching imperial and colonial narratives through the study of energy systems reframes existing narratives and opens up the possibility of constructing new ones. First of all, charting the desires for new energy resources permits historians to locate supplementary explanations for the colonial drive of many empires and use that as an anchor to explore their economics, politics, and culture. For instance, the pursuit of hydroelectricity in British colonies such as Canada or New Zealand provides insight into how the British Colonial Office used energy systems as a measure for circumventing inconvenient land legislation and stripping Indigenous land rights in the name of providing cheaper energy.[4] Furthermore, the ubiquity and universal scale of energy systems allow for colonial narratives to cross traditional temporal and geopolitical boundaries. For example, the study of the categorisation and extraction of coal in China during the nineteenth century presents a narrative in which many European, Chinese, and American imperialisms intersected and created a "new imperialism" that is best described by the shared use of coal between the countries.[5] And lastly, insisting on a connection between energy regimes and empire historians disrupts existing assumptions about the "natural" or obvious succession of energy sources, and the empires themselves.[6] The adoption of energy sources does not follow a teleological plan; the decision to use one over another must be understood as a sociocultural phenomenon in which energy systems were made to fit. The transition from animal and water power to fossil fuels in Great Britain during the early nineteenth century was not a "natural next step." Instead the transition is better understood as a politically and econom-ically advantageous decision made by industrialists and investors hoping to consolidate labour and stockpile resources.[7] One could make a similar argument using many different energy systems (i.e., natural gas, petroleum, or rivers). However, given current conversations about the embeddedness of fossil fuels and electric power, I find it more salient to focus on recent his-toriographic contributions concerning coal and electrification in nineteenth- to twenty-first-century histories of empire.

Energy and power

The next step in addressing environmental crises will come from the humanities and social sciences ... What we need to do is, first, grasp the full intricacies of our imbri-cation with energy systems (and with fossil fuels in particular), and second, map out other ways of being, behaving, and belonging in relation to both old and new forms of energy.[8]

Imre Szeman and Dominic Boyer (2017)

The crisis of climate change and utility of energy as a metric for historical analysis has given rise to ever-increasing lists of "energy histories," in both academic and popular presses, such as Imer Szeman and Dominic Boyer's *Energy Humanities: An Anthology* (2017) or Richard Rhodes's *Energy: A Human History* (2018), both dealing with empire explicitly.[9] As Szeman and Boyer suggest in *Energy Humanities*, understanding how certain energy systems are embedded in soci-eties and cultures will be a necessary step in affecting change to existing exploitative energy regimes. The establishment of fossil fuel and electric power regimes in the nineteenth century coincides with the largest global imperial and colonial acquisitions in history. The transition to these new systems of power production and political organisation is not coincidental. The utilisation of these resources played a role in fabricating imperial and colonial expansion in the nineteenth century, thus embedding practices of resource manipulation, environmental neglect, and labour exploitation into the foundations of those energy infrastructures, many of which persist. Therefore, in an answer to the call for humanists to take up the study of energy, scholars of colonialism and imperialism—not just those of the nineteenth century—are well suited to make substantial contributions to the effort to grasp the intricacies of our imbrication with energy systems.

Many deployments of "energy" or "energy systems" as a framework for historical inquiry have appeared in the last few decades as means of studying politics, economics, and culture, including, but not limited to, Daniel Yergin's *The Prize* (1980), Thomas Hughes's *Networks of Power* (1983), David Nye's *Electrifying America* (1990), Richard White's *The Organic Machine* (1995), and Timothy Mitchell's *Carbon Democracy* (2011).[10] Any one of these books, or many other unnamed titles, could establish the advantage of studying energy in history; however, for discussing the relationship between energy and empire, *The Organic Machine* serves as a useful starting point because of the model Richard White provides in his treatment of energy. In this work, now considered a classic in most environmental history departments (and likely many other specialties), White challenges readers to see the Columbia River, which is often only seen as an inactive resource, as an active part of human interaction through time. This accomplishes two things. First, as with most environmental history, his argument makes the Columbia River, a source of energy, an active participant in the interplay between people in the space. Secondly, and most importantly for this chapter, the river allows White to speak across time, space, and cultures. At different periods of the Columbia River's history many different groups of people utilised the river for food, irrigation, hydroelectricity, and nuclear power. Such framings provide historians with tools to craft colonial narratives around the energy source and measure the ways in which energy inspired, directed, and expanded empires.

Coal

This order is now bound to the technical and economic conditions of machine production which today determine the lives of all the individuals who are born into this mechanism, not only those directly concerned with economic acquisition, with irresistible force. Perhaps it will so determine them until the last ton of fossilized coal is burnt.[11]

Max Weber (1905)

As indicated by the connection Weber made between energy and the economic regimes of Europe and America, as well as China, the extraction and usage of coal influenced many imperial agendas of the nineteenth and twentieth centuries. Increasingly coal, and indeed the fossil fuel regime, has received scholarly attention due to the growing awareness of the role fossil fuels play in spurring climate change. In their histories of coal and empire, Shellen Wu, Peter Shulman, and Steven Gray make coal the centre of their narratives about Chinese, American, and British imperialism.[12] The utilisation of coal, at least on an industrial scale, requires a vast human–machine network in which the material is painstakingly mined, processed, transported, and burnt. Even though these three cases deal specifically with China, the United States, and Great Britain, the adoption of coal power during this period spanned the globe. The surge in coal mining and usage during the mid nineteenth century did not arise as an afterthought of colonial expansion or imperialist ideas; it forged them.

The use of coal did not begin during the Industrial Revolution. Europeans burnt coal for thousands of years, from the Roman occupation through the Middle Ages. They recognised that the mineral generated intense heat in cooking, warming, and forging. What marks the significance of coal during the Industrial Revolution is not the use of the resource itself, but the attachment of the resource to economic ideas about the maintenance and expansion of empires and its use in the steam engine. This action ignited both the engines of industry and the birth of the fossil fuel economy. In *Fossil Capital* (2016), Andreas Malm argues that the decision to begin using coal is best understood as a decision made, generally by the economic elite, to utilise *stock* instead of *flow* energy resources, which meant that energy resources could be stockpiled, centralised in urban spaces, and made to turn the growing engines of industry, rather than relying on the more geographically contingent methods of power production like water, wind, or human power. The usage of coal manufactured the British Empire's reliance on a new network of supply, materials, and labour. During the nineteenth century, coal not only fuelled the British, American, and Chinese Empires but acted as a formative agent in directing their imperial policies.[13] As the journalist Archibald Hurd suggested in 1898, the Colonial Office and Royal Navy knew the significance of coal:

> Coal rules the destiny of nations ... Coal is the source of commercial prosperity and the secret of our naval superiority ... Coal is the requisite of empire. Without coal it [the British Empire] will be in the position of a man with a pipe and matches and no tobacco.[14]

Accordingly, historians have also acknowledged the importance of coal in powering ships or railways and global commerce, but we must also show how coal determined the course of colonial acquisition and imperial policy. The entrance of coal into the production and transportation spheres of British society during the eighteenth and nineteenth century reshaped life on the isles. However, the concerns spanned Great Britain and its global empire. The study of coal usage in the British Navy represents one of the most fruitful areas of historical research on the connection between energy and empire. The same steam engine that restructured the British economy proved instrumental in re-forging the larger, faster, and more battle-ready ships of the British Navy. Following the conclusion of the Crimean War (1856), in which the last British Navy battle with sailing ships took place, the steamship became the prime mover of British military authority. Even though steamships expanded the capabilities of seafaring vessels, they also introduced new logistical challenges, namely the strategic placement of fuelling stations, physical barriers for the large ships, and maintenance of an increasingly complex supply network. In short, coal may have increased the speed and range of the Royal Navy, but it also tied it to land and relied on new industries.

Even though the British Navy's transition to steam power took decades, beginning with the HMS *Devastation* (1871), the supply network, location of coaling stations, and preferred quality of coal controlled the movement of the British Empire. In Steven Gray's *Steam Power and Sea Power*, coal is both the fuel source for the British Navy and the impetus behind the British Empire's global spread.[15] The logistics of coaling proved a major point of anxiety for British Navy officers and the Colonial Office. Gray, and many other scholars, utilise a passage from the Carnarvon Commission, a Colonial Office report written by Lord Carnarvon 1882, which states:

> The best means ... of providing for the defence and protection of Our Colonial Possessions and commerce ... special attention being given to necessity of providing safe coaling, refitting and repairing stations ... in time of war.[16]

Among many other concerns facing the expansive military and commercial British Empire, coal maintained a prominent place in imperial policy. Following this report, we can observe a "coal consciousness," as the goal of maintaining a maritime hegemony produced anxiety within the Colonial Office. In order to combat their own anxiety and maintain global naval sovereignty, the Colonial Office and Royal Navy sold their high-quality Welsh coal, the most superior naval coal, to their foreign imperial competitors such as Russian, German, Austrian, Italian, Swedish, French, and Spanish governments.[17] This was not foolish oversight but an intentional method of brokering their geopolitical might by threatening to withhold coal or deny foreign governments access to the British collieries. Just as historians have highlighted the effectiveness of British technologies or arms of government to create their empire, so too must we wrestle with the ways the British Empire used coal.

The dominant legacy of steamships aligns with the "tools of empire" narrative which suggests that superior Western technologies led to imperial successes.[18] However, coal is also a useful tool for historians to demonstrate the ineffectiveness, or at least weaknesses, of the British Empire (and other empires). This is not petty revisionism, but rather an attempt to combat deterministic assumptions about the superiority of fossil fuels and related technologies. Too often, the development and success of steamship and rail technology are used primarily as means of explaining the success of the British, which overshadows the moments when coal exposes failure and inadequacy. Coal spurred anxiety among economists, politicians, and Royal Navy officers, not only because they feared supply complication but also because the limits of steam ships, and therefore coal, were at times exposed throughout the British Empire. In *Steamboats on the Indus*, Clive Dewey outlines the ways steam ships proved to be technological and financial failures for the British Empire on the Indus River because of their insistence on their technological superiority, much of which was attached to coal.[19] Once again focusing on the usage, and occasional problems, with coal narratives like this proves instructive in deconstructing narratives of universal British technological superiority.

Besides the British Empire, coal figured prominently in American imperialism at the turn of the twentieth century. In Peter Shulman's book, *Coal and Empire: The Birth of Energy Security in Industrial America*, he makes three arguments centred on coal. First, the American focus on fossil fuels did not begin with oil but with coal during the nineteenth century. Second, the supposed "need" for distant coaling stations did not lead to the establishment of the American island empire during the 1890s; rather the creation of that empire created an unprecedented need for coal. And finally, the technological shift to fossil fuel technologies, fuelled by both coal and oil, defined the emerging United States' global role. Much like Gray's argument about the Royal Navy's dependence on coal, Shulman demonstrates the formative role energy consideration plays in addressing international and colonial policies. The American Empire did not just happen upon coaling stations throughout the Pacific; their growing awareness of the centrality of fossil fuels demanded that an international network of coal figure in their political and military policies.[20]

In addition to enriching the history of Western imperialisms, following the development of coal usage provides insight into the growth of non-Western empires. The one-sided exploitation of resources is a key marker of colonialism and imperialism.[21] During the nineteenth century, this style of exploitation existed in non-Western empires too. For instance, in Shellen Wu's *Empires of Coal: Fueling China's Entry Into the Modern World Order* she compares and contrasts the usage of fossil fuels in China with its Western counterparts—not as a means of suggesting Chinese imperialism mirrors British or American imperialism, but as a way of suggesting how shared patterns of energy usage led to the creation of similar patterns of resource exploitation and economic change. As Wu argues, the Qing government spurred on industrialisation and found coal a readily available solution to meet their manufacturing and transportation plans. In response to Western geological surveys and engineering, primarily with the input of Ferdinand von Richthofen and German engineers, Qing imperial policy connected China's wealth of mineral resources to industry, modernity, and a metaphor sovereignty. By capturing this rhetoric surrounding coal, Wu usefully explains that the exploitation of energy resources as part of imperial ideology is not a Western phenomenon, but a global one. Even though the German Empire informed Qing resource policies, the Chinese operated with essentially the same motivation well into the twentieth century. The embrace of coal as necessity unified imperial activity around the world. Acknowledging the breadth of this phenomenon reforms how we measure imperial success and strengthens the association of fossil fuels with exploitation.[22]

Electricity

> What are the advantages of these electrical exhibitions? They undoubtedly denote progress, and they instil and spread knowledge. They enable the public to see with their own eyes what is being done in special fields of manufacture and enterprise.[23]
>
> *William Preece (1882)*

Colonisers, and some subsequent historians, treated electric power systems, much like coal, as a tool of empire. During the late nineteenth century, for many colonial powers, electric power indicated progress. Thus electricity has served as a marker of modernity through technological progress, light, communication, and motive force.[24] Such portrayals obfuscate colonial ideas about and justification for electrification, rather than explicitly linking electrification and colonialism. It is true that electrical technologies provided light and electromotive force, but they also served to expand colonial desires to control the environment, affect political control through infrastructure, and perform the racial ideologies of the colonisers. The historiography of electrification is both well researched and under-researched. Histories of national electrification and the cultural import of electricity abound but these histories do not adequately account for the connections between electrification and colonialism. This is not to suggest that they ignore colonisation completely but that their focus on large technical systems and the social impact of electricity in colonial metropoles has left much to explore. Building on the tools of earlier electrical history, the following represents a sampling of recent scholarly efforts to explore the relationship between electrification and colonialism and encourages further study of the deployment of electric power in colonial spaces.

Technological infrastructures shape, change, and ignore political and geographic boundaries. "Technology was not only the instrument but also the substance of state power."[25] After the mid-nineteenth century, Western states used, or at least hoped to use, electrical infrastructures to shorten the distance between their colonies and political centres to redefine their boundaries around the world. The telegraph is often treated as a precursor to electrical power technology; however colonial powers like the British Empire used the telegraph to manage their colonial holds and blueprint the electric power grid. The expansion of the global telegraph network is a useful subject for exploring how competing colonial powers utilised communication networks to not only increase the speed of communication for its citizens, but also consolidate state management within a technological infrastructure. For instance, in *Communication and Empire*, Dwayne Winseck and Robert Pike argue that the global expansion of telegraphy is not just about technological competition between individual empire but one about the intrusion of Western systems (communication companies and public works departments) into colonial spaces using the rhetoric of civilisation and modernity.[26] Similarly to the earlier observations about coal, the establishment of electrical communication infrastructures did not follow the expansion of the British and American Empires during the nineteenth century. The channels of electric communication redefined physical boundaries by installing new nodes of power and increasing the distance from which political, military, and commercial power were allowed to flow.

During the 1870s and 1880s, much of the success of electric lighting technology depended upon the testing of lights in colonial spaces. Many colonial powers sought to use the electric light to illuminate their harbours, streets, and subjects. However, electric lights were not installed for purely beneficent reasons. They were not only technologies of relief. As much as lighting might have offered comfort or a sense of modern technological capability, so too did it increase colonial access to harbours, control over electrical infrastructure, and bureaucratic management of utilities. The spread of electric lighthouse technology provides one of the clearest representations of this. Despite the relatively small number of electric lighthouses constructed during the late nineteenth century, the electric lighthouse became a symbol of technological modernity. In their articles on nineteenth-century electric lighthouses, Michael Schiffer and Robert Bickers show how colonial ports served as laboratories for these "mundane" technologies in which scientists and engineers to demonstrate the effectiveness of the electric light and facilitate colonial control and commerce.[27] Schiffer explores how British, French, and American scientists and engineers developed electric lighthouses to prove the utility of different kinds of electric light. At the same time, the light communicates each empire's standing as a maritime power. Bickers examines the installation of lighthouses on the Chinese coast between the 1860s and 1930s. Even though the language used by proponents of electric lighthouses justified their existence as accident avoidance or as means to increase maritime trade, most insisted their utility lay in potential advantage to the empire as a whole. Much like the telegraph and coaling stations, the British Empire, particularly the Admiralty, did not build these lighthouse networks as an afterthought of colonisation but explicitly used them to create their informal East Asian empire.

Besides providing physical infrastructures to create and uphold colonial rule and imperial policy, electric lighting highlighted and extended racial boundaries in both settler and subject colonies. Historians can use the construction of these electrical infrastructures to see the ways infrastructures reflect and enact colonial racial inequalities. For instance, Moses Chikowero's article "Subalternating Currents" argues that the electrification of Bulawayo, Zimbabwe between 1894 and 1939 shows how the rapid acceptance of domestic electric technology solidified the centralised administration of power sources, thereby furthering the colonial government's hold on the urban community. The arrangement of Bulawayo's electrical grid then accentuated urban racial tensions.[28] Even though colonial officials insisted that electricity would alleviate economic disadvantage and brighten the colony for everyone, access to electric power followed racist and classist lines which reinforced colonial ideas about European superiority. The electric power system effectively created new slums and furthered social inequalities by leaving certain areas of town disconnected and enforcing high utility rates that limited access to electricity. By closely following narratives of electrification in the colonies, such as the one in colonial Zimbabwe, colonial historians can see the ways in which electricity, the emblem of many empires' civilising or modernisation projects, often worked to bolster the spatial and racial divides of colonial populations.

Electric power also provides historians with a means of exploring the colonial foundations of technological infrastructures from the height of European imperial expansion well into the twenty-first century. For instance, in the British Empire, many electrification projects began in colonies but concluded in mandates or dominions. This transition to both a new source of power and a new government organisation allows us to chart the legacy of colonialism in these new states, especially as former empires attempted to decolonise and distance themselves from colonialism. For example, in *Current Flow: The Electrification of Palestine*, Ronen Shamir examines how the British, and later Mandate, efforts to electrify increasingly marginalised Arab communities, while favouring the Jewish population and economy after the fall of the Ottoman Empire. As with many electrification stories, the end-result of a state-wide electric grid is known, but how electric power became a function of the state is frequently left uncontextualised. In the case of Palestine, the British government granted companies and individuals exclusive rights to control water sources and generate electric power. British officials specifically mobilised an electric infrastructure that would benefit their commercial interests in the region, effectively crippling local political influence over energy production. While this study focuses on Palestine's electrification, it makes an excellent case for energy humanists to look for the ways that early-twentieth-century technological infrastructures (including engineers, politicians, consumers) implemented and mirrored colonial behaviours at the local and global level.[29]

In the twenty-first century, as we transition away from fossil fuels and develop new means of production, it is useful to consider the legacy of colonialism within contemporary debates about electric power. It is no coincidence that electrification began at the height of Western imperial expansion. Electric power systems achieved the promises of many empires' modernisation projects, such as increasing bureaucratic management of technological systems, displaying Western visions of technological superiority, and powering the consolidated centres of production. Unfortunately, the colonial foundations of electric power production sometimes reappear in the expansion of alternative fossil fuel alternatives, such as wind power, despite being designed to halt the climatic damage and human suffering. For instance, in Cymene Howe and Dominic Boyer's book *Wind Power in the Anthropocene*, they explore the development, and ultimate failure, of the wind industry on the isthmus of Tehuantepec.[30] Alongside analysing the reasons for the wind parks' failure and difficulties, their most interesting insight is that the behaviour of the wind company officials and international non-governmental organisations (NGOs) mimics earlier European colonial practices in Mexico. Although the proponents of the project claimed that wind power offered a solution to other environmentally harmful sources, they adopted the same land acquisition and profiteering practices of other extractive industries. The population of the isthmus, especially the Indigenous population, rightly saw the encroachment of the wind companies as an external force stripping them of their resources without fair compensation for their land and environment, much as various colonial powers had done for centuries. This is not to suggest that wind power itself is colonial; rather, any energy production method which requires the displacement of people, distributes profits unequally, and causes environmental damage is derived from colonial ideas about energy resources. We must begin reconfiguring our characterisation of energy sources to move away from binary constructions of "good" and "bad" energy resources. We must reconcile with the fact that energy resources alone didn't create the Anthropocene; rather, the Anthropocene is the result of oppressive power systems, like colonialism, extracting those sources and exploiting the people to which it belongs.

Conclusion

There are many connections between energy and empire. "Coal" and "electric power" represent a small portion of the scholarship on energy and empire. Nuclear power, petroleum, and solar power have also become useful topics for scholars interested in highlighting connections between energy resources, technologies, and infrastructures in colonial spaces. Historians such as Gabrielle Hecht and Traci Brynne Voyles explore the colonial legacy of nuclear power through the tragic human and environmental costs of uranium mining and hegemonic nuclear infrastructures.[31] Timothy Mitchell and Andreas Malm track the history of the petroleum industry and its relationship with economic and political policies of expansion and exploitation.[32] Only in the last decade or so have humanists, such as George Gonzalez and Benjamin Sovacool, begun to make explicit connections between colonialism and solar power, which remains a rich subject for future research.[33]

As indicated by the examples above, energy is a useful category of analysis for historians of empire because the search for energy resources and subsequent application of those resources motivated so many colonial and imperial movements. Furthermore, and perhaps most importantly, studying the connection between empire and energy systems reveals the ways in which certain systems, especially those that remain popular today, came to dominate because of colonial and imperial systems of economic, social, and environmental exploitation. Even as empires touted decolonisation policies throughout the twentieth century, governments and energy firms remained entrenched in the same energy infrastructures put in place during the late nineteenth century, most especially the oil industry, hydroelectric bureaucracies, and coal mining. Historians, sociologists, and other energy humanists are well suited to expose this relationship and make clear how empire is embedded in numerous energy systems of the past and present, not only to expose the exploitation but to look for moments in which people use energy systems to actively decolonise.

Notes

1 William Stanley Jevons, *The Coal Question: An Inquiry Concerning the Nation and the Probable Exhaustion of Our Coal Mines*, 2nd ed. (London: MacMillan and Co., 1866), 4.
2 Vaclav Smil, *Energy Transitions: History, Requirements, Prospects* (Oxford: Praeger, 2010), 17.
3 Ute Hasenöhrl and Jan-Henrik Meyer, "The Energy Challenge in Historical Perspective," *Technology and Culture* 61, no. 1 (2020): 295–306.
4 Daniel Macfarlane and Peter Kitay, "Hydraulic Imperialism: Hydroelectric Development and Treaty 9 in the Abitibi Region," *American Review of Canadian Studies* 46, no. 3 (2016): 380–97.
5 Shellen Wu, *Empires of Coal: Fueling China's Entry into the Modern World Order, 1860–1920* (Stanford, CA: Stanford University Press, 2015).
6 A critique of this long-standing historiography is thoroughly detailed and deconstructed in J. van der Straeten and U. Hasenöhrl, "Connecting the Empire: Neue Forschungsperspektiven auf das Verhältnis von (Post)Kolonialismus, Infrastrukturen und Umwelt," *NTM International Journal of History and Ethics of Natural Sciences, Technology and Medicine* 24, no. 4 (2016): 355–91.
7 Andreas Malm, *Fossil Capital: The Rise of Steam Power and the Roots of Global Warming* (New York: Verso Books, 2016).
8 Imre Szeman and Dominic Boyer, *Energy Humanities: An Anthology* (Baltimore, MD: Johns Hopkins University Press, 2017), 3.
9 Richard Rhodes, *Energy: A Human History* (New York: Simon Schuster, 2018).
10 Daniel Yergin, *The Prize: The Epic Quest for Oil, Money, and Power* (New York: Simon and Schuster, 2008); Thomas Hughes, *Networks of Power: Electrification in Western Society, 1880–1930* (Baltimore, MD: Johns Hopkins University Press, 1984); David E. Nye, *Electrifying America: Social Meanings of a New Technology* (Cambridge, MA: MIT Press, 1992); Richard White, *The Organic Machine: The Remaking of the Columbia River* (New York: Hill and Wang, 1995); Timothy Mitchell, *Carbon Democracy: Political Power in the Age of Oil* (London: Verso Books, 2011).
11 Max Weber, *The Protestant Work Ethic and the Spirit of Capitalism*, 3rd ed., trans. Talcott Parsons (New York: Charles Scribner's Sons, 1950), 181.
12 Wu, *Empires of Coal*; Peter Shulman, *Coal and Empire: The Birth of Energy Security in Industrial America* (Baltimore, MD: Johns Hopkins University, 2015); Steven Gray, *Steam Power and Steam Power: Coal, the Royal Navy, and The British Empire, c. 1870–1914* (New York: Palgrave MacMillan, 2018).
13 Malm, *Fossil Capital*.
14 Archibald Hurd, "Coal, Trade, and the Empire," *The Nineteenth Century: A Monthly Review* 44, no. 261 (1898): 718–23.
15 Gray, *Steam Power*.
16 *London Gazette*, September 12, 1879.
17 Gray, *Steam Power*, 74.
18 Daniel Headrick, *Tools of Empire: Technology and European Imperialism in the Nineteenth Century* (Oxford: Oxford University Press, 1981).
19 Clive Dewey, *Steamboats on the Indus: The Limits of Western Technological Superiority in South Asia* (Oxford: Oxford University Press, 2014).
20 Shulman, *Coal and Empire*.
21 D.K. Fieldhouse, *The Colonial Empires: A Comparative Survey from the Eighteenth Century* (London: Weidenfeld and Nicolson, 1966), 386.
22 Wu, *Empires of Coal*.

23 W.H. Preece, "Electrical Exhibitions," *Telegraphic Journal and Electrical Review*, 30 December 1882, 505–6.

24 Ute Hasenöhrl, "Rural Electrification in the British Empire," *History of Retailing and Consumption* 4, no. 1 (2018): 10–27.

25 Gyan Prakash, *Another Reason: Science and the Imagination of Modern India* (Princeton, NJ: Princeton University Press, 1999), 160.

26 Dwayne Winseck and Robert Pike, *Communication and Empire: Media, Markets, and Globalization, 1860–1930* (Durham, NC: Duke University Press, 2007).

27 Michael Bryan Schiffer, "The Electric Lighthouse in the Nineteenth Century: Aid to Navigation and Political Technology," *Technology and Culture* 46, no. 2 (2005): 275–305; Robert Bickers, "Infrastructural Globalization: Lighting the China Coast, 1860s–1930s," *The Historical Journal* 56, no. 2 (2013): 431–58.

28 Moses Chikowero, "Subalternating Currents: Electrification and Power Politics in Bulawayo, Colonial Zimbabwe, 1984–1939," *Journal of South African Studies* 33, no. 4 (2007): 287–306.

29 Ronen Shamir, *Current Flows: The Electrification of Palestine* (Stanford, CA: Stanford University Press, 2013).

30 Cymene Howe and Dominic Boyer, *Wind Power in the Anthropocene* (Durham, NC: Duke University Press, 2019).

31 Gabrielle Hecht, *Being Nuclear: African and the Global Uranium Trade* (Cambridge, MA: MIT Press, 2009); Traci Brynne Voyles, *Legacies of Uranium Mining in Navajo Country* (Minneapolis: The University of Minnesota Press, 2015).

32 Timothy Mitchell, *Carbon Democracy: Political Power in the Age of Oil* (New York: Verso Books, 2011); Andreas Malm, *Fossil Capital: The Rise of Steam Power and the Roots of Global Warming* (New York: Verso Books, 2016).

33 George A. Gonzalez, *Energy and Empire: The Politics of Nuclear and Solar Power in the United States* (Albany, NY: SUNY Press, 2012); Benjamin Sovacool, *Visions of Energy Futures: Imagining and Innovating Low-Carbon Transitions* (New York: Routledge, 2019).

11

SCIENCE, EMPIRE, AND THE OLD SOCIETY OF JESUS, 1540–1773

Maria Pia Donato and Sabina Pavone

Since their foundation in 1540, and well after their first suppression by Pope Clement XIV in 1773, Jesuits have been key actors in the intertwined goals of the competing Catholic empires: colonisation and evangelisation. Both goals implied a momentous effort in collecting, producing, and transmitting knowledge on the natural world, which the Society of Jesus, a missionary order, and shortly after its founding also a teaching order, employed in practically all the regions of the world. As the Jesuits became fixtures throughout the Catholic empires, these attributes came to particularly define the order.

The peculiar place of Jesuits in the historiography of early-modern science and empire has been established for more than a century. In the nineteenth and early twentieth centuries, the master historical narrative was about their importance in disseminating science as part of the Western civilising mission. A sub-narrative was often interwoven with this main one, namely the harmonious coexistence of science and religion, and the positive influence of Christianity in world history. The Society of Jesus, which had been restored in 1814 and had progressively resumed its pastoral and educative activities in many European countries and their overseas territories, played a very active part in both debates. As the disciplinary histories of science were being written and shaped, methodologically it meant retracing Jesuit contributions to various branches of science; for this purpose, the fact that one was a Jesuit was to an extent secondary to his being a man of science.

Within such a framework, national sub-narratives extolled the non-European locations of the development of science, pinpointing the Jesuits' work in local contexts, or, conversely, imputing backwardness to the Society's role in education.[1] The national paradigm was articulated differently in the former Iberian colonies, which became independent nation-states in the nineteenth century and claimed their share in Western modernity, and the East, where the (re)-discovery of, and admiration for, ancient and possibly superior traditions of learning was pursued by Asian scholars and European and American historians alike. In the wake of Joseph Needham, sinologists working on the Jesuits highlighted the Order's important role in facilitating East–West intellectual exchanges, an insight which paved the way towards de-centring the history of science—even at the cost of overlooking the ultimate religious goal of their missions.

In past decades, while historians have come to study early-modern missions as a field for cross-cultural encounters, the positivistic take on Jesuit science has been replaced by a subtler understanding of the circulation of knowledge, and a more interconnected stance on the history of science. The revision of Jesuit science in the early-modern world gained momentum in the 1990s, when Jesuit studies underwent what Fabre and Romano aptly termed *désenclavement*—the opening up of the history of the Society beyond small circles of specialists,[2] just as "science and empire" was consolidating as a research subfield. It has blossomed into a mainstream line of academic investigation for both Jesuit and non-Jesuit scholars, which the Society, in their commitment to inter-religious dialogue and their role as a global teaching order, promotes worldwide. Reflecting a growing body of scholarship, recent compendia include chapters on science. Conversely, the notion of *missionary science* and *savoirs missionnaires* has entered the history of science as a convenient way to describe a wide range of activities.[3] Incidentally, studies on Jesuit science have contributed to the reassessment of the place of the Iberian Empires in the global history of science.

As a matter of fact, scholarly interest in Jesuit science has been fuelled by the Society itself for a very long time. A bureaucratic and relatively centralised organisation, it has always been careful in recording its own activity, for the sake of internal governance and for propaganda *vis-à-vis* "heretics" and, no less importantly, the courts and rival orders. The Jesuits thus produced a tremendous number of letters and reports, narratives and descriptions of remote lands and peoples, which, while praising the fathers' endeavours, vindicated their usefulness in empire building. In the late nineteenth century, in a period when the role of the Catholic Church in society came under attack by liberal governments and secular intelligentsia in several countries, monumental editions of sources began to see the light too, in a period when, in the words of the leading Jesuit historian John W. Malley, "the publication of critical editions of historical sources was becoming an international industry" as the best means to "refute the calumnies hurled at the Society."[4] These collections were themselves embedded in imperial culture and in ethnically and racially determined visions of progress. Missionary histories of individual countries mostly shared a Eurocentric and Catholic triumphalism disparaging Indigenous peoples.

Non-Jesuit historians working on science and empire have therefore been able to draw upon a wealth of published sources—in actuality often paying limited attention to the typologies and intended audiences of missionary documentation. In doing so, both Jesuit and non-Jesuit historians of science have long inadvertently reproduced a civilisational understanding of the spread of Western science, one of the "founding blocks" of the history of science, and which is ultimately a by-product of colonial and Jesuit propaganda. Today, though, these materials and well-ordered archives are available for new questions and interpretations in regard to cross-fertilisation of knowledge and Indigenous agency.

The Jesuits' ability to fashion their own story distinguishes them from other congregations and must be taken into account as its own historical problem. Dominicans, Augustinians, Franciscans, and Discalced Carmelites were engaged in evangelisation and colonisation earlier (and, in some territories like New Spain and the Philippines, more systematically) but Jesuits developed a more effective communication network. Other distinctive features include their up-to-date humanist education, according greater importance to the arts and sciences than the friars' scholastic mindset, and their global mobility. The Society was organised in large "imperial" and regional entities (*assistancies* and provinces), each under the care of a superior, who responded to the Superior General in Rome; fathers and brothers, however, were not canonically bound to any specific place and could be dispatched anywhere their superiors considered them to be more useful and appropriate. Issues of mobility recurrently raised tensions between the Society and the secular powers on which the *assistancies* depended, yet missionaries could still be dispatched anywhere. Both features helped the Jesuits enter complex information networks.[5] As Steven J. Harris has argued, the strength, longevity, and flexibility of the Jesuit scientific tradition owes much to an organizational structure that effectively combined spatially distributed networks (the Society's overseas missions) with multiple nodal points or nexuses (Jesuit colleges and universities). ... Its production of natural knowledge arose within and simultaneously helped sustain that institutional configuration.[6]

In this context, there are many ways of addressing science, empire, and the Jesuits. One is to deal with each empire separately. After all, although Jesuits circulated, the political circumstances, imperial infrastructures, and patterns of colonial settlement shaped their action nonetheless. However, intra-imperial and national histories tend to reify boundaries that were in fact uncertain and contested, and to discount both the discrepancies that existed between imperial and religious agendas as well as the Superior General's and the missionaries' local agency. In Brazil, for instance, Jesuits had to comply and compete with four empires, three of which were rival Catholic powers. Tensions also recurrently rose between the order and the papacy, and between the papacy and secular powers, sometimes making the Jesuits in the field either the pawns or the scapegoats of such conflicts. The Roman congregation de Propaganda Fide, created in 1622, was intended by the papacy as a means to simultaneously (though hardly effectively) control the missionaries and the crowns, and it repeatedly reprimanded the Jesuits' lack of compliance. The Old Society's story is in actuality one of conflicts and setbacks, eventually culminating in the order's dissolution.

Another common way of telling this story is a discipline-based one, i.e., of Jesuits' contribution to astronomy, natural history, anthropology, etc.[7] Once the old diffusionist paradigm was replaced by a multipolar one, as in recent scholarship, such an approach allows for an in-depth appraisal of the circulation of knowledge in the early-modern world. Focusing on the scientific output, however, tends to downplay the shortfalls, hazards, and ultimately the relations of power embedded in knowledge making. It is worth noting incidentally that historians of Jesuit science still tend not to address issues of domination overtly, and rather emphasise the globalisation of knowledge.[8]

This chapter seeks to combine these narratives by adopting a tentative situational approach; that is, by looking at Jesuits within intra- and trans-imperial configurations and interconnected structures of governance. Such an approach, we posit, helps avoid the pitfalls of treating science, empire, and the Jesuits as obvious and unitary entities straightforwardly conjoined with one another. In our view, an essential tension still pervades the history of, and historiography on, the Society of Jesus, generating some persistent confusion between imperialism and Catholic universalism. Although historically imbricated with each other and mutually dependent both ideologically and practically, these projects were nonetheless distinct: looking at them from the standpoint of the history of science has sometimes enhanced the confusion. Likewise, looking at empire building through Jesuits' science (and Jesuit sources) might entail overrating the Society of Jesus's actual width, strength, and coherence. Therefore, adopting an approach that distinguishes between configurations of power in which Jesuits operated might help put both projects in perspective without concealing their inherently power-driven nature.

In other words, by highlighting a few underlying ambivalences and blind spots in the historiography on Jesuit science in imperial contexts and, conversely, scrutinising the heuristic value of "science and empire" in relation to the Jesuits' project and its implementation, this chapter aims to contribute to the current scholarly effort towards a better situated understanding of both science making and empire making in the early-modern world.

Jesuits as imperial agents, or science as an appendage to imperialism

As early as 1547, a few years after the first Jesuit, Francisco Xavier (1506–52), had left the port of Belem under the patronage of the Portuguese monarchy (*Padroado*) to reach the Eastern possessions and outposts of the still expanding *Estado da India,* Ignatius urged missionaries to send information about "climate, diet, customs and character of the places and the peoples."[9] Ever since, the Society's missionaries provided descriptions of the circumstances and environments in which they operated. Indeed, by the 1570s they were already active in Portuguese Goa, Brazil, the Golden Coast in Africa, as well as in New Spain, New Castile, and the Philippines. Their writings encompassed many genres, in printed and manuscript form: reports and letters for internal use, histories of provinces, costumes books, travel journals, *lettres curieuses,* and dictionaries, to list only the most common genres. Missionary accounts gained a place of their own in the expanding book and news market, providing European readerships with a taste of the exotic and a powerful, if indirect, argument for imperialism.[10]

Miguel de Asúa has highlighted the central features of Jesuit natural histories of Spanish America. He shows how early chronicles dwelt on the marvels of nature and always kept sight of the history of salvation, to which the chronicle authors' subordinated the political facet of European expansion. Many chronicles used aboriginal names, evidence of Jesuit preoccupation with local languages for the purpose of preaching.[11] Other missionaries and colonial officials also described the wonders of the newly conquered lands, but Jesuits conceived a comprehensive vision of the New World. José de Acosta's *Historia natural y moral de las Indias* is an instance of, and a model for, Jesuit natural history and discourse on *maravilloso Americano.*[12] In the eighteenth century, the indiscriminate accumulation of materials gave way to more sober but not less appreciative regional monographs, like Pedro Lozano's *Descripción … del Gran Chaco,* published in Cordoba (present-day Argentina) in 1733, or José Gumilla's *El Orinoco ilustrado,* written in Rome while Gumilla served as Procurator for New Granada and published in Madrid in 1741.[13]

The European courts invested in compiling botanical knowledge, and the Jesuits played their part in imposing European categories on local resources while relying on native informants. *Materia medica* (the curative properties of natural substances) received special attention. As soon as they ventured to exotic, potentially dangerous environments in the East and the West, Europeans were eager to learn about local remedies to protect themselves. Physicians like Garcia de Orta and Cristovão da Costa (Acosta)—Acosta had worked in a Jesuit-run hospital in Goa—were imitated by missionaries.[14] Offering medical services was instrumental to conversion as it helped missionaries demonstrate the superiority of Christianity, but hospitals and pharmacies were in fact sites of exchange, places where different medical traditions were shared and local knowledge was put into practice.[15] Plants and recipes circulated among the order's missions. Some appeared in print, like the *Remedios fáciles para diferentes enfermedades,* by the pharmacist, teacher, and later Provincial Superior of Manila, Bohemian Pavel Klein, or the *Florilegio medicinal* by his compatriot at Colegio Maximo in Mexico Johannes Steinhöfer, both published in 1712.[16]

Other forms of knowledge were relevant in the spiritual and material conquest of new territories. Jesuit mapmaking has sustained a particularly large body of scholarship.[17] Maps were crucial for the control of the missionary space, as well as powerful representations of the order's achievements.[18] Though mainly drawn for missionary purposes, Jesuit maps circulated and contributed to display the empire's mastery of space and peoples.[19] In several instances, missionaries (and their unacknowledged local guides) explored territories that the crowns claimed but had not occupied. The 1607 expedition South of Rio Grande de Norte and into the Ibiapaba

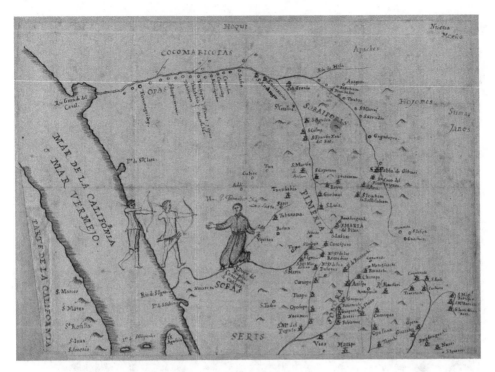

Figure 11.1 Map of New Mexico and California, drawing by Francis Xavier Saetta SJ, c. 1696. © Archivum Romanum Societatis Iesu, Rome, *Grandi Formati*, cassetto 1, n. 24. With permission of the Society of Jesus.

Mountains, Luis Figueira's 1636 mission into the Amazon, and Eusebio Kino's (Chini) exploration of Arizona and Baja California are cases in point.[20] On occasions, they joined official expeditions, like Jacques Marquette's exploration of the Mississippi valley with Louis Jolliet on behalf of Louis XIV of France in 1673—one of those enterprises that attracted the attention of early historians of cartography, and was later the focus of much discovery literature.[21] Another such instance is the famous 1746 Spanish naval expedition to Patagonia.[22]

Jesuit science overseas clearly reflects the institutionalisation of the natural sciences as global disciplines, embedded in a Janus-faced project of charting and taming the natural world. As Andrés I. Prieto has emphasised, it also mirrors the consolidation of the order's educational facilities.[23] The Society typically established itself in larger colonial towns, where they opened their colleges, and from which they radiated into the hinterlands. The order's colleges and universities, like those in Mexico City, Cordoba, Santiago, Quito, Manila, Macao, Goa, Bahia, as well as Rio de Janeiro (established 1567, after the Jesuits helped the Portuguese defeat the French), were instrumental in educating colonial élites and criollo fathers, and as such were essential both for the Society's evangelical project and for empire building. By the mid seventeenth century, 250 college towns around the world were sites for the printing of Jesuit titles in the natural sciences.[24]

Science was often taught by priests educated in Europe. Italian, Austrian, and Bohemian mathematicians and astronomers in the Americas, like Valentin Stensel in Salvador de Bahia and Nicolò Mascardi in remote Chiloe, are cases in point.[25] The French Empire usually only allowed French nationals in the Jesuit missions, although Roman-born Francesco Giuseppe

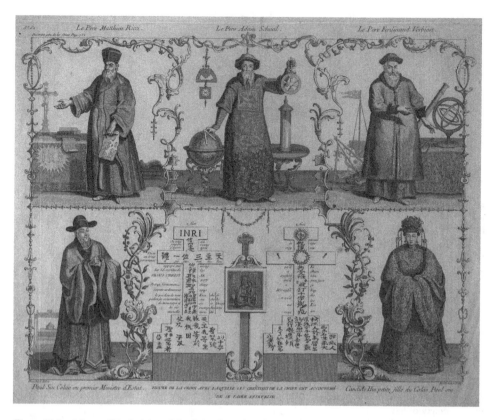

Figure 11.2 Matteo Ricci, Adam Schaal, Ferdinand Verbiest and others in China, 1735. *Le Père Matthieu Ricci, le Père Adam Schaal, le Père Ferdinand Verbiest, Paul Siu Colao ou premier Ministre d'Estat, Candide Hiu petite fille du Colao Paul siun*, Engraving in *Description géographique, historique, chronologique, politique et physique de l'empire de la Chine et de la Tartarie chinoise*, vol. 3 (Paris 1735). © Archivum Romanum Societatis Iesu, Rome, *Grandi Formati*, cassetto 12, nr. 3. With permission of the Society of Jesus.

Bressani was the one who narrated the Huron mission.[26] David Buisseret has demonstrated that Jesuit mapmaking in America was connected to the rise of mixed mathematics and drawing in the curriculum in Europe.[27] Still, by the mid eighteenth century, a process of creolisation was noticeable in the Spanish Americas, and to a much lesser extent in Brazil.[28] Future research should assess how creolisation impacted the imperial nature of Jesuit science. Regrettably, with exceptions, scientific output and teaching are still often treated by distinct historiographies; we therefore lack a fine-grained understanding of their intertwined evolution over time.[29]

To advance the field, a finer comparative analysis of the flow of information between the colonies and their metropolis would also be welcome. In fact, several scholars firmly point to Rome as the central hub of the Society's universal project; in pinpointing the role of prominent Jesuit scholars teaching and working in the order's institutions in Rome, including Antonio Possevino, Daniello Bartoli, Athanasius Kircher, and Roger Boscovich, they cast doubts on the usually assumed imperial characterisation of Jesuit science.[30] A subtler geography of Jesuit publications, as well as a more careful appraisal of the production of prints and manuscripts in their respective chronologies and geographies, would shed light on this point.

As for the exploitation of natural resources, Jesuits were actively involved in the acclimatisation of economically valuable plants. Tobacco remained a major asset for the Society well into the nineteenth century.[31] In Bahia, Jesuits experimented with acclimating cinnamon and pepper in an attempt to transplant plants to Brazil, following territorial losses in Asia.[32] The Society's involvement in the trade of the Peruvian bark and yerba mate is well known.[33] Although excellent studies are available on the Jesuit economy, most do not address its interplay with science.[34]

Still, downsizing the religious part of Jesuit engagement with nature, as it is still common in the historiography of science, particularly in regard to the eighteenth century, can be misleading. Just like Acosta's *Historia* was part of a larger evangelisation project, Jesuit naturalists penned chronicles (some of which subtly criticised civil authorities), as well as narratives of the spiritual *conquista* and catechisms. Lozano was not only a naturalist, he also authored the *Historia de la Compañía de Jesús en la Provincia del Paraguay* (Madrid, 1755). After the order's expulsion from the Portuguese, Spanish, and French dominions, science and apology combined afresh. Before composing his comprehensive *Paraguay natural ilustrado* and *El Paraguay cultivado* during his exile in Ravenna, José Sánchez Labrador completed a bulky *Paraguay Catholico*: these works have been the object of distinct historiographies,[35] but are in fact the two facets of a universalist vision that cannot be entirely subsumed under the category of imperialism.

Science as a concealment of imperialism: Jesuits as transimperial agents

In 1552, after his second journey to Japan, Francis Xavier wrote to Rome that Jesuits sent there should have some knowledge of astronomy, since the Japanese were interested in the motions of the heavens and natural phenomena.[36] The features that made Jesuits good candidates for seeking contacts with the Eastern empires were already clear to European courts by the middle of the sixteenth century. Well educated as Jesuits were, they were deemed particularly apt at approaching local elites, whose esteem for learning was proverbial. Spain tried to infiltrate Jesuit missionaries from the Philippines into China. Portugal went even further, sponsoring Jesuit missions to Madagascar, Tamil Nadu, China, Japan, and Indonesia, comprised of both national and "neutral" fathers. The papacy and the Order's hierarchy favoured for these global missions especially learned Italian Jesuits, who were supposedly unburdened by competing imperial allegiances.

This section considers Jesuits as transimperial agents and examines how they used science for two—not always compatible—goals: winning the Eastern empires over to Roman Catholicism and establishing diplomatic and commercial relations between these and the European powers that sponsored and financed them.

It was in China and later Siam that Jesuit missionaries found that astronomy and science more generally were precious resources for winning the confidence of elites, obtaining permission to celebrate Christian rites and to preach, and finally, gaining access to court. The first Italian Jesuit sent to China by the Portuguese *Padroado* in 1582, Michele Ruggieri (Luo Mingjiang), while writing the first Chinese catechism (*Thianzu shilu*, 天主實錄, *A True Record of the Lord of Heaven*), drew maps of China's provinces which he based on local administrative charts. Chinese visitors went to the Jesuit house in Zhaoqing (Guangdong) to admire the big clock and astronomical instruments. Shortly thereafter Matteo Ricci translated ancient and modern scientific texts into Chinese with the help of Sabatino de Ursis and the convert Xu Guangqi; fashioning himself the reputation of a man of great learning, Ricci came into contact with Ming imperial officials and eventually even collaborated with them.[37]

The second generation of Jesuits in China—most of them Italians (Niccolò Longobardo, Giulio Aleni, Alfonso Vagnoni, Luigi Buglio), but also Portuguese (Gabriel de Magalhães),

Flemish (Ferdinand Verbiest), and German (Adam Schall von Bell)—followed Ruggieri and Ricci's lead. This new generation managed an impressive activity of translation, collection, and adaptation of European and Chinese scholarship, while using their networks to get access to instruments and information.[38] Explanations of Galileo's telescope in Chinese were printed by Manoel Dias as early as 1615.[39] In the late 1620s, Jesuits were involved in official calendar computations, as well as foundry, herborising, mapmaking, and engineering. Schall became the leading figure of the Astronomical Bureau and increased his influence in the Ming–Qing dynastic transition. In 1669, under the protection of emperor Kangxi and his endeavour to conciliate Chinese and Western learning, Verbiest replaced his rival Yang Kuang-hsien as the head of Bureau.[40]

Later in the seventeenth century, French Jesuits similarly used science as a means to advance the French imperial agenda. The *mathématiciens du Roi,* six Jesuits led by Jean de Fontaney who while still in Europe had joined the Royal Academy of Science, were instructed to exploit their talents and Paris-made instruments to enhance diplomatic relations and start commerce between China and France.[41] In the following years, in spite of acute rivalries between French- and Portuguese-sponsored priests, Jesuits of both groups, including Tomas Pereira, Prospero Intorcetta, and Philippe Couplet, served as physicians, surgeons, and translators.[42] Jesuits were involved in Qing projects of mapping "Tartary" and surveying land of an empire that, since Ricci's time, had tripled in size.[43] French Jesuits stationed in Pondicherry similarly made astronomical observations and studied Indian astronomy, and the Raja of Jaipur Sawai Jai Singh invited Claude Stanislas Boudier, active in Chandernagore (near Kolkata), to collaborate in new astronomical tables.[44]

Scholars have spilled much ink on the Jesuits in Asia, in particular their strategy of accommodation and relations with local learned elites. Traditionally Jesuit missions in Asia have been considered within national histories or the history of religion. More recently, Asian scholars have questioned and overtly challenged some Western assumptions. The history of Jesuit science in Asia has become a battlefield of contrasting interpretations and approaches for understanding the cultural influence of the Jesuits, with both Asian and Western scholars using three different interpretive models: positivistic, civilisational, and culturalist.[45]

For instance, there has been much debate about the modernity of astronomy brought by Jesuits to the East, and European astronomy's long-term impact on Chinese science.[46] More recently, historians of science, drawing on earlier scholarship by sinologists and historians of religion, have placed greater emphasis on the cultural and religious aspects of the East–West exchange in astronomy, time computation, and cosmography—and on the mutual misunderstandings in regard to the role of the learned and the place of philosophical inquiry in political and religious orthodoxy.[47]

Likewise, debates on Jesuit cartography in Asia, especially China, have moved from a positivist standpoint (that is, following criteria of accuracy and novelty) to a culturalist focus. This literature now emphasises the ideological implications of mapmaking, the worldviews embedded in the production and use of maps, as well as a finer analysis of individual figures, their peculiar discursive strategies, and collaborations.[48]

More generally, in the context of increased academic cooperation and a better cross-cultural knowledge of European and Asian sources, historians of science are deepening their understanding of the manifold adaptations (technical as well as philosophical), rather than adoption, of European ideas and practices.[49] While scrutinising the work of local go-betweens and Asian literati, they shed new light on the bi-directional flow of information, texts, and objects which were mediated by the Jesuits.[50] Newer scholarship has studied these transformations from the standpoint of Chinese literati. This work has considered the place of Jesuits within Ming and Qing agendas, especially Emperor Kanxi's consolidation of the Manchu multi-ethnic empire through the "theory of the Chinese Origin of Western Learning."[51] Furthermore, a new generation of scholars has analysed the complex, multi-ethnic, multi-religious, and transimperial circuits of circulation and appropriation of Jesuit-Asian science in Tibet, Korea, India, and the greater Persian world; in so doing, they have deconstructed Asia, and even China, as a homogeneous and whole entity.[52] Civilisational blocks, and the very idea of "East," ultimately appear to be by-products of colonial and Jesuit propaganda.[53]

This point leads to the thorniest of questions: do the importance and prominence of the history of Jesuit science in Asia in modern scholarship on East–West relations ultimately reproduce the concealment of imperialism that science was meant to serve in the early-modern age? Is it still consistently dependent on a narrative that was crafted at the height of nineteenth- and twentieth-century European colonialism in Asia?

As indicated, it should not be overlooked that, in the eyes of the European powers, Jesuit missions to Asian empires and kingdoms were instrumental to commercial penetration, if not military invasion. At the end of the sixteenth century, Jesuits conceived of Japan as a target for Portuguese military invasion just as seriously as an arena for evangelism; Spanish Jesuits like Alonso Sánchez considered China the next step in Spain's global expansion. Portuguese and Spanish officials patronised Italian missionaries, reputed for their humanist education, precisely because it helped in the concealment of imperial designs under the guise of culture and religion. The chronology and itineraries of Jesuit missions corresponded to acute phases of competition among rival Catholic, and later Protestant and Orthodox, empires. In the seventeenth century, rival seaborne empires, including the "heretic" Dutch and English nations, shattered Portugal's *Estado da India*, depriving the Society of many outposts and opening a window for France's overseas ambitions. France changed tactics, but not strategy: since the beginning, French missionaries acted as agents of the French state and overtly aimed at sustaining Louis XIV's power politics, thus irritating Portugal and keeping the Holy See and the order's hierarchy busy in trying to settle the conflict among these rival Catholic powers for years.

Likewise, the Czars of Russia used Jesuits in seeking contact with China since the 1670s. In the mid eighteenth century, the Academy of Science in Saint Petersburg corresponded with Jesuits in Beijing, including the imperial astronomer Antoine Gaubil. Under Catherine II, a purportedly scientific mission to the Qing court was planned, and the Jesuit architect and astronomer Gabriel Gruber signed a letter promising to bring scientific instruments to China. The document was in fact penned by Russian officials, whose aim was to counter Britain's commercial expansion.[54] After 1773, when the Society was suppressed everywhere but in Russia, Gruber resumed the project and tried unsuccessfully to send three Jesuit scientists (Giovanni Grassi, Norbert Korsak, and Jan Stürmer) to China in order to restart evangelisation.[55] Eventually, Grassi was sent to catechise Maryland.

This last episode raises a further issue. Asia is not only a standpoint from which to observe how Jesuit scientists were instrumental in concealing imperialism; it also provides a good vantage point to assess the discrepancies between imperial visions, on the one hand, and the order's and the papacy's religious priorities, on the other. Last but not least, it exposes the precariousness of Western attempts at infiltrating Eastern empires. The history of early-modern Jesuit missionaries in Asia is ultimately one of defeat and retreat.[56] So much for their scientific reputation! It is not just that they concealed imperialism in service to the European Crowns; they were repeatedly denied entry, expelled, and put on trial. Just after celebrating Galileo's discoveries, Dias and his brethren and converts were expelled from Nanjing. In 1664, Schall and Verbiest were arrested and temporarily banned. Shogun Toyotomi Hideyoshi expelled Jesuits from Japan in 1587 and again in 1597, and in 1635 Tokugawa Ieyasu expelled missionaries and foreigners altogether, largely in response to Christian evangelisation. As of today, these episodes are still regarded as minor accidents in the narrative of Jesuit science. In spite of relinquishing an overtly Eurocentric stance, modern scholarship tends to focus on "achievements" and breakthroughs of Jesuit science.

Conclusion: "reinventing" Jesuit science in the nineteenth century

The Society of Jesus was dissolved progressively between 1757 and 1773. Its suppression was largely a matter of empire. In the middle of the eighteenth century, the European powers sought to reinvigorate their economies through a better exploitation of their colonies and more direct forms of sovereignty, governance, and tax collection, in a bid to counter British expansion. At that time, the Society became an obvious target, regardless of the true extent of the order's political and economic influence.

The spark occurred in the Uruguay and Parana regions, which had been the setting for much linguistic, ethnographic, and naturalistic endeavours by the Jesuits, yet it was a region where the limits of considering the Jesuits solely as imperial agents are exposed.[57] The Jesuit reductions enjoyed a quasi-autonomous status. Their latifundia had been enlarged, and conflicts with settlers and slave merchants had increased. To defend them, the Jesuits went as far as arming the Indians against both Spain and Portugal to oppose the implementation of the 1750 Treaty of Madrid—to which Jesuits had contributed their cartographical expertise.[58]

Expulsion of the Society of Jesus from most Catholic countries and their colonial territories resulted in a wide circulation of formerly Jesuit texts and objects; manuscripts, books, records, and specimens were confiscated by lay institutions like the Bibliothèque du Roi in Paris, and provided nourishment to the scientific institutions of modern imperialism. In India, information-hungry British officials and collectors appropriated Jesuit archives and Orientalist knowledge.[59]

Expelled fathers went to Europe, especially to the Papal States, where many conducted intense scholarly activity. Spanish subjects actually received a stipend to write in praise of Spanish colonialism, which many did, but praised the Society nonetheless. Former criollo Jesuits interpreted a nascent American consciousness, and their writings were later appropriated by the new independent states as national heritage.[60]

Several made a living as teachers, men of science, and travellers. Jesuits in China, for instance, continued their work at court, casting themselves as dispassionate representatives of the best that European civilisation had to offer. Even amidst anti-Jesuit polemics, France actively supported French missionaries while searching for new go-betweens to replace them.[61] Maria Theresa of Habsburg reduced the Society's control over universities, but continued to provide employment for ex-Jesuits. In 1772, shortly before the dissolution of the order, the director of the Jesuit Observatory in Vienna, Joseph Liesganig, was sent to Galicia, the newest Habsburg annexation. During the following decades, Liesganig prepared maps and cadastral surveys and continued his astronomical observations at the Observatory in Lviv.[62] Prague-educated Marcin Poczobut served as director of the observatory in Vilnius, where he remained after the suppression of the order; he later re-entered the Society in Russia, the only country where it survived.[63] Russia actually became a vast field of investigation for former Jesuits: they travelled to the southern regions (Saratov, Odessa, Astrakhan, Mozdok) and into remote Siberia, penning ethno-geographic observations.[64] Adam František Kollár coined the term *ethnologia* in 1783 as "the science of nations and peoples ... [an inquiry into] the origins, customs, languages and institutions of various nations, and finally into their fatherlands and ancient seats."[65]

In short, this suppression changed the way in which many fathers experienced their multiple identities as Jesuits, Catholic missionaries, men of learning, and citizens of nation-states.[66] After the formal reestablishment of the Society in 1814, the Jesuits resumed their overseas evangelisation activities progressively from the mid-1830s onward. The world, however, had profoundly changed in the meantime: the Iberian Empires had vanished with the independence of their former American colonies, and the expansion of the British rule in Asia drastically downsized the role of Catholic missionaries. Moreover, the Jesuits now tended to align themselves with the reactionary sectors within the Catholic Church (although with more nuances than historiography has long conceded), and they thought of missions as compensation for the position they had lost in Europe. Although the Society continued to invest massively in education, the reactionary shift considerably lessened their impact on contemporary science. It cannot therefore come as a surprise that, as we have indicated at the beginning of this chapter, in the nineteenth century the Society also spent much effort in crafting retrospectively its role in the dissemination of Western science and civilisation in the early age of the European expansion.

In conclusion, rather than a fact, science and empire is more aptly used as heuristic device to help put both Jesuit science and early-modern imperialism in perspective. To this goal, recent cross-cultural approaches to missions help reassess Jesuit scientific activity on the field, whereas isolating Jesuit science as a self-evident object risks reproducing an Eurocentric ideology and understanding of world history.

Notes

1 Henrique Leitão and Francisco Malta Romeiras. "The Role of Science in the History of Portuguese Anti-Jesuitism," *Journal of Jesuit Studies* 2, no. 1 (2015): 77–99.

2 Pierre-Antoine Fabre and Antonella Romano, "Les Jésuites dans le monde moderne. Nouvelles approches historiographiques," *Revue de Synthèse* 120 (1999): 247–54.

3 See for example Charlotte De Castelnau-L'Estoile et al., eds., *Missions d'évangélisation et circulation des savoirs. XVIe-XVIIIe siècle* (Madrid: Casa de Velazquez, 2011); Guillermo Wilde, ed., *Saberes de la conversión: jesuitas, indígenas e imperios coloniales en las fronteras de la cristiandad* (Buenos Aires: Editorial SB, 2011).

4 John W. O'Malley, "The Historiography of the Society of Jesus," in *The Jesuits: Cultures, Sciences and the Arts, 1540–1773*, ed. John W. O'Malley (Toronto: University of Toronto Press, 1999), 3–37, here p. 14.

5 For studies of these networks, and in relation to the Republic of Letters, see Luke Clossey, *Salvation and Globalization in the Early Jesuit Missions* (Cambridge: Cambridge University Press, 2008); Markus Friedrich, *Der lange Arm Roms? Globale Verwaltung und Kommunikation im Jesuitenorden 1540–1773* (Frankfurt: Campus, 2011); Mordechai Feingold, ed., *Jesuit Science and the Republic of Letters* (Cambridge, MA: MIT Press, 2003); Paula Findlen, "How Information Travels: Jesuit Networks, Scientific Knowledge, and the Early Modern Republic of Letters, 1540–1640," in *Empires of Knowledge: Scientific Networks in the Early Modern World*, ed. Paula Findlen (London: Routledge, 2018), 57–105.

6 Steven J. Harris, "Mapping Jesuit Science: The Role of Travel in the Geography of Knowledge," in O'Malley, *The Jesuits*, 215.

7 For recent instances of such an approach, see Augustin Udias, *Jesuit Contribution to Science: A History* (Heidelberg: Springer, 2015); Alberto Gómez Gutiérrez and Jaime Bernal Villegas, *Scientia xaveriana: los jesuitas y el desarrollo de la ciencia en Colombia, siglos XVI–XX* (Bogotá: Pontificia Universidad Javeriana, 2008).

8 See for example Joseph A. Gagliano and Charles E. Ronan, eds., *Jesuit Encounters in the New World: Jesuit Chroniclers, Geographers, Educators and Missionaries in the Americas, 1549–1767* (Rome: Istituto Storico, 1997); Alexandre Chen Tsung-min, ed., *Catholicism's Encounters with China: 17th to 20th century* (Leuven: Ferdinand Verbiest Institute, KU Leuven, 2018). See also the presentations at the upcoming conference *Engaging the World: The Jesuits and Their Presence in Global History*, Lisbon, Portugal, organised in June 2020 and postponed to June 2021 the Institute for Advanced Jesuit Studies at Boston

College and Brotéria in Lisbon and the Catholic University Portugal https://jesuitportal.bc.edu/news/february-2020-program-for-2020-international-symposium-on-jesuit-studies-engaging-the-world-the-jesuits-and-their-presence-in-global-history/.

9   "De aëris temperie, victu, moribus et ingeniis locorum ac hominum," referring to India, in *Monumenta Ignatiana, Epistolæ et instructiones*, vol. 1 (Madrid: Lopez del Horno, 1903), 649.

10   On this point, with a special focus on translations, see Galaxis Borja González, *Jesuitische Berichterstattung über die Neue Welt* (Göttingen: Vandenhoeck & Ruprecht, 2011).

11   Miguel de Asúa, "Natural History in the Jesuit Missions," in *The Oxford Handbook of the Jesuits*, ed. Ines G. Županov (Oxford: Oxford University Press, 2019), 708–36. On Jesuits and the learning of native languages, see Aliocha Maldavsky, *Vocaciones inciertas. Misión y misioneros en la provincia jesuita del Perú en los siglos XVI y XVII* (Lima: CIS, 2012), esp. 267–87.

12   Jacques Lafaye, *Quetzalcóatl et Guadalupe. La formation de la conscience nationale au Mexique (1531–1813)* (Paris: Gallimard, 1974); Luis Milliones Figueroa and Domingo Ledezma, eds., *El saber de los jesuitas, Historias naturales y el nuevo mundo* (Vervuert: Iberoamericana, 2005); Jorge Cañizares-Esguerra, *Nature, Empire, and Nation: Explorations of the History of Science in the Iberian World* (Stanford, CA: Stanford University Press, 2006). On the differences between the accounts from Spanish and French America, see Marc André Bernier, Clorinda Donato, and Hans-Jürgen Lüsebrink, eds., *Jesuit Accounts of the Colonial Americas: Intercultural Transfers, Intellectual Disputes and Textualities* (Toronto: University of Toronto Press, 2014). On the Philippines, see José Pardo-Tomás, "Las primeras historias naturales de las Filipinas (1583–1604)," *Nuevo Mundo Mundos Nuevos*, http://journals.openedition.org/nuevomundo/76534, doi.org/10.4000/nuevomundo.76534.

13   Pedro Lozano, *Descripción chorographica de terreno, ríos, arboles, y animales de los dilatadísimas provincias del Gran Chaco, Gualamba, y de los ritos y costumbres de las inumerables naciones de barbaros e infideles que le habitan. Con un cabal relación histórica de lo que en ellos han obrado para conquistarlas algunos gobernadores y ministros reales, y los misioneros jesuitas para reducirlos a la fe del verdadero Dios* (Córdoba: Joseph Santos Balbas, 1733); José Gumilla, *El Orinoco ilustrado: Historia natural, civil y geographica de este gran rio...* (Madrid: Manuel Fernandez, 1741). On the pitfalls of clear-cut partitions between baroque and Enlightenment natural philosophies, see Margaret Ewalt, *Peripheral Wonders: Nature, Knowledge, and Enlightenment in the Eighteenth-Century Orinoco* (Bucknell, PA: Bucknell University Press, 2008); Carlos Del Cairo and Esteban Rozo, "El salvaje y la retórica colonial en El Orinoco ilustrado (1741) de José Gumilla S. J.," *Fronteras de la Historia* 11 (2006), doi.org/10.22380/20274688.533.

14   Ines G. Zupanov, "Drugs, Health, Bodies and Souls in the Tropics: Medical Experiments in Sixteenth-Century Portuguese India," *The Indian Economic and Social History Review* 39, no. 1 (2002): 1–45; Allan Greer, "The Exchange of Medical Knowledge between Natives and Jesuits in New France," in Figueroa and Ledezma, *El saber de los jesuitas*, 135–146. Medicine was excluded from the curriculum of Jesuit colleges in Europe, but was considered an essential part of missionary activity.

15   Sabine Anagnostou, *Missionspharmazie. Konzepte, Praxis, Organisation und wissenschaftliche Ausstrahlung* (Stuttgart: Franz Steiner Verlag, 2011). On Jesuit "spiritual medicine," see Eliane Fleck, "'Da mística ás luzes': medicina experimental nas reduçoes jesuítico-guaranis da Provincia Jesuítica do Paraguai," *Revista Complutense de Historia de América* 32 (2006): 158–78; Ines G. Županov, "Conversion, Illness and Possession: Catholic Missionary Healing in Early Modern South Asia," in *Divins remèdes; Médecine et religion en Inde*, ed. Ines G. Županov and Caterina Guenzi (Paris: EHESS, 2008), 263–300.

16   Pablo Clain [Pavel Klein], *Remedios fáciles para diferentes enfermedades* (Manila: Corea, 1712); Johannes Steinhöfer [Juan de Esteyneffer], *Florilegio medicinal de todas las enfermedades: sacado de varios, y clasicos authores, para bien de los pobres, y de los que tienen falta de medicos, en particular para provincians remotas ...* (Mexico: Herederos de Jaun Joseph Guillena Guarrascoso, 1712); Margarita Artschwager Kay, "The Florilegio Medicinal: Source of Southwest Ethnomedicine," *Ethnohistory* 24, no. 3 (1977): 251–9; Oldřich Kašpar, *Los jesuitas checos en la Nueva España, 1678–1767* (Mexico: Universidad Iberoamericana, 1991); Timothy Walker, "The Medicines Trade in the Portuguese Atlantic World: Acquisition and Dissemination of Healing Knowledge from Brazil (c. 1580–1800)," *Social History of Medicine* 26 (2013): 411–14.

17   For an excellent overview, see Robert Batchelor, "Historiography of Jesuit Cartography," in *Jesuit Historiography Online*, doi.org/10.1163/2468-7723_jho_COM_212546.

18   Brian J. Harley, "The Map as Mission: Jesuit Cartography as an Art of Persuasion," in *Jesuit Art in North American Collections*, ed. Jane B. Goldsmith (Milwaukee, WI: Patrick and Beatrice Haggerty Museum of Art, Marquette University, 1991), 28.

19 Maps drawn for internal use might eventually be exploited by colonial officials and merchants; see Carl Kupfer and David Buisseret, "Seventeenth-Century Jesuit Explorers' Maps of the Great Lakes and Their Influence on Subsequent Cartography of the Region," *Journal of Jesuit Studies* 6, no. 1 (2019): 7–70; Manonmani Restif-Filliozat, "The Jesuit Contribution to the Geographical Knowledge of India in the Eighteenth Century," *Journal of Jesuit Studies* 6, no. 1 (2019): 71–84.

20 Pablo Ibáñez-Bonillo, "Rethinking the Amazon Frontier in the Seventeenth Century: The Violent Deaths of the Missionaries Luis Figueira and Francisco Pires," *Ethnohistory* 65, no. 4 (2018): 575–95; Herbert Bolton, *The Rim of Christendom: A Biography of Eusebio Francisco Kino* (New York: Macmillan, 1936); Gabriel Gómez Padilla, *En la 'isla' más grande del orbe: Kino en California* (Hermosillo, Mexico: Secretaría de Educación y Cultura del Estado de Sonora, 2008); Salvador Bernabéu Albert, ed., *El gran norte mexicano. Indios, misioneros y pobladores entre el mito y la historia* (Seville: CSIC, 2009). Not all Jesuit explorations were successful, of course. For instance, several attempts at sailing up the Paraguay river failed: Patricio Fernández, *Relación historial de las misiones de indios Chiquitos* (Madrid: Manuel Fernandez, 1726).

21 See for instance Alfred Hamy, *Au Mississippi: la première exploration, 1673: le Père Jacques Marquette de Laon, prêtre de la Compagnie de Jésus (1637–1675), et Louis Jolliet* (Paris: Honoré Champion, 1903); Joseph P. Donnelly, *Jacques Marquette S.I., 1637–1675* (Chicago: Loyola Institute, 1932).

22 Guillermo Furlong, *El Padre José Quiroga* (Buenos Aires: Peuser, 1930); Carmen Martínez Martín, "La expedición del Padre Quiroga, S.J. a la costa de los Patagones (1745–46)," *Revista Complutense de Historia de América* 17 (1991): 121–37.

23 Andrés I. Prieto, *Missionary Scientists: Jesuit Science in Spanish South America, 1570–1810* (Nashville, TN: Vanderbilt University Press, 2011).

24 Harris, "Mapping Jesuit Science," 224.

25 Carlo Z. Camenietzki, "Baroque Science between the Old and the New World: Father Kircher and his Colleague Valentin Stansel (1621–1705)," in *Athanasius Kircher: The Last Man Who Knew Everything*, ed. Paula Findlen (New York: Routledge, 2004), 311–28; Prieto, *Missionary Scientists*, 116–40. Astronomy was probably the field in which the mobility of Jesuits was more consistent. Astronomical observations and calculations did serve imperial pursuits (they were crucial to navigation), but the exchange of information was routine in this domain. By the seventeenth century a multipolar network of observation was established, and it grew into properly international campaigns in the following century. The Jesuits scattered in the overseas territories cooperated in such enterprises, largely regardless of the underpinning "systems of the world": see Luís Miguel Carolino, "Astronomy, Cosmology, and Jesuit Discipline, 1540–1758," in Zupanov, *Oxford Handbook of the Jesuits*, 670–707. Furthermore, numerous Bohemian and German Jesuits served as pharmacists; see note 16 above, and Renée Gicklhorn, *Missionapotheker. Deutsche Pharmazeunten in Latinoamerika des 17. and 18. Jahrhunderten* (Stuttgart: Wissenschaftliche Verlagsgesellschaft, 1973), esp. 37–44, 51–6.

26 Francesco Giuseppe Bressani, *Breve relatione d'alcune missioni … nella Nuova Francia* (Macerata: Per gli heredi d'Agostino Grisei, 1653); Lucien Campeau, SJ, *La mission des Jésuites chez les Hurons, 1634–1650* (Rome: Institutum Historicum S. I., 1987).

27 David Buisseret, "Jesuit Cartography in Central and South America,'" in Gagliano and Ronan, *Jesuit Encounters*, 113–62.

28 David Block, *Mission Culture on the Upper Amazon: Native Tradition, Jesuit Enterprise, and Secular Policy in Moxos, 1660–1880* (Lincoln: University of Nebraska Press, 1994), 103–24; Dauril Alden, *The Making of an Enterprise. The Society of Jesus in Portugal, Its Empire, and Beyond 1540–1750* (Stanford, CA: Stanford University Press, 1996), 255–87; José Sala Catalá, *Ciencia y técnica en la metropolización de América* (Madrid: Doce Calles, 1994). The problem of native clergy is the object of a distinct body of literature.

29 Teaching is still largely the province of historians of education, who commonly adopt a national or local scale, and several Jesuit colleges and universities have their own historians. See, however, Luís Miguel Carolino and Carlos Ziller Camenietzki, eds., *Jesuítas, Ensino e Ciência, sécs. XVI–XVIII,* (Lisbon: Caleidoscópio, 2005); Ugo Baldini, "The Jesuit College in Macao as a Meeting Point of the European, Chinese and Japanese Mathematical Traditions: Some Remarks on the Present State of Research, Mainly Concerning Sources (16th–17th centuries)," in *The Jesuits, the Padronado and East Asian Science (1552–1773)*, ed. Luis Saraiva and Catherine Jami (Singapore: World Scientific Publishing, 2008), 33–80.

30 See for instance Antonella Romano, *Impressions de Chine. l'Europe et l'englobement du monde (XVIe–XVIIIe siècle)* (Paris: Fayard, 2016), 20; Adriano Prosperi, "Introduzione," in *Istoria della Compagnia di*

Gesù, l'Asia, by Daniello Bartoli (Turin: Einaudi, 2019), I: xxi–lxxxi; Findlen, *Athanasius Kircher*; Piers Bursill-Hall, ed., *R.J. Boscovich: vita e attività scientifica/His Life and Scientific Work* (Rome: Istituto della Enciclopedia Italiana, 1993).

31 The monopoly of the sale of betel leaves and tobacco bestowed to the Jesuits in Manila led to the rise in price of these commodities; see Omri Bassewitch Frenkel, "Transplantation of Asian Spices in the Spanish Empire 1518–1640" (PhD diss., McGill University, 2017); Thomas Murphy, SJ, *Jesuit Slaveholding in Maryland, 1717–1838* (New York: Routledge, 2011).

32 José Alberto do Amaral Lapa, "O Brasil e as drogas do Oriente," *Studia* 18 (1966): 7–40; Luis Fernande de Alemida, "Aclimatação de plantas do Oriente no Brasil durante os séculos XVII e XVIII," *Revista Portuguesa de Historia* 15 (1975): 339–481.

33 Matthew James Crawford, *The Andean Wonder Drug: Cinchona Bark and Imperial Science in the Spanish Atlantic, 1630–1800* (Pittsburgh, PA: Pittsburgh University, 2011); Claudio Ferlan, *Sbornie sacre, sbornie profane. L'ubriachezza dal vecchio al nuovo mondo* (Bologna: il Mulino, 2018).

34 Reference works on Jesuit economy include Alden, *The Making*; German Colmenares, *Las haciendas de los jesuitas en el nuevo reino de Grenada* (Bogotà: Universidad del Valle, 1969); Nicholas P. Cushner, *Lords of the Lands: Sugar, Wine and Jesuit Estates of Coastal Perù, 1600–1767* (Albany, NY: SUNY Press, 1980); *Jesuit Ranches and the Agrarian Development in Colonial Argentina, 1650–1767* (Albany, NY: SUNY Press, 1983); Charles J. Borges, *Economics of the Goa Jesuits 1542–1759: An Explanation of their Rise and Fall* (New Delhi: Concept, 1994); Stephen Lenik, "Mission Plantations, Space, and Social Control: Jesuits as Planters in French Caribbean Colonies and Frontiers," *Journal of Social Archaeology* 12, no. 1 (2012): 51–71; Andrew Dial, "The 'Lavalette Affair': Jesuits and Money in the French Atlantic" (PhD diss., McGill University, 2019).

35 José Sánchez Labrador, *Los indios pampas, puelches y patagones,* ed. Guillermo Furlong (Buenos Aires: Viau y Zona, 1936); José Sánchez Labrador S.J., *Peces y aves del Paraguay natural ilustrado,* 1767, ed. Mariano Castex (Buenos Aires: Fabril Editora, 1968); Hector Sainz Ollero et al., *José Sánchez Labrador y los naturalistas jesuitas del Rio de la Plata: la aportación de los misioneros jesuitas del siglo XVIII a los estudios medioambientales* (Madrid: MOPU, 1989).

36 Letter from Francis Xavier to Ignacio de Loyola, 9 April 1552, in *Monumenta historica Japoniæ, II: Documentos del Japon, 1547–1557* (Rome: Instituto Historico de la Compañía de Jesús, 1990), 358. On the Jesuit "clock diplomacy" in Japan, see Ryuji Hiraoka, "Jesuits and Western Clock in Japan's 'Christian Century' (1549–c.1650)," *Journal of Jesuit Studies* 7, no. 2 (2020): 204–20.

37 Ruggieri worked on an Atlas of China from 1579 (when he arrived in Macao) to 1606 but never published it. The first edition was edited by Eugenio Lo Sardo: Michele Ruggieri, *Atlante della Cina* (Rome: Istituto Poligrafico dello Stato, 1993). On Ricci, in an extremely large and still growing body of scholarship, see Ronnie Po-chia Hsia, *A Jesuit in the Forbidden City: Matteo Ricci 1552–1610* (Oxford: Oxford University Press, 2010).

38 Buglio translated chapters of the *Ornithologiæ sive de avibus historiæ* by the Bolognese naturalist Ulisse Aldrovandi into Chinese. And see Gabriel de Magalhães, *Nouvelle Relation de la Chine, contenant la description des particularitez les plus considerables de ce grand empire* (Paris: Barbin, 1688).

39 Massimo Bucciantini, Michele Camerota, and Franco Giudice, *Il telescopio di Galileo. Una storia europea* (Turin: Einaudi, 2012), 253–65.

40 Ho Peng-Yoke, "The Astronomical Bureau in Ming China," *Journal of Asian Studies* 3, no. 2 (1969): 137–57; George Dunne, *Generation of Giants: The Story of the Jesuits in China in the Last Decades of the Ming Dynasty* (London: Kessinger Publishing, 1962); Isaia Iannaccone, *Johann Schreck Terrentius: le Scienze Rinascimentali e lo Spirito dell'Accademia dei Lincei nella Cina dei Ming* (Naples: Istituto universitario orientale, 1998); Claudia von Collani and Erich Zettl, eds., *Johannes Schreck-Terrentius. Wissenschaftler und China-Missionar (1576–1630)* (Stuttgart: Franz Steiner, 2016); Malek Roman, ed., *Western Learning and Christianity in China: The Contribution and Impact of Johann Adam Schall von Bell SJ (1592–1666)* (Nettetal: Steyler, 1998); John W. Witek, ed., *Ferdinand Verbiest (1623–1688): Jesuit Missionary, Scientist, Engineer and Diplomat* (Nettetal: Steyler, 1994).

41 The other five Jesuits were Joachim Bouvet, Jean-François Gerbillon, Claude de Visdelou, Louis-Daniel Lecomte, and Guy Tachard. Bouvet would later tutor the emperor in mathematics; see Claudia von Collani, *Eine wissenschaftliche Akademie für China* (Stuttgart: Steiner, 1989). During the journey to China, they stopped in Siam, welcomed by the king Phra Narai. Tachard remained there while the other five left and the king sent him back to France on a diplomatic trip. Tachard wrote two different books on his expeditions, *Voyage de Siam* (1686) and *Seconde voyage au royaume de Siam* (1689). See Florence C. Hsia, *Sojourners in a Strange Land: Jesuits and Their Scientific Missions in Late Imperial China*

(Chicago: University of Chicago Press, 2009). A second group of Jesuits arrived in Siam in 1687 but the plan to build an observatory failed following an internal political conspiracy that led to regime change.

42 Luis Filipe Barreto, ed., *Tomas Pereira, S.J. (1646–1708), Life, Work and World* (Lisbon: Centro Cientifico e Cultural de Macao, 2010); Arthur K. Wardega, SJ, and António Vasconcelos de Saldanha, eds., *In the Light and Shadow of an Emperor: Tomás Pereira, SJ (1645–1708), The Kangxi Emperor and the Jesuit Mission in China* (Newcastle upon Tyne: Cambridge Scholars Publishing, 2012); Nicolas Standaert, ed., "Networks and Circulation of Knowledge: Encounters between Jesuits, Manchus and Chinese in Late Imperial China," special issue, *East Asian Science, Technology, and Medicine* 34 (2011); Beatriz Puente Ballesteros, "Jesuit Medicine at the Kangxi Court (r. 1662–1722): Imperial Networks and Patronage," *East Asian Science, Technology, and Medicine* 34 (2011): 86–162; Beatriz Puente Ballesteros, "Isidoro Lucci S.J. (1661–1719) and João Baptista Lima (1659–1733) at the Qing Court: The Physician, the Barber-Surgeon, and the Padroado's Interests in China," *Archivum Historicum Societatis Iesu* 82 (2013): 165–216. Also see Huiyi Wu, *Traduire la Chine au XVIIIe siècle: les jésuites traducteurs de textes chinois et le renouvellement des connaissances européennes sur la Chine (1685–ca. 1740)* (Paris: Honoré Champion, 2017). On Jesuit court artists, see Elisabetta Corsi, *La fábrica de las ilusiones: los jesuitas y la difusión de la perspectiva lineal en China, 1698–1766* (Mexico: Colegio de Mexico, 2004); Petra Ten-Doesschate Chu and Ning Ding, eds., *Qing Encounters: Artistic Exchanges between China and the West* (Los Angeles: Getty Research Institute, 2015); Marco Musillo, *The Shining Inheritance: Italian Painters at the Qing Court, 1699–1812* (Los Angeles: Getty Research Institute, 2016).

43 Laura Hostetler, *Qing Colonial Enterprise: Ethnography and Cartography in Early Modern China* (Chicago: University of Chicago Press, 2001); Mario Cams, *Companions in Geography: East–West Collaboration in the Mapping of Qing China (c.1685–1735)* (Leiden: Brill, 2017).

44 The Jesuit involvement in astronomy in India is less known than that in China, though as early as the first mission to the Mughal court of Akbar in 1580, Antonio de Monserrate (1536–1600) registered geographical and astronomical observations. For these and later developments, see Rajesh K. Kocchar, "Secondary Tools of Empire: Jesuit Men of Science in India," in *Discoveries, Missionary Expansion and Asian Cultures*, ed. Teotonio R. de Souza (New Delhi: Concept Publishing Company, 1994), 175–83; Augustin Udias, *Searching the Heavens and the Earth: The History of Jesuit Observatories* (Dordrecht: Kluwer, 2003), 99–102; Will Sweetman and Ines G. Županov, "Rival Missions, Rival Science? Jesuits and Pietists in Seventeenth and Eighteenth-Century South India," *Comparative Studies in Society and History* 61, no. 3 (2019): 624–65. On artistic hybridisations of Jesuit art in India, see Gauvin A. Bailey, *The Jesuits and the Grand Mogul: Renaissance Art at the Imperial Court of India, 1580–1630* (Washington, DC: Freer Gallery of Art, Arthur M. Sackler Gallery & Smithsonian Institution, 1998).

45 The shift of focus to East–West exchanges and the adaptive cultural strategies of Jesuit missionaries gained momentum in the late 1980s and was at once an impulse for, and a product of, the new stance on Jesuit missions and the interdisciplinary dialogue beyond area studies, both encouraged by the Society of Jesus. See for instance Charles E. Ronan and Bonnie B.C. Oh, eds., *East Meets West: The Jesuits in China, 1582–1773* (Chicago: Loyola University Press, 1988); Thomas H.C. Lee, ed., *China and Europe, Images and Influences in Sixteenth to Eighteenth Centuries* (Hong Kong: Chinese University Press, 1991); Federico Masini, ed., *Western Humanistic Culture Presented to China by Jesuit Missionaries: XVII–XVIII Centuries* (Rome: Institutum historicum S.I., 1996); Catherine Jami and Hubert Delahaye, eds., *L'Europe en Chine: interactions scientifiques, religieuses et culturelles aux XVIIe et XVIIIe siècles* (Paris: Collège de France, 1993).

46 Joseph Needham and Wang Ling, *Science and Civilization in China*, vol. 3, *Mathematics and the Sciences of the Heavens and the Earth* (Cambridge: Cambridge University Press, 1959), 437–61, suggested that the Jesuits imported the incorrect Ptolemaic–Tychonic astronomy and cosmology, and were therefore indirectly responsible for China's "backwardness," to which Pasquale D'Elia, *Galileo in China: Relations Through the Roman College Between Galileo and the Jesuit Scientist-Missionaries (1610–1640)* (Cambridge, MA: Harvard University Press, 1960), contended that Jesuits were in fact secretly Copernicans. The debate on Jesuit scientific culture was rekindled in the following decades by, among others, Nathan Sivin, "Copernicus in China," *Studia Copernicana* 6 (1973): 63–122; Isaia Iannaccone and Adolfo Tamburello, eds., *Dall'Europa alla Cina: contributi per una storia dell'astronomia* (Naples: Istituto universitario orientale, 1990). More recently, historians of science like Ulrich Liebrecht and Catherine Jami have argued that the Jesuit *mathémathiciens du Roi* imported a rather different science than their predecessors.

47 Catherine Jami, *The Emperor's New Mathematics: Western Learning and Imperial Authority during the Kangxi Reign (1662–1722)* (Oxford: Oxford University Press, 2012).

48 There were for instance significant differences between Ricci's and Martino Martini's maps. Although Ricci put China at the centre of his world map, he was in fact diminishing Ming ideas that their Empire embraced the whole world. In his *Novus Atlas Sinensis* (1655), Martini introduced elements of the Chinese cosmography and toponyms of the provinces. For a critical discussion of Ricci's supposed Sino-centrism, see Forin-Stefan Morin, "The Westerner: Matteo Ricci's World Map and the Quandaries of European Identity in the Late Ming Dynasty," *Journal of Jesuit Studies* 6, no. 1 (2019): 14–30.

49 Wu Xiaoxin, ed., *Encounters and Dialogues: Changing Perspectives on Chinese–Western Exchanges from the Sixteenth and Eighteenth Centuries* (Nettetal: Steyler Verlag, 2005).

50 Bianca Maria Rinaldi, *The "Chinese Garden in Good Taste": Jesuits and Europe's Knowledge of Chinese Flora and Art of the Garden in the 17th and 18th Centuries* (Munich: Meidenbauer, 2005). Noël Golvers, *Libraries of Western Learning for China: Circulation of Western Books between Europe and China in the Jesuit Mission (ca. 1650–ca. 1750)* (Leuven: K.U. Leuven, 2013–2015); Harold J. Cook, *Translation at Work: Chinese Medicine in the First Global Age* (Leiden: Brill, 2020).

51 Benjamin Elman, *On Their Own Terms: Science in China, 1550–1900* (Cambridge, MA: Harvard University Press, 2005); Qi Han: "Between the Kangxi Emperor (r. 1662–1722) and Leibniz: Joachim Bouvet's (1656–1730) Accommodation Policy and the Study of the Yijing," in *Beyond Borders: A Global Perspective of Jesuit Mission History*, ed. Shinzo Kawamura and Cyril Veliath (Tokyo: Sophia University Press, 2009), 172–81; Qi Han, "Chinese Literati's Attitudes toward Western Science: Transition from the Late Kangxi Period to the Mid-Qianlong Period," *Historia Scientiarum, International Journal of the History of Science Society of Japan* 24, no. 2 (2015): 76–87; Qi Han, "The Legitimization of the Transmission of Western Science: Xu Guangqi's Proposal for the Calendar Reform," in *The Generation of Giants 2: Other Champions of the Cultural Dialogue between Europe and China*, ed. Luisa M. Paternicò (Trento: Centro Studi Martino Martini, 2015), 19–25; Minghui Hu, *China's Transition to Modernity: The New Classical Vision of Dai Zhen* (Seattle: University of Washington Press, 2015). See further Thierry Meynard, "Recent Chinese Translations (2000–2010) of Early Texts and Modern Studies Related to Missionaries in China in the Seventeenth and Eighteenth Centuries," in *Acta Pekinensia: Western Historical Sources for the Kangxi Reign* (Macau: Macau Ricci Institute, 2013), 435–59.

52 Lim Jongtae, "Matteo Ricci's World Maps in Late Joseon Dynasty," *The Korean Journal for the History of Science* 33, no. 2 (2011): 277–96; Lim Jongtae, "Learning Western Astronomy from China: Another Look at the Introduction of the *Shixian li* Calendrical System into Late Joseon Korea," *The Korean Journal for the History of Science* 34, no. 2 (2012): 197–217; Yunli Shi, "Calculating the Fate of Chinese Dynasties with the Islamic Method: Chinese Study and Application of Arabic Astrology in the 17th Century," *Micrologus. Natura, scienze e società medievali* 25 (2016): 311–15; Lobsang Yongdan, "A Scholarly Imprint: How Tibetan Astronomers Brought Jesuit Astronomy to Tibet," *East Asian Science, Technology, and Medicine* 45 (2017): 91–117; Wu Huiyi, Alexander Statman, and Mario Cams, "Displacing Jesuit Science in Qing China," special issue, *East Asian Science, Technology, and Medicine* 46, no. 1 (2017); Richard A. Pegg, "The Star Charts of Ignatius Kögler (1680–1746) in the Korean Court," *Journal of Jesuit Studies* 6, no. 1 (2019): 44–56. For the Qing theory that Western sciences originated in China, you may wish to access Jiang Xiaoyuan and Shilun Qingdai, "Xixue Zhongyuan shuo," *Ziran kexue shi yanjiu* 7–2 (1983):101–8.

53 Sylvia Murr, *L'Inde philosophique entre Bossuet et Voltaire* (Paris: Ecole française d'Extrême-Orient, 1987); Donald Lach, *Asia in the Making of Europe* (Chicago: University of Chicago Press, 1994); Joan-Pau Rubiés, "Reassessing 'the Discovery of Hinduism': Jesuit Discourse on Gentile Idolatry and the European Republic of Letters," in *Intercultural Encounter and the Jesuit Mission in South Asia (16th–18th centuries)*, ed. Anand Amaladass and Ines Županov (Bangalore: ATC, 2014), 113–55; Michela Catto, "The Generalate of Claudio Acquaviva: The Birth of the Jesuit Myth of China," in *The Acquaviva Project: Claudio Acquaviva Generalate SJ (1581–1615) and the Emergence of Modern Catholicism*, ed. Pierre-Antoine Fabre and Flavio Rurale (St. Louis–Rome: The Institute of Jesuit Sources-Institutum Historicum Societatis Iesu, 2017), 129–47.

54 Gregory Afinogenov, "Jesuit Conspirators and Russia's East Asian Fur Trade, 1791–1807," *Journal of Jesuit Studies* 2, no. 1 (2015): 56–76.

55 Stanislav Južnič, "Central-European Jesuit Scientists in China, and Their Impact on Chinese Science," *Asian Studies* 3, no. 2 (2015): 89–118.

56 Liam Matthew Brockey, *Journey to the East: The Jesuit Mission to China, 1579–1724* (Cambridge, MA: Harvard University Press, 2009).

57 We borrow the expression from Kenneth Maxwell, "The Spark: Pombal, the Amazon and the Jesuits," *Portuguese Studies* 17 (2001): 168–83. See Jeffrey D. Burson and Jonathan Wright, eds., *The Jesuit Suppression in Global Context: Causes, Events, and Consequences* (New York: Cambridge University Press, 2015).

58 Miguel de Asúa, *Science in the Vanished Arcadia: Knowledge of Nature in the Jesuit Missions of Paraguay and Río de la Plata* (Leiden: Brill, 2014), especially 194–9; Girolamo Imbruglia, *The Jesuit Missions of Paraguay and a Cultural History of Utopia (1568–1789)* (Leiden: Brill, 2017).

59 Ângela Barreto Xavier and Ines G. Županov, *Catholic Orientalism: Portuguese Empire, Indian Knowledge (16th–18th Centuries)* (New Delhi: Oxford University Press, 2015), 287–329.

60 Miguel Batllori, *La cultura hispano-italiana de los jesuitas expulsos; españoles, hispano-americanos, filipinos, 1767–1814* (Madrid: Editorial Gredos, 1966); Jorge Cañizares Esquerra, *How to Write the History of the New World: Histories, Epistemologies, and Identities in the Eighteenth-Century Atlantic World* (Stanford, CA: Stanford University Press, 2001).

61 Sabina Pavone, "Ricostruire la Compagnia partendo da Oriente? La comunità gesuita franco-cinese dopo la soppressione," in *Missioni, saperi e adattamento tra Europa e imperi non cristiani*, ed. Vincenzo Lavenia and Sabina Pavone (Macerata: Eum, 2015), 129–64; Ronnie Po-chia Hsia, "Jesuit Survival and Restoration in China," in *Jesuit Survival and Restoration: A Global History, 1773–1900*, ed. Robert Maryks and Jonathan Wright (Leiden: Brill, 2015), 245–60.

62 Gerhard Geissl, *Joseph Liesganig: Die wiener Meridianmessung und seine Arbeiten im Gebiet von Wiener Neustadt* (Wiener Neustadt: Verein Museum und Archiv für Arbeit und Industrie im Viertel unter dem Wienerwald, 2001); Madalina Veres, "Scrutinizing the Heavens, Measuring the Earth: Joseph Liesganig's Contribution to the Mapping of the Habsburg Lands in the Eighteenth Century," *Journal of Jesuit Studies* 6, no. 1 (2019): 85–98. For another case in point, see Per Pippin Aspaas and László Kontler, *Maximilian Hell (1720–92) and the Ends of Jesuit Science in Enlightenment Europe*, Jesuit Series (Leiden: Brill, 2019).

63 Mark O'Connor, "Oświecenie katolickie i Marcin Poczobut SJ", in *Jezuici a kultura polska*, ed. Ludwik Grzebień and Stanisław Obirek (Krakow: Wydawn. WAM, Księża Jezuici, 1993), 41–9; Sabina Pavone, *Una strana alleanza* (Naples: Bibliopolis, 2008), 35–6, 189, 303–54.

64 Like in the Americas, many of these observations were published in the nineteenth century, e.g., Augustin Carayon, S.I., *Documents inédits concernant la Compagnie de Jésus*, vol. 20, *Missions des Jésuites en Russie (1804–1824)* (Poitiers: Henri Oudin, 1869); Anna Peck, "Between Russian Reality and Chinese Dream: The Jesuit Mission in Siberia 1812–1820," *The Catholic Historical Review* 87 (2001): 17–33.

65 Adam František Kollár, *Historiae iurisque publici regni Hungariae amoenitates* (Vienna: Bavmeisterianis, 1783), 1:80, quoted in Paul Shore, "Ex-Jesuit Librarian-Scholars Adam František Kollár and György Pray: Baroque Tradition, National Identity, and the Enlightenment among Jesuits in the Eastern Habsburg Lands," *Journal of Jesuit Studies* 6, no. 3 (2019): 470.

66 Leonor Correa Etchegaray, Emanuele Colombo, and Guillermo Wilde, eds., *Las misiones antes y después de la restauración de la Compañía de Jesús: continuidades y cambios* (Mexico City: Pontificia Universidad Javeriana/Universidad Iberoamericana de México, 2014); Maryks and Wright, *Jesuit Survival*.

12

NETWORKS OF KNOWLEDGE IN THE INDO-PACIFIC, 1600–1800

Dorit Brixius

Using networks as a tool of historical analysis has proved useful to understand the exchange, circulation, and production of knowledge at a global scale and across national and cultural boundaries.[1] This essay explores the historiography of scientific networks in the context of the seventeenth- and eighteenth-century Indian and Pacific Oceans. In bringing together the historiographies of these two oceans I do not suggest conceptualising one connected world of the Indo-Pacific in a Braudelian sense, but rather to emphasise the many horizons of those worlds, in Sugata Bose's words.[2] Ever since Fernand Braudel's study on the distinct worlds of the Mediterranean, historians have sought to investigate regional connections and interactions in oceanic contexts.[3] Transoceanic contexts have been applied in Atlantic history and they have also entered the distinct fields of the Indian and Pacific Ocean worlds.[4] Regional connections and interactions in the context of oceans have proved fruitful when looking at nuanced Indo-Pacific cross-cultural encounters.[5] These studies underline the potential of the Indo-Pacific to those who write about knowledge networks. I recognise the distinct histories and historiographies of those oceans while suggesting that using the Indo-Pacific as a regional framework might help to seriously examine extra-European knowledge and cross-cultural exchange.[6]

This essay has a historiographical objective in the first instance, to show how the Indo-Pacific past has been framed with regard to the history of scientific networks. It also demonstrates where its historiography stands now and how writing within Indo-Pacific contexts might help to overcome the challenges of global approaches to the history of science by privileging Indigenous and cross-cultural networks. In terms of scope, it does not merely take the common view that the Indian Ocean world stretches from the Cape of Good Hope to the South China Sea, but that it includes also Pacific islands, namely the Philippines and those of the Malay world. And it comprises European as well as non-European empires and merchant networks.

I argue that a reconceptualisation of the Indo-Pacific as an interregional space of study can help to overcome the challenges that global historians of science continue to face. In so doing I build the argument in four steps. First, I bring to light the many connections and boundaries of the Indo-Pacific. Second, I examine the historiography of scientific networks in the Indo-Pacific by identifying debates on the polycentric nature of science production. In a third step, while insisting on the limits of recent approaches, I critically review cross-cultural studies that emerged after the global turn in the history of science.[7] In particular I seek to question what so-called "global networks" mean in more recent historiography concerned with global pre-modern science in the Indo-Pacific. And finally, and in conclusion, this essay examines the possibilities and limits of Indo-Pacific methodologies and the history of science, elaborating on the question if historians of science will be able to write an inclusive history of Indo-Pacific knowledge networks.

The Indo-Pacific: lessons to be learned for historians of science

What historians of the Indian Ocean have been attempting recently is similar to the development of Atlantic history starting four decades ago, when scholars began to bring the Atlantic world into focus.[8] There is now a large literature which uses oceanic histories to frame and write about non-European cultures.[9] Historians of area studies of both the Indian and the Pacific Oceans have long recovered Indigenous agency and non-European histories while questioning narratives of European historical trajectories. The Indo-Pacific is arguably more diverse than the Atlantic or Mediterranean, "with greater variations in languages, cultures, geography, and political units in the countries spread around its littoral from South Africa to Indonesia and China."[10] Seeking to find unifying factors of this oceanic space, Kirti Chaudhuri argued a few decades ago that we can see a unified Indian Ocean when looking at trade routes, particularly those of the Silk Road connecting the maritime world with the mainland.[11]

More recently, Indian Ocean scholarship opened up to plural ideas of this ocean's worlds.[12] Indeed, such a perspective reveals the diversity of traders and merchant networks. Already the history of merchant networks in the fifteenth century shows the diversity of the Indo-Pacific traders, ranging from Indian merchants sailing regularly to the Red Sea and the Persian Gulf, Arab traders dominating the Arabian Sea, the area between China and the Malay Archipelago (the South China Sea) under the control of the Chinese, and Indonesia included in Japanese and Malay merchant networks.[13] The work underlines the pluralised worlds of the Indian Ocean, in which traders, mariners, and merchants of South Asia, China, and East Africa, as well as Islamic missionaries, members of the Armenian diaspora, and European colonisers, all interacted and intersected with each other, blown together by the natural phenomenon of the monsoon.[14]

Historians have shed light on the Indian Ocean, what most prominently Sugata Bose called an "interregional arena" of economic, political, cultural, and as I detail below, scientific histories.[15] Using the Indian Ocean as an analytical framework has the potential to explain its worlds outside regional or colonial identities imposed by Western geographical categories, with a greater depth of economic and cultural meaning.[16] As a response to colonial frontiers limiting trans-colonial exchanges, global macro-models have helped us to see the larger picture, yet, as Bose observes, these area studies were too focused on national identities and have created mostly large-scale comparisons.[17] However, the work of Sanjay Subrahmayam has shown that micro-historical area studies can recover local agency while surveying historical connections across the Indian Ocean.[18]

By the same token, historical studies of the Pacific have been collectively marked by the in-depth examination of Indigenous history, culture, and knowledge, heavily influenced by postcolonial approaches. The field has done a great deal to argue against the assumed backwardness and isolation of the Pacific region that, as Epeli Hau'ofa famously argued, went hand in hand with European conceptions which imposed colonial and racial boundaries between the worlds of the Pacific.[19] First but foremost, migration in the Pacific Ocean began long, long before the first European ship entered these waters.[20] The efforts of scholars to resurrect Indigenous history of the Pacific have been very fruitful, also thanks to a revaluation of archives by insisting on the validity of oral history, genealogies, and cosmologies as valuable sources to

Figure 12.1 The power of south wave, Ranca Buaya Beach, West Java, Indonesia. Photograph by Fikry Anshor, Unsplash.

reconstruct the past.[21] As a result, as Alison Bashford observes, "Pacific historiography is marked by a cross-cultural epistemology, different ways of knowing culture, nature and history."[22]

As recent studies have shown, the Indian Ocean and the Pacific as an interregional geographic entity, was—and is—a place for telling different histories. These histories foster national identities while simultaneously revealing interactions among communities. There is no physical border between the Indian and the Pacific Oceans, and although there is an older tradition of keeping these regions separate, their histories cannot be detached. No single mode of writing the history of the Indian Ocean and the Pacific has been persistent, but the former has given rise to histories of islands, bays, and passages.[23] As Michael Pearson underlined, it is "people, not water, that created unity and a recognisable Indian Ocean that historians can study,"[24] and this also applies to the shared histories of the Indo-Pacific. Pluralising Indo-Pacific worlds, and examining their knowledge networks, indicates both division and unity. The plural worlds of the Indo-Pacific spoke to and competed with each other.

And what about Indo-Pacific histories and the history of science? Local interactions in the Indo-Pacific in the seventeenth and eighteenth centuries have received much less attention than the question how native knowledge was mobilised and received in Europe or used by European imperial attempts.[25] To overcome this challenge, the Indo-Pacific offers historians of science opportunities to embrace new, pluralised methodologies advocated by Indo-Pacific scholars. Historians of science will be able to write local, or as I call it "context-based," Indo-Pacific histories by utilising comparative, "trans-local" and "transoceanic" approaches to geographic spaces and human encounters, which examine communication between areas and people.[26] Above all, a reconceptualisation of the Indo-Pacific as an interregional space of study can help to overcome the marginalisation of Africa and Southeast Asia in the history of science. The history of science in the Indo-Pacific worlds was a network of knowledge and materials, where the unifying factors were migration—both forced and unforced— as well as trade, cultural exchange, and religious expansion.

Using a network and context-based approach to the history of science in the Indo-Pacific will help to identify and recognise the myriad of scientific actors—ranging from European and non-European learned men to European and non-European figures of non-privileged backgrounds such as dealers, servants, slaves, women, artisans, and beyond. With such a diversity of people, culture-based practices mingled, meshed, and clashed. Here, we must certainly be aware of the danger of drawing an idyllic picture of those encounters since they were embedded in power relations: movement was tied to warfare, commercial goals, forced migration, and violence. Instead of regarding the Indo-Pacific as a free-flowing or bundled world, we must be aware of the plurality of different worlds tied together by networks that sometimes interacted, while at other times remained distinct because of disconnections and boundaries.

Knowledge networks in the Indo-Pacific: de-centred views

Looking at the circulatory property of knowledge has challenged the institutionalisation and social and national bounds of knowledge production. De-centred approaches make it possible to look at knowledge networks from three angles: first, exchange, commerce, and science all overlap; second, the more nuanced human interactions connected histories and go-betweens of scientific practice;[27] and third, local networks generated knowledge. There is by now a sizable literature on the practices that shaped science in the early-modern period, including de-centred views on science production in the early-modern world.[28] Yet, the historiography concerned with de-centred views on knowledge production in European empires is still highly unbalanced.

Recent studies have emphasised that there was no single network type, and there were a myriad of scientific agents. Knowledge networks cannot be seen abstractly; they had many different facades, components, and underlying motivations. To be sure, European scientific institutions, such as the academies, royal societies, and scientific periodicals, offered important frameworks for knowledge exchange and epistemological networking.[29] The study of intellectual and merchant networks is an important way to incorporate new commercial, material, and global approaches to scientific networks in order to understand how those were formed and how they intersected. Scholars have started to argue that the process and management of knowledge production were shaped by multi-centred dynamics, in which networks were fragile, uncertain, and mingled. Networks were more than an intellectual affair and more complex than a bilateral exchange between European metropolitan centres and the European colonial peripheries in the Indo-Pacific. They were both material and immaterial, and included people, artefacts, and ideas constituting long-distance and short-distance networks that characterised the early-modern world of the Indo-Pacific and elsewhere. Rather than being constructed around a single centre, networks should be considered polycentric.[30] The centres of these networks could be administrative or merchant hubs, and in other cases were built around intellectual issues and problems, even while including cultural heterogeneous actors.

Religious networks fit very well into polycentric narratives. From the sixteenth century onwards, European trading networks began reaching out to the commercial, political, and intellectual worlds of the Indo-Pacific, and mingled with the large-scale missionary networks and those of the Society of Jesus in particular.[31] By the mid sixteenth century, religious networks at a global scale began to intersect with networks of intellectual exchange and that rapidly encouraged, as Paula Findlen has observed, "densely layered, overlapping networks of correspondents who exchanged books, specimens, drawings, instruments, ideas, information, and news."[32] Religious networks are a tool to examine de-centred histories because religious orders were transnational by default and thus were also cross-imperial and cross-national networks.[33] Jesuit networks have gained particular attention amongst historians of science in order to examine the circulation of materials and knowledge as a global phenomenon. In her monograph on Jesuits at the court of Beijing, Antonella Romano traces the production and exchange of geographical knowledge between the "places of knowledge" of Rome, Madrid, Manila, Macau, Mexico City, and Beijing.[34] While relying explicitly on European-language sources (despite the richness of South Asian archives), Ângela Barreto Xavier and Ines Županov's study on Catholic missionaries and the construction of "Oriental knowledge," produced by Catholic agents of the Portuguese Empire, unravels the constituting practices of Catholic Orientalism at the geographical scale of South Asia, Italy, Portugal, and France.[35] Jesuit networks have also proved particularly fruitful for exploring the importance of trans-oceanic botanical networks. A representative example is that of the Bohemian Jesuit father Georg Joseph Kamel (1661–1706) stationed in Manila under Spanish authority whose correspondence networks reached from Southeast Asia to London and beyond.[36] These studies bring to light the existence of knowledge hubs within the cross-national and cross-cultural construction of scientific networks.[37]

Historians of science have long thought of the Society of Jesus as forming networks, and these studies continue to bear fruit. There were many other networks operating in the Indo-Pacific world, built around commercial and institutional connections. Historians of colonial botany and medicine, for instance, have argued for the overlaps between botany, empire, and commerce.[38] As Emma Spary has stressed, "botanical networks had a more literal connection with European economies than other sorts of collecting networks."[39] The link between the travel of plants for profit and the rise of early-modern European empires has gained significant scholarly attention, with a primary concern for European colonies in the Atlantic.[40] For colonial botany in the pre-modern Indo-Pacific, and particularly in the French and Spanish colonies, concrete and coherent studies are still lacking.

Previously, scientific networks have been considered as central to Europe, with European institutions as the main market for scientific data and research. Studies of global and colonial scientific networks have explored, from a top-down perspective, the accumulation process as always being centred on a European metropolitan institution or broader European commercial goals.[41] Early-modern French science has been regarded as particularly centralised in Paris.[42] Bruno Latour famously developed his actor–network theory (ANT) of centralised science production using the example of the French expedition to the Pacific island of Sakhalin conducted by the navigator Jean-François de Galaup, comte de La Pérouse.[43] Latour claims that European scientific establishments were "centres of calculations" where new forms of scientific knowledge were accumulated, with Europe as the main consuming market. According to Latour, La Pérouse extracted local geographical knowledge about Sakhalin and transmitted it to the imperial centre in Paris. Here this knowledge was accumulated, calculated, and eventually translated into European knowledge systems, namely into the language of cartography.

More recent bottom-up approaches argue for a focus on de-centred, local interactions. In applying such approaches, Latour's theory has been challenged by historians of science when looking at various other examples. First, the argument of long-distance networks must be re-thought when assuming that science production was polycentric yet hierarchical, which can be illustrated through the examples of acclimatisation gardens.[44] On the French colonial island that we call Mauritius, to name one example, the garden became a centre of accumulation itself, strongly built upon local networks interacting between islands of the Indo-Pacific.[45] Secondly, ANT has drawn extensive critique not only because of its hierarchicalisation of centres and peripheries. but also because of neglecting local and Indigenous agency. By the same token, scholars have pointed to the fact that Latour's theory is not nuanced when it comes to power relations and knowledge claims. In direct response to the Latourian macro-perspective of La Pérouse's expedition, Michael Bravo and Julie Cruikshank investigate La Pérouse's interaction with the Ainu and Tlingi people.[46] Examining a series of encounters on the ground, they argue for the importance of local knowledge, created in diverse and highly complex cross-cultural negotiations. These studies are part of recent scholarship that has brought to light that European-made knowledge relied to a great extent on non-European information.[47]

De-centred approaches do allow exploring the plurality and diversity of agents and sites in the process of knowledge production. These networks were the basis for large-scale and massively complex interactions that were, however, very often exploitative and hierarchical, and they had European end-points: encyclopaedic works, cabinets of curiosities, botanical gardens, and astronomical observations, to name the most representative examples. De-centred approaches to scientific networks include the European dependence on Indigenous knowledge. This knowledge was appropriated, very often without any reimbursement or credit, the latter of which was erased in printed accounts of the time.[48] For scholars today, intensive archival work and close readings of manuscripts have been the keys to rediscover non-European agency. Next, let us examine more explicitly how these approaches serve to include cross-cultural exchanges.

Local exchanges and global circulation: cross-cultural perspectives

As the previous section shows, recent studies have given us a chance to understand networks not only as linking European scholars but also as a more complex web of linkages in Indo-Pacific contexts. This section reveals how historians of science have tried to understand these complex webs in more depth. Even though "science and empire" has been studied for some years now, an inclusive history challenging the common view that science was a tool of imperialist domination has only slowly become part of methodological approaches. Two decades ago Roy MacLeod observed that, even though European perspectives should be questioned, scholars' efforts to see colonial science through the lens of "multiple engagements" have been rarely achieved.[49] Ever since MacLeod's observation, the discipline has developed significantly. Over the last 20 years, new approaches have arisen which include: the study of different scientific disciplines, new interpretations between geographical and territorial spaces, and the historiographical treatment of the "local."[50]

Historians of science interested in the Indo-Pacific histories have used concepts like "contact zones" and "middle grounds" in which "go-betweens" acting as cultural brokers mediated between different cultural and knowledge traditions.[51] These concepts are essential for writing histories of science outside exclusive European frameworks. Kapil Raj's study on cross-cultural science production was certainly ground breaking in exploring the production of various types of knowledge in European-Indian contexts.[52] Exploring in-depth cases of geographical and botanical surveys in early colonial India, he unveiled the interplay between local scientists, mediators, and Europeans, and the discourse which geographical surveys created.

The term "cross-cultural" has helped to describe the exchanges between different cultures and knowledge systems which were rooted in complex hierarchies and power relations. These debates have given rise to a so-called "global" history of science. Sujit Sivasundaram once observed that the term global "does not denote a total or singular history" but that writing about global science is about the connections and disconnections within a web of linkages.[53] Global history that reaches beyond a mere "global scale" or the globalisation of the world in the past can only move forward by incorporating micro-historical tools of analysis.[54] Only by pluralising sites and agents can scholars de-centralise European networks of science, something

Figure 12.2 Zanzibar, Tanzania. Photograph by Kaspars Eglitis, Unsplash.

that has been advocated by scholars like Kapil Raj, Sujit Sivasundaram, and Simon Schaffer.[55] As Fa-Ti Fan observed, in order to critically analyse power relations in science, we must look at disruptions rather than a presumed smooth circulation of knowledge.[56]

More balanced approaches to science in colonial contexts seek to situate local exchanges that recognise the asymmetrical character of power relations between Indigenous agents, mediators, and Europeans. In this context, the concept of "hybrid knowledge" has been criticised for implying a static view of culture.[57] Hybridity emerged as a key analytic in postcolonial theorising.[58] Instead of using the term "hybrid" to indicate a (smooth) fusion of two entities, recent considerations in Atlantic history suggest looking at "entanglements" between cultures, that is to say, their continuous and unresolved processes of cross-cultural interactions. As Karen Graubart has beautifully put it, "Rather than name an outcome, entanglement suggests ongoing confrontations, shifts, and revisions: a state of mutual learning and pushback which does not dissolve into a final product."[59] In this sense, entanglement implies the highly complex and multifaceted nature of cross-cultural interactions and that produced multiple possible reactions, ranging from resistance to cooperation. Also, "entangled" encounters are not restricted to a particular time or place but can be applied in possibly every context of human (and non-human) interaction.

The history of plant knowledge has proved particularly useful because scholars can incorporate it while writing cross-cultural histories of knowledge production in various colonial contexts, even though they do not necessarily use the term "entangled." Even though she uses the term "hybrid," Anna Winterbottom in her study of Indian flora demonstrates that Madras served as a lively hub for botany and medicine, and was critical for the construction of knowledge in the English East India Company world of the seventeenth century. Winterbottom shows how resident surgeons sought to mobilise collectors across the Indo-Pacific, expanding their networks from Manila, Batavia, and Cape Town.[60] Indian flora became a particular interest of Dutch and Portuguese naturalists in the late sixteenth and the late seventeenth century. For the former case, in the Malayalam-speaking parts of southwest India, botanical surveys resulted in the twelve-volume *Hortus Malabaricus*, published from 1678 to 1693 under the direction of the colonial administrator of the Dutch East India Company and naturalist Hendrik van Rheede tot Drakenstein (1636–91). Richard Grove, in particular, has argued that this text was fundamentally shaped by South Asian plant knowledge, through the help of three Tulu Brahmin physicians.[61] By the same token, Ines Županov and Ângela Barreto Xavier have highlighted the Portuguese attempts to bioprospect in South India and beyond, reinforcing Richard Grove's arguments for the significance of Southwest Indian knowledge for the Dutch and Portuguese construction of "tropical" nature.[62]

Indo-Pacific islands and their networks have gained particular attention in light of cross-cultural encounters in the context of managing, exploiting, and experimenting with their natural environments. For instance, Mauritius, over the entire course of European colonisation, was a place to examine the impact of deforestation, while also serving as a site for accumulating plants from Africa and Asia. Richard Grove has vividly demonstrated in his study of early environmentalism the tension between "edenic" visions of Mauritius and the island's vulnerability.[63] Because of heavy deforestation throughout the seventeenth and first half of the eighteenth centuries, difficult (and dangerous) environmental and climate conditions were the colonial reality. Particularly because of cyclones, the island remained extremely vulnerable.[64] French colonial settlers sought to balance the natural difficulties by applying a new, local, and environmental-orientated management strategy that included the planting of trees imported from Africa and Asia and the introduction of insect-eating birds.[65] The acclimatisation of non-native plants was a matter of intersecting horticultural knowledge from Southeast Asia, India, China, Madagascar, and East Africa.[66]

Acclimatisation gardens played significant roles in the construction of botanical knowledge, not only on Mauritius but also on Ceylon (today's Sri Lanka) where intermediaries acted as part of a network linking localities and knowledge traditions.[67] This observation has also been made for the Dutch interest in the natural world of the Indonesian islands. For example, Genie Yoo has examined the role of Malay go-betweens in the construction of Dutch natural knowledge, exemplified by the works of the Dutch naturalist Georg Everhard Rumphius (1627–1702).[68] Here, Yoo provides insights into the knowledge networks of the Maluku world. In so doing, she investigates not only the role of cultural go-betweens in networks of knowledge but also how the networks of local informants of Muslim elites worked amongst themselves.

These studies explore the specifically local character of individual encounters and their connections to increasingly global networks. European imperial networks were in constant interaction and exchange with each other, and so were Indigenous networks, in which Indigenous actors interacted amongst themselves and with Europeans who came to the region. By examining how these networks shaped the conditions and outcomes of cross-cultural exchanges, recent studies link various parts of the globe and simultaneously provide extensive details of local encounters.

Global histories of science, including cross-cultural or entangled histories where Indigenous brokers *contributed to* European scientific practices, have had an important role in re-writing the history of science. Despite this, they are limited because their circulatory approach fails to understand practices in local contexts. In the future, scholarship on the Pacific and the Indian Ocean might help to move beyond studies of how non-European knowledge travelled or how it influenced developments elsewhere in shaping what we call science today. Otherwise, and as Michael McDonnell has recently argued, global history as it is done in an ever-expanding field will "inevitably push us farther and farther away from indigenous peoples"[69] and their networks.

The history of Indo-Pacific knowledge networks: *quo vadis?*

In order to understand local knowledge production, we need to observe more than the interplay between global and local narratives; we need to develop micro-historical studies grounded in Indo-Pacific cultures. In conclusion, let us examine alternative ways of developing local histories of knowledge networks in the Indo-Pacific.[70] Recently, there have been important and outstanding studies that have innovatively pushed for methodological interventions that permit scholars to approach the histories of science without traditional analytical framing.

In the past, because of an inability to read and interpret sources written in local and Indigenous languages, or because these sources no longer exist, historians of science have investigated local knowledge using sources recorded through the eyes of Europeans. In the Mexican context, to name a case from Atlantic history, Iris Montero Sobrevilla claims that the *Florentine Codex*, a European source written in mostly European languages and mostly in a European fashion, allows historians of science to study the natural world as Indigenous peoples experienced and studied it.[71] Nonetheless, in many cases, European written sources are frustratingly silent about local, non-European knowledge, let alone in the context of slavery, even as recent research has tried to do justice to this silence.[72] We must also consider that European colonial archives are a product of the colonial past in which the agency of local and non-elite actors was often downplayed, and Indigenous knowledge systems were neglected in relation to European science. European colonial archives sponsored by the imperial states hence were (and still largely are) the result of the record keeping of European bureaucracies.[73]

Can we do better than relying on these European sources? The answer is yes. The archives of Indo-Pacific cultures might indeed help to overcome such a dilemma since there are the archives of non-European cultures, such as the Mughals, and the populations of India, Sri Lanka, the Malay world, and China.[74] The most representative example of this from the Indian Ocean is probably Sivasundaram's study of late-eighteenth- and early-nineteenth-century Ceylon in which he used Indigenous palm-leaf manuscripts to write the island's history, without just relying on British colonial archives.[75] Touching on aspects of botany, medicine, and other types of knowledge, Sivasundaram tells the history of British colonisation from a different angle, namely that British colonisers learned from Indigenous customs. In doing so, he re-ascribed the making of social and territorial space as a product of local, dynamic realities in the British encounters in South Asia.

For a new generation of historians of science who are guided by a vision of inclusiveness, who seek to incorporate values of multi-situated intellectual engagements, and who move beyond traditional archival sources, one way of dealing with archives is to cross-contextualise them. Appealing to use such "cross-contextualization," Sivasundaram argued that "by stretching the category of science across cultural perspectives and across different genres of recording and thinking, it is possible to appreciate the distinctive features of the history of science in the Pacific, and to take indigenous knowledges seriously."[76] Future research must be open to sources of historical evidence such as oral histories and material culture. This will overcome the rigid binary and grand divide between Indigenous and European perspectives and pluralise the notion of science, knowledge, and archive.[77] Our goal should not be to write a total or global history of science in the Indo-Pacific. We must look at the relevance of local knowledge networks beyond their Indo-Pacific location. And in so doing, thinking and researching local Indo-Pacific networks will help us escape the straitjacket of a Western orientation.

Notes

1 For a recent edited volume, see Paula Findlen, ed., *Empires of Knowledge: Scientific Networks in the Early Modern World* (New York: Routledge, 2019).

2 Fernand Braudel, *The Mediterranean and the Mediterranean World in the Age of Philip II*, vol. 1 (Berkeley: University of California Press, 1995); Sugata Bose, *A Hundred Horizons: The Indian Ocean in the Age of Global Empire* (Cambridge, MA: Harvard University Press, 2006).

3 Braudel, *The Mediterranean*; David Armitage, Alison Bashford, and Sujit Sivasundaram, eds., *Oceanic Histories* (Cambridge: Cambridge University Press, 2017).

4 On the Atlantic see Bernard Bailyn, *Atlantic History: Concept and Contours* (Cambridge, MA: Harvard University Press, 2005). On the Indian Ocean, see Bose, *A Hundred Horizons*; Michael N. Pearson, *The Indian Ocean* (London: Routledge, 2003). On the Pacific, see Matt K. Matsuda, *Pacific Worlds: A History of Seas, Peoples, and Cultures* (Cambridge: Cambridge University Press, 2012); David Armitage and Alison Bashford, eds., *Pacific Histories: Ocean, Land, People* (New York: Palgrave Macmillan, 2014).

5 See Armitage, Bashford, and Sivasundaram, *Oceanic Histories*; Jerry H. Bentley, "Sea and Ocean Basins as Frameworks of Historical Analysis," *Geographical Review* 89 (1999): 218–24.

6 I reinforce the arguments made in Šebestián Kroupa, Stephanie Mawson, and Dorit Brixius, "Introduction: Science and Islands in the Indo-Pacific Worlds," special issue, *British Journal for the History of Science* 51, no. 4 (2018): 541–58.

7 See "Focus: Global Histories of Science," special issue, *Isis* 101, no. 1 (2010).

8 Bose, *A Hundred Horizons*, 4–5.

9 On oceanic histories, see Armitage, Bashford, and Sivasundaram, *Oceanic Histories*.

10 Michael Pearson, "Introduction," in *Trade, Circulation and Flow in the Indian Ocean World*, ed. Michael Pearson (Houndmills: Palgrave Macmillan, 2015), 1–14, here p. 3. In Atlantic contexts there are emerging studies that focus on the place of Africa, the slave trade, and non-European agents outside non-elite subjects, also with regard to the history of science. See James Delbourgo and Nicholas Dew, eds., *Science and Empire in the Atlantic World* (New York: Routledge, 2008); Neil Safier, "Itineraries of Atlantic Science: New Questions, New Approaches, New Directions," special issue, *Atlantic Studies* 7, no. 4 (2010); Pablo F. Gómez, *The Experiential Caribbean: Creating Knowledge and Healing in the Early Modern Atlantic* (Chapel Hill, NC: UNC Press, 2017). See also Judith Ann Carney, *Black Rice: The African Origins of Rice Cultivation in the Americas* (Cambridge, MA: Harvard University Press, 2009); John Thornton, *Africa and Africans in the Making of the Atlantic World, 1400–1800* (Cambridge: Cambridge University Press, 1998).

11 Kirti N. Chaudhuri, *Trade and Civilisation in the Indian Ocean: An Economic History from the Rise of Islam to 1750* (Cambridge: Cambridge University Press, 1985).

12 Sujit Sivasundaram, "The Indian Ocean," in Armitage, Bashford, and Sivasundaram, *Oceanic Histories*, 31–61.

13 Stephan Conermann, "South Asia and the Indian Ocean," in *Empires and Encounters, 1350–1750*, ed. Wolfgang Reinhard (Cambridge, MA: Belknap Press, 2015), 391–554, here p. 511. On the early modern South China Sea, see Eric Tagliacozzo, "The South China Sea," in Armitage, Bashford, and Sivasundaram, *Oceanic Histories*, 113–33, here pp. 120–3. On Armenian networks, see Sebouh David Aslanian, *From the Indian to the Mediterranean. The Global Trade Networks of Armenian Merchants from New Julfa* (Berkeley: University of California Press, 2011), 44–65. On Ottomans and Mughals, see Suraiya Faroqhi, *The Ottoman and Mughal Empires: Social History in the Early Modern World* (London: I.B. Tauris, 2019).

14 Bose, *A Hundred Horizons*; Engseng Ho, *The Graves of Tarim: Genealogy and Mobility Across the Indian Ocean* (Berkeley: University of California Press, 2006); Abdul Sheriff, "Globalisation with a Difference: An Overview," in *The Indian Ocean: Oceanic Connections and the Creation of New Societies*, ed. Abdul Sheriff and Engseng Ho (London: Hurst & Company, 2014), 11–41; Edward A. Alpers, *The Indian Ocean in World History* (Oxford: Oxford University Press, 2014).

15 Bose, *A Hundred Horizons*, 4–6.

16 For the history of medicine, see also Anna Winterbottom and Facil Tesfaye, eds., *Histories of Medicine and Healing in the Indian Ocean World*, 2 vols. (New York: Palgrave Macmillan, 2015).

17 Bose, *A Hundred Horizons*, 7.

18 Sanjay Subrahmayam, *Explorations in Connected History: From the Tagus to the Ganges* (Oxford: Oxford University Press, 2005).

19 Epeli Hau'ofa, "Our Sea of Islands," *The Contemporary Pacific* 6 (1994): 147–61; Margaret Jolly, "Imagining Oceania: Indigenous and Foreign Representations of a Sea of Islands," *The Contemporary Pacific* 19 (2007): 508–45.

20 Matsuda, *Pacific Worlds*, 9–36.

21 See for instance Damon Salesa, "The Pacific in Indigenous Time," in Armitage and Bashford, *Pacific Histories*, 31–52; Hau'ofa, "Our Sea of Islands," 152; Nicholas Thomas, *Islanders: The Pacific in an Age of Empire* (New Haven, CT: Yale University Press, 2010).

22 Alison Bashford, "The Pacific Ocean," in Armitage, Bashford, and Sivasundaram, *Oceanic Histories*, 62–84, here p. 79.

23 Sivasundaram, "The Indian Ocean," 60.

24 Michael Pearson, *The Indian Ocean* (New York: Routledge, 2003), 27.

25 The Pacific has been seen as particularly important for eighteenth-century expeditions: the French voyages of Louis Antoine de Bougainville and the Comte de La Pérouse, the British expeditions of James Cook and Joseph Banks, the expedition of the Danish cartographer and explorer Vitus Bering in the service of Russia, and the expedition of Italian explorer and navigator Alessandro Malespina for the Spanish Crown, to name the most prominent examples. On the practices of natural history during the Cook voyages, see Anne Mariss, *"A World of New Things" Praktiken der Naturgeschichte bei Johann Reinhold Forster* (Frankfurt am Main: Campus Frankfurt, 2015).

26 Matt Matsuda suggested the concept of "trans-local," Matsuda, *Pacific Worlds*, 5–6. On the concept of "transoceanic," see Barbara Watson Andaya, "Oceans Unbounded: Transversing Asia across 'Area Studies,'" *The Asia-Pacific Journal* 5 (2007), 15.

27 Kapil Raj, "Beyond Postcolonialism … and Postpositivism: Circulation and the Global History of Science," *Isis* 104, no. 2 (2013): 342–3; Nicholas B. Dirks, "Colonial Histories and Native Informants: Biography of an Archive," in *Orientalism and the Postcolonial Predicament: Perspectives on South Asia*, ed. Carol A. Breckenridge and Peter van der Veer (Philadelphia: University of Pennsylvania Press, 1993), 279–313.

28 László Kontler et al., eds., *Negotiating Knowledge in Early Modern Empires: A Decentred View* (Houndmills: Palgrave Macmillan, 2014). For an Indian Ocean case, using manuscript collections in eighteenth-century India, see Stéphane Van Damme, "Capitalizing Manuscripts, Confronting Empires: Anquetil-Duperron and the Economy of Oriental Knowledge in the Context of the Seven Years' War," in Kontler et al., *Negotiating Knowledge in Early Modern Empires*, 109–28.

29 See for instance Alice Stroup, *A Company of Scientists: Botany, Patronage, and Community at the Seventeenth-Century Parisian Royal Academy of Sciences* (Berkeley: University of California Press, 1990).

30 As Sven Dupré and Sachiko Kusukawa claimed, there was also no global Republic of Letters but a series of "intersecting networks and communities" in which the hubs were characterised by the need to communicate with each other via agents. Sven Dupré and Sachiko Kusukawa, "Introduction: The Circulation of News and Knowledge in Intersecting Networks," special issue, *History of Universities* 23, no. 2 (2008): 4.

31 See for instance Ines G. Županov, *Disputed Mission: Jesuit Experiments and Brahmanical Knowledge in Seventeenth-Century India* (New Delhi and New York: Oxford University Press, 1999).

32 Paula Findlen, "Introduction," in Findlen, *Empires of Knowledge*, 1–22, here p. 14.

33 For critical reflections on the concept of "transnational" passively implying the restrictive and often anachronistic concept of "nation," see Christopher A. Bayly et al., "Conversation: On Transnational History," *American Historical Review* 111, no. 5 (2006), 1441–64.

34 Antonella Romano, *Impressions de Chine: l'Europe et l'englobement du monde, XVIe-XVIIe siècle* (Paris: Fayard, 2016). On Jesuit translations of scientific Chinese texts, see Huiyi Wu, *Traduire la Chine au XVIIIe siècle: les jésuites traducteurs de textes chinois et le renouvellement des connaissances européennes sur la Chine (1685–ca. 1740)* (Paris: Éditions Honoré Champion, 2017).

35 Ângela Barreto Xavier and Ines G. Županov. *Catholic Orientalism: Portuguese Empire, Indian Knowledge* (New Delhi: Oxford University Press, 2015).

36 Sebastian Kroupa, "Ex Epistulis Philippinensibus: Georg Joseph Kamel SJ (1661–1706) and his Correspondence Network," *Centaurus* 57 (2015): 229–59; Raquel Reyes, "Botany and Zoology in the Late Seventeenth-Century Philippines: The Work of Georg Josef Camel SJ (1661–1706)," *Archives of Natural History* 36 (2009): 262–76.

37 Christopher A. Bayly, *Empire and Information: Intelligence Gathering and Social Communication in India, 1780–1870* (Cambridge: Cambridge University Press, 1999); David N. Livingstone, *Putting Science in its Place: Geographies of Scientific Knowledge* (Chicago: University of Chicago Press, 2003); James A. Secord, "Knowledge in Transit," *Isis* 95, no. 4 (2004): 654–72. For the Dutch case, see also Fokko J. Diksterhuis, Andreas Weber, Huib Zuidervaart, eds., *Locations of Knowledge in Dutch Contexts* (Leiden: Brill, 2019).

38 Pamela H. Smith and Paula Findlen, eds., *Merchants and Marvels: Commerce, Science and Art in Early Modern Europe* (New York: Routledge, 2001); Londa Schiebinger and Claudia Swan, eds., *Colonial Botany: Science, Commerce, and Politics in the Early Modern World* (Philadelphia: University of Pennsylvania Press, 2005); Harold J. Cook, *Matters of Exchange: Commerce, Medicine, and Science in the Dutch Golden Age* (New Haven, CT: Yale University Press, 2007); Alette Fleischer, "(Ex)Changing Knowledge and Nature at the Cape of Good Hope, Circa 1652–1700," in *The Dutch Trading Companies as Knowledge Networks*, ed. Siegfried Huigen, Jan L. de Jong, and Elmer Kolfin (Leiden: Brill, 2010), 243–65.

39 Emma C. Spary, *Utopia's Garden: French Natural History from Old Regime to Revolution* (Chicago: University of Chicago Press, 2000), 56. See also Emma Spary, "Botanical Networks Revisited," in *Wissen im Netz: Botanik und Pflanzentransfer in europäischen Korrespondenznetzen des 18. Jahrhunderts*, ed. Regina Dauser, Stefan Hächler, and Michael Kempe (Berlin: Akademie-Verlag, 2008), 1–18.

40 See for instance Samir Boumediene, *La colonisation du savoir: une histoire des plantes médicinales du "Nouveau monde" 1492–1750* (Vaulx-en-Velin: les Éditions des mondes à faire, 2016).

41 For the British case, see Lucile Brockway, *Science and Colonial Expansion: The Role of the British Royal Botanic Gardens* (New York: Academic Press, 1979) and for the French case, see Spary, *Utopia's Garden*, 61–88.

42 But French histories of centralised science continue to be written, including James E. McClellan III and François Regourd, *The Colonial Machine: French Science and Overseas Expansion in the Old Regime* (Turnhout: Brepols, 2011). McClellan and Regourd's work has been criticised and challenged for their top-down approach: Loïc Charles and Paul Cheney, "The Colonial Machine Dismantled: Knowledge and Empire in the French Atlantic," *Past & Present* 219 (2013): 127–63; Lissa Roberts, "'Le Centre de Toutes Choses': Constructing and Managing Centralization on the Isle de France," *History of Science* 52 (2014): 319–42; Dorit Brixius, "A Hard Nut to Crack: Nutmeg Cultivation and the Application of Natural History between the Maluku Islands and Isle de France (1750s–1780s)," special issue, *British Journal for the History of Science* 51, no. 4 (2018): 585–606.

43 Bruno Latour, *Science in Action: How to Follow Scientists and Engineers Through Society* (Cambridge, MA: Harvard University Press, 1987), 215–58.

44 Matthew Sargent, "Recentering Centers of Calculation: Reconfiguring Knowledge Networks within Global Empires of Trade," in Findlen, *Empires of Knowledge*, 297–316, here pp. 308–12.

45 Roberts, "'Le Centre de Toutes Choses'" and Brixius, "A Hard Nut to Crack."

46 Michael Bravo, "Ethnographic Navigation and the Geographical Gift," in *Geography and Enlightenment*, ed. David N. Livingstone and Charles W.J. Withers (Chicago: Chicago University Press, 1999), 199–235; Julie Cruikshank, *Do Glaciers Listen? Local Knowledge, Colonial Encounters, and Social Imagination* (Vancouver: UBC Press, 2014), 127–53.

47 Kapil Raj, *Relocating Modern Science: Circulation and the Construction of Knowledge in South Asia and Europe, 1650–1900* (Basingstoke: Palgrave Macmillan, 2007); Anna Winterbottom, *Hybrid Knowledge in the Early East India Company World* (Houndmills: Palgrave Macmillan, 2016). See also Neil Safier, "Global Knowledge on the Move: Itineraries, Amerindian Narratives, and Deep Histories of Science," *Isis* 101, no. 1 (2010): 133–45.

48 See also examples from Atlantic history, Neil Safier, *Measuring the New World: Enlightenment Science and South America* (Chicago: University of Chicago Press, 2008) and Boumediene, *La colonisation du savoir*.

49 Roy M. MacLeod, "Introduction," in "Nature and Empire: Science and the Colonial Enterprise," special issue, *Osiris* 15 (2000): 6.

50 "Focus: Global Histories of Science," special issue, *Isis* 101, no. 1 (2010).

51 Richard White suggested the term "middle ground": Richard White, *The Middle Ground: Indians, Empires, and Republics in the Great Lakes Region, 1650–1815* (Cambridge: Cambridge University Press, 1991). On works deploying "contact zone" see for instance Kapil Raj, "The Historical Anatomy of a Contact Zone: Calcutta in the Eighteenth Century," *Indian Economic & Social History Review* 48 (2011): 55–82; Lissa Roberts, "Situating Science in Global History: Local Exchanges and Networks of Circulation," *Itinerario* 33 (2009): 9–30. On go-betweens and cultural brokers in the history of science see Schaffer et al., *The Brokered World*; Alida C. Metcalf, *Go-Betweens and the Colonization of Brazil, 1500–1600* (Austin: University of Texas Press, 2005); and for the Pacific in particular see Bronwen Douglas, "Agency, Affect, and Local Knowledge in the Exploration of Oceania," in *Indigenous Intermediaries: New Perspectives on Exploration Archives*, ed. Shino Konishi, Maria Nugent, and Tiffany Shellam (Canberra: ANU Press, 2015), 103–30.

52 Raj, *Relocating Modern Science*.

53 Sujit Sivasundaram, "Introduction," *Isis* 101, no. 1 (2010): 96.

54 Romano, *Impressions de Chine*, 17; Francesca Trivellato, "Is There a Future for Italian Microhistory in the Age of Global History?", *California Italian Studies* 2, no. 1 (2011).

55 Raj, *Relocating Modern Science*; Sujit Sivasundaram, "Sciences and the Global: On Methods, Questions, and Theory," *Isis* 101, no. 1 (2010): 146–58; Simon Schaffer et al., *The Brokered World: Go-Betweens and Global Intelligence, 1770–1820* (Sagamore Beach, MA: Science History Publications, 2009).

56 Fa-Ti Fan, "The Global Turn in the History of Science," *East Asian Science, Technology and Society* 6 (2012): 249–58, here p. 252.

57 Ralph Bauer and Marcy Norton, "Introduction: Entangled Trajectories: Indigenous and European histories," special issue, *Colonial Latin American Review* 26, no. 1 (2017): 1–17, referencing Nicholas Thomas, *Entangled Objects: Exchange, Material Culture and Colonialism in the Pacific* (Cambridge, MA: Harvard University Press, 1991).

58 As the key text for the theoretic development of hybridity, see Homi Bhabha, *The Location of Culture* (London: Routledge, 1994).

59 Karen B. Graubart, "Shifting Landscapes. Heterogeneous Conceptions of Land Use and Tenure in the Lima Valley," *Colonial Latin American Review* 26, no. 1 (2017): 62–84.

60 Winterbottom, *Hybrid Knowledge*. See also Minakshi Menon, "Making Useful Knowledge: British Naturalists in Colonial India, 1784–1820" (PhD diss., University of California, 2013); Pratik Chakrabarti, "Medical Marketplaces Beyond the West: Bazaar Medicine, Trade and the English Establishment in Eighteenth-Century India," in *Medicine and the Market in England and Its Colonies, c.1450–c.1850*, ed. Mark S.R. Jenner and Patrick Wallis (London: Palgrave Macmillan, 2007), 196–215.

61 Richard Grove, "Indigenous Knowledge and the Significance of South-West India for Portuguese and Dutch Constructions of Tropical Nature," *Modern Asian Studies* 30, no. 1 (1996): 121–43. On the *Hortus*, see also Minakshi Menon's ongoing project "Hortus Indicus Malabaricus: The Eurasian Life of a Seventeenth-Century 'European' Botanical Classic," which will soon produce a single-volume critical edition of the *Hortus'* plant descriptions.

62 Ines G. Županov and Ângela Barreto Xavier, "Quest for Performance in the Tropics: Portuguese Bioprospecting in Asia (16th–18th Centuries)," *Journal of the Economic and Social History of the Orient* 57 (2014): 511–48; Grove, "Indigenous Knowledge."

63 Richard H. Grove, *Green Imperialism: Colonial Expansion, Tropical Island Edens and the Origins of Environmentalism, 1600–1860* (Cambridge: Cambridge University Press, 1995).

64 On cyclones and observation on Mauritius, see Martin Mahony, "The 'Genie of the Storm': Cyclonic Reasoning and the Spaces of Weather Observation in the Southern Indian Ocean, 1851–1925', special issue, *British Journal for the History of Science* 51, no. 4 (2018): 607–33.

65 Grove, *Green Imperialism*; Roberts, "'Le Centre de Toutes Choses'"; Etienne Stockland, "Policing the Oeconomy of Nature: The Oiseau Martin as an Instrument of Oeconomic Management in the Eighteenth-Century French Maritime World," *History and Technology* 30, no. 3 (2014): 207–31.

66 Brixius, "A Hard Nut to Crack."

67 Sujit Sivasundaram, "Islanded: Natural History in the British Colonization of Ceylon," in *Geographies of Nineteenth-Century Science*, ed. David Livingstone and Charles W. Withers (Chicago: University of Chicago Press, 2011), 123–48.

68 Genie Yoo, "Wars and Wonders: The Inter-Island Information Networks of Georg Everhard Rumphius," special issue, *The British Journal for the History of Science* 51, no. 4 (2018): 559–84. For natural knowledge in Indonesia, see Marya Rosenberg Leong, "Global Trade and Local Knowledge: Gathering Natural Knowledge in Seventeenth-Century Indonesia," in *Intercultural Exchange in Southeast Asia: History and Society in the Early Modern World*, ed. Tara Alberts and David Irving (London: I.B. Tauris, 2013), 144–60; Peter Boomgaard, "Introduction: From the Mundane to the Sublime: Science, Empire, and the Enlightenment, 1760s–1820s," in *Empire and Science in the Making: Dutch Colonial Scholarship in Comparative Global Perspective, 1760–1830*, ed. Peter Boomgaard (New York: Palgrave Macmillan, 2013), 23.

69 Michael A. McDonnell, "Facing Empire: Indigenous Histories in Comparative Perspective," in *The Atlantic World in the Antipodes: Effects and Transformations Since the Eighteenth Century*, ed. Kate Fullagar (Newcastle: Cambridge Scholars Publishing, 2012), 225.

70 There are important works on their way, including Minaski Menon, ed., "Indigenous Knowledges and Colonial Sciences in South Asia," special issue, *South Asian History and Culture* (forthcoming 2021).

71 Iris Montero Sobrevilla, "Indigenous Naturalists," in *Worlds of Natural History*, ed. Helen Anne Curry et al. (Cambridge: Cambridge University Press, 2018), 112–30.

72 See Gómez, *The Experiential Caribbean*. On Atlantic slavery and science see Londa L. Schiebinger, *Secret Cures of Slaves: People, Plants, and Medicine in the Eighteenth-Century Atlantic World* (Stanford, CA: Stanford University Press, 2017). For the Indian Ocean, see Dorit Brixius, "From Ethnobotany to Emancipation: Slaves, Plant Knowledge, and Gardens on Eighteenth-Century Isle de France," *History of Science* 58 (2020), 51–75.

73 Engseng Ho, *The Graves of Tarim: Genealogy and Mobility Across the Indian Ocean* (Berkeley: University of California Press, 2006), xxiii; Marie Houllemare, "La fabrique des archives coloniales et la naissance d'une conscience impériale (France, XVIIIe siècle)," *Revue d'Histoire Moderne et Contemporaine* 2 (2014), 7–31; Anne Laura Stoler, *Along the Archival Grain: Epistemic Anxieties and Colonial Common Sense* (Princeton, NJ: Princeton University Press, 2010); James C. Scott, *Against the Grain: A Deep History of the Earliest States* (New Haven, CT: Yale University Press, 2017).

74 For the nineteenth-century Indian Ocean see also Sujit Sivasundaram, "The Oils of Empire," in Curry et al., *Worlds of Natural History*, 379–98.

75 Sujit Sivasundaram, *Islanded: Britain, Sri Lanka, and the Bounds of an Indian Ocean Colony* (Chicago: University of Chicago Press, 2013).

76 Sujit Sivasundaram, "Science," in Armitage and Bashford, *Pacific Histories*, 239.

77 Greg Dening, *Performances* (Melbourne: Melbourne University Press, 1996); Thomas, *Entangled Objects*; Neil Safier, "Masked Observers and Mask Collectors: Entangled Visions from the Eighteenth-Century Amazon," *Colonial Latin American Review* 26, no. 1 (2017): 104–30; Sivasundaram, *Islanded*. See also Adriana Craciun and Simon Schaffer, eds., *The Material Cultures of Enlightenment Arts and Sciences* (Houndmills: Palgrave Macmillan, 2016).

13

BETWEEN TRANSIMPERIAL NETWORKING AND NATIONAL ANTAGONISM

German scientists in the British Empire during the long nineteenth century

Ulrike Kirchberger

In the nineteenth century, a large number of the scientists who explored the non-European world in British services came from German-speaking central Europe. Young men, often with degrees in the natural sciences from German universities, participated in prestigious expeditions that were organised and funded by the Royal Geographical Society, the East India Company, the Foreign Office, and other British government bodies and scientific institutions. Other German scientists were employed in the colonial administrations in Africa, Asia, and Australia, in the Indian Forest Service, and in the botanical gardens that were established all across the British Empire. Exact numbers are hard to establish. However, historians of British colonial science often notice the presence of these Germans. For the Australian colonies, for example, the historian R.W. Home has observed that there was an "overrepresentation of Germans in positions of scientific leadership" in the middle of the nineteenth century.[1]

Historians who deal with the German scientists in the British Empire often choose a biographical approach.[2] In recent years, these biographies have been largely influenced by postcolonial theory. Whereas older studies presented narratives of heroic Humboldtians who ventured into the unknown alone and were only interested in their scientific research,[3] new biographies question this perspective. They emphasise that the German explorers depended to a considerable degree on the knowledge and cooperation of local experts and intermediaries. The German scientists in the contact zones of the British Empire, they argue, were participants in multi-ethnic enterprises. Different interest groups had different motives to join or support an expedition. Multiple ethnic hierarchies determined the progress of the journeys.[4]

On the global level, the German scientists in Britain and its colonies are defined as contributors to transimperial networks. They produced specific forms of knowledge in the African, Asian, and Australian contact zones that was then transferred and changed through different imperial sites and scales.[5] Currently historians are widening the perspective, and they are integrating plants, animals, and natural objects as active participants in these transimperial networks. Applying concepts from actor–network theory and multispecies studies, they argue that non-humans developed their own dynamics and had considerable impact on the knowledge production in the contact zones and beyond.[6]

In this way, new approaches from postcolonial theory and the global history of science have found their way into the biographies of individual German scientists in the British Empire. As a group, however, they have still received little attention so far, although they developed a certain transoceanic coherence. Some of them corresponded regularly about their research and exchanged data, biota, and publications across different continents. Others shared diasporic experiences when they joined German ethnic life in the British settler colonies or propagated the foundation of German overseas colonies.

Motives

German scientists sought employment in the British Empire for different reasons. Most of them were academics who were keen on exploring the non-European world. They practised what historians often call "Humboldtian science,"[7] and strove for a comprehensive, all-encompassing overview of the world. An important method for achieving such a universal perspective was the accurate description of nature, by travelling and doing fieldwork, and then classifying, taxonomising, and comparing the observations. Following the example of the German naturalist Alexander von Humboldt, many young academics believed overseas travel would allow for the collection of data that would contribute to the knowledge of the world.

However, Germans not only sought employment in British services because they wanted to rise to the scientific paradigm of Humboldtian research. They had many other personal, political, and religious reasons. For example, Ernst Dieffenbach, who travelled to New Zealand as a scientific adviser of the British "New Zealand Company" in 1839, was a political refugee. Others discovered "terra incognita" as members of British missionary societies. In many respects, the Germans who explored the non-European world in British services were a heterogeneous group of people with different motivations for joining the British imperial project.[8]

There were, however, unifying factors amongst these German researchers and scholars. Most importantly, the German confederation of states had no overseas colonies. Britain, by contrast, had a powerful colonial empire. Therefore, German Humboldtians turned to Britain. Starting in the late eighteenth century, when enlightenment scholars from Göttingen joined British expeditions to explore the Pacific region, German travellers realised their overseas interests by using British influence outside of Europe.[9] Britain's diplomatic, trading, and military outposts all over the globe provided the infrastructure the scientists needed to organise and conduct their journeys. German scientists found employment in Britain's colonial administrations in Africa, Asia, Australia, and the Americas that were not available in German realms until the late nineteenth century. Furthermore, the large overseas ports in London and Liverpool were important departure points for the transoceanic journeys of German overseas travellers in the middle of the nineteenth century. Many German Humboldtians reached their non-European destinations via London. They often used this intermediate stop to do research in the city's libraries and museums, and to establish contacts with influential British peers and patrons of science before they began their transoceanic passage.

Other structural preconditions that explain the relatively high number of German scientists in British colonial services can be found in the different university systems in Britain and Germany in the first half of the nineteenth century. In Prussia and other German states, university reforms began in the early decades of the nineteenth century. As a result, universities focused on free research and developed a specialisation in subjects such as philology, linguistics, and natural sciences. At that time, the English universities in Oxford and Cambridge still followed a traditional concept of providing a broad-based liberal education for members of the Anglican Church. Reforms were first introduced in the middle of the nineteenth century. Therefore, the process of specialisation and professionalisation that turned Enlightenment naturalists and gentleman scholars into experts of specific scientific disciplines in the course of the nineteenth century began slightly earlier in Prussia than in Oxford and Cambridge. In the first half of the nineteenth century, influential British academics not only admired Alexander von Humboldt but they were also interested in the Prussian education system as a model for their own reform initiatives.[10] As they viewed Humboldtian science and the Prussian reformed universities in a positive light, they were willing to employ the graduates of German universities who had studied botany, geography, geology, or forestry at a time when there was a high demand for educated young men to explore the non-European world and to provide the data and information necessary for British economic and colonial expansion. In spite of occasional tensions, educated Germans were welcome contributors to the British "subimperialism of science."[11]

Central to organising the transfer of German scientists to the British Empire in the middle of the nineteenth century was a close-knit group of British and German scholars and diplomats in London and Berlin. Alexander von Humboldt was not only famous for his books and journeys, he also was, in the last decades of his life, an influential patron of science. At the royal court in Berlin, Humboldt supported many young German men who wanted to do research in Africa and Asia, and he established connections for them to join British expeditions. His most important contact for that purpose was the Prussian envoy in London, Christian von Bunsen. During Bunsen's tenure in the 1840s and early 1850s, the Prussian legation was a central meeting place for British and German cultural and academic elites. Bunsen helped to place German scientists in a number of spectacular British expeditions in Africa and Asia. He was himself a man of learning as well as a Protestant theologian. He combined scientific interest in overseas exploration with the mission of spreading Protestantism and civilisation to the so-called "heathen." Although Anglo-German scholarly networking was increasingly dominated by the figure of the secularised "scientific expert" in the course of the nineteenth century, a sense of shared commitment to Protestantism, though often overshadowed by national conflicts, remained a link binding many German and British scholars together.[12] From the middle of the century onwards, this sense of belonging to the same Protestant religion was intertwined with racialised notions of a common "Anglo-Saxon" identity.[13]

Imperial hierarchies and multi-directional networks

Throughout the nineteenth century, the German scientists in the British Empire were part of the core–periphery system of imperial science. Both British and German scientists in the British colonies had to report the results of their research to the Royal Geographical Society, the Royal Botanic Gardens, and other organisations of metropolitan science. The scientists in the colonies were expected to collect the data that was then organised and categorised in London. The naming and classifying of "newly discovered" non-European species of plants and animals, for example, were done in London. The official floras and faunas of different colonial regions were edited and published in London. The German scientists in the British colonies were part of this centralised system.

However, these imperial core–periphery hierarchies slowly lost significance during the second half of the nineteenth century. Conflicts between metropolitan and colonial scientists increased. Multipolar, transimperial networks grew more important.[14] New centres of global science emerged in the non-European peripheries of European empires. The German scientists in the British Empire contributed to this process of decentring the core–periphery hierarchies of imperial science. Although they worked for the British Empire, they imagined themselves as collaborating with a larger project of discovery and improvement. They communicated across oceans and imperial borders, and exchanged letters, publications, plants, and animals regardless of national allegiances.

In Melbourne, for example, the director of the botanical gardens and government botanist of Victoria, Ferdinand von Mueller, built such a communication network with scientists all over the world in the course of the second half of the nineteenth century. At the same time, his relations with the directors of the Royal Botanic Gardens in Kew, the centre of imperial botany, became increasingly strained. He grew unhappy with his role of being a mere collector for Kew and began to keep his own herbaria and named and classified plants in Melbourne without consulting the botanists in Kew. He also took offence when the directors in Kew did not make him co-editor of the official *Flora Australiensis* that was published in London in several volumes in the middle of the nineteenth century. In 1882, he produced his own flora of Australia with Melbourne as the place of publication.[15]

In the Indian Ocean region, German scientists cultivated networks that connected them with their British, French, and Dutch peers in Asia, Africa, and Australia. At a time when transoceanic travelling and communication were facilitated by the introduction of the steamship and the telegraph, the Germans contributed to the increasing intercolonial transfer across the Indian Ocean while the traditional way of corresponding between the colonies via London lost significance. For example, Dietrich Brandis, a high-ranking official of the Indian Forest Service, still sent biota from British India to Australia via Kew in the 1870s, but he also began to dispatch duplicates directly to Australia across the Indian Ocean.[16] Mueller also engaged in direct exchange of seeds with colonial institutions in South Africa and British India, such as, for example, with the Agri-Horticultural Society of India in Calcutta (Kolkata).[17] This increase in intercolonial and transimperial networking across the Indian Ocean was enhanced by the rise of important scientific centres in the region, such as, for example, the botanical gardens and research station of Buitenzorg (Bogor) in Dutch Java. At the end of the nineteenth century, Buitenzorg turned into a leading institution for tropical botany and agronomy and attracted scientists from all over the world, among them many German botanists and foresters in the colonies surrounding the Indian Ocean.[18]

At the same time, German scholars developed a German scientific network spanning the Indian Ocean. For example, Ferdinand von Mueller in Melbourne and the director of the botanical gardens in Adelaide, Richard Schomburgk, began to exchange research and biota with the German-born members of the Indian Forest Service, including Dietrich Brandis, William Schlich, and Berthold Ribbentrop. After the German protectorates were established in Africa in the mid-1880s, German–German exchanges between the German scientists in the British colonies and in German East Africa began. Both Mueller and Brandis engaged in species transfers to the German colonies. In German East Africa, foresters began to cultivate Australian eucalypts and, encouraged by Brandis, they experimented with South Asian bamboo to "improve" economy and civilisation in the new colony.[19] At the same time, the German scientists in the Indian Ocean region began to develop contacts with their peers in the United States. Mueller, for example, transferred Australian eucalypts and acacias to California in the middle of the nineteenth century.[20] Brandis played an advisory role in the making of the North American forest services from the 1870s.[21]

Compared to the institutionalised core–periphery relations of imperial science, the intercolonial German–German networks were often informal in character. Like many communications between individual scientists all over the world, they were free and equal cooperations based on mutual scientific interests and trust. However, the German scientists in the British Empire were in different places and positions and could also be competitors rather than friends. Mueller, for example, remained in Melbourne until his death. He participated in the ethnic life of the German minority. His psychological distance from Kew increased. Brandis, by contrast, spent the last decades of his life in Bonn and London. He cultivated a lifelong friendship with the Hooker family and criticised Mueller in his letters to Kew.[22] The Germans shared scientific interests and the same homeland, but they also developed their own networks and itineraries. Some communicated with each other, whereas others had no intercolonial connections to their German peers.

Colonial propaganda

The German scientists contributed to British economic and colonial expansion in many ways. They fulfilled British expectations by providing data and information about regions where cotton and other cash crops could be cultivated and about potential sites for the establishment of settler colonies.[23] They shared the racist ideology that legitimised British and European expansionism. Convinced of European superiority over non-European societies, they usually despised their African and native Australian collaborators and ignored their contributions in their publications.[24]

In this way, German scientists were committed to the goals of British expansionism. At the same time, however, they engaged in the colonial movement in Germany. In the 1840s and early 1850s, when German overseas emigration reached an unprecedented peak, the German scientists in the British Empire were important supporters of the numerous emigration schemes and colonial projects that emerged in Germany during these years. Ernst Dieffenbach, for example, advocated initiatives to establish a German colony on the Chatham Islands near New Zealand in the early 1840s.[25] After his return to Germany in 1844, Dieffenbach fought for a German colonial empire in the revolution of 1848–9.[26] Others advertised regions they had explored in Africa and South America as suitable places for German emigrants' colonies.[27] Christian von Bunsen backed plans for German overseas colonies in the 1840s, during his tenure as Prussian envoy in London.[28]

In the 1850s, after the revolution had failed and emigration figures decreased, German interest in overseas colonisation declined as well. Politicians in the German states concentrated on intra-German affairs. The German scientists in the British Empire, however, continued to propagate for the creation of a German overseas empire. In 1861, for example, the astronomer Georg von Neumayer envisioned a unified Germany with a strong navy in a speech he gave at the German Society in Melbourne.[29] Ferdinand von Mueller championed colonial ideas in his correspondences after 1865. He referred to New Guinea, Abyssinia, the Pacific islands, and other places in the southern hemisphere as destinations for German colonisation.[30]

The German scientists in the British Empire not only identified with the British imperial project, they also were a dynamic force in the German colonial movement throughout the nineteenth century. Even when political interest in Germany was low, they advertised German colonial ideas from the periphery of the British Empire. As transnational intermediaries in the non-European periphery, their colonial ambitions were not always in sync with metropolitan politics.

1884 as a watershed?

In the last decades of the nineteenth century, the conditions for overseas exploration changed. In 1884–5, Germany established colonies in Africa and the Pacific world. Anglo-German political relations began to suffer from Wilhelmine "Weltpolitik." Furthermore, the end of the nineteenth and early twentieth centuries witnessed a paradigm shift in the history of science. Humboldtian fieldwork slowly lost prestige and significance, and was replaced by the chemical and physical disciplines which were characterised by laboratory-based research. We might conclude that, as a consequence of these changes, German scientists no longer sought British support to prepare for their overseas expeditions, and the overall number of German scientists who moved to Britain and its colonies declined. There are, however, no hard figures that would confirm such a trend. Therefore it is a debatable question in how far these changes marked a turning point for German scientists in the British Empire.

On the one hand, with the rise of the colonial movement in the 1870s, the scientific exploration of the non-European world gained a strong political lobby in Germany. Newly founded associations, such as, for example, the *Centralverein für Handelsgeographie und Förderung deutscher Interessen im Ausland* and the *Deutsche Kolonialgesellschaft*, propagated the scientific exploration of the non-European world as part of German colonial policy and economic expansion. The botanical gardens in Berlin/Dahlem emerged as a new centre for global botany. Politicians, aristocrats, merchants, missionaries, and scientists now formed a broad coalition that supported German exploratory enterprises as projects of national prestige. German scientists did not depend on British support to the same extent as before.

Anglo-German scientific cooperation was indeed given up or reduced in the late nineteenth and early twentieth centuries. The trainee programme for the candidates of the Indian Forest Service, for example, shifted its focus from Continental Europe to Britain when a new school of forestry was opened in Coopers Hill in the 1880s.[31] Initiatives by Prussian and Hanoverian diplomats to bring German scientists to the Royal Botanic Gardens in Kew and other institutions of British imperial science also decreased in the last decades of the nineteenth and early twentieth centuries.[32]

On the other hand, however, Anglo-German cooperation continued in many fields and in manifold ways in the years before the First World War.[33] Germans could still be found in the staff of the botanical gardens, natural history museums, and other institutions of science in the British world. They were present in the new science departments of British universities, and they were appointed to professorships in the universities of Australia and New Zealand. In the Indian Forest Service, Germans, like for example Berthold Ribbentrop, remained in high-ranking positions. The forestry school in Coopers Hill was founded by the German-born forest official William Schlich.[34]

As cooperation continued, so did the national rivalry that had always accompanied Anglo-German joint ventures in the context of scientific imperialism. Not only in the decades before the First World War but throughout the nineteenth century did conflict and national prejudice flare up when Germans participated in important British overseas expeditions or were appointed to posts in the colonial administrations. British scientists accused the Germans who joined British expeditions of keeping the results of their research from the British side and, instead, sharing their findings with their German compatriots first.[35] They complained that Germans would benefit from opportunities that were denied to British scientists who were equally qualified to participate in overseas expeditions or hold offices in the British colonial administrations.[36] In the decades before the First World War, German scientists who had been in British services also came under attack from German imperialists. Carl Peters, for example, argued that the generation of German scientists of the middle of the nineteenth century had been too focused on their scientific interests and had neglected their tasks as civilisers and state builders.[37]

However, such outbursts of national hostility never completely undermined transnational cooperation. Even when Anglo-German political relations deteriorated before the First World

War, many German scientists were still well integrated into the established elites in the British world. They were married to British women and received decorations from British learned societies for their scientific achievements. In the summer of 1914, when the First World War had already begun and many students and young academics in Britain and Germany left their universities for the battlefields in Western Europe, the British Association for the Advancement of Science held one of their regular meetings in Australia. Following a long-standing tradition, scientists from many different countries participated in the conference, and the German participants were treated with professional respect. The late Ferdinand von Mueller's contributions to Australian botany were acknowledged.[38] A German-born physicist, Arthur Schuster, was appointed as president of the association. In their diaries and memoirs, the German participants highlighted that many British peers expressed hope that international scientific cooperation would continue after the war. They also wrote, however, that they were treated as "enemy aliens" in Australia and on their arrival in Britain.[39] Although Anglo-German political relations had collapsed, scientific cooperation still continued to a certain extent at a global level. The scientists who helped to realise the aims of British and German imperialism pursued their own scientific interests, which did not always follow the chronologies of national politics.

Conclusion

The analysis of the role of German scientists in the British Empire shows the ambiguities of the relationship between science and imperialism. In the nineteenth century, German Humboldtians were attracted to the British Empire because it offered opportunities to explore the non-European world that the German confederation of states could not provide, at least not until the end of the nineteenth century. The British academic and political elites needed the personnel to pursue their imperial project and therefore invited German scientists to join their overseas expeditions and appointed them to high-ranking posts in British colonial administrations.

In this way, Germans were part of the British subimperialism of science. They provided metropolitan science with information and biota. Many of them were well integrated into British academia and society. At the same time, they participated in transimperial networks that undermined the core–periphery hierarchies of British imperial science. Furthermore, they established German–German intercolonial networks, and they were involved in diasporic exchanges about a German colonial empire throughout the nineteenth century. Although they contributed to imperial knowledge production, their colonial interests and initiatives did not always follow the rhythms of national politics in London and Berlin. At different times and sites of Anglo-German scientific imperialism, they developed their own specific dynamics of know-ledge production, colonialism, and ethnic identity.

Notes

1 R.W. Home, "Science as a German Export to Nineteenth Century Australia," Working Papers in Australian Studies, 104, Sir Robert Menzies Centre for Australian Studies, University of London, 1995, 17–18.

2 See for example these studies from a broad range of biographical material, Moritz von Brescius, *German Science in the Age of Empire: Enterprise, Opportunity and the Schlagintweit Brothers* (Cambridge: Cambridge University Press, 2019); Andrew Wright Hurley, *Ludwig Leichhardt's Ghosts: The Strange Career of a Traveling Myth* (Rochester, NY: Camden House, 2018); Heike Hartmann, ed., *Der Australienforscher Ludwig Leichhardt. Spuren eines Verschollenen* (Berlin: Bebra Wissenschaftsverlag, 2013); Pauline

Payne, *The Diplomatic Gardener: Richard Schomburgk: Explorer and Botanic Garden Director* (Adelaide: Jeffcott Press, 2007); Herbert Hesmer, *Leben und Werk von Dietrich Brandis 1824–1907. Begründer der tropischen Forstwissenschaft, Förderer der forstlichen Entwicklung in den USA, Botaniker und Ökologe* (Opladen: Westdeutscher Verlag, 1975); R.W. Home, A.M. Lucas, Sara Maroske, D.M. Sinkora, and J.H. Voigt, eds., *Regardfully Yours: Selected Correspondence of Ferdinand von Mueller*, 3 vols. (Bern: Peter Lang, 1998, 2002, 2006).

3 See for example Hanno Beck, "Geographie und Reisen im 19. Jahrhundert," *Petermann's Geographische Mitteilungen* 10, no. 1 (1957): 1–14; Franz-Josef Schulte-Althoff, *Studien zur politischen Wissenschaftsgeschichte der deutschen Geographie im Zeitalter des Imperialismus* (Paderborn: Schöningh Verlag, 1971).

4 Moritz von Brescius, "Humboldt'scher Forscherdrang und britische Kolonialinteressen. Die Indien- und Hochasien-Reise der Brüder Schlagintweit (1854–1858)," in *Über den Himalaya. Die Expedition der Brüder Schlagintweit nach Indien und Zentralasien 1854 bis 1858*, ed. Moritz von Brescius et al. (Cologne: Böhlau Verlag, 2015), 31–88; Anja Schwarz, "Schomburgk's Chook: The Entangled South Australian Collections of a German naturalist," *Postcolonial Studies* 21, no. 1 (2018): 20–34.

5 Bernhard C. Schär, "From Batticaloa via Basel to Berlin: Transimperial Science in Ceylon and Beyond around 1900," *Journal of Imperial and Commonwealth History* 48, no. 2 (2020): 230–62.

6 See for example Stephanie Zehnle, "Animal Skinners: A Transcolonial Network and the Formation of West African Zoology," in *Environments of Empire: Networks and Agents of Ecological Change*, ed. Ulrike Kirchberger and Brett M. Bennett (Chapel Hill: University of North Carolina Press, 2020), 151–75; Ulrike Kirchberger, "Temporalising Nature: Chronologies of Colonial Species Transfer and Ecological Change across the Indian Ocean in the Age of Empire," *International Review of Environmental History* 6, no. 1 (2020): 101–25.

7 The somewhat controversial term was introduced by Susan Cannon, *Science in Culture: The Early Victorian Period* (New York: Dawson, 1978). For a more recent interpretation see for example Ian F. McNeely, "The Last Project of the Republic of Letters: Wilhelm von Humboldt's Global Linguistics," *Journal of Modern History* 92 (June 2020): 241–73.

8 Ulrike Kirchberger, *Aspekte deutsch–britischer Expansion. Die Überseeinteressen der deutschen Migranten in Großbritannien in der Mitte des 19. Jahrhunderts* (Stuttgart: Franz Steiner Verlag, 1999), 247–77, 332–8.

9 Stephen Conway, *Britannia's Auxiliaries: Continental Europeans and the British Empire, 1740–1800* (Oxford: Oxford University Press, 2017); Andreas Daum, "German Naturalists in the Pacific around 1800: Entanglement, Autonomy, and a Transnational Culture of Expertise," in *Explorations and Entanglements: Germans in Pacific Worlds from the Early Modern Period to World War I*, ed. Hartmut Berghoff, Frank Biess, and Ulrike Strasser (New York: Berghahn Books, 2019), 79–102.

10 John Davis, "Higher Education Reform and the German Model: A Victorian Discourse," in *Anglo-German Scholarly Networks in the long Nineteenth Century*, ed. Heather Ellis and Ulrike Kirchberger (Leiden: Brill, 2014), 39–62.

11 Robert A. Stafford, *Scientist of Empire: Sir Roderick Murchison, Scientific Exploration and Victorian Imperialism* (Cambridge: Cambridge University Press, 1989), 223.

12 Kirchberger, *Aspekte*, 348–413.

13 Kirchberger, *Aspekte*, 384.

14 For general studies on the transimperial networks of science, see for example Brett M. Bennett and Joseph M. Hodge, eds., *Science and Empire: Knowledge and Networks of Science across the British Empire, 1800–1970* (Basingstoke: Palgrave Macmillan, 2011); James Beattie, Edward Melillo, and Emily O'Gorman, eds., *Eco-Cultural Networks and the British Empire: New Views on Environmental History* (London: Bloomsbury Academic, 2016); Benedikt Stuchtey, ed., *Science across the European Empires, 1800–1950* (Oxford: Oxford University Press, 2005). For the tensions between colonial and metropolitan science, see for example Hilary Howes, *The Race Question in Oceania: A.B. Meyer and Otto Finsch between Metropolitan Theory and Field Experience, 1865–1914* (Frankfurt am Main: Peter Lang, 2013).

15 Ferdinand von Mueller, *Systematic Census of Australian Plants* (Melbourne: M'Carron, Bird & Co., 1882).

16 James Sykes Gamble, Ass. to the Inspector General of Forests, to W. Thiselton-Dyer, Bot. Gardens, Kew, Simla, 16/5/1877; Dietrich Brandis to Thiselton-Dyer, Office of the Inspector General of Forests, Simla, June 1877; Directors' Correspondence DC/153/118, Dietrich Brandis to Joseph Hooker, Simla, 22 July 1877, in Miscellaneous Reports, Victoria (MR/412), Archives Royal Botanic Gardens Kew (hereafter RBGK), London.

17 *Journal of the Agricultural and Horticultural Society of India*, vol. 6, 1882; Monthly Proceedings of the Society, Thursday, the 27th February, 1879, x; Monthly Proceedings of the Society, Thursday, the 22nd May, 1879, xiii, in Agri-Horticultural Society of India, Library, Kolkata.

18 For a history of Buitenzorg, see for example Andrew Goss, *The Floracrats: State-Sponsored Science and the Failure of the Enlightenment in Indonesia* (Madison: University of Wisconsin Press, 2011).

19 Kaiserlicher Landeshauptmann von Togo, Köhler to the Reichskanzler, Fürst zu Hohenlohe Schillingsfürst, 3 January 1896, Eucalyptus in Togo; Kaiserliches Gouvernement von Togo to the Auswärtige Amt, Kolonial-Abteilung, Lome, 26 December 1904, in R 1001/7740, Bundesarchiv Berlin-Lichterfelde (BArch), Berlin; H.E. Gast an das Kolonialwirtschaftliche Komittee in Berlin, 3 July 1914, in R 1001/7736, BArch, Berlin; Adolf Engler, "Bemerkungen über Schonung und verständige Ausnützung der einzelnen Vegetationsformationen Deutsch-Ostafrikas," in *Berichte über Land- und Forstwirtschaft*, ed. Kaiserliches Gouvernement von Deutsch-Ostafrika Dar-es-Salaam, vol. 2 (Heidelberg: Carl Winter's Universitätsbuchhandlung, 1904–1906), 5; Dietrich Brandis, "Anbau der großen Bambusen in Deutschafrika," *Der Tropenpflanzer* 3 (1899): 440; Dietrich Brandis, "Zur Bambuskultur in Deutschafrika," *Deutsches Kolonialblatt* 11 (1900): 473–6.

20 Ian Tyrell, "Peripheral Visions: Californian–Australian Environmental Contacts, ca. 1850s–1910," *Journal of World History* 8, no. 2 (1997): 275–302.

21 Hesmer, 327–84.

22 See for example Dietrich Brandis to Thiselton-Dyer, Simla, 18 August 1877, Directors' Correspondence, DC/153/120, RBGK, London.

23 See for example Robert Hermann Schomburgk, "Remarks on Anegada," *Journal of the Royal Geographical Society* 2 (1832): 152–70; Ernst Dieffenbach, "An Account of the Chatham Islands," *Journal of the Royal Geographical Society* 11 (1841): 195; Berthold Seemann, "Remarks on a Government Mission to the Fiji Islands," *Journal of the Royal Geographical Society* 32 (1862): 50–62.

24 See for example Heinrich Barth, *Reisen und Entdeckungen in Nord- und Central-Afrika in den Jahren 1849 bis 1855 von Dr. Heinrich Barth. Tagebuch seiner im Auftrag der Britischen Regierung unternommenen Reise*, vol. 1 (Gotha: Justus Perthes Verlag, 1857), vii–ix.

25 Ernst Dieffenbach, "Beschreibung der Chatham Inseln," *Warrekauri*, 17–37.

26 Ernst Dieffenbach, "Neuseeland und die Colonisation," *Beilage zur Allgemeinen Zeitung* 191 (July 10, 1846): 1523, 1524; "Promemoria an die Nationalversammlung in Frankfurt," *Der deutsche Auswanderer. Centralblatt der deutschen Auswanderung und Colonisierung* 23 (2 June 1848): 354–8.

27 Robert Hermann Schomburgk, *Geographisch-statistische Beschreibung von Britisch-Guiana, seine Hilfsquellen und seine Ertragsfähigkeit, der gegenwärtige Zustand der Kolonie und deren Aussichten. Aus dem Englischen von O. A. Schomburgk* (Magdeburg: Schmillinsky, 1841), vi, vii; Heinrich Barth to Karl Sieveking, 9 July 1846, in *Heinrich Barth. Er schloss uns einen Weltteil auf. Unveröffentlichte Briefe und Zeichnungen des großen Afrika-Forschers*, ed. Rolf Italiaander (Hamburg: Pandion, 1970), 144, 145; Barth, *Reisen und Entdeckungen*, I: vii–ix.

28 Friedrich Nippold, ed., *Christian Carl Josias Freiherr von Bunsen. Aus seinen Briefen und nach eigener Erinnerung geschildert von seiner Witwe*, vol. 2 (Leipzig: Brockhaus, 1869), 112, 244, 245, 344.

29 Georg von Neumayer, "Ein Vortrag im Deutschen Verein in Melbourne, Victoria (1861)," in *Auf zum Südpol! 45 Jahre Wirkens zur Förderung der Erforschung der Südpolar-Region 1855–1900* (Berlin: Vita Deutsches Verlagshaus, 1901), 19–33.

30 Ferdinand von Mueller to August Petermann, 26 September 1865, 26 November 1865, 6 November 1870, 28 February 1872, and May 1874, in *Die Erforschung Australiens: Der Briefwechsel zwischen August Petermann und Ferdinand von Mueller 1861–1878*, ed. Johannes H. Voigt (Gotha: Justus Perthes Verlag, 1996), 74, 75, 98, 99, 109, 121.

31 S. Ravi Rajan, *Modernizing Nature: Forestry and Imperial Eco-Development 1800–1950* (Oxford: Oxford University Press, 2006).

32 See the numerous letters in German Letters, 1841–1855, RBGK, London, compared to the fewer number for the years before the First World War, in German & Austrian Letters, 1901–1914, RBGK, London.

33 See for example the exploration of Antarctica around 1900, the international dimension of which is highlighted by Pascal Schillings, *Der letzte weiße Flecken. Europäische Antarktisreisen um 1900* (Göttingen: Wallstein Verlag, 2016).

34 Ulrike Kirchberger, "Deutsche Naturwissenschaftler im britischen Empire: die Erforschung der außereuropäischen Welt im Spannungsfeld zwischen deutschem und britischem Imperialismus," *Historische Zeitschrift* 271 (2000): 639–42.

35 Howard Saunders to Norton Shaw, 2 April 1857; Keith A. Johnston to Howard Saunders, 12 June 1854; August Petermann, African Discovery: A Letter Addressed to the President and Council of the Royal Geographical Society of London, 1854, in Archives of the Royal Geographical Society, London.

36 Sir John Kirk to Joseph Hooker, 5 November 1882, Kilimanjaro Expedition, East Africa Archives, Kilimanjaro Expedition, RBGK, London; Georg von Neumayer to Justus von Liebig, Melbourne, 14 September 1857, in R.W. Home and Hans-Jochen Kretzer, "The Flagstaff Observatory Melbourne: New Documents Relating to its Foundation," *Historical Records of Australian Science* 8 (1991): 234, 235.

37 Carl Peters, *Afrikanische Köpfe. Charakterskizzen aus der neueren Geschichte Afrikas* (Berlin/Vienna: Ullstein, 1915), 18, 19.

38 *Report of the Eighty-Fourth Meeting of the British Association for the Advancement of Science, Australia, July 28–August 31, 1914* (London: John Murray, Albemarle Street, 1915), 560.

39 Albrecht Penck, *Von England festgehalten. Meine Erlebnisse während des Krieges im britischen Reich* (Stuttgart: Engelhorn, 1915), 25.

14

IBERIAN SCIENCE, PORTUGUESE EMPIRE, AND CULTURES OF INQUIRY IN EARLY-MODERN EUROPE

Hugh Cagle

In Lisbon, on a sunlit day in June of 1515, in a yard alongside the Casa da Índia, an Asian elephant and a rhinoceros from Cambay squared off to the delight of King Manuel I, his court, and a festive crowd. To believe the engraving that has survived, the duel was vigorous and heated. In one moment, the elephant bent its left foreleg, lunged with its right, and drove a tusk at the head of its stocky opponent. The rhino dropped low. The tusk flew wide of its mark. It would not be a match to the death—not, at least, on this occasion. The spectacle ended when the elephant fled the yard, plunged headlong into the narrow streets of the imperial capital, and found its way northward, back to the calm of the royal stables in Rossio. The creature was probably in good company. More than a dozen elephants ambled ashore across Lisbon's wharves in the sixteenth century. On ceremonial occasions, Dom Manuel would mount his favourite pachyderm to lead a procession from the Paço da Ribeira up the winding streets to Lisbon's cathedral.[1] It was a fitting performance for a ruling family that claimed dominion over "either side of the sea in Africa" and styled itself "Lord of Guinea, and of the Conquest, Navigation, and Commerce of Ethiopia, Arabia, Persia, and India."[2]

At the convergence of global networks of exchange and accumulation, Lisbon in the sixteenth and early seventeenth centuries swelled with the rare, the exotic, and the marvellous. Finely carved African ivory, silken Indian headdresses, brightly glazed martabans from Pegu, Chinese porcelain, Japanese armour, and silver-ornamented coconuts from the Maldives filled shops along Rua dos Mercadores.[3] Civet cats scampered across the grounds of the royal menagerie. Baboons scaled its trees. Grey parrots from Guinea, parakeets from South Asia, and macaws from the far side of the Atlantic all spread their wings in the aviary of the Alcaçova palace—their plumage spanning the rainbow from luminous yellows to regal blues and reds.[4]

Gardens, meanwhile, greened with the leaves of exotic flora. Bananas and plantains from Guinea, ornamental plants from the Konkan Coast of South Asia, and tobacco from the Americas grew on the estates of imperial ministers, royal factors, and returned colonial governors.[5] Abbeys and apothecaries alike stocked tamarind and senna from West Africa, along with Asian *materia medica* ranging from amber to zedoary. Dispensaries sold them to the sick. Infirmaries served them to the poor. Contrary to Sidney Mintz's influential findings, in early-sixteenth-century

Portugal, the spoils of empire delighted the senses, filled the bellies, and fortified the souls of even the kingdom's unlikeliest subjects.[6] In Portugal, as across much of early-modern Europe, new institutions emerged to variously manage, contain, and study the booming traffic in spices, medicines, objects, and animals. Cabinets of wonders filled. Collections of curiosities multiplied. An overseas world that seemed unaccounted for in the texts of antiquity materialised across early-modern Europe. Partly as a consequence, the authority of texts began to ebb while the authority of evidence from experience and the senses grew, helping give rise to new visions of the world and new philosophies of knowledge about that world.

In ways such as these, the empires of Spain and Portugal helped propel the social, cultural, and intellectual transformations now collectively termed the Scientific Revolution. Yet that is not how this history has often been told. Instead, scholarship on the history of science has long tended to sideline, downplay, and ignore—or simply condescend to—the history of natural inquiry in Spain, Portugal, and their empires. Science in the Iberian world, scholars have often argued, was hamstrung by Inquisitorial censorship, lacked an institutional basis, and was plagued by dogmatic adherence to the authority of texts and ancient authors. Much of this conventional account dates to the middle of the eighteenth century, when it was built out from the Black Legend, an image of still older vintage that portrayed Iberians, and especially Spaniards, as violent and rapacious, intellectually brutish, and mired in religious obscurantism.

It was against the lasting influence of the Black Legend that late-twentieth-century scholars began to develop a framework for reassessing the contributions of what came to be termed Iberian science. And it was precisely the story of institutional development, the proliferation of *Kunst-* and *Wunderkammern*, and the ontological and epistemic shifts they accompanied within Spain and Portugal, which the concept of "Iberian science" was meant to highlight.

This scholarship was a long time in coming. Scattered works by historians such as Charles Boxer, Simon Schwartzman, David Goodman, and the sociologist John Law appeared in the middle and late twentieth century.[7] In 1983, in a foundational survey of colonial Latin America, James Lockhart and Stuart Schwartz noted the importance of naturalists and learned societies to the region but stressed that these remained as yet little studied.[8] All of these works built on large and sophisticated Spanish and Portuguese literatures in the history of science. But these enjoyed only limited circulation beyond Iberia, in part because they were rarely translated. Then, in 2004, that scholarship was influentially brought to the attention of a wider audience by Jorge Cañizares-Esguerra, who provocatively asked "how much longer" Iberian science would be ignored by mainstream, largely anglophone, historians of science.[9]

A collective research endeavour had begun to crystallise. Historians Víctor Navarro Brotóns in Spain and William Eamon in the United States assembled a multinational group of scholars for what turned out to be a formative conference in Valencia in 2005 and that led in turn, in 2007, to a dual-language (Spanish and English) volume titled *Más allá de la Leyenda Negra: España y la Revolución Científica* (*Beyond the Black Legend: Spain and the Scientific Revolution*).[10] Conferences and publications on the topic have since proliferated.[11]

The scholarship on Iberian science has not only demonstrated the ways in which Spanish and Portuguese seaborne endeavours revealed new terrestrial and maritime theatres of empire. It has illustrated the ways in which they too vitally participated in the intellectual, acquisitive, and investigative practices that would help define the new sciences of the seventeenth and eighteenth centuries. Ongoing scholarship inspired by the Iberian science framework has led to other important shifts in historical understanding as well. Challenging the dominant narrative of a single, coherent, and unified scientific revolution, they have revealed a more plural history of epistemic transformation—one that embraces and explores numerous disparate and innovative investigative programmes. They have highlighted the importance of contingency and improvisation to the history of investigative method. They have drawn attention to the ways in which race and gender helped condition knowledge. And they have disclosed colonial histories in which the enslaved and other marginalised inhabitants of the Iberian world fashioned the vernacular epistemologies that not only sustained colonial communities from Asia to the Americas for centuries, but that also patterned the development of theories and practices of knowledge within metropolitan Europe.[12]

At the outset, however, architects of the Iberian science framework had narrower and more specific concerns. Countering the legacies of the Black Legend often meant not only highlighting the contributions to investigative theory and practice made by Spaniards and Portuguese at home and abroad but also showing that they (rather than their northern European counterparts) were the earliest and most forceful protagonists of early-modern intellectual transformation. Iberian science scholars sought not to call into question the very idea of a Scientific Revolution as to place Spaniards and Portuguese in its vanguard—as pathbreaking advocates of the evidence of the senses over texts, of experimentalism and inductive reasoning over ancient authority and the deductive logic of the syllogism.

More recently, a number of scholars from a range of disciplinary backgrounds—including many of those who helped establish it—have begun to offer critiques of the Iberian science framework. They have variously challenged the elitism, triumphalism, nationalism, eurocentrism, and teleology implicit in the original approaches.[13]

In what follows, I build on those critiques. Yet my concern here lies less with the conceptual underpinnings of the Iberian science approach than with the substance of some of its claims. If a foundational insight of contemporary science studies holds true—if natural knowledge is a product of the particular networks that enable it, the entangled interests that focus investigative attention and motivate collaboration, and the distinct institutional arrangements that sustain those efforts—then there is every reason to examine more closely the claims that the Iberian science framework makes about two sets of relationships. One is the set of relationships it envisions between Iberia and other parts of Europe. The other is the set of relationships it imagines to have taken shape within Iberia, between the investigative projects of metropolitan Spain and Portugal.

The Iberian science framework embraces a vision of epistemic culture in which procedures for the production of knowledge were more or less unified within Iberia, and which distinguished Iberia from northern Europe. By sketching connections and comparisons largely within sixteenth- and seventeenth-century Europe, I make the case, first, that this vision exaggerates differences between Iberia and northern Europe and, second, that it elides important differences in the ways in which natural inquiry was organised within metropolitan Spain and Portugal. Iberia was more similar to Europe, and metropolitan Spain and Portugal were more different from one another, than the Iberian science framework recognises. I argue for a more expansive, kaleidoscopic perspective in which metropolitan Spain, Portugal, and their empires were all integral parts of a wider cultural and intellectual world that encompassed much of early-modern imperial Europe—and in which a range of heterogeneous and variously theorised epistemic practices coexisted, circulated, and intermittently jostled for adherents and authority. In contrast to a great deal of comparative work on Iberian empires, I also suggest that important differences between the cultures of inquiry in metropolitan Spain and Portugal are best understood not as expressions of essential cultural difference or idealised imperial models, but as consequences of institutional arrangements that took shape in response to the contingencies of early-modern colonialism.

An exploration of the similarities and differences between metropolitan epistemic cultures not only highlights the heterogeneity and entanglement of early-modern ways of knowing even within early-modern Europe, but is a reminder, as well, of the inadequacy of the national and imperial boundaries so often used to describe them and trace their histories. If metropolitan ways of knowing can be understood as more varied and less monolithic, then it becomes easier to discern their entanglements with the ostensibly distinct, colonial epistemologies that they are so often imagined to have isolated, overtaken, and vanquished.

Defining Iberian science

The constituent features of the Iberian science perspective and the historical trajectory envisioned by many of its practitioners can be sketched easily enough. Religion and royal authority are understood as central. The Iberian vision of science and empire was at once patrimonial, messianic, and providential: the arrival of the Portuguese to Asia and of Spaniards to the Americas seemed evidence to Iberian Catholic leaders that their endeavours enjoyed divine favour, and would usher in a new era of world-historical transformation. Those values in turn suffused natural inquiry. As Juan Pimentel has written,

> the organization and development of scientific activity was marked by a determination to build an empire [that achieved both] a real and symbolic appropriation of the world, whereby knowledge gleaned from nature in the context of baroque, courtly, and Catholic society would serve the mission that Providence had reserved for the monarchy.[14]

The leaders of the two Iberian empires sought the appropriation of nature both symbolically (such that it became an emblem of political power and courtly splendour) and instrumentally (such that it became a productive resource and hence the basis of material wealth and economic power).

Such a vision was operationalised through a shared institutional infrastructure. The Casa de Contratación in Seville and the Casa da Mina in Lisbon collected and ordered reports of lands and peoples overseas, and compiled cartographic information to create atlases of their respective empires. Standardised formularies disciplined this labour and produced commensurate bodies of knowledge about otherwise disparate overseas locations. The movement of personnel back and forth between these institutions reinforced their common culture of inquiry and shared techniques.

Inter-imperial rivalry and a shared patrimonial ethos meant that secrecy and symbols of possession became common priorities. This was a culture of inquiry that eschewed publication and other generalised forms of participation in the production of new knowledge. The sovereign was the steward and final authority over newly encountered nature and peoples, and the final arbiter of knowledge about those same places. Through appointees to royal institutions, the Crown authorised and sustained epistemologies fashioned to make sense of the previously unknown. The aim was not to amplify the production or spread of new knowledge, but instead to render a carefully crafted vision of the widening extent of Iberian Catholic sovereignty over newly disclosed worlds.

At the same time, in the wake of arrivals in coastal Africa, Asia, and the Americas, Spaniards and Portuguese overseas grappled with the tensions between the known and the novel. Far from home and bereft of familiar analogues, they variously embraced empirical and utilitarian epistemologies to make both sense and use of the strange things they found (avocados, allspice, and manatees in South America; neem and hooded cobras in South Asia) in lands that otherwise remained distinctly unfamiliar. A voracity for things unfamiliar and a penchant for epistemic improvisation combined with an institutionalised and utilitarian approach to the production of natural knowledge in the service of empire are taken as hallmarks of Iberian science.

News of the initial achievements of Iberian science—its new institutions and its improvised epistemologies—ricocheted through metropolitan Europe. Spanish and Portuguese rulers, their emissaries, their imperial investigative institutions and techniques acquired epistemic and cultural authority. Imperial and epistemic successes, together, inspired similar programmes elsewhere in Europe.

But that changed, in the Iberian science account, beginning in the seventeenth century. Iberian institutions and investigative pursuits persisted but, especially after the end of the union of the Crowns (Portugal and its empire were part of the Spanish Habsburg world from 1580 to 1640), the dream of an Iberian Catholic universal monarchy seemed increasingly untenable. Meanwhile, amid the political reconfigurations of France and England (the dramatic expansion of the French imperial state under Louis XIV; the upheavals of the English Civil War and its consequences), new institutions with novel publishing programmes and more expansive communities of participants took shape (the Royal Society in England and the Académie Royale in France), and new forms of learning about the natural world gained prominence (experimentalism, mechanical philosophy, and the mathematisation of natural phenomena). These variously constituted the "new philosophy." To quote Pimentel again, within Iberia "the forms of science being promoted [secrecy, symbolic appropriation, and the centralization of authority on matters of truth about nature] remained attached to forms of representation [a divinely sanctioned universal monarchy] that had lost credibility."[15] Within metropolitan Europe, institutional and epistemic authority had shifted decisively away from Iberia.

Changes within metropolitan Europe were accompanied by the redistribution of expertise across European empires. The weakening of the Iberian Crown as a consequence of the dissolution of the dual monarchy led as well to a decline of patronage of metropolitan investigative institutions and the centre of investigative gravity shifted to the Iberian colonial world. Influence over investigative programmes accrued to colonial religious orders, viceregal administrations, and creole elites.

Cultures of inquiry across early-modern Europe

The political, epistemic, and institutional transformations highlighted in that narrative are of course real and important, and reflective of regionally differentiated approaches to the production of knowledge about nature. However, cultures of collection and inquiry across early-modern Europe were more thoroughly interconnected; investigative sensibilities—especially an emphasis on the inspection of particulars and an appreciation for subtle differences—were more widely shared; and the reasons for challenging textual authority and the works of ancient authors were more varied than the Iberian science perspective allows.

Of course, it was not only in Iberia but across western Christendom that the exchange and display of nature performed the work of patronage, symbol, and emblem—and provided the very stuff of investigation. Indeed, it was for those reasons that much of what came to Lisbon in the sixteenth and seventeenth centuries travelled onward. The victorious rhino of the *terreiro do paço* would be immortalised in Dürer's well-known engraving before the creature was gifted to Pope Leo X and sank with its ship—clad in felt and gold chain—while crossing the Mediterranean. Live parrots from Brazil burst from meat pies at a royal dinner in Brussels. "Liquid amber" from Asia found its way into the collection of Francesco Calzolari of Verona. Unicorn horns from Africa and Asian bezoars embellished Roman homes (a range of *naturalia* including rhinoceros horns and narwhal tusks circulated under the designation of unicorn horns in this period). Tupi featherwork sent from Jesuits in Brazil appeared in Milan in the collection of the cleric Manfredo Settala. And the German Jesuit Athanasius Kircher kept the large bill of a Brazilian toucan.[16]

Family connections often sustained these interlinked metropolitan itineraries and the shared epistemic sensibilities that underwrote them. For Portugal, perhaps nowhere was the breadth of these connections more apparent than in networks that resulted from the marriage of Dom Manuel's successor, D. João III, to Catherine of Austria. This drew together Portugal's House of Aviz and Habsburgs spread between Lisbon, Madrid, Vienna, Prague, Brussels, Innsbruck, Graz, and Munich.[17] The queen herself was Lisbon's foremost collector of rarities from gilded snakeheads to scorpions' tongues (petrified shark teeth, as it turned out). In Europe both above and below the Pyrenees, material collections of nature variously served ends practical, political, spiritual, investigative, and philosophical.[18]

So, too, did patrimonial, spiritual, and messianic sensibilities pattern attitudes toward overseas peoples and nature not only among Iberians but among Europeans generally. Catholics and Protestants alike understood travel and discovery in terms that were deeply religious and often tied to national destinies. They had in common the inheritance of a crusading ideology that legitimated colonial violence committed in the service of the Crown and Christianity. Colonists from both Iberia and northern Europe alike conceived of New World peoples in particular as ensnared in Satanic misdirection; they understood overseas flora and fauna as elements of a natural world that was often as numinous as it was unfamiliar and threatening. Scholars and naturalists of all nations sought to reconcile unanticipated encounters overseas with both prevailing Biblical interpretations and with knowledge drawn from ancient texts.[19]

While colonial cultures of natural inquiry overseas were not reducible to the epistemic tools and preoccupations of metropolitan contemporaries,[20] colonial projects were nevertheless presented to patrons in Europe in ways that were carefully calibrated to appeal to widely shared metropolitan sensibilities. When Ambrósio Fernandes Brandão, the New Christian planter and author of a manuscript now published under the title *Dialogos das grandezas do Brasil*, sought the patronage of an Austrian nobleman in Lisbon at the turn of the seventeenth century, he framed his inquiries much as the German natural philosopher Johann Joachim Becher (1635–82) would do some decades later in Vienna, as councillor to the Holy Roman Emperor Leopold I: by positioning himself as an intermediary between the landed wealth of the titled nobility and the commercial wealth of overseas empire. Sugar plantations were, according to Brandão, like the titled estates of the northern Habsburg realm—bulwarks of communitarian prosperity and producers of food and goods set against the individualism of merchant capitalists whose speculative investments created nothing but more money. Never mind the high-risk loans that financed sugar plantations and the tenuous nature of their possession. The sugar planter from Pernambuco understood perfectly well the cultural world of an Austrian Habsburg and the kinds of solicitations required of him during his audience in Lisbon.[21]

That Columbian and other overseas encounters influenced a range of epistemic practices in Europe beginning in the sixteenth century is foundational to the Iberian science perspective. And indeed, historians of science focusing on many regions of early-modern Europe have compellingly shown this to have been the case using a range of archival materials. But the claim that Iberian endeavours overseas uniquely propelled epistemic transformations requires substantial qualification. Both the timing and the extent of Iberian influence have long been open to question. In response to what Anthony Grafton would call the "shock of discovery," J.H. Elliott argued that the construction of the Columbian encounters as a seismic cultural and intellectual event took decades.[22] More recent scholarship has shown that other more or less contemporaneous economic and cultural processes were also vitally important.

Perhaps the most salient of these other transformative influences was the intensification of global trade. Both an increasingly varied gift economy and an emergent merchant capitalism demanded more discerning taste from their participants. Gifting and especially the profitable buying and selling of exotica (mace and nutmeg, say, or silk) demanded a keen eye for subtle differences between seemingly similar kinds of vegetable goods or manufactures. Knowledge of those differences required a direct somatic encounter. Commoditisation, in other words, propelled empiricism.[23]

Recent studies have shown as well that Old World encounters not only in Africa and Asia but much closer to home also compelled the attention of European eyes, hands, mouths, and noses. Here the story is of shifts in medical education and the emergence of a multinational, polyglot, and multi-confessional community of naturalists within sixteenth-century Europe. By the early sixteenth century the writings of Theophrastus and Dioscorides had become standard references for medical students throughout Europe and pharmacy (and, along with it, a knowledge of medicinal plants) was becoming more integral to the curricula of university-trained physicians. Of course, these ancient authors focused their texts on the plant and animal life surrounding the Mediterranean. But as students of medical faculties such as that at the University of Leiden ventured into the countryside trying to identify northern European plants with Mediterranean descriptions, it became evident that even within Western Christendom the catalogue of nature was incomplete. The inadequacy of ancient observations grew more apparent with the intensification of medical travel. Students from throughout western Christendom who travelled to such medical faculties as those of Padua or Bologna for their degrees might not have observed discrepancies between texts and observations. But as those same students returned to Lisbon, Antwerp, or Frankfurt, they began to recognise that the flora they learned as students differed from the plant life of the regions in which they lived and practised. New descriptions were needed and generations of sixteenth- and seventeenth-century naturalists—some with formal schooling but many without it—began to compile their own sprawling compendia based on observations of their own and of a widening community of practitioners.[24] Here again, the particulars of things, their significance, and their uses gained salience quite apart from Iberian or other encounters overseas.

Imperial arrangements and the location of expertise

Less evident but perhaps more important, and almost certainly more contentious, are differences in the institutional arrangements through which the investigation of nature took place in Spain, Portugal, and their empires. While the notion of a common Iberian science surely captures a common patrimonial and evangelical approach to empire—one that served both utilitarian and symbolic appreciations of the natural world—that did not mean that metropolitan Spain and Portugal shared a common culture of inquiry, common objects of inquiry, or organised that inquiry in the same ways. Both empires located the accumulation of information, and the production, management, and redeployment of natural knowledge within metropolitan institutions. And the development of those institutions was closely linked—even to the point of sharing personnel. But in both cases, on-the-ground contingencies overseas gave rise to particular kinds of empire, inspired distinct institutional arrangements, and located expertise in idiosyncratic ways.

In Seville, the Casa de Contratación not only managed trade by, among other things, levying tolls and collecting taxes, but also accumulated information on winds, tides, currents, and coasts to create and then improve sailing instructions and sea charts. Beginning in 1508 with Amerigo Vespucci—recently returned from his voyages to Brazil in the employ of Portugal's King Manuel

I—Ferdinand of Spain appointed the first *piloto mayor* to instruct and examine pilots on the use of quadrants and astrolabes. Not two decades later, in Madrid in 1524, the Consejo de Indias came into being, and it was by licence of the Consejo that unfamiliar New World *naturalia* was commercialised in Spain. In the service of tighter administration, the Consejo compiled natural histories by men like Gonzalo Fernández de Oviedo on the climate and geography of the Americas. They dispatched in 1571 Francisco Hernández to provide a survey of New World nature, and through the person of Juan López de Velasco orchestrated, among other projects, the *Relaciones Geográficas* in 1577.[25]

A similar institutional infrastructure took shape in metropolitan Portugal. The centralisation of commercial and navigational activity took place with the Casa da Guiné and the Armazém da Guiné, both of which had their origins in institutions in Lagos but had in the early 1480s—with the construction of the São Jorge da Mina castle on the Gold Coast—been transferred to the square just behind the wharves in Lisbon. By 1510 the Casa da Guiné was referred to as the Casa da Índia. It enforced royal trade regulations, dispatched and received ships and trade goods, and levied and collected taxes. Officials there also attempted to prevent fraud and the counterfeiting of Asian spices. Damião de Góis recalled one such occasion when the sweet smell of a counterfeit spice—wrongly shipped to Lisbon as cinnamon from Ceylon—was set ablaze in the Terreiro do Paço. The Casa was not only a clearing and counting house that channelled textiles, copper, ivory, slaves, drugs, and spices into and out of Lisbon. It also assured the accuracy of the massive scales used for trade at the Mina castle, which were periodically disassembled, packed up, shipped home, recalibrated, and dispatched anew to the Lower Guinea Coast.[26]

Those were not the only mechanical calibrations that underwrote the empire. A parallel institution, the Armazém da Índia, tended to the navigational instruments and cartographic demands of the empire. As early as 1494—with the appointment of the ageing Bartolomeu Dias as *recebedor* (now a decade after his rounding of the Cape)—the Armazém had begun not only to ensure pilots had the skills and instrumental familiarity for open-water sailing; it had also begun to instruct them on how to correct sea charts provided expressly for that purpose. Those corrected maps formed the basis of global views, such as the Cantino planisphere. When in 1547 King João III formalised instruction of pilots with the appointment of the mathematician Pedro Nunes to the post of *cosmógrafo-mor*, it was attached to the Armazém.[27]

Yet those similarities obscure as much as they reveal. The Portuguese Casa da Índia may have been the model upon which the Spanish Casa de Contratación was built.[28] But Spanish and Portuguese imperial institutions developed along divergent trajectories and undertook what were often dramatically different imperial investigative projects. Not until at least the eighteenth century did the Portuguese Crown undertake a project like that of the *Relaciónes Geográficas*.[29] In Portugal, occasional royal inquiries did generate inventories of drugs and other *naturalia* from the *Estado da Índia*. Two of the earliest surviving lists are those of the apothecary Tomé Pires in 1516 from Malacca and of apothecary Simão Alvares in 1548 from somewhere along the Konkan Coast. But such reports on overseas nature appear always to have been drafted in direct response to discrete enquiries, rather than as part of concerted programmes for the accumulation of natural knowledge. Reportage on Brazil was similarly ad hoc but took much longer to materialise. They include the accounts of Pero de Magalhães Gândavo's *História da província de Santa Cruz* (1576) and the *Tratado descritivo* by Gabriel Soares de Sousa (1587). Individual authors wrote accounts of overseas flora, fauna, and native peoples if and when they saw fit to do so, and for their own purposes. In the rare event that they did put pen to page, that work found its way back to metropolitan Portugal in circumstances that remain unknown. And of those, the works that were collected into an imperial archive appear to have remained—much like the work of Brandão cited above, a manuscript by Francisco de Buytrago on West Central Africa, and the lavishly hand-illustrated catalogue of Amazonian nature by the cleric Frei Christovão de Lisboa—unknown for centuries.[30] Of course, Jesuit missionaries in Brazil—as in the Spanish Americas—were prolific and often controversial writers.[31] But their letters and reports were not produced at the request of the Crown and the imperial state did not formally collect them.

A number of writings from Portugal's empire first found their way into print in Italy (as was often the case with abridged versions of Jesuit letters from Brazil) and northern Europe (as was the case with the Jesuit Fernão Cardim's account of Brazilian flora and fauna), where they and many other unpublished works on Portuguese dominions overseas also appear to have circulated in manuscript (as did the work of Ambrosio Brandaõ, discussed above).[32] Yet while proponents of the Iberian science framework have made much of an ostensible policy of Iberian secrecy, it was, at best, put into practice much more thoroughly and systematically in Spain. At the instigation of the Crown, the Portuguese may have carefully guarded natural knowledge. But in the case of what was presumably some of the most sensitive material—imperial maps—it seems that, as one team of researchers put it, if "an overall policy of silence did exist and did apply to cartography, the least that could be said is that it was completely ineffective."[33]

Nor was it the case that empirical practices were centralised or otherwise institutionalised in the ways they were in the Casa de Contratación. Unlike in metropolitan Spain, no institution in metropolitan Portugal collected, compared, or was commissioned to test unfamiliar *materia medica*. This perhaps fits within the broader pattern of decentralisation within Portugal and long evident in studies of its handling of early Atlantic exploration and settlement: the ivory and slave trades in West Africa and the dyewood trade of Brazil were managed through subcontracts rather than conducted on the royal account. Within Portugal—and, again, rather unlike Spain—medical practice and the provision of *materia medica* to metropolitan pharmacies were rarely the subjects of state oversight. Indeed, although specialists of Spain and its colonies frequently assert institutional similarities, even the *protomedicato* (which oversaw the medical profession in Habsburg Spain) had no parallel in Portugal until well into the eighteenth century and well after the era of Spanish Habsburg rule.[34]

The culture of inquiry that predominated in metropolitan Portugal and that patterned imperial approaches to the study of nature was one in which expertise was delegated to inhabitants throughout the Portuguese colonial world. Some were agents of the Crown. Some were Portuguese abroad. Many were neither of those things. It is a seemingly idiosyncratic geography of imperial knowledge and is easily dramatised by a pair of episodes from the sixteenth century. The first surfaces in an exchange of letters between Dom Manuel I's successor João III of Portugal and the Catholic King of Kongo, Afonso I. Well known is the 1529 letter to Dom João III sent from Central Africa on behalf of D. Afonso, in which Afonso asks for the speedy dispatch of Portuguese physicians. Less well known is that this letter was part of an exchange of letters that year in which D. João III agreed to send physicians but asked Afonso if perchance "there be unicorn horns in your kingdom" and—if the answer were no—whether Afonso "might [nevertheless] know where" João could find some. And if, Dom João added, King Afonso were unsure whether certain horns were indeed those of a unicorn, he encouraged his Kongolese counterpart to convey a few of them northward anyway.[35] The second episode comes from Goa. When in about 1542 the Portuguese physician Garcia de Orta and the apothecary Simão Alvares were queried by then Governor Martim Affonso de Sousa about the true identify of pepper—whether black and white pepper were the same and whether they came from the same plant—the physician and the apothecary disagreed.[36] Citing the usual suspects (Dioscorides, Pliny, Galen, Isidore of Seville, and "all the Arabs"), Orta argued (wrongly) that they were different things and came from distinct plants. Alvares insisted that they were the same thing and came from the same plant, and offered his own experience in handling pepper to make the case.

No one in either of these exchanges worked on the assumption that a standing institution in Lisbon could, would, or—importantly—should sort out such basic issues. Solutions, instead, were variable. The befuddled governor sought clarification from the king and court of Cochin. Alvares, meanwhile, sought verification for his claims from a coterie of apothecaries in Lisbon.[37] And even though Dom João solicited the opinion of Dom Afonso, he also made clear that if the latter were uncertain of the identity of purported unicorn horns, there were specialists he could consult in Portugal as well.

Colonial encounters, imperial infrastructures, and investigative sensibilities

Both of the Iberian kingdoms sponsored exploratory voyages that by the 1490s had set their sights on a notional Asia, lured there in part by its mythic trade in long-known but rare and expensive drugs and spices. Outcomes, however, differed markedly. The Spaniards unwittingly found their way to the Caribbean where, despite their initial hopes, commercial networks did not yield the kinds of lucrative trade goods they had hoped. The Portuguese managed to make their way to India and on to island Southeast Asia, enabling sustained commerce in precisely those things they had hoped to find, with pepper, nutmeg, and cinnamon among them.

These contingencies—rather than idealised imperial models devised in advance of exploration and colonisation[38]—patterned the divergent development of imperial investigative infrastructures. In the Spanish Americas—and especially before the discovery of silver in northwestern New Spain (Zacatecas) and the Andean highlands (Potosí)—what drove natural inquiry was a preoccupation with finding New World substitutes for the Asian trade goods they had hoped to monopolise but could not. Building markets for new, unfamiliar substitutes demanded, first, assembling people and plant matter and, second, adjudicating the viability of such substitutions. The Consejo de Índias came into being to undertake just that translation. Portuguese officials, by contrast, had perhaps good reason to believe that they could do without such institutions. Although in Asia, too, the nature encountered was unexpectedly varied, familiar commodities with well-known uses and existing markets were ready to hand. Similarly, in Portuguese America colonial economies centred on—and imperial revenue derived from—the old and familiar rather than from the new and unexpected. The sustained cultivation of sugarcane, an imported crop, began in the 1530s with imported labour, expertise, and machinery.

Instead of an infrastructure that prioritised the mobilisation of exotic nature, the Portuguese state elaborated one that was purpose-built to focus on the acquisition and marketing of familiar drugs and spices in Asia and, in the Americas, on the perpetuation of a plantation system centuries in the making. This manner of locating expertise in the colonies rather than in Lisbon would prove lasting and become a matter of concern among at least some of the members of the Concelho Ultramarino (itself a metropolitan Portuguese institution inspired by the example of the Spanish Consejo). One councillor, João Bernardes de Morais, summed up these worries in a letter published in the front matter of a book about health and medicine in Brazil. According to Morais, the Portuguese might claim sovereignty over Brazil but the Dutch, with their more thorough knowledge of its flora and fauna (he had in mind the compendious *Historia naturalis Brasiliae*, co-authored by the German Georg Marcgraf and the Dutchman Willem Piso during the Dutch occupation of Pernambuco), more effectively possessed it.[39]

In the late 1670s, the Portuguese diplomat to the court of Louis XIV, Duarte Ribeiro de Macedo, grappled with precisely this idiosyncratic location of expertise when he devised a plan to transplant Asian drugs and spices to Brazil. Macedo described a culture of inquiry that he believed now separated Lisbon from its colonies. Curiosity in metropolitan Lisbon gave rise to wondrous pleasures; its gardens delighted. But what Macedo described as "rustic curiosity" lay, he claimed, in the colonies. Instead of state-led investigations of nature that might allow the transplantation of South Asian flora to South American plantations, he advocated that local "rustics" be transported alongside the plants they knew best.

Expertise on matters of nature was spread across Portugal's colonies overseas rather than assembled into institutions within metropolitan Lisbon. To recognise that arrangement is not to diminish it or to suggest that it was deficient. There is no reason to take the more centralised institutional configurations of sixteenth- or seventeenth-century Spain, France, or England as normative, much less optimal. Observers in metropolitan Portugal did not see the arrangement as disadvantageous until the eighteenth century, when expeditions orchestrated within the wealthier and more powerful British and French Empires made large, expensive, and more centrally administered investigative endeavours seem more efficacious. In both Spain and Portugal, programmes of imperial reform entailed the expansion of state-sponsored science. These reforms—Bourbon and Pombaline alike—are almost always described as a renewal of older efforts to centralise the collection of natural knowledge. But that's not quite right. In the Portuguese case, it amounted to an attempt to install such a programme for the very first time. José Celestino Mutis in colonial Colombia could (and did) compare himself to Francisco Hernáandez two and a half centuries earlier.[40] But Portuguese contemporaries could only set themselves against the Dutch, who, by knowledge, if not by their actual presence on the ground, seemed to have surer possession of Portuguese dominions overseas.

In conclusion, the concept of "Iberian science" has rightly focused scholarly attention and catalysed investigative energy. But it is at once too narrow and too capacious. It fails to capture the very real and meaningful linkages between Spain, Portugal, and the wider world of sixteenth- and seventeenth-century Europe. It simultaneously elides critical differences in the ways in which the two Iberian empire-states orchestrated their investigations of nature. Inhabitants of the two Iberian empires participated in the wide-ranging cultural and intellectual transformations underway in early-modern Europe. But the particular ways in which they did so differed markedly and changed over time as a result of the particularities of unforeseen encounters and the material possibilities of improvised empires.

Notes

1 Annemarie Jordan Gschwend, "A Procura Portuguesa por Animais Exóticos" [The Portuguese Demand for Exotic Animals], in *Cortejo Triunfal com Girafas: Animais Exóticos ao Service do Poder* [Triumphal Procession with Giraffes: Exotic Animals at the Service of Power] (Lisbon: N.p., n.d.), 33–77.

2 Vasco da Gama, *Em Nome de Deus: The Journal of the First Voyage of Vasco da Gama to Índia*, trans. and ed. Glenn J. Ames (Leiden: Brill, 2009), 157.

3 Donald F. Lach, *Asia in the Making of Europe*, vol. 2, *A Century of Wonder. Book One: The Visual Arts* (Chicago: University of Chicago Press, 1970), 10–16; Annemarie Jordan Gschwend, "Catarina de Áustria: Colecção e *Kunstkammer* de uma Princesa Renascentista," *Oceanos* 16 (1993): 62–70; Annemarie Jordan Gschwend, "As Maravilhas do Oriente: Colecções de Curiosidades Renascentista em Portugal" [The Marvels of the East: Renaissance Curiosity Collections in Portugal], in *A Herança de Rauluchantim* [The Heritage of Rauluchantim], ed. N.V. Silva (Lisbon, 1996), 82–127.

4 Gschwend, "A Procura Portuguesa"; Palmira Fontes da Costa, "Secrecy, Ostentation, and the Illustration of Exotic Animals in Sixteenth-Century Portugal," *Annals of Science* 66 (2009): 59–82.

5 According to Clements Markham one Dom Francisco de Castelo Branco, the *camareiro-mor* to João III, kept a banana plant at his "country house." See Garcia de Orta, *Colloquies on the Simples and Drugs of India*, trans. and ed. Clements Markham (London: Henry Southern and Company, 1913), 200, n. 1; Carolus Clusius, *Rariorum aliquot stirpium per Hispanias* (Antwerp: Christopher Plantin, 1576), 131, 254, 280, 289, 299, 444; Rose Standish Nichols, *Spanish and Portuguese Gardens* (London: Constable and Company, 1922), 225–6; and Damião de Gois, *Chronica do felicissimo rei Dom Emanuel* (Lisbon: Francisco Correa, 1566–7), pt. I, ch. 56, 52–52v.

6 Lach, *Century of Wonders*, 1, 11–12; Isabel M.R. Mendes Drummond Braga, *Assistência, saúde pública e prática médica em Portugal: Séculos XV-XIX* (Lisbon: Universitária Editora, 2001); Lisbeth de Oliveira Rodrigues and Isabel dos Guimarães Sá, "Sugar and Spices in Portuguese Renaissance Medicine," *Journal of Medieval Iberian Studies* 7 (2015): 176–96. For a contrary interpretation see Sidney W. Mintz, *Sweetness and Power: The Place of Sugar in Modern History* (New York: Penguin, 1986).

7 C.R. Boxer, *Two Pioneers of Tropical Medicine: Garcia d'Orta and Nicolás Monardes* (London: Hispanic and Luso-Brazilian Councils, 1963); Simon Schwartzman, *A Space for Science: The Development of the Scientific Community in Brazil* (University Park: The Pennsylvania State University Press, 1991 [1979]); David Goodman, *Power and Penury: Government, Technology, and Science in Philip II's Spain* (New York: Cambridge University Press, 1988); John Law, "On the Methods of Long-Distance Control: Vessels, Navigation and the Portuguese Route to India," in *Power, Action and Belief: A New Sociology of Knowledge*, ed. John Law (London: Routledge, 1986), 234–63.

8 James Lockhart and Stuart B. Schwartz, *Early Latin America: A History of Colonial Spanish America and Brazil* (New York: Cambridge, 1983).

9 Jorge Cañizares-Esguerra, "Iberian Science in the Renaissance: Ignored How Much Longer?" *Perspective on Science* 12 (2004): 86–124.

10 Víctor Navarro Brotóns and William Eamon, eds., *Más allá de la Leyenda Negra: España y la Revolución Científica (Beyond the Black Legend: Spain and the Scientific Revolution)* (Valencia: CSIC, 2007). Despite the implications of the title, the volume included work on the Portuguese.

11 The Spring 2015 meeting of the Association of Spanish and Portuguese Historical Studies (ASPHS) held at Johns Hopkins, titled "History of Iberian Science and Medicine," was meant to survey the terrain since 2005 and included many of the participants from Valencia. The network LAGLOBAL, based at the Institute of Latin American Studies at the University of London and led by Mark Thurner, has hosted a number of conferences on Iberian science and related topics since 2016.

12 For a brief survey and references, please see Hugh Cagle, "Objects and Agency: Science and Technology Studies, Latin American Studies, and Global Histories of Knowledge in the Early Modern World," *Latin American Research Review* 54, no. 4 (2019): 976–91.

13 Critical reappraisals were collected in a special issue of *History of Science* themed "Iberian Science: Reflections and Studies," and published in the wake of the 2015 ASPHS conference noted above. See especially the essays by Juan Pimentel and José Pardo-Tomás, "And Yet, We Were Modern: The Paradoxes of Iberian Science after the Grand Narratives," *History of Science* 55 (2017): 133–47; and John Slater and Maríaluz López-Terrada, "Being Beyond: The Black Legend and How We Got Over It," *History of Science* 55 (2017): 148–66.

14 Juan Pimentel, "The Iberian Vision: Science and Empire in the Framework of a Universal Monarchy, 1500–1800," *Osiris* 15 (2000): 17–30, here p. 23.

15 Pimentel, "Iberian Vision," 23.

16 Daston and Park, *Wonders*, 154–5; Amy Buono, "Crafts of Color: Tupi *Tapirage* in Early Colonial Brazil," in *The Materiality of Color: Production, Circulation, and the Application of Dyes and Pigments, 1400–1800*, ed. Andrea Feeser, Maureen Daly Goggin, and Beth Fowkes Tobin (New York: Routledge, 2012), 240, 242; Miguel de Asúa and Roger French, *A New World of Animals: Early Modern Europeans and the Creatures of Iberian America* (London: Routledge, 2019), 175.

17 Almudena Pérez de Tudela and Annemarie Jordan Gschwend, "Luxury Goods for Royal Collectors: Exotica, Princely Gifts, and Rare Animals Exchanged Between the Iberian Courts and Central Europe in the Renaissance (1560–1612)," in *Exotica: Portugals Entdeckungen im Spiegel fürstlicher Kunst- und Wunderkammern der Renaissance*, ed. Helmut Trnek and Sabine Haag (Mainz am Rhein: Philip von Zabern, 2001), 1–127.

18 In addition to the Iberian examples cited above, the topic has been covered extensively for many parts of early modern Europe. See for example Daston and Park, *Wonders*, especially Chapters 6 and 7; Paula Findlen, *Possessing Nature: Museums, Collecting, and Scientific Culture in Early Modern Italy*

(Berkeley: University of California Press, 1994); Brian Ogilvie, *The Science of Describing: Natural History in Renaissance Europe* (Chicago: University of Chicago Press, 2006).

19 Jorge Cañizares-Esguerra, *Puritan Conquistadors: Iberianizing the Atlantic, 1550–1700* (Stanford, CA: Stanford University Press, 2006); Richard Drayton, "Knowledge and Empire," in *The Oxford History of the British Empire*, vol. 2, *The Eighteenth Century*, ed. Peter J. Marshall (New York: Oxford University Press, 1998), 231–52; Lorraine Daston and Katharine Park, *Wonders and the Order of Nature, 1150–1750* (New York: Zone Books, 2001).

20 This is an issue I address in Hugh Cagle, *Assembling the Tropics: Science and Medicine in Portugal's Empire, 1450–1700* (New York: Cambridge University Press, 2018).

21 Cagle, *Assembling the Tropics*, 233–51; Pamela Smith, *The Business of Alchemy: Science and Culture in the Holy Roman Empire* (Princeton, NJ: Princeton University Press, 1994), ch. 4.

22 J.H. Elliott, *The Old World and the New, 1492–1650* (New York: Cambridge University Press, 1970); Anthony Grafton with April Shelford and Nancy Siraisi, *New Worlds, Ancient Texts: The Power of Tradition and the Shock of Discovery* (Cambridge, MA: Harvard University Press, 1995).

23 Pamela H. Smith and Paula Findlen, eds., *Merchants and Marvels: Commerce, Science, and Art in Early Modern Europe* (New York: Routledge, 2001); Harold J. Cook, *Matters of Exchange: Commerce, Science, and Art in the Dutch Golden Age* (New Haven, CT: Yale University Press, 2007).

24 Ogilvie, *Science of Describing*; Alix Cooper, *Inventing the Indigenous: Local Knowledge and Natural History in Early Modern Europe* (New York: Cambridge University Press, 2007).

25 María M. Portuondo, *Secret Science: Spanish Cosmography and the New World* (Chicago: University of Chicago Press, 2009), especially 72–102; Antonio Barrera-Osorio, *Experiencing Nature: The Spanish American Empire and the Early Scientific Revolution* (Austin: University of Texas Press, 2006), 13–28, 35–55; Barbara E. Mundy, *The Mapping of New Spain: Indigenous Cartography and the Maps of the Relaciones Geográficas* (Chicago: University of Chicago Press, 1996), 17–23; David C. Goodman, *Power and Penury: Government, Technology and Science in Philip II's Spain* (Cambridge: Cambridge University Press, 1988).

26 This institution grew out of the Casa de Ceuta in Lagos, then in the mid fifteenth century successively became Casa da Guiné, Casa da Mina, and then by c. 1509 the Casa da Guiné, Mina, e Índias, or simply Casa da Índia. On these and other institutional arrangements, see George D. Winius, *Portugal the Pathfinder: From the Medieval toward the Modern World, 1400–1600* (Madison, WI: Hispanic Seminary of Medieval Studies, 1995).

27 Maria Fernanda Alegria, Suzanne Daveau, João Carlos Garcia, and Francesc Relaño, "Portuguese Cartography in the Renaissance," *The History of Cartography*, vol. 3, *Cartography in the European Renaissance* (Chicago: University of Chicago Press, 2007), 975–1068. A chair in mathematics meant to offer courses in navigation—the *aula da ésfera*—was created at the Jesuit Collégio de Santo Antão in 1559. See Henrique Leitão, *A ciência na Aula da Esfera do Colégio de Santo Antão, 1590–1759* (Lisbon: Comissariado Geral das Comemorações do V Centenário do Nascimento de S. Francisco Xavier, 2007); Luis de Albuquerque, "A 'aula da esfera' do Colégio de Santo Antão no século XVII," *Anais da Academia Portuguesa de História* 21 (1972): 337–91; Bailey W. Diffie and George D. Winius, *Foundations of the Portuguese Empire, 1415–1580* (Minneapolis: University of Minnesota Press, 1977), 316–17.

28 Influence moved in both directions. Not until 1642, shortly after the Union of the Crowns had ended, did João IV create the Conselho Ultramarino—an institution that would be similar in function to the Spanish Consejo de Indias. See Erik Lars Myrup, *Power and Corruption in the Early Modern Portuguese World* (Baton Rouge: Louisiana State University Press, 2015).

29 A classic but somewhat disorganised study of these efforts is that by William Joel Simon, *Scientific Expeditions in the Portuguese Overseas Territories (1783–1808): The Role of Lisbon in the Scientific-Intellectual Community of the Late Eighteenth Century* (Lisbon: Instituto de Investigação Científica Tropical, 1983).

30 Brandão, *Dialogos*; Francisco de Buytrago, *Arvore da vida, Thezouro descuberto da Arvore irmaã daque se fez a cruz da nossa Redempção. Para livrar dos maleficios do Demonio, p.a vida e saude dos enfeitiçados ou vexados do mesmo Demonio, e outras muitas enfermidades E muitos e singulares remedios para muitos achaques aprovado tudo com muitas experiencias prodigiozas, como hé publico e se vera dos daquelles tractados*, Fundo Reservados 437, códice 13114, Biblioteca Nacional de Portugal, Lisbon; Cristovão de Lisboa, "História dos animais e árvores do maranhão," Livros do Maranhão e Grão Pará, códice 1660, Arquivo Histórico Ultramarino, Lisbon.

31 Cagle, *Assembling the Tropics*, ch. 6.

32 Ibid.

33 Maria Fernanda Alegria, Suzanne Daveau, João Carlos Garcia, and Francesc Relaño, "Portuguese Cartography in the Renaissance," *The History of Cartography*, vol. 3, *Cartography in the European Renaissance* (Chicago: University of Chicago Press, 2007), 1007.

34 An excellent discussion of the situation in Portugal is Laurinda Abreu, "A organização e regulação das *profissões médicas* no Portugal Moderno: entre as orientações da Coroa e os interesses privados," in *Arte Médica e Imagem do Corpo: de Hipócrates ao final do século XVIII*, ed. Adelino Cardoso, António Braz de Oliveira, and Manuel Silvério Marques (Lisbon: Biblioteca Nacional de Portugal, 2010), 97–122.

35 Brásio, António, ed., *Monumenta Missionaria Africana: África Ocidental*, series 1, 15 vols. (Lisbon: Agência Geral do Ultramar, 1952–88), 1: 521–39, especially p. 531.

36 The episode is notable since out of dozens of dialogues Orta described in his *Colóquios* (1563), this is the only one that is has appeared in other surviving documentation.

37 Orta's *Colóquios* (2: 248) provides the physician's explanation and his account of Alvares's reply. The document bearing the title "Emformação que me dey symão allũez buticayro mor del Rey noso sõr do naçymento de todelas droguas que vão pera o Reyno o quoal ha XXXIX Anos q serue nestas partes da Imdia seu o ficio home gramdemente curyoso destas cousas" was first transcribed by Jaime Walter, "Simão Alvares e o seu rol das drogas da Índia," *Studia* 10 (1962): 136–49.

38 See the discussion in, for example, Michael Adas and Hugh Cagle, "Age of Settlement and Colonisation," in *The Ashgate Companion to Modern Imperial Histories*, ed. Philippa Levine and John Marriott (Burlington, VT: Ashgate, 2012), 41–74; and Eric Hinderaker and Rebecca Horn, "Territorial Crossings: Histories and Historiographies of the Early Americas," *Willian and Mary Quarterly* 67 (2010): 395–432. For a carefully considered, alternative viewpoint see Sanjay Subrahmanyam, "Holding the World in Balance: The Connected Histories of the Iberian Overseas Empires, 1500–1640," *American Historical Review* 112, no. 5 (2007): 1359–85.

39 João Ferreira da Rosa, *Trattado unico a constituiçam pestilencial de Pernambuco offerecido a el Rey N. S.* (Lisbon, 1694), xxij–xxiij.

40 J.R. Marcaída and J. Pimentel, "Green Treasures and Paper Floras: The Business of Mutis in New Granada," *History of Science* 52, no. 3 (2014): 277–96.

15

THE DYNAMIC TRAJECTORY OF FRENCH COLONIALISM AND SCIENCE

Michael A. Osborne

This chapter examines science in the French colonies in the nineteenth and twentieth centuries.[1] It engages little with the important topics of religion, missionaries, and imperialism but follows in chronology the chapter in this volume by Maria Pia Donato and Sabina Pavone on "Science, empire, and the old Society of Jesus, 1540–1773," which does treat religion and science prior to the French Revolution (Chapter 11). The nineteenth century was an exciting and important time in the history of colonialism, imperialism, science, and technology. Biology, chemistry, medicine, and physics underwent paradigmatic shifts. This was the century of Charles Darwin, the birth of thermodynamics, the extension of railways, and the expanded use of steam ships. It was also the century of cell theory, germ theory, and our modern conceptions of digestion and physiology.

The history of French colonial science is less a narrative of the implantation of French science overseas than an *asymmetric co-evolution* of relationships which grew in concert with two world wars, decolonisation, and uneven postcolonial scientific and technological cooperation, financial assistance, and economic interdependence. All too often, however, the former French colonies now struggle to create and sustain scientific institutions and to nurture sustainable and self-reproducing scientific communities. Many former colonies now find themselves in positions of scientific and technological dependency even as they have sought the assistance and expertise of the international community beyond France.

In 1900 the French colonial empire was second in extent only to that of Britain. This was significant for France itself as well as for peoples colonised by France. France suffered extreme demographic stagnation after World War I because about one in ten of the male population died in the conflict and three times that many came home wounded. By 1930 about 65 million colonial subjects, more than the population of France, lived under the French flag overseas. This is only slightly less than the entire population of France today (2021), estimated at about 67 million. Colonialism's legacy continues as nearly 6 per cent of French citizens trace their origins to the former North African French colonies of Algeria, Morocco, and Tunisia. Thus France's relationships with its colonies and former colonies is a two-way street with multiple reciprocal influences.

In 1950 the geographical extent of the empire was 22 times greater than that of France itself. Military, commercial, and to some extent French scientific societies and scientific interests built the empire. It was also the work of municipal imperialism as French cities and ports such as

Bordeaux, Marseille, and Toulon had commercial and cultural ties with the colonies dating back to before the French Revolution.[2] The army and navy were key actors. The French Revolution and the Napoleonic wars that followed effectively merged advanced engineering and cartography with military activities and left an imprint on French science and the French Empire. At the defeat of the Napoleonic Empire in 1815, the army emerged as the strongest military force. The French navy fulfilled its role as a guardian of the coasts, provided transport to and from the colonies, and provided colonial medical personnel in support of French forces and the few French nationals who settled in the colonies. French mining companies established the first railway in France in 1828 and the rail system expanded during the Second Empire (1852–70). The French exported this important technology to Algeria in 1862 when a line opened between the city of Algiers and the city of Blida to the southwest. Railway development spread among French colonies with natural resources. Cochinchina (later incorporated into Indochina) opened its first tramway in 1881. It connected the Chinese district of Cholon on the Saigon River (now incorporated into Ho Chi Minh City) with the city of Saigon. Another line of about 20 kilometres in length opened in 1885 connecting the shipping lanes of the Mekong Delta at the port of Mỹ Tho with those of Saigon. A similar pattern of railway development can be seen in Senegal in West Africa where railways hauled peanuts and other commodities to the port at Dakar.

French scientists and physicians conducted formal surveys of colonies and potential colonies. Notable surveys of the nineteenth century included those of Napoleon's Egyptian campaign (1798–1801) and a scientific expedition to Algeria (1839–42). French explorers also established a basic cartography for French activities in Asia and sub-Saharan Africa. The French government funded most of these activities and sometimes shared costs with individual colonies. Neither France nor the French Empire were stable entities. Both added and lost territory; in the case of Alsace-Lorraine, France lost and then regained a considerable swath. During the Third Republic (1870–1940), the French Ministry of Public Instruction funded hundreds of overseas study missions. Yet colonial missions to North Africa and Indochina accounted for only 10 per cent of the total expenditure. About 12 per cent of all funding went to medicine, while the physical sciences of geology and astronomy accounted for about 7 per cent. A majority of the money was spent on European missions.[3] These multi-disciplinary teams examined cultural monuments, health conditions, race and ethnicity, languages, the possibilities for colonial botany and agriculture, cartography, mineral resources, and other topics. A few well-to-do explorers and adventurers also produced useful surveys. For example, the wealthy nineteenth-century naturalist Alfred Grandidier (1836–1921) made three largely self-funded trips to the East African island of Madagascar in 1865, 1866, and 1868. After returning to Paris he and his son published 40 volumes on the natural history of the island to establish a scientific baseline still used today.[4] The island became a French protectorate in 1897 and a colony the next year.

Historians of the French Empire often divide the empire into old colonies, in other words those pre-revolutionary colonies such as New France in Canada and holdings in India and the Caribbean, and the new colonies formed after the Revolution such as Algeria, Indochina, and Madagascar. French Saint-Domingue in the Caribbean was by far the largest and most lucrative of France's old colonies. A plantation colony founded on slave labour, it exported to France cane sugar, coffee, indigo, and cotton. In 1790 about 500,000 slaves resided there and together freed people of colour and the white population totalled about 50,000. The other French Caribbean colonies were smaller, with the population of the islands of Guadeloupe and Martinique totalling about 20 per cent and 17 per cent, respectively, of that of Saint-Domingue.[5] French influence in the Americas declined after Saint-Domingue broke away from France in 1804 to become Haiti, a year after France sold the Louisiana Purchase to the United

States. After 1820 only 3 per cent of the populations of the Caribbean colonies were French, and for reasons of space my discussion of this region is brief.[6]

This simple scheme of old and new colonies, while useful, obscures some important historical factors. Sometimes the "old" colonies became "new" colonies because French intervention intensified. This was the case in West Africa where commercial companies chartered by the French crown had been active since the 1630s. The British removed the French from the region as a result of the Seven Years War (1756–63), but France regained title to the area in 1817. In the twentieth century, French West Africa and particularly Senegal became a major source for African troops who policed the empire and defended France in two world wars. The city of Saint-Louis in Senegal became the site of the first sub-Saharan bacteriological laboratory in 1896.

Medicine, science, and the colonies

Not all sectors of French science and technical education were deemed exportable to the colonies. Some colonial institutions conducted "pure" scientific research in astronomy and geophysics but on balance these activities were mixed with practical meteorology and climatology.[7] French colonial scientific activity inclined toward the applied sciences of botany, agronomy, medicine, engineering, and extractive technologies, including mining and export agriculture. There was also, as there was in France and Germany, widespread interest in hydrotherapy at numerous colonial spas as a palliative treatment for tropical fevers and ills contracted in the colonies.[8] Much depended on what was envisioned for the colony in terms of scientific development, the density of French presence, the era the colony entered the French orbit and also its proximity to France. Nearest to France, of course, were the North African colonies and they became important sites of science and medicine and had some opportunities to train for scientific work.

French forces invaded the city of Algiers in 1830 and began a process of colonisation which would last until 1962. The conquest expelled the Ottoman rulers of the country and is notable for savage pacification campaigns by French forces, outbreaks of disease, and famines. Algeria became the major French settlement colony. The relation of a given colony to France is important for the development of its science and cultural institutions and colonies were governed in diverse ways. In the case of Algeria, it was quickly integrated into the French departmental system of governance. This is the case today for the Caribbean colonies of Guadeloupe, Martinique, French Guiana, and for the Indian Ocean island of Réunion. In contrast, New Caledonia in the South Pacific, which is rich in nickel and was also a former penal colony, is part of an overseas collectivity of France and has a complicated political history. In 1906, for example, the French made several islands in the Pacific part of an Anglo-French Condominium (joint-protectorate). During World War II Japan claimed sovereignty over the region but failed. What this means is that some former colonies like Algeria, and present-day Guadeloupe, Martinique, French Guiana, and Réunion, were or now are integrated into France. They are technically part of France and not mere outposts. This integration into France confers some rights to governance and provides for the creation of educational institutions, and it is why, for example, the 118 islands of French Polynesia do not have the right to run their own court system or to create universities. This also means that there are asymmetries of development and science between the various colonies as well as between the individual colonies and France. Algeria itself would be administered by the army until the Third Republic. France later extended rule across North Africa by forming protectorates in Tunisia (1881) and Morocco (1912). When the French colonial system collapsed in the 1950s and 1960s France was ill prepared for the influx of some 800,000 refugees, the so-called Pieds-Noirs, who fled back to France from Algeria.

The military arts were well institutionalised around the Ottoman Empire and the Regency of Tunis adopted aspects of Western science after the fall of Algiers. For example, Ahmad I (1806–55), the provincial governor (Bey) of Tunis after 1837, founded a modern military school, the École Polytechnique du Bardo. He also travelled to France in 1846 on a study and diplomatic mission. His successor Sadoq Bey (1813–82) founded the Collège Sadiki in 1875. This institution offered instruction in foreign languages, mathematics, physics, physical geography, mineralogy, botany, medicine, and other scientific disciplines. After France established its protectorate in 1881, instruction was conducted in both French and Arabic.

The sciences of colonial botany and horticulture were corner stones of French colonisation, and aside from field hospitals and military technology, the first French scientific institution in Algeria was an experimental garden and demonstration farm at Hamma near Algiers. Its foundation in 1832 reveals the plans France had for the colony and the state of French botanical science. Initially, the French hoped to transform Algeria into a producer of agricultural crops, including fruits and spices not grown in France. The Jardin d'Essai at Hamma became the administrative hub of about 20 gardens around Algeria, many of which were staffed by men trained at a horticultural school at the Muséum d'Histoire Naturelle in Paris.[9] The project was bound up with the collective memory of the loss of Saint-Domingue, an anchor of the French triangle trade and the Old Regime supplier of sugar and coffee. But the French underestimated the harshness of the North African climate and the fertility of the soil even as they promoted the idea that North Africa had once been the granary of Rome and French rational agriculture and rule would resurrect the region to its former greatness.[10] In the end, the region became a producer of wine and grain, two products France produced in abundance. In 1900, for example, wine constituted about half of all Algerian exports and constituted about one-third of its gross domestic product.[11] The methods of agricultural science changed too from one infused with Lamarckian ideas of adaptation, acclimatisation, and horticultural "know how" to a more modern experimental approach undergirded by Mendelian genetics and focused on plant selection. In 1908 the French founded the empire's first modern agricultural experiment station at Tunis, and after World War I similar experiment stations replaced the older *jardins d'essais*.[12]

The physician and critic of colonialism Frantz Fanon (1925–61) has pointed to medicine as an important factor in the domination and exploitation of colonised peoples. French authorities regarded medicine as an essential tool of colonisation, although it was not at first intended to serve Indigenous populations. The encounter of Western medicine and colonised peoples continues to animate the history of colonialism, and a number of scholars have written on the "Arab mind," colonial and postcolonial psychiatry, psychology, and the racially coded sciences of anthropology and ethnology.[13] Readers who wish to know more about psychiatry are invited to consult Matthew Heaton's essay in this volume on "Colonial psychiatry" (Chapter 6). About 25 per cent of France's military medical corps accompanied the invasion of Algeria.[14] By 1845 38 French hospitals, not all of them military, dotted the country. Malaria and other fevers were prevalent throughout the coastal region and in 1832 in the eastern Algerian city of Bône (Annaba) two army physicians, Jean André Antonini and François Clément Maillot, distinguished malaria from typhoid fever by experimenting with doses of quinine sulphate and discovered that the drug was an effective treatment for malaria. Later, the French adopted the drug as a prophylactic against malaria. Another notable medical discovery occurred in the city of Constantine in northeastern Algeria. There, in 1880, the French army physician Charles-Louis-Alphonse Laveran (1845–1922) discovered the parasite that causes malaria. He subsequently quit the army to devote himself to medical parasitology at the Pasteur Institute in Paris and in 1907 won the Nobel Prize in Physiology or Medicine.

French medical and scientific education expanded slowly in the middle of the nineteenth century. Following the 1848 founding of the colony's first *lycée*, the Collège d'Alger, the French opened the Algiers Preparatory School of Medicine and Pharmacy in 1857 at a civil hospital with a staff of eight professors and four assistants. It would become a full faculty of medicine in 1909 with the right to issue the MD degree. Although the colony was officially a department of France, graduates did not have the right to practise medicine in France. The first woman to graduate from the medical faculty, Aldjia Noureddine Benallègue (1909–2015) became a distinguished paediatrician. By 1947 the school enrolled more than 1,000 students, of whom 282 were women. In 2016 about 1,000 professors and instructors at the faculty served about 15,000 postgraduate and undergraduate students in medicine, pharmacy, and dentistry. Preparatory schools for law, science, and letters opened in 1880 and as with medicine they would be elevated to full faculty status in 1909. An École des Sciences opened in the 1880s, staffed by four professors and three assistants who taught physics, mathematics, astronomy, agronomy, and zoology. Professors conducted research on local flora, fauna, mineral resources, and the industrial sciences. In 1895, only 30 students followed classes in science while 177 were registered in medicine and pharmacy. Algerian students continued to prefer the study of law over either medicine or science.

The French astronomer, Charles Bulard, who had worked at the Observatoire de Paris, opened an astronomical observatory in Algiers in the 1850s. The observatory relocated in 1889 to a hill in what is now the Algiers suburb of Bouzaréah. In the 1890s the astronomer Charles Trépied directed the observatory. He participated in compiling a new sky atlas and the institution took more than 1,000 solar photographs and conducted spectroscopy research. In an independent Algeria, Bouzaréah is now home to the Algeria Space Agency and cooperates with France, Argentina, India, and China. Algeria is a seismologically active region and in 1906 Ferdinand de Montessus de Ballore published the first seismic study of the region, and shortly thereafter the French created a Service of Seismology.[15]

Other observatories and geophysical institutions were founded in the Sahara and elsewhere, and in 1907 the weather service separated from astronomical activities. Weather services, including climatological work, were common to many French colonies. In Tunisia, for example, meteorological services were organised in 1885. By the 1920s Gaston Ginestous, who held a doctorate from Paris, had organised a network of reporting stations. He and his successor Charles Bois worked on rainfall and hydrology, critical to agricultural needs. In Morocco, the first resident general, Louis-Hubert-Gonzalve Lyautey (1854–1925), hired the geographer Louis Gentil (1868–1925) as his science advisor. This resulted in the 1920 creation of the Institut Scientifique Chérifien, the country's oldest modern university research institution. In the 1930s the French built the Averroès Observatory south of Casablanca and a research station at Ifrane in the Atlas mountain chain.

France is a leader in atomic and nuclear physics and has been so for many years. In 1903 Marie Curie won her first Nobel Prize in Physics which she shared with her husband Pierre Curie and the physicist A. Henri Becquerel for work on radioactivity. Currently about 75 per cent of France's power needs is supplied by nuclear reactors. Atomic and later nuclear weapons development have had continuing environmental impacts on France's former colonies and subjects. In 1958 Charles de Gaulle created programmes to investigate atomic and nuclear weapons. Later, as president of France, de Gaulle oversaw the creation of a "strike force" of sea-, land-, and air-based nuclear weapons. Between 1960 and 1966, France conducted four above-ground and 13 underground atomic tests in southern Algeria. Although the country was independent of France in 1962, a secret agreement allowed French underground testing to continue for four more years. France also conducted at least 181 nuclear tests, a number of them above ground, in French Polynesia. In 2010 the French government agreed to compensate people exposed to radiation in the two regions.

The Pasteur Institutes and science in the colonies

The chemist and bacteriologist Louis Pasteur was the most celebrated French scientist of the nineteenth century. In his massive encyclopaedia of the French colonies, published in conjunction with the Paris Colonial Exposition of 1931, the historian Gabriel Hanotaux praised Pasteur as the "grand master of modern colonization."[16] Pasteur's studies of microbes, vaccines, and industrial processes were replicated worldwide and would have significant impact on a number of infectious diseases in France, its colonies, and beyond. Members of the Pasteur Institute of Paris, founded in 1887, embraced the role of health consultants for the colonies and did so in a series of missions to North and West Africa, Southeast Asia, the Mediterranean basin, and elsewhere.

The Pasteur Institute in Paris was founded as a private institution separate from the French university system, but it was important for research on a variety of infectious diseases, the dispersion of the germ theory of disease, parasitology, and tropical medicine. Generations of military and civil physicians who worked throughout the empire trained there in microscopy and bacteriology in a course initially taught by Émile Roux who was assisted by Alexandre Yersin, the co-discoverer of the bacterium causing bubonic plague. In 1894 the French Ministry of Colonies gained independence from the navy. The Pastorians embraced the role of health consultants for the empire and investigated an astounding array of human and animal diseases, including bubonic plague, malaria, sleeping sickness, typhus, yellow fever, and cholera. The Pastorians were called in when epidemics of yellow fever struck Saint-Louis and Dakar in Senegal in 1900 and 1912 and investigated another epidemic of 1908–9 which struck the Caribbean colony of Martinique.[17] Pasteur's colleague Albert Calmette, who had followed some of the microbiology course taught by Roux and Yersin, founded the first overseas laboratory associated with the Paris Pasteur Institute in Saigon in 1892. Others followed in Indochina at Nah Trang (1895), Hanoi (1925), and Dalat (1936). The network of overseas Pasteur Institutes or associated laboratories extended as well to China and to many sites across Africa where, in addition to the Saint-Louis, Senegal site—subsequently relocated to Dakar where it became the Dakar Pasteur Institute (1924)—institutions were founded in Tunis (1893), Algiers (1894), Tangier (1911), Casablanca (1911), the French West Africa town of Kindia (1925), Brazzaville (1908) in French Equatorial Africa, and Tananarive, Madagascar (1898).[18]

French science and medicine first influenced the principalities of Indochina (Cambodia, Laos, and Vietnam) through contacts with the Vietnamese court when the naval surgeon Jean Marie Despiau (d. 1824) became one of Emperor Gia Long's personal physicians. Despiau convinced Gia Long to start a medical service for his subjects and was active in bringing smallpox vaccination to Vietnam. Later natural history, mapping, and the search for natural resources and transportation routes drove Henri Mouhot (1826–61) and others to explore the Mekong River basin. Others such as Francis Garnier (1839–73) explored widely in Cambodia, which became a French protectorate in 1863. The French naturalist Auguste Pavie (1847–1925) explored Laos and became a close friend of King Oun Kham.

The first scientific bureau in Indochina, a meteorological service, was founded in 1884. In the twentieth century, the French established an Oceanographic Institute at Nha Trang (1922). Scientific research was mainly of a practical nature and centred on agricultural technologies for rubber, palm oil, coffee, and tobacco, and snake bite anti-venom. The Vietnamese court, however, was interested mainly in medicine.[19] In 1905, two years after the first French "native medical aid" programme was organised in Madagascar, the French created an *Assistance médicale indigène* (Indigenous Medical Service) to serve the health needs of Indochina and extended activity into the Cambodian sector in 1907. A medical school founded in Hanoi in 1902 began training Khmer peoples in Western medicine in 1910, but by the 1920s most applicants were Vietnamese, and the Cambodian health service was mainly staffed by Vietnamese.[20] Shortly after World War II, the French physician Pierre Huard became Dean of Hanoi Medical University. Huard and Maurice Durand published one of the earliest articles to insert the history of French colonial science into the general history of science.[21] The authors signalled the technological insufficiency of the Vietnamese as an initial source of French victory and noted how they later embraced Western science and technology as a path to independence.

Several scholars are investigating the environmental history of the French Empire and its legacy. In Indochina, for example, the French tried to improve transportation in the Mekong Delta and its tributaries with dredgers similar to those used on the Rhône in France. But their efforts were thwarted by the dense mud of the Mekong Delta and its many canals as well as misunderstandings of land tenure, political alliances, and state making. Alexandre Yersin was one of the early investors in Indochinese rubber plantations, and newer scholarship investigates the agricultural, ecological, industrial, labour, and health dimensions of plantation culture. By the 1930s, investigators were researching scientific and rational systems of rubber tapping, plant physiology, and plant genetics.

Imperial hydrology and agricultural practices have also attracted the attention of historians of North Africa. In the Third Republic, after the opening of the Suez Canal in 1869, the geographer François Elie Roudaire and the builder of the Suez Canal, Ferdinand de Lesseps, both imbued with the utopian ideas of Saint-Simon and his disciple Prosper Enfantin, resurrected a scheme to flood the northern Sahara with sea water. The novelist Jules Verne later immortalised this techno-scientific dream in his last novel, *The Invasion of the Sea*. The French colonies offered a sort of scientific and technical laboratory for modernisation projects which could not have been accomplished in France. However, in both instances, Indochina and North Africa, the French often misread the local environment and colonised peoples and what could be accomplished.[22] Readers wishing to investigate further environmental issues in Southeast Asia are invited to consult Timothy Barnard's essay in this volume titled "Empire, cultivation and the environment in Southeast Asia since 1500" (Chapter 19).

The Pasteur Institutes were important for the transfer of Western medical and scientific knowledge to the colonies and they often addressed local needs and issues. Directors and institute personnel knew the empire in detail and often rotated between colonial postings and Paris. In Tunis, for example, the founding director, Pasteur's nephew Adrien Loir (1862–1941), studied viniculture and human and animal diseases, including rabies. Replacing Loir in 1902 was the mercurial Charles Nicolle (1866–1936) who discovered the asymptomatic carrier state. He later won the Nobel Prize in Physiology or Medicine (1928) for his work on typhus. Algiers obtained a Pasteur Institute in 1894 and professors at the medical school used its laboratories for their own research. Subsequent to founding the bacteriological laboratory in Saigon, Albert Calmette became director of the Algiers Pasteur Institute in 1909. He was assisted by the Algiers-born Edmond Sergent (1876–1969) who, like his brother Étienne (1878–1948), had trained at the Algiers school of medicine. The two became towering figures in Algerian science and medicine and worked on malaria, tuberculosis, mycology, and many human and animal diseases.[23] In Morocco, the two Pasteur Institutes at Tangiers and Casablanca developed research programmes on hygiene, typhus, dengue, anthrax, leprosy (Hansen's disease), and other afflictions. In West Africa the Pasteur Institute at Dakar performed fundamental work leading to an effective vaccine for yellow fever and is now co-managed with the Senegalese government. Training for health officers and medical assistants to staff an *Assistance médicale indigène* was organised in 1905 and in 1960 Dakar's school of medicine transformed into a full faculty able to award the MD degree.

Conclusion: the scientific legacy of empire

With decolonisation, the former French colonies lost scientific personnel and scientific institutions suffered. For example, Algeria had about 350 physicians and 400 scientists at independence in 1962. Most European scientists and engineers soon fled the region. Morocco and Tunisia had far fewer scientific and medical personnel. All three nations recognised the need to produce their own engineers and physicians and they were assisted in this by the World Bank, UNESCO, and other international agencies. Still, a "brain drain" of scientific personnel has persisted. France has continued to fund the four major research institutes in Algeria, and in 1969 the two countries created the Organisme de Coopération Scientifique to administer them. France has been particularly keen to keep North African petroleum flowing to France. Algeria and Morocco have reached out to the United States, China, Spain, and other countries, and Algeria has announced plans to become an international hub of biotechnology by 2020.

These plans have been incompletely realised after the political upheavals of the Arab Spring. Although the *Koran* is not an anti-scientific document, a resurgence of Islamic fundamentalism has pressed the scientific trajectory of the region ever more toward utilitarian and applied endeavours. In addition, harsh reactions to Westernisation and Western business interests operating in North Africa continue to endanger the scientific and technical communities. For example, in 2013 a terrorist attack on a natural gas facility at Ameras near the Algerian border with Libya killed 40 people, and another attack in 2016 at Krechba in the Sahara south of Algiers caused the Norwegian petroleum company Statoil and British Petroleum to withdraw their employees.[24] Further, a study of African scientific publications between 1987 and 1990 calculated that the summed per capita scientific and technical publications of Algeria, Morocco, and Tunisia were about half that of Nigeria, Africa's third most productive country in terms of scientific publication.[25]

The story differs of course depending on which former colonies are assessed, but French colonial policies throughout the empire tended to distance colonial subjects from pure scientific and higher medical activities. Scientific underdevelopment or asymmetry *vis-à-vis* the West is one legacy of colonialism. Possibly the adoption of new technologies like cell phones and solar power will enable the former colonies to bypass outdated technological steps taken by the West. This will require the sustained investment, assistance, and support of the international scientific community, but as Frantz Fanon observed, "The people who take their destiny into their own hands assimilate the most modern forms of technology at an extraordinary rate."[26]

Notes

1 This essay draws on Michael A. Osborne, "Science and the French Empire," *Isis* 96 (2005): 80–7 and Michael A. Osborne, "Maghreb of North Africa," in *The Cambridge History of Science*, vol. 8, *Modern Science in National, Transnational, and Global Context*, ed. Hugh Richard Slotten, R.L. Numbers, and D.N. Livingstone (Cambridge: Cambridge University Press, 2020), 476–94. Detailed references for much of the information presented here and more will be found in these two publications. For further reading about science, technology, and medicine in several empires, and which includes significant bibliographic information, see Roy MacLeod, ed., "Nature and Empire: Science and the Colonial

Enterprise," special issue, *Osiris* 15 (2000). For a comprehensive and detailed treatment of internationalism, science, technology, and medicine across several modern empires, see Slotten, Numbers, and Livingstone, *The Cambridge History of Science*, vol. 8, *Modern Science in National, Transnational, and Global Context*. I thank Anita Guerrini and Andrew Goss for careful reading and comments.

2 Michael A. Osborne, *The Emergence of Tropical Medicine in France* (Chicago: University of Chicago Press, 2014).

3 Michael J. Heffernan, "A State Scholarship: The Political Geography of French International Science during the Nineteenth Century," *Transactions of the Institute of British Geographers*, n.s., 19 (1994), table 3.

4 Jehanne-Emmanuelle Monnier, *Profession explorateur, 1836–1921* (Rennes: Presses Universitaires de Rennes, 2017), 128–43.

5 James E. McClellan III, *Colonialism and Science: Saint Domingue in the Old Regime* (Baltimore, MD: Johns Hopkins University Press, 1992), 51.

6 Stuart McCook, "Greater Caribbean," in Slotten, Numbers, and Livingstone, *The Cambridge History of Science*, 8: 782–98.

7 Lewis Pyenson, *Civilizing Mission: Exact Sciences and French Overseas Expansion, 1830–1940* (Baltimore, MD: Johns Hopkins University Press, 1993), 87–153. For short biographies of colonial physicians born before 1888, see B. Brisou and M. Sardet, eds., *Dictionnaire des médecins, chirurgiens et pharmaciens de la Marine* (Paris: Service Historique de la Défense, 2010).

8 Eric T. Jennings, *Curing the Colonizers: Hydrotherapy, Climatology, and French Colonial Spas* (Durham, NC: Duke University Press, 2006).

9 Michael A. Osborne, *Nature, the Exotic, and the Science of French Colonialism* (Bloomington: Indiana University Press, 1994), 145–71.

10 Will D. Swearingen, *Moroccan Mirages: Agrarian Dreams and Deceptions, 1912–1986* (Princeton, NJ: Princeton University Press, 1987); Diana K. Davis, *Resurrecting the Granary of Rome: Environmental History and French Colonial Expansion in North Africa* (Athens: Ohio University Press, 2007).

11 Giulia Meloi and Johan Swinnen, "The Rise and Fall of the World's Largest Wine Exporter—And Its Institutional Legacy," *Journal of Wine Economics* 9 (2014): 3–33.

12 Christophe Bonneuil and Mina Kleiche, *Du Jardin d'essais colonial à la station expérimentale, 1880–1930* (Paris: Centre de Coopération International en Recherche Agronomique pour le Développement, 1993).

13 Alice L. Conklin, *In the Museum of Man: Race, Anthropology, and Empire in France, 1850–1950* (Ithaca, NY: Cornell University Press, 2013); Kathleen Kilroy-Marac, *An Impossible Inheritance: Post-Colonial Psychiatry and the Work of Memory in a West African Clinic* (Berkeley: University of California Press, 2019); Warwick Anderson, Deborah Jenson, and Richard Keller, eds., *Unconscious Dominions: Psychoanalysis, Colonial Trauma, and Global Sovereignties* (Durham, NC: Duke University Press, 2011); Richard C. Keller, *Colonial Madness: Psychiatry in French North Africa* (Chicago: University of Chicago Press, 2007); Spencer D. Segalla, *The Moroccan Soul: French Education, Colonial Ethnology, and Muslim Resistance, 1912–1956* (Lincoln: University of Nebraska Press, 2019).

14 Michael A. Osborne, "Maghreb of North Africa," 476–94.

15 Assia Harbi, Amal Sebaï, and Mohamed S. Boushacha, "A Glimpse at the History of Seismology in Algeria," in *The Geology of the Arab World—An Overview*, ed. Abderrahamne Bendaoud et al. (Cham: Springer Nature Switzerland, 2019), 341–79.

16 Gabriel Hanotaux, "Conclusion: L'oeuvre médicale de la France dans les colonies africaines," in *Histoire des colonies françaises et de l'expansion de la France dans le monde*, ed. Gabriel Hanotaux and Alfred Martineau (Paris: Société de l'histoire nationale et Librairie Plon, 1929–1933), 4: 592–3.

17 Michael A. Osborne, "The Several Meanings of Global Health History: The Case of Yellow Fever," Rockefeller Archive Center Research Reports, 2017, http://rockarch.issuelab.org/resources/29167/29167.pdf.

18 Anne-Marie Moulin, "Patriarchal Science: The Network of Overseas Pasteur Institutes," in *Science and Empires: Historical Studies about Scientific Development and European Expansion*, ed. Patrick Petitjean, Catherine Jami, and Anne-Marie Moulin (Boston: Kluwer Academic Publishers, 1992), 307–22; Aro Velment, *Pasteur's Empire: Bacteriology and Politics in France, its Colonies, and the World* (Oxford: Oxford University Press, 2020).

19 C. Michelle Thompson, "Indochina," in Slotten, Numbers, and Livingstone, *The Cambridge History of Science*, 8: 593–608.

20 Sokhieng Au, *Mixed Medicines: Health and Culture in French Colonial Cambodia* (Chicago: University of Chicago Press, 2011).
21 Pierre Huard and Maurice Durand, "The Beginnings of Science in Viet Nam," in *History of Science*, ed. René Taton, 4 vols. (New York: Basic Books, 1963–6), vol. 3, pp. 579–84.
22 Mitchitake Aso, *Rubber and the Making of Vietnam: An Ecological History, 1897–1975* (Chapel Hill: University of North Carolina Press, 2018); David A. Biggs, *Quagmire: Nation-Building and Nature in the Mekong Delta* (Seattle: University of Washington Press, 2010); Christophe Bonneuil, "Crafting and Disciplining the Tropics: Plant Science in the French Colonies," in *Science in the Twentieth Century*, ed. John Krige and D. Pestre (Amsterdam: Harwood Academic Publishers, 1997), 77–96; Michael J. Heffernan, "A French Colonial Controversy: Captain Roudaire and the Saharan Sea, 1872–83," *Maghreb Review* 13, nos. 3–4 (1988): 145–59; Swearingen, *Moroccan Mirages*; Davis, *Resurrecting the Granary of Rome*.
23 Jean-Pierre Dedet, *Edmond et Étienne Sergent et l'épopée de l'Institut Pasteur d'Algérie: double biographie* (Péznas: Domens, 2013).
24 Clifford Krauss, "BP and Statoil Pull Employees from Algeria Gas Fields after Attack," *New York Times*, 21 March 2016.
25 Yvon Chatelin and R. Waast, "L'Afrique scientifique de la fin des années 1980," in *Les Sciences du Sud: états des lieux*, ed. R. Waast (Paris: Office de la Recherche Scientifique et Technique Outre-Mer, 1996), 73–90.
26 Frantz Fanon, "Medicine and Colonialism," in *A Dying Colonialism*, trans. Haakon Chevalier (New York: Grove Press, 1967), 145. Originally published in 1959.

16

ANOTHER EMPIRE

Science in the Ottoman lands

Daniel A. Stolz

An overview of the literature on science and empire should reckon with the ways in which attention to imperial context has sometimes reinforced the prevailing Eurocentric bias in the historiography of science. Much of the original impetus for the turn to imperial context derived from a critique of science as an instrument of European colonisers.[1] This critique, for all its merit, did not challenge the assumption that Europeans carried science on their ships, fully formed, when they left the metropole. A subsequent generation of scholarship moved away from this understanding of science as a tool that colonisers wielded, and toward an understanding of empire as a context within which modern science itself took form.[2] Such studies drew attention to the role of globally circulating knowledge and actors in the emergence of putatively "Western" disciplines. Yet the focus remained on empires whose metropoles lay within a narrow slice of the world: the Iberian, French, Dutch, and British examples being most common. We came to understand modern "science" as the result of a multi-directional process of global circulation, yet "empire"—even when understood in similarly multi-directional terms—remained tethered to northwest Europe.[3] This form of Eurocentricity is one reason historians of science and empire have not adequately accounted for the heterogeneity of "empire" as a category.

Ottoman history can help correct this tendency by calling attention to contexts that overlap with, yet differ from, conventional geographies of science and empire. Like other early-modern empires of Europe, the Ottomans sought to expand their knowledge of the seas as they struggled to control Indian Ocean and Mediterranean trade. Like other modernising states of Europe in the nineteenth century, the Ottomans mobilised science in projects of conscription, education, and industrialisation. Yet Ottoman science was not simply European science centred on the southeast corner of the continent. The Ottomans largely wrote and spoke different languages; their material culture emerged from a different matrix of trade, diplomatic exchange, and warfare; their guilds, marketplaces, and property regimes operated under different norms; and Ottoman political economy emerged from distinctive methods of waging war, providing food, and apportioning wealth. To contextualise science in *Ottoman* Empire is less about recovering "non-Western knowledge" than it is about attending to another meaning of empire.

Ottoman history calls attention to the heterogeneity of empire not only because of Ottoman difference from other European empires, but also because the nature of Ottoman imperialism changed markedly over time. This essay proceeds from the view that Ottoman imperialism can usefully be divided into three major periods: the era of initial state formation and expansion in the fourteenth to sixteenth centuries; the renegotiation of imperial–provincial relations in the seventeenth and eighteenth centuries; and the period of reformist movements from 1774 to 1922. Working within this periodisation, the essay begins by situating the astral sciences, cartography, and horology in the context of imperial "worldmaking" in the first three Ottoman centuries. The second part of the essay briefly explores how the sciences, like other aspects of Ottoman society, increasingly took shape within specifically regional or urban contexts in the seventeenth and eighteenth centuries. The final section of this essay will highlight the transformative role accorded to the sciences during the turbulent period from 1774 to the end of empire, as scientific expertise was newly militarised, institutionalised, and ultimately fashioned into a ruling ideology.

One challenge of attending to the heterogeneity of "empire" across time and space is that the meaning of "science" was equally variable, if not more so. In its modern sense, the category only came together in Ottoman society (as elsewhere) in the nineteenth century. Moreover, older actors' categories that are familiar to historians of science were not used by the Ottomans, or not used in the same way. "Natural philosophy," for example, provides a cramped window on to the scope of Ottoman science, and even Islamic categories, such as the distinction between "rational sciences" and "transmitted sciences," are of limited use (among other faults, privileging the writing of a particular class of scholars over artisanal knowledge, not to mention non-Muslim discourses).[4] A rigorous subtitle for this essay might be, "various forms of Ottoman naturalistic inquiry, some of which are now considered part of the genealogy of science, while others were specifically excluded from the making of 'modern science' as a category." I have chosen a broad lens advisedly, guided in part by the dearth of synthetic accounts of Ottoman science across multiple periods.[5] Rather than attempting to track a specifically defined category of Ottoman science, or to focus on a specific time or place, I aim to show how shifting forms of empire privileged different kinds of inquiry—and how the specific circumstances of the nineteenth century shaped the Ottoman institutionalisation of "science" in its modern sense.

Ottoman worldmaking: astral science, cartography, and timekeeping in the fourteenth to sixteenth centuries

In his evocative portrait of Ottoman temporal culture in the eighteenth century, the historian Avner Wishnitzer remarks that the Ottomans conceived of time in ways that "bound together heaven and earth, society and nature, and the fate of humans with the course of planets. ... [B]y claiming correlation with divine rhythms, hegemonic temporal culture served to legitimise and reaffirm the very mundane social order."[6] This elaborate nesting of cyclical phenomena, which placed the revolutions of politics within the epicycles of planets and the rotation of watch hands, derived from the work of several centuries of Ottoman "worldmaking," as I will call it. By "worldmaking," I mean both astral sciences and cartography, and I seek to underline the connection between these disciplines and the making of an Ottoman world empire between the fourteenth and sixteenth centuries.

Beginning in 1300, the Ottoman household and its network rapidly transitioned from a small beylicate on the northwest fringe of Anatolia to an empire that wielded power on three continents. The conquest of Constantinople, in 1453, is well known as a marker of the Ottoman displacement of Byzantine rule in the Balkans and western Anatolia. But the expansions of the sixteenth century also defined the Ottoman polity. In 1517, defeat of the Mamluk Sultanate brought Ottoman rule into the Arabic-speaking symbolic heartlands of Islam in Egypt, Syria, and the Hijaz. This expansion established the Ottomans as the dominant power on both the

Eastern Mediterranean and the Red Sea. By the middle of the sixteenth century, the conquests of Suleiman the Magnificent (Kanuni Suleiman) had extended Ottoman rule westward across most of North Africa and eastward all the way to Basra, giving Ottoman ships a second opening on to the Indian Ocean.

The transformation of the Ottoman state into a world empire provided a new political framework for the sciences, including institutional forms and pathways of exchange that fostered a dynamic literary and material culture. The astral sciences were perhaps the most well established at the Ottoman court, with sultans routinely employing their practitioners since at least the mid fifteenth century. The term "astral sciences" here denotes a range of practices associated with the position of the *müneccim*, literally a practitioner of *ilm-i nücûm*, the "science of the stars." Such practices included determining the horoscope of a ruler or other prominent figure; publishing an astrological forecast (*takvim*) and offering advice on the regulation of the several Ottoman calendars; explicating a treatise on the arrangement of the celestial spheres; designing or explaining the use of an instrument such as an astrolabe or sundial; or interpreting an astronomical table (*zij*) to predict a celestial occurrence. Muslim scholars had, by the thirteenth century, articulated a disciplinary distinction between astronomy (*'ilm al-hay'a*) and astrology (*'ilm ahkam al-nujum*),[7] but the reality, at least in Ottoman practice, was blurry. What is clear is that the Ottoman ruling class valued the full range of astral sciences. The late-fifteenth-century sultan Bayezid II appears to have been the first to spend lavishly on them, employing many scholars in the position of *müneccim*.[8] These practitioners may have aided Bayezid in the political rivalries that characterised his reign, both with the Mamluk sultans and with his brother Cem. Patronising the astral sciences allowed Bayezid and later sultans to fashion themselves as philosopher-kings, while the *müneccim* also offered practical advice on the timing of military campaigns and major buildings.

The sultans patronised the astral sciences even when they were not of immediate political use. Ottoman scholars did new work in the specific subfield of *'ilm al-hay'a*, sometimes called *al-hay'a al-basita*, which explored the form and motion of the celestial spheres. This cosmological field had emerged as a distinct genre in the Islamic world between the eleventh and thirteenth centuries, as scholars working primarily in Arabic and Persian critiqued the models articulated in the Ptolemaic tradition of the *Almagest*.[9] As early as the mid fifteenth century, Ottoman scholarly circles in Anatolia began to assimilate key masterworks of this genre, including al-Jaghmini's *al-Mulakhkhas fi al-hay'a al-basita* and Nasir al-Din al-Tusi's *al-Tadhkira fi 'ilm al-hay'a*, both of which would enjoy long traditions of translation and commentary in Ottoman scholarship.[10] Mehmed II spent lavishly to acquire the service of 'Ali al-Qushji, one of the leading astronomers of the day and a former member of the Samarqand Observatory circle.[11] Qushji wrote an original *hay'a* work on the occasion of Mehmed II's successful campaign against the Aq Qoyunlu in 1473, and subsequently took up residence in Istanbul. Such patronage for cosmological work should be understood in multiple contexts: the sultans' interest in fashioning themselves in the princely image of the patrons of the great observatories at Maragha and Samarqand; a broader Ottoman assimilation of Islamic scholarly genres in both Persian and Arabic; and, simultaneously, the dramatic rise in Ottoman patronage of the Islamic institution of prestigious learning, the *medrese* (more on which below).

The strength of the astral sciences at the Ottoman court, and among provincial elites as well, can be seen in the enthusiasm for mechanical timepieces that swept through Ottoman society beginning in the sixteenth century. European diplomats and others seeking the favour of the sultans quickly learned that a novel clock was always a welcome gift, and both English and Continental clockmakers developed a whole line of "eastern market" pieces that catered to Ottoman taste.[12] (Some of their finest specimens can still be seen in Istanbul's Topkapı Palace

Museum.) But the Ottomans also had their own horologists, most famously Taqi al-Din ibn Ma'ruf, an Egyptian-born polymath appointed by Selim II to oversee an observatory in Istanbul in the mid-1570s.[13] Some Ottoman scholars crafted timepieces themselves. Others, who had trained in the Islamic discipline of *'ilm al-miqat* (the science of timekeeping), specialised in the astronomical techniques necessary to set mechanical clocks according to the Ottoman convention of reckoning time from sunset.[14] This practice, called *gurubi saat* ("sunset time"), was sufficiently ubiquitous and distinctive to be adopted as a marker of identity in the nineteenth century, when telling time *alla turca* came to denote one's scepticism toward Westernising (*alla franca*) elite cultural tastes.[15] The interaction that timepieces mediated between scholarly knowledge and material culture illustrates the importance of the astral sciences in Ottoman society, as well as the dynamism of these sciences over several centuries.

Arguably the most pragmatic area in which Ottoman imperial ambitions converged with sciences of worldmaking was cartography. Here, initial Ottoman efforts were directly connected to developments among their Venetian, Genoese, and Iberian counterparts, and only secondarily to assimilation of the Arabic and Persian cartographic traditions. The Ottomans possessed early and masterful examples of portolan charts and "Catalan" maps in the fifteenth century, with some of the former category being of Ottoman production.[16] After the conquest of the Mamluk Sultanate, the Ottomans sponsored intensive efforts to garner information regarding the Empire's new frontiers in terms of both physical and human geography. Ibrahim Pasha, Sultan Suleiman's powerful Grand Vizier, enlisted an eclectic cast of naval captains and corsairs to explore the possibilities for Ottoman expansion around the Indian Ocean littoral.[17] One result of these efforts was Piri Reis's encyclopaedic *Book of the Sea,* which offered a state-of-the-art account of the geography of the Mediterranean, information on the voyages of Columbus and Da Gama, and the first Ottoman Turkish account of the Indian Ocean. The emergence of an Ottoman imperial geography is epitomised in the work of Selman Reis, who visited ports around the Indian Ocean and reported to Ibrahim Pasha with his observations on the physical, political, and economic pathways for continued Ottoman expansion.[18]

"Worldmaking" sciences illustrate the many sites of learning and pathways of exchange within which Ottoman sciences took shape in the fourteenth to sixteenth centuries. On the one hand, Arabic and Persian scholarship became fundamental for most Ottoman elite learning in this period. Areas such as *hay'a* illustrate the extent to which Ottoman scholarship built upon the dynamic translation and appropriation of Greek, Syriac, and Indian learning undertaken in the Islamic empires since the ninth century. Likewise, Islamic institutional forms of learning flourished in Ottoman society. Ottoman elites, including women, frequently used the legal device of the *waqf* to consolidate wealth within a family while patronising a favoured pious cause, such as a hospital or *medrese* (Arabic *madrasa*). The latter was a site of advanced learning where students gathered in circles to study texts with master scholars, with both students and teachers benefiting (to varying degrees) from endowed stipends. The extent to which fields such as astronomy and mathematics were regularly taught in the *medrese* remains an open question; Islamic jurisprudence, exegesis, and Arabic language sciences were generally privileged.[19] Regardless of their place in the *medrese*, however, it was possible—and, in some contexts, typical—for pursuits such as the astral sciences to be conducted through individual scholar–student relationships.

In addition to assimilating learned languages, genres, and institutions from Arab- and Perso-Islamic society, the Ottomans benefited from, and promoted, Mediterranean pathways of knowledge exchange in the fourteenth to sixteenth centuries. As we saw in the case of cartography, it sometimes makes more sense to understand Ottoman science in the context of other early-modern Mediterranean empires than it does to place them in their Islamic context. A web of ties bound Ottoman learned culture to Venice, Genoa, and even Lisbon. The threads on this web included diplomatic exchange, the trade in captives from naval warfare, and the travel of Renaissance humanists and other European scholars in search of knowledge.[20] Ottoman religious minorities also played a role, especially as the Spanish persecution and expulsion stimulated the growth of Jewish communities around the Mediterranean, including in Ottoman cities like Salonica. Men like Moses ben Judah Galeano, a Jewish physician of Crete who spent time at the court of Bayezid II and had connections to Padua, helped create an Eastern Mediterranean world of knowledge circulation. Recent scholarship has argued that such Ottoman intermediaries may even have transmitted into Latin scholarship certain mathematical models that originated in Arabic and Persian, and which Copernicus later used while developing his heliocentric planetary models.[21]

While this story makes for a striking episode in the history of the "Scientific Revolution," it is also part of a larger pattern. As the Ottomans established themselves as an imperial state, their efforts set in motion calendars, clocks, maps, instruments, and even cosmological treatises. By such work, Ottoman scholars, sailors, timekeepers, and stargazers forged close ties to the ruling elite, who in turn gained an understanding of the world that both legitimated and extended their power.

Provincial contexts of science in the seventeenth and eighteenth centuries

In the two centuries that began with the "Long War" (1593–1606), the Ottoman Empire experienced new military, fiscal, and environmental challenges. Technical changes in warfare, the expansion of European empires, and droughts associated with the "Little Ice Age" destabilised the order established in the previous era.[22] As the need for more infantry led the Janissary Corps to grow, it put down social and economic roots in urban society and became a political force to rival the palace. The recruitment of irregular infantry, combined with repeated crop failures, stimulated the growth of "brigandage" in the countryside. The currency lost value. Expansion slowed, and on some fronts came to a halt. In the eighteenth century, these pressures were exacerbated by the rise of a modernising and expansionist state in Russia, which inflicted a catastrophic defeat on the Ottomans in the war of 1768–74. The state adapted by increasing its fiscal and administrative reliance on provincial elites, diminishing the power of imperial officials.

The crises of the seventeenth and eighteenth centuries were once understood within a narrative of imperial "decline," and thereby merged with a broader interpretation of post-classical Islamic history as an era of cultural and intellectual stagnation. Since the rise of economic and social history, however, and more recently with the contribution of approaches ranging from material culture and book history to environmental and global history, a different picture has emerged.[23] Thus, to take one example, rather than focusing on why the Ottomans did not develop a thriving print culture in this period, recent scholarship has shown that Ottoman cities witnessed the birth of new forms of non-elite reading, authorship, and book production in manuscript form.[24] Even the notorious provincial elites, rather than serving as evidence of imperial weakness, are now seen as part of the dynamic renegotiation of power that characterised the "Age of Revolutions" in the late eighteenth century.[25]

Natural inquiry hardly disappeared from Ottoman society in the seventeenth and eighteenth centuries, but rather took shape within its new political and socioeconomic realities. To be sure, some of these realities were grim: by the second half of the seventeenth century, currency debasement had impoverished the endowed teaching positions in Istanbul's *medreses*. Harun Küçük argues that this harsh economy undermined the accumulation of scholastic knowledge in the capital (with the exception of Islamic jurisprudence, which could still lead to remunerative state employment as a judge). By the same token, however, Istanbul became fertile ground for what Küçük calls "practical naturalism," including a thriving medical marketplace—as befit one of the most plague-ridden cities of the early-modern era.[26] Astrological prognostication formed an integral part of this marketplace, apparently meeting little objection from the pietistic movements that sometimes gained influence in the capital in the late seventeenth century. Paracelsian "new medicine," too, found a home in some early-modern Ottoman cities, facilitated by a tradition of alchemical learning in Sufi circles.[27]

A congruent picture has emerged from research on provincial centres of learning in this period. In Cairo, mathematics, astral sciences, and other practices of divination became the favoured pursuit of scholars who did not hail from the city's established scholarly or merchant families, and who turned to the "uncommon sciences" to forge ties with the city's military-political elite.[28] Yet at least some of these scholars, in contrast to Küçük's analysis of the Istanbul context, continued to situate their studies within textual genealogies ultimately rooted in Greek philosophy. Indeed, early-modern Ottoman-Arab scholars broke new ground in Aristotelian logic, sometimes articulating their work as part of a broader ethos of "verification."[29] Location-specific declines in the material conditions for scholarship should not be taken for a total disappearance of dynamism in scholarly genres. But the larger point here is that any discussion of "Ottoman science" must consider how the transformation of empire into a web of more-horizontal relationships altered the significance of the term "Ottoman" as an analytical category. If what made the sciences of worldmaking "Ottoman" in the earlier period was their connection to defining an imperial Ottoman world, the seventeenth and eighteenth centuries were a time when the culture of learning in Ottoman cities began to be defined more by particular urban contexts and provincial patronage.

Science to save the empire: militarising, institutionalising, and objectifying science

The relationship between science and empire in the Ottoman lands underwent profound transformation beginning in the late eighteenth century, as factions in Istanbul and the provinces sought to revise the empire's military, fiscal, and administrative capabilities. The "New Order" (*nizâm-ı cedid*) advocates under Selim III (r. 1789–1807), the bureaucratic leaders of the mid nineteenth century Reforms (*tanzimat*), the autocratic sultan Abdülhamid II (r. 1876–1909), and the revolutionary architects of the Second Constitutional Era (1908–13) each sought changes that were different in both kind and degree. Some focused more on the military-fiscal challenges faced by the imperial state, and the overlapping problem of relations between imperial and provincial elites; other efforts had more to do with the diplomatic context, in which the pursuit of great power alliances became increasingly imperative. The legal status of non-Muslim communities attracted scrutiny and reform, as did the relationship between the empire's different ethnic communities. The state invested in new industrial infrastructure and experimented with new land management techniques and tenure regimes, both in response to the demand for revenue and the need to resettle large populations from the empire's lost territories. These efforts bespeak not a single attempt to "Westernise," but rather a variety of shifting efforts to redefine empire in a period of rapid economic, political, technological, and demographic change.

Yet at least two common themes ran through these efforts. First, all sought to preserve *some* Ottoman order, though not the status quo; in other words, they were efforts to salvage empire.[30] Second, reform programmes increasingly placed their hopes in new sciences. Thus, the sciences in late Ottoman society were intimately linked with evolving efforts to maintain an imperial order. In the remainder of this essay, I argue that the imperial context was crucial to late Ottoman science in three ways: militarisation, institutionalisation, and an ideological process that I liken to "objectification." To be clear, while late Ottoman military and institutional reform are often described in terms of centralisation, I offer a different analysis here. Late Ottoman science had multiple centres, and even within these centres, it became a field of competition between different factions of the elite. Moreover, non-elite classes took up the new sciences as a means of resistance to state power. By situating late Ottoman science in terms of militarisation, institutionalisation, and objectification, I try to move beyond "centralisation" and offer an analysis that attends to these multi-directional dynamics.

Military personnel and locations played a leading role in late Ottoman science. As early as the 1770s, the dominant performance of the Russian navy led the Ottoman government to make new investments in nautical science, translating up-to-date manuals from French and establishing a new imperial military engineering school (*mühandishâne-i berrî*) in Istanbul by the end of the century. The military itself was conceived of as an object of scientific reform, both in the "New Order" army of Selim III and later the infantry established by Mahmud II after 1826.[31] As advisers to such new military units, a steady stream of European physicians, engineers, and other specialists became influential figures in late Ottoman governments. From writing about logarithms to implementing health surveillance, sailors and soldiers were often at the cutting edge of late Ottoman science—as its agents, in part, but also as its objects of discipline. They translated texts from foreign languages (usually French), taught the new techniques in new schools, and deployed them in the barracks, on the training grounds, in hospitals, at sea, and in battle. Such efforts unfolded not only in Istanbul, but also in the province of Egypt, which Mehmed Ali Pasha sought to turn into a semi-autonomous possession for his household beginning around 1820. His project depended in part upon a new and sprawling military-technical complex—an army medical school, a naval school and shipyards, engineering and translation schools—whose legacy can still be felt in the close association between the scientific professions and the security state in Egypt.[32]

The novelty in all this was not the tight relationship between science and political power—that connection had existed at the Ottoman court in the fifteenth century. After 1774, however, investment in new scientific education and technological capacity altered the political position and social composition of the armed forces in particular. Rather than a semi-autonomous and representative social institution, the new armies became a vanguard of social change and instruments of central state power both in Istanbul and Cairo. It is true that the 80-odd years from the destruction of the Janissary Corps (1826) to the Young Turk Revolution are often noted for the diminished role of the military in Ottoman politics, in contrast to the frequent Janissary rebellions of the previous century. What this account can obscure, however, is the extent to which the strengthening of government forces through scientific training and technological capacity—and the marginalisation of armed force that lay outside government control—ultimately created an army that both identified with the interests of the central state and was able to exercise power over its direction. (In Cairo, the 'Urabi Revolution revealed something of this emerging dynamic at the end of the 1870s, though British intervention was decisive in that case.) In other words, the return of the armed forces to political influence after the Young Turk Revolution of 1908 should be seen as emerging from the socio-political changes of the nineteenth century, even as it marked a departure from the relationship between the government and the army for most of that period.

To be sure, the militarisation of science meant more than the accumulation of new powers in the central state. Even where late Ottoman science was most intimately bound up in the imperial administration's efforts to make new inroads in the provinces—as was the case, for example, in the increasingly techno-political conception of population re-settlement, public health, and resource management—these efforts were, by definition, shaped in places like Mecca or Jidda in addition to Istanbul.[33] Moreover, a broad range of society participated in the new scientific techniques and rhetoric to make new political claims and even to resist the state's entrance into their lives. As Khaled Fahmy has demonstrated, for example, Egyptian peasants embraced forensic medicine and the authority of the police doctor as a means of resisting the arbitrary exercise of power by the large landholder families who dominated the mid-nineteenth-century Egyptian countryside.[34] Meanwhile, the military-technical career was an important lane in the new road of social mobility that military education offered late Ottomans, opening up the imperial administration to a new class of men born to ordinary provincial families.[35] We might still describe this process in terms of "centralisation," but only if we understand "centralising" as a dynamic whereby the state was remade from the provinces and by competing claimants to power.

Militarisation played a major, but not exclusive, role in shaping the new institutional forms of late Ottoman science. The military-technical academies established in Istanbul beginning in the reign of Selim III and in Egypt under Mehmed Ali Pasha were the first formal sites of learning to compete with the *medrese* in Ottoman society. These academies provided an alternative set of expectations and credentials that could define what it meant to be learned. After 1869, however, the military academies became only one branch of an expansive, nearly empire-wide system including civil preparatory schools that fed into prestigious training institutes for the bureaucracy.[36] In addition, Christian missionary organisations operated schools throughout the empire (as did the Alliance Israélite Universelle), and these missionary schools often emphasised science education, for reasons rooted in both natural theology and a pragmatic view of how to enrol students.[37]

Older locations and modes of study also participated in late Ottoman science. Indeed, Muslim scholars (*ulema*) were often the pioneering translators, teachers, and interpreters of new sciences and technology in late Ottoman society. While much has been made of *ulema* resistance to the new sciences, this resistance had more to do with the association of the new subjects with increasing bureaucratic regulation of the *medrese* than with theological or legal objections to their substance. In fact, much as imperial reformists embraced the sciences to save the empire, Ottoman religious reformists emphasised the importance of the sciences for reviving a pious Islam that could preserve the *umma* from the political and cultural onslaught of European imperialism.[38]

Islamic discourse also played a broader role in learning and debating new sciences among the late Ottoman bureaucratic and intellectual elite. To the extent that imperial reform was a conservative project—oriented toward preserving the empire and cultivating loyalty toward the state—a conservative ethos imbued a great deal of the rhetoric around the new scientific learning, not only among *ulema*. For bureaucrats, literary figures, and—perhaps especially—for advocates of further reforms, like the Young Ottomans, new scientific learning was necessarily linked with an emphasis on morality and virtue.[39]

Yet the association of science with new locations of learning and new social classes had destabilising consequences. Arguably the most radical novelty was the very conception of "science" as a unitary and privileged body of knowledge. I refer to this understanding as the objectification of science, much as scholars of religion have identified the objectification of Islam in modernity.[40] Just as it became possible to ask, "What does Islam say?", it became possible (and, for many, just as desirable) to ask, "What does science say?" The parallelism is not coincidental. New methods of schooling, legal codification, rising print and literacy, and (not least) the scrutinising gaze of European colonial powers led Muslim thinkers increasingly to point to "Islam" as a single, total, and fixed entity, rather than as the lived consensus of diverse communities. If we replace legal codification with other new forms of textualising knowledge (see below), much the same factors allowed late Ottomans to fashion "science" as a single, fixed, external authority, rather than looking to diverse kinds of knowledge.[41]

An essential part of this story was the rise of Ottoman print culture in the second half of the nineteenth century. Though official "literacy" remained both low and predominantly male, late Ottoman society had a burgeoning periodical literature in Turkish, Arabic, and other Ottoman languages, and including prominent female authors and women's journals. Newspapers and other printed text were read aloud at home and in coffeehouses. In keeping with the global history of print culture in this period, science popularising played a major role in the growth of the late Ottoman press. General-interest periodicals often included articles that celebrated the march of scientific progress, and some journals, such as the Turkish *Mecmua-i Fünûn* and the long-running Arabic monthly *al-Muqtataf*, specialised in science and related topics such as medicine and industry. Such journals propagated an understanding of science as an overarching, authoritative way of thinking about the world.

Late Ottoman science popularisers almost always preached the virtues of reform, whether in terms of imperial administration, education, or religion. Print culture thus connected the celebration of scientific knowledge with an appeal to science as an ideological programme—in other words, with scientism. Even in the mid nineteenth century, elite technical academies, particularly the Imperial Medical College, were becoming breeding grounds for Feuerbachian *Vulgarmaterialismus* and Comtean positivism.[42] Materialism enjoyed a thorough airing in the late Ottoman press, mostly in Turkish periodicals—although the Syrian-Egyptian doctor and journalist Shibli Shumayyil gained notoriety for translating Büchner's commentaries on Darwin into Arabic and advocating the elimination of most education outside of science.

Such radical views were marginal within late Ottoman society, and hardly representative even of the elite. But scientism held sway among a small coterie of military officers, doctors, and literary figures who formed the Ottoman Committee of Union and Progress (CUP), which became the nucleus of the "Young Turk" opposition to Abdülhamid II in the 1890s and early 1900s. The CUP, whose very name nodded to Comte, orchestrated the revolution that ended the sultan's authoritarian rule in 1908. At least until 1913, CUP leaders tended to downplay their scientism as they sought common cause with *ulema* reformists.[43] Ironically (for men who had set out to salvage the empire), the force of their commitment to the scientific reform of society was unleashed only in the transition from empire to nation-state, where it laid the groundwork for the radical social engineering of the early Turkish Republic under Mustafa Kemal (Atatürk). Kemalists sought to fashion their ideology as an *ex nihilo* creation of the new nation-state, but much of their programme had its roots in the late Ottoman context.[44] Atatürk's own saying, "The truest guide in life is science," which remains aphoristic in Turkish, is very much a relic of the late Ottoman elite who believed that science would save the empire.

Conclusion

This essay has argued that the imperial context provides a meaningful lens with which to interpret certain aspects of Ottoman science. The Ottoman case thus expands the way in which historians of science have approached the study of "science and empire," by highlighting the heterogeneity of "empire" across both space and time. To focus on the imperial context, however, is to marginalise others. This analysis has privileged centres of governance (though not only Istanbul), and it has followed a periodisation derived from the governance of the empire as a whole, rather than the experience of specific regions, cities, or peoples. Needless to say, it is possible to situate Ottoman science in other ways, yielding different narratives with different inflection points.

Even a focus on the imperial context, however, has shown that the sciences in Ottoman society were never made only at the centre. Just as British and French science emerged from circulations of people and objects along the conduits of their respective empires, Ottoman science emerged from its own imperial patterns of circulation. Even in the nineteenth and early twentieth centuries, Istanbul's attention to new sciences, though intended to consolidate power in the imperial administration, in fact owed much to rival centres within the empire (like Cairo), to the aspirations of ordinary provincial families (who enrolled their sons in the new schools), and to a burgeoning press culture that increasingly linked science with a radical social programme.

This essay has focused on the significance of the imperial context for the sciences in Ottoman society specifically, rather than following Ottoman science as it circulated (or circulated back) to other parts of the world. Such circulations certainly merit the attention they have begun to receive in the early-modern case, where it is increasingly clear that Ottoman imperialism, by linking Arab- and Perso-Islamic learning with Latin Europe, played a role in the "Scientific Revolution." In the nineteenth century, too, Ottoman imperialism shaped patterns of circulation that helped fashion modern science. Although this dynamic is not yet well understood, we might begin by considering the thousands of ambitious late Ottoman subjects who ventured abroad to pursue advanced training at their government's behest. Further research may show that the story of modern Europe's polytechnics, laboratories, and observatories is in part an Ottoman story.[45]

Notes

1 Daniel Headrick, *The Tools of Empire: Technology and European Imperialism in the Nineteenth Century* (New York: Oxford University Press, 1981). A similar premise arguably underlies the influential work of David Arnold, *Colonizing the Body: State Medicine and Epidemic Disease in Nineteenth-Century India* (Berkeley: University of California Press, 1993).

2 Kapil Raj, *Relocating Modern Science* (London: Palgrave, 2006); Simon Schaffer et al, eds., *The Brokered World: Go-Betweens and Global Intelligence, 1770–1820* (Sagamore Beach, MA: Science History Publications, 2009).

3 My argument develops the critique in Daniel Stolz, *The Lighthouse and the Observatory: Islam, Science, and Empire in Late Ottoman Egypt* (Cambridge: Cambridge University Press, 2018). My analysis is partly inspired by the example of historians of Qing science such as Laura Hostetler, *Qing Colonial Enterprise: Ethnography and Cartography in Early Modern China* (Chicago: University of Chicago Press, 2001); see also the essay by James Flowers in this volume (Chapter 18). For a powerful reminder of heterogeneity among continental empires, see Deborah R. Coen, *Climate in Motion: Science, Empire, and the Problem of Scale* (Chicago: University of Chicago Press, 2018).

4 Harun Küçük, *Science without Leisure: Practical Naturalism in Istanbul, 1660–1732* (Pittsburgh, PA: University of Pittsburgh Press, 2020).

5 Cf. Miri Shefer-Mossensohn, *Science among the Ottomans: The Cultural Creation and Exchange of Knowledge* (Austin: University of Texas Press, 2016), an accessible introduction.

6 Avner Wishnitzer, *Reading Clocks, Alla Turca: Time and Society in the Late Ottoman Empire* (Chicago: University of Chicago Press, 2015), 18.

7 George Saliba, "Islamic Astronomy in Context: Attacks on Astrology and the Rise of the Hay'a Tradition," *Bulletin of the Royal Institute for Inter-Faith Studies* 4, no. 1 (2002): 25–46.

8 Ahmet Tunç Şen, "Reading the Stars at the Ottoman Court: Bayezid II (r. 886/1481–918/1512) and His Celestial Interests," *Arabica* 64 (1017): 557–608.

9 For a concise overview of the development of *hay'a* through the thirteenth century, see F.J. Ragep, *Nasir al-Din al-Tusi's Memoir on Astronomy*, vol. 1 (New York: Springer-Verlag, 1993), 29–35.

10 On the early Ottoman assimilation of *hay'a*, see Hasan Umut, "Theoretical Astronomy in the Early Modern Ottoman Empire" (PhD diss., McGill University, 2019), 136–52.

11 Umut, *Theoretical Astronomy in the Early Modern Ottoman Empire*, 98–105.

12 There is a large literature on early modern Ottoman timepieces. Notable for including extensive images are Kemal Özdemir, *Ottoman Clocks and Watches* (Istanbul: Creative Yayıncılık, 1993), and Ian White, *English Clocks for the Eastern Markets* (Sussex: Antiquarian Horological Society, 2012).

13 Avner Ben-Zaken, *Cross-Cultural Scientific Exchanges in the Eastern Mediterranean* (Baltimore, MD: Johns Hopkins University Press, 2010).

14 Daniel Stolz, "Positioning the Watch Hand: 'Ulama' and the Practice of Mechanical Timekeeping in Cairo, 1737–1874," *International Journal of Middle East Studies* 47 (2015): 489–510.

15 Wishnitzer, *Reading Clocks, Alla Turca*.

16 Giancarlo Casale, *The Ottoman Age of Exploration* (New York: Oxford University Press, 2010), 18–20.

17 Casale, *Ottoman Age of Exploration*, 34–52.

18 Casale, *Ottoman Age of Exploration*, 39.

19 On this debate, see Sonja Brentjes, *Teaching and Learning the Sciences in Islamicate Societies (800–1700)* (Turnhout, Belgium: Brepols, 2018); and Shefer-Mossensohn, *Science among the Ottomans*, 61–63.

20 Sonja Brentjes, *Travellers from Europe in the Ottoman and Safavid Empires, 16th–17th Centuries: Seeking, Transforming, Discarding Knowledge* (Aldershot, UK: Ashgate, 2010).

21 Robert Morrison, "A Scholarly Intermediary between the Ottoman Empire and Renaissance Europe," *Isis* 105 (2014): 32–57.

22 On the seventeenth and eighteenth centuries, see Rifa'at Abou-El-Haj, *The Formation of the Modern State* (Albany, NY: SUNY Press, 1991); Karen Barkey, *Bandits and Bureaucrats: The Ottoman Route to State Centralization* (Ithaca, NY: Cornell University Press, 1994); Sam White, *The Climate of Rebellion in the Early Modern Ottoman Empire* (New York: Cambridge University Press, 2011).

23 For an accessible overview of such trends in early modern Ottoman studies, see the introduction by Dana Sajdi in *Ottoman Coffee, Ottoman Tulips: Leisure and Lifestyle in the Eighteenth Century* (London: I.B. Tauris, 2007).

24 Nelly Hanna, *In Praise of Books: A Cultural History of Cairo's Middle Class, Sixteenth to Eighteenth Centuries* (Syracuse, NY: Syracuse University Press, 2003); Dana Sajdi, *The Barber of Damascus: Nouveau Literacy in the Ottoman Levant* (Stanford, CA: Stanford University Press, 2013); Nir Shafir, "The Road to Damascus: Circulation and the Redefinition of Islam in the Ottoman Empire, 1620–1720" (PhD diss., UCLA, 2016).

25 Ali Yaycioglu, *Partners of the Empire: The Crisis of the Ottoman Order in the Age of Revolution* (Stanford, CA: Stanford University Press, 2016).

26 Küçük, *Science without Leisure*.

27 Feza Günergun, "Convergences in and around Bursa: Sufism, Alchemy, and Iatrochemistry in Turkey, 1500–1750," in *Entangled Histories: Materials, Practices, and Knowledge across Eurasia*, ed. Pamela Smith (Pittsburgh, PA: University of Pittsburgh Press, 2019), 227–57; Tuna Artun, "Hearts of Gold and Silver: The Production of Alchemical Knowledge in the Early Modern Ottoman World" (PhD diss., Princeton University, 2013).

28 Jane H. Murphy, "Locating the Sciences in Eighteenth-Century Egypt," *British Journal for the History of Science* 43 (2010): 557–71; Jane H. Murphy, "Ahmad al-Damanhuri and the Utility of Expertise in Early Modern Ottoman Egypt," *Osiris* 25 (2010): 85–103.

29 Khaled El-Rouayheb, *Relational Syllogisms and the History of Arabic Logic, 900–1900* (Leiden: Brill, 2010); Khaled El-Rouayheb, "Was There a Revival of Logical Studies in Eighteenth-Century Egypt?" *Die Welt des Islams* 45 (2005): 1–19.

30 On the "Young Turks" as preservers of empire (rather than Turkish nationalists), see M. Şükrü Hanioğlu, *Preparation for a Revolution: The Young Turks, 1902–1908* (New York: Oxford University Press, 2001).

31 On science and military discipline in the Nizâm-ı Cedid, see Ali Yaycioglu, "Guarding Traditions and Laws—Disciplining Bodies and Soul: Tradition, Science, and Religion, in the Age of Ottoman Reform," *Modern Asian Studies* 52 (2018): 1542–1603. On the naval sciences, see Tuncay Zorlu, *Innovation and Empire in Turkey: Sultan Selim III and the Modernisation of the Ottoman Navy* (London: I.B. Tauris, 2008).

32 Pascal Crozet, *Les sciences modérnes en Égypte: transfer et appropriation, 1805–1902* (Paris: Geuthner, 2008); Khaled Fahmy, *All the Pasha's Men* (Cambridge, UK: Cambridge University Press, 1997);

Khaled Fahmy, "Dissecting the Modern Egyptian State," *International Journal of Middle East Studies* 47 (2015): 559–62.

33 Michael Christopher Low, "Ottoman Infrastructures of the Saudi Hydro-State: The Techno-Politics of Pilgrimage and Potable Water in the Hijaz," *Comparative Studies in Society and History* 57 (2015): 942–74; Chris Gratien, "The Ottoman Quagmire: Malaria, Swamps, and Settlement in the Late Ottoman Mediterranean," *International Journal of Middle East Studies* 49 (2017): 583–604.

34 Khaled Fahmy, *In Quest of Justice: Islamic Law and Forensic Medicine in Modern Egypt* (Berkeley: University of California Press, 2018).

35 Michael Provence, *The Last Ottoman Generation and the Making of the Modern Middle East* (Cambridge, UK: Cambridge University Press, 2017), 10; Pascal Crozet, "La trajectoire d'un scientifique égyptien au XIXe siècle," in *Entre réforme sociale et mouvement national: Identité et modernisation en Égypte* (Cairo: CEDEJ, 1995), 285–309.

36 Provence, *Last Ottoman Generation,* 18–29; Benjamin Fortna, *Imperial Classroom: Islam, the State, and Education in the Late Ottoman Empire* (Oxford: Oxford University Press, 2002).

37 Marwa Elshakry, "The Gospel of Science and American Evangelism in Late Ottoman Beirut," *Past and Present* 196 (2007): 701–30.

38 Marwa Elshakry, *Reading Darwin in Arabic* (Chicago: University of Chicago Press, 2013); Stolz, *The Lighthouse and the Observatory.*

39 M. Alper Yalçınkaya, *Learned Patriots: Debating Science, State, and Society in the Nineteenth-Century Ottoman Empire* (Chicago: University of Chicago Press, 2015).

40 Dale Eickelman and James Piscatori, *Muslim Politics* (Princeton, NJ: Princeton University Press, 1996).

41 Peter Harrison's argument that the modern dichotomy of "science and religion" only emerged in the nineteenth century can be extended to Ottoman society, in my view. Peter Harrison, " 'Science' and 'Religion': Constructing the Boundaries," *Journal of Religion* 86 (2006): 81–106.

42 M. Şükrü Hanioğlu, "Blueprints for a Future Society: Late Ottoman Materialists on Science, Religion, and Art," in *Late Ottoman Society: The Intellectual Legacy,* ed. Elisabet Özdalga (London: RoutledgeCurzon, 2005), 28–116.

43 M. Şükrü Hanioğlu, *A Brief History of the Late Ottoman Empire* (Princeton, NJ: Princeton University Press, 2008).

44 On the late Ottoman context for the early Turkish Republic, see Erik Zürcher, *The Young Turk Legacy and Nation-Building: From the Ottoman Empire to Atatürk's Turkey* (London: I.B. Tauris, 2010). For specific connections between Kemalism and CUP ideology, see M. Şükrü Hanioğlu, *Atatürk: An Intellectual Biography* (Princeton, NJ: Princeton University Press, 2011).

45 See the discussion of Ottoman-Egyptians at the Paris Observatory in Stolz, *The Lighthouse and the Observatory,* 89–92.

17

THE PLANTING OF "COLONIAL" SCIENCE IN RUSSIAN SOIL

Anna Kuxhausen

The eighteenth century witnessed the birth of academic and practical science in Russia. Before the reign of Peter the Great (1696–1725), Russia had not yet begun to participate in the European scientific world. In the two centuries before Peter took the throne, trade had opened with the West, and European culture had made incursions, mainly through its delicacies and amenities, into the homes of Russia's elite. Yet Russia remained mostly isolated from the intellectual richness of Western Europe. While first the Renaissance flourished and then the era of scientific discovery began to blossom in Europe, not even one university or academy grew in Russia. Until the eighteenth century, skilled physicians were exceedingly scarce and, like other luxuries, imported from abroad. The few existing boarding schools tended to be modest on curriculum and offered inadequate preparation for admission to foreign universities. Families of means with the rare ambition to educate their sons had to hire foreign tutors to prepare their progeny to matriculate at universities in Leiden, Halle, and Strasbourg. The complete ecclesiastical monopoly on printing presses hampered the growth of a reading public and its corollary social spaces until well into the eighteenth century. With limited opportunities for a secular, intellectual elite to grow, Russia remained until Peter's reign, in the words of the renowned philosopher Gottfried Wilhelm Leibniz, "*tabulam rasam*."[1]

This article is about the planting of academic and practical science in Russian soil—an empty field, as it were. Two primary institutions were the main greenhouses: the Medical Chancellery and the Academy of Sciences.[2] Peter, serving as lead botanist, handpicked his master gardeners to tend his seedlings, nurture them into mature plants, and then to propagate them. But here we must depart from placid and gentle metaphors, because the story of Peter's reign, and the subsequent eighteenth century in Russia, is anything but. The European men of science who answered Peter's call were relentless cultural colonists driven by professional ambition and a taste for adventure. Whether in the relative wilds of St. Petersburg or further afield in uncharted Siberian territory, the blank slate of the Russian frontier offered them the thrilling opportunity to fashion themselves as heroes of science. While they steered, local men served as the ploughs that forcibly opened this new frontier for cultivation. In some cases, bringing science to Russia led to fame beyond their lifetimes. In others, answering opportunity's call proved lethal. In all, their missions did not rely upon wealthy patrons or the whims of the crown: these scientists were servitors of a physically and intellectually expanding state. I argue that the "cultivation of science" in Russia and the priorities of the expanding state were symbiotic. Peter's colossal will to seed scientific institutions in Russia—and to have the state reap their benefits—led to massively expensive, state-sponsored cultural colonisation by hired Western European agents.[3]

Symbiosis of science and colonisation

To conceptualise the symbiotic nature of the colonisation process we must recognise its spatial significance and its different geographical centres. In the continued imperial expansion and colonisation of non-Russian and Indigenous peoples, Moscow and St. Petersburg constituted the "metropole" while the conquered territories, contiguous lands to the south toward the Black and Caspian seas, and eastward across Siberia to the Pacific Ocean, became the Russian "colonies."[4] In the process of scientific colonisation, as Sverker Sörlin has argued, rather than geopolitical power determining the axis, academic reputation established the predominant authorities from which scientific expansion issued.[5] The agents of this colonising process in Russia came from universities in northern Europe and German lands. These colonisation processes were mutually supportive throughout the eighteenth century. As the tools for ruling and exploiting colonised regions became more sophisticated, thanks to the work of these scientists, the state in turn increased investment in the institutions of science.[6]

Two luminaries of the Enlightenment proved influential in the scientific expansion in Russia. Dr Herman Boerhaave (1668–1738), at Leiden University in the Netherlands, and Carl Linnaeus (1707–78), at Uppsala University in Sweden, became beacons of scientific Enlightenment in Russia. Boerhaave, one of the founders of physiology and early promoters of the teaching hospital, attracted students from all over Europe, including Russia.[7] Boerhaave's emphasis on clinical medicine—the treating of symptoms, the privileging of observation and experience over theory—was well suited to Russian conditions, in which the need for practitioners was acute and access to theoretical study problematic.[8] While Boerhaave's disciples proclaimed clinical medicine, Linnaeus's "apostles," as he dubbed them, spread the gospel of describing the entire natural world according to his system. His colleagues and pupils began arriving in St. Petersburg before his binomial taxonomy had become the dominant classification system, and their presence in Russia proved a tremendous benefit to Linnaeus's scientific authority.[9] As his network of agents expanded eastward, it worked to establish his system as the authoritative one in Russian science. This network of agents also created a conduit for specimen collections to reach Linnaeus from the Russian steppe, Siberia, and Central Asia.[10] In eighteenth-century Russia, where medicine and natural science were in their absolute nascence, the very foundations of these fields—their methodologies, values, and infrastructures—were shaped by the disciples of Boerhaave and Linnaeus.

Practical science: the origins of the imperial medical chancellery

Peter's lifelong obsession with empire building stimulated his interest in the practical uses of science. The tsar's fervent quest for territory on the Baltic, Black, and Caspian seas led him to wage war against Sweden (1700–21), the Ottoman Empire (1695–6, 1710–12), and the Persian Empire (1722–3). Seeking a better trade route via the rumoured Northeast Passage and hoping to discover mineral wealth, Peter invested in expeditions to regions unexplored by Europeans or Russians. Peter's empire-building and colonising activities presented expensive logistical challenges. Moving troops overland to distant theatres of war involved mobilising an enormous quantity of horses, carts, cannons, arsenals, munitions, provisions, and supporting personnel. Expeditions similarly required immense organisation to execute, including planning for journeys that began over land and then moved to waterways. This involved transporting the materials for constructing water vessels by ground and then building them on site.[11] Having introduced a navy in 1696 and a large standing army in 1699, Peter's military budget consumed 80 to 85 per cent of state revenues. The burden of these expenditures would drive Peter's reforms in the realm of science and knowledge for his empire. To invest such capital in preparing for battles only to lose fighting power to illness, scurvy (which could disable an entire crew), and lack of even primitive aid for treatable injuries struck Peter as untenable.[12]

During his diplomatic mission to Western Europe, known as the Grand Embassy (1697–8), the tsar dispatched envoys to recruit physicians and barber-surgeons to serve in his military. The tsar also fed his own appetite for training in medicine; at Leiden University, he attended lectures in the anatomical theatre, became personally well acquainted with Boerhaave, and even participated in an autopsy. Peter's lifelong fascination with anatomy and curiosity regarding surgery aided his interest in establishing medical science as an institution in his homeland. He acquired medical instruments, textbooks, and various teaching aids during his Grand Embassy. Some of these souvenirs became favourite tools of the tsar; he had a fondness for extracting the rotting teeth of his friends and underlings, which he then had processed, engraved, and added to his anatomical collection.[13]

Peter's emissaries managed to recruit many medical workers from abroad, expanding on the tradition of previous tsars to recruit court physicians from Western Europe. The admiral of Peter's newly constructed marine fleet persuaded more than 50 European surgeons to serve in the tsar's navy.[14] Peter had imagined that Russians could train on the job, so to speak, while working at the elbows of these foreign specialists. Alas, in many cases, the doctors could not speak Russian, the Russians could not understand orders in other languages, and translators were in even shorter supply than physicians. (Indeed, the first Russian-born physician was drafted into diplomatic service because of his linguistic skills![15]) Continuing to recruit foreign medical personnel in order to staff every ship and battalion of the growing military would be inefficient and expensive. These positions carried substantial risk of mortality, making long-term contracts a hard sell. Hence, the tsar decided to establish a Russian training programme and address the problem of domestic supply.

Peter's initial goals were expedient: train Russians to serve as low-level medics and barber-surgeons and dispatch them as soon as possible to the front. Without a university, plans for establishing a medical faculty would have to wait. Peter commissioned his own physician at the time, the Dutch doctor Nicolaas Bidloo, to design and run what would become the first teaching hospital in Russia and the first school to offer medical courses according to Western European academic methods. Bidloo earned his medical degree at Leiden University, where one of his mentors also worked with Boerhaave. Opened in 1707, the Moscow hospital school's first cohorts of students were solicited from the Holy Synod's Slavic–Latin–Greek Academy, an enrolment decision that at least diminished the language barrier. The state treasury paid their tuition.[16] Most of the graduates from the Moscow hospital school became barber-surgeons in the military, while a few of the most talented matriculated at Western European universities to earn medical doctorates—at the state's expense. Russian ties to Leiden University (and Boerhaave's approach) continued throughout the eighteenth century. Through public sponsorship and private family trusts, increasing numbers of Russian graduates pursued advanced degrees in Leiden (the wealthy brought their serfs, who often enrolled alongside their lords). Throughout the eighteenth century, the recruitment of Bidloo's successors from Leiden fostered the ongoing exchange of students and professors between the two institutions.[17]

In 1715–16, Peter established two more training hospitals based on Bidloo's Moscow school, one designated for the ground forces and one for the navy. (In total, seven hospitals were built on the tsar's orders.[18]) In his Military Charter from 1716, in Chapters 33 and 34, Peter addressed the organisation of medical care in the military. Peter stipulated that the medical workers in service to the military would not charge for their services and "must treat everyone in the regiment ..., from the top to even the bottom, without payment, because they receive their salaries" from the state's treasury.[19] According to some historians, this marked the future trajectory of the distribution and management of medical care: from the very beginning of institutionalised medicine in Russia, the practice, established by charter, was for the Russian state to pay the physicians' salaries directly. The state maintained the right to draft civilian medical practitioners, as well as to recall retired military physicians, surgeons, and medics to service, all of which served colonisation purposes.[20]

Still, supply could not meet the demands of Peter's imperial war machine and further administrative reforms followed. In 1721, Peter reorganised the Apothecary Chancellery in Moscow into the Medical Chancellery, now charged with the management of apothecary and medical practitioners throughout the empire. With the stroke of his pen, all personnel and related establishments came under the purview of the state. Peter's drive to prevent disability and maintain the health of his troops, for the purpose of enlarging the empire, further stimulated the expansion of the state medical administration. The new office continued to license medical practitioners and apothecaries, while also assuming oversight of existing and future hospitals, in anticipation of their establishment throughout the empire. Perhaps most significantly, especially for those ensnared by this new rule, any practitioner licensed by the state could be dispatched by the Medical Chancellery to any region of the empire—no matter how distant. This administrative reform laid the groundwork for institutionalised medicine to develop in Imperial Russia primarily according to the state's dictates rather than demands of the private market.[21] Archival records of the Medical Chancellery confirm that the office posted physicians and surgeons to provincial towns in remote regions of the empire. Some of them expected this imperial arrangement to work in the reverse direction, and petitioned the Medical Chancellery to send more resources.[22] Crucially, these men provided medical care while also acting as imperial agents; when queried about epidemics and other issues, they sent reports back to St. Petersburg.

Academic science: "debarbarising" Russia and the creation of the Academy of Sciences

During his journeys to Western Europe universities, Peter's curiosity about scientific research drew him to visit with professors and scholars of the Enlightenment. Every encounter became a hands-on learning experience for Peter. He met Antonie van Leeuwenhoek and viewed organisms through his microscope. He watched dissections by the professor of anatomy Dr Frederik Ruysch, learned how to suture a wound from him, participated in an autopsy, and later purchased Ruysch's anatomical collection.[23] In 1711, while in Saxony, Peter became acquainted with the German philosopher-scientist Leibniz, who had been pursuing patronage from the tsar for a decade. According to Leibniz, who had been promoting transnational cooperation in scientific inquiry and exploration, the tsar favourably received his ideas for taking scientific "measurements in his vast lands."[24] He also praised Peter's interest in spreading Enlightenment, noting the tsar's desire "to debarbarize [*débarbariser*] this vast empire."[25]

A year later, Leibniz became a salaried advisor to the tsar, having suggested that a scientific academy would be a worthy project to pursue.[26] Peter eventually appointed the physician of the royal family, Lavrentii Blumentrost, who studied in Leiden with the famous Boerhaave, to draw up plans for the Academy of Sciences. The academy would crown St. Petersburg, Russia's new capital on the Baltic Sea, with the glory of scientific discovery and achievement. Blumentrost's plans, which presumably manifested Peter's vision, intentionally adapted the Western European models for scientific academies to the deficits of the Russian context. Without other educational institutions supporting it, the Russian Academy of Sciences would have to serve both research and instructional purposes.[27] In Blumentrost's words, it was essential to "consider the conditions of this country in regard to instructors and students" and therefore mandate that scholars take on teaching duties in addition to conducting research. Not only laurels to the state would issue from the discoveries of its scientists; the academy would "benefit the people in the future through instruction in the sciences and dissemination of scientific knowledge."[28] Thus, from its original design, the Academy of Sciences in St. Petersburg expanded on the models of academies in London, Paris, and Berlin by hosting a secondary school and a university to train the next generation of Russian scientists.

One of Leibniz's pupils, the celebrated German philosopher Christian Wolff, made Peter's short list of candidates for vice president of the academy. Peter tasked Johann Schumacher, of the Medical Chancellery, to convince Wolff to accept the position and relocate to St. Petersburg. Schumacher admitted to Wolff that there were drawbacks to the location—"the climate in St. Petersburg is rough"—but he heaped praise on the city for its high calibre of elites, freedom of religion, and overall degree of refinement. "St. Petersburg is such a polished and civilized city," he wrote, "[It] cannot be excelled by any in Germany."[29] By Schumacher's assessment, Peter's efforts to "debarbarize" his realm were beginning to see results—at least for the elite strata in his new capital. Wolff declined Peter's offer to relocate but agreed to leverage his prominence and networks to recruit prominent faculty from Europe.[30]

Thanks to Wolff's influence and compensation above Western European standards, the first wave of European scholars rolled on to the banks of St. Petersburg in 1725. Some endured as many as seven weeks of travel to reach its shores. Blumentrost and Schumacher greeted the brave souls with hospitality that must have been reassuring; the esteemed scholars were treated to celebratory parties and a month of catered meals. Regrettably, they also ended up guests at Peter's funeral that same year. The tsar's premature death did not threaten the future of the Academy, as some had feared, and its opening proceeded as planned. This speaks to how firmly the Academy had been planted; its future financing and management had been laced to the state rather than to the person of the tsar.[31]

Master gardeners and their seedlings

Among the ranks of the first professors were prominent scholars of the age, along with those just embarking on their academic careers. In addition to fully furnished housing, made cosier with free firewood and a cook supplied by the Academy, these scholars had access to significant research budgets and light teaching loads (owing partly to a dearth of students initially).[32] Nonetheless, most luminaries completed their five-year contracts and soon after departed for familiar and lucrative positions in Western Europe. The established scholars tended to be in their 30s and 40s when they came to Russia. The first cohort included a friend of Leibniz, Jakob Herrmann, a Swiss mathematician who had studied under the famous Jacob Bernoulli. Wolff assisted the philosopher Georg Bernhard Bilfinger in escaping a dustup in Tubingen for a position at the Russian Academy. After five years, Hermann and Bilfinger moved on to other prestigious positions in Basel and Paris respectively. Theophilus Siegfried Bayer, a German specialist in antiquities with expertise in Chinese linguistics and artefacts, served for 11 years until his death. More unusual was the path of French astronomer Joseph-Nicolas Delisle, who arrived from Paris at the age of 37. Perhaps partly because he enjoyed expeditions to less inhabited regions of the Russian Empire, and the collecting of artefacts for his own cabinet of curiosities that they afforded, the esteemed Delisle spent 22 years at the Russian Academy of Sciences.[33]

Some very young academics joined the Russian Academy of Sciences during its early years. Another member of the famous family of mathematicians, Daniel Bernoulli, arrived at the age of 24. He recruited his younger friend, the mathematician Leonard Euler, to join him in St. Petersburg. The two shared an accommodation, but only Euler took to the environment and stayed beyond his initial contract. His legendary memory allowed him to learn Russian quickly, and this in turn gave him access to experiences beyond St. Petersburg. Like Delisle, Euler embraced the adventure offered by his surroundings. As a respite from academic life, the young Euler even enlisted for a short tour of duty as a medic in the Russian navy.[34] The Swiss botanist Johann Amman, who had trained at Leiden, came to Russia at the age of 26. Soon after Amman arrived, Linnaeus began to write to him, soliciting plants that grew in Russia, which Amman would send on the next ship heading west.[35]

These scholars and others at the Academy worked in the first decade to build its local facilities and international reputation. Early on, regular meetings and conferences were organised. Beginning in 1728, the Academy published the research of its members in Latin in an annual journal, *Commentarii academiae scientarum petropolitanae*, which found an audience in the intellectual community of Western Europe.[36] Personal connections facilitated other gains for the Academy. The ownership of manuscript, artefact, and specimen collections sometimes passed to the Academy upon a scholar's death, as with Bayer's manuscripts. Peter's personal library became the foundation for the Academy's, and his cabinet of curiosities, the Kunstkamera, became the repository of ever-growing collections of artefacts, specimens, and minerals in addition to its public role as a museum. The academy campus grew to incorporate a botanical garden, an astronomical observatory with powerful telescope, an anatomical theatre, and various laboratories.[37] In Western Europe, the reputation of the upstart academy in the hinterlands continued to grow: Christian Wolff referred to it as "the paradise of scholars."[38]

Propagation: the Academy of Sciences and Expeditions

As in early European explorations of colonial territories, expeditions into the far reaches of the Russian Empire primarily aimed to create maps and inventories of natural resources. In contrast to the range of private and public expeditions in the Americas, the state sponsored all expeditions in Russia. Peter had long dreamt of state-led expeditions advancing his imperial goals. Peter intended for exploratory expeditions to discover better routes through as yet unmapped waterways, scout sites for establishing fortresses that could grow into reinforced Russian settlements, and, ultimately, to reroute the silk route so that it passed through Russia on its way to Europe.[39] And if silver and gold could be found, so much the better for Russia's trade balance, in the mercantilist calculus of the era. Career captains and hard-bitten crewmembers staffed Peter's early scouting expeditions. Though impressive navigators and hardy warriors, they lacked the scientific training to produce usable maps and to inventory mineral deposits. For this work, Peter needed geodesists, cartographers, and mineralogists.[40]

The shift to including scientists on expeditions happened at the behest of the Scotsman Robert Erskine, who served as Peter's personal physician, consultant, and superintendent of the Kunstkamera. As the museum's collection of artefacts and specimens was expanding quickly, Erskine recognised the need to engage a specialist to process, classify, and organise the collections. Erskine recruited the German physician and scholar Daniel Messerschmidt, who had extensive curatorial experience. Their correspondence reveals that Erskine foresaw sending Messerschmidt on exploratory expeditions.[41] In 1720, Messerschmidt departed for Siberia with orders to study the region's geography, natural history, and epidemic diseases, and to collect specimens of healing botanicals and any other "rarities" he encountered. Commandeering the expedition was a formerly prisoner of war, the Swedish officer who had been taken as a prisoner of war and deported to Siberia. An able cartographer, Philip von Strahlenberg, had surveyed portions of the region while in exile. More than seven years later, Messerschmidt returned to St. Petersburg with 14 horse-drawn wagons of botanical samples, fossils, taxidermised animals, and artefacts taken from Indigenous communities. Messerschmidt's expedition journals produced five volumes of notes, which together with his maps proved invaluable to future expeditions and to the classification of 359 plants unique to Russia.[42]

Planning for a larger-scale exploratory expedition of Siberia and the northern Pacific coast of North America began during the last years of Peter's life.[43] To his dying day, the tsar never stopped dreaming of imperialism and colonialism on a grandiose scale, believing that science and technology together with soldiers, cannons, and gunpowder, could achieve anything he was capable of imagining.[44] As with his endless military campaigns, he imagined epic transformations that disregarded the cost to human lives.[45] A sign of his confidence in imperial benefits of these missions, the tsar committed roughly one-fifth of the state's revenues toward "empirical" expeditions. Peter had learned how essential scientists were to acquiring usable data. He tasked the Academy of Sciences with the planning and execution of all measurements and data collection, and retained military commanders to handle the logistics, security, and transportation.[46]

In 1724, Peter hired a Russianised Dane, Vitus Bering, to lead the first expedition to Kamchatka, the large peninsula off northeastern Siberia. Bering, a naval officer during the Great Northern War, sailed with his crew from Kamchatka to discover the strait later named for him. After fog and ice forced Bering to turn back before reaching land, he began planning for the Second Kamchatka Expedition. Requiring years of preparation, this initiative launched in 1733 under Bering's command. Begun during the reign of Empress Anna (1730–40) and concluded under that of Elizabeth, the Great Northern Expedition (1733–43), as it came to be called later, was one of the most expensive and grand exploratory expeditions of the era. It involved more than 500 scientists, navigators, shipbuilders, carpenters, illustrators, sailors, and soldiers, with an additional 1,500 crew members and Cossacks hired for limited contracts en route.[47]

The Academy of Sciences received detailed instructions regarding their responsibilities. The academic detachment of the expedition should survey the topography of Siberia, make measurements and accurate maps of the route from Kamchatka to the Pacific coast of North America, describe and inventory all of the natural resources, and provide descriptions of the Indigenous peoples and their cultures. Specimens, samples, and artefacts were to be collected and shipped to St. Petersburg. Empress Anna spared no expense; by imperial order, the academic detachment enjoyed *carte blanche*, including the authority to requisition anything they required from administrative and military posts along their route.[48] In return for this unbounded support, crown and state expected that all scientific data, knowledge, specimens, and artefacts would belong to the Russian Academy of Sciences.[49]

Given the youth of the Academy itself, Western European scientists led the academic detachment of the Great Northern Expedition, and Russian scientists-in-training assisted. Three young German scholars became the *de facto* leaders of the multi-disciplinary, international detachment: Johann Georg Gmelin, Gerhard Friedrich Müller, and Georg Wilhelm Steller (who joined later at Okhotsk and sailed on Bering's ship). Gmelin and Steller were physician-naturalists and Müller a historian. Russian graduate students accompanied them, serving as translators, interpreters, and assistants. One Russian in particular, Stepan Krasheninnikov, became an indispensable junior partner to Gmelin and Müller. Gmelin sent Krasheninnikov ahead to Kamchatka to prepare conditions for the scientists. This included supervising the construction of dwellings and making some preliminary surveys of the region—no small order given the inhospitable terrain (swamps, cliffs, geysers, active volcanoes, and mountains). Krasheninnikov collected so much data during his years in the field, he was able to defend a dissertation based on his field research not long after his return to St. Petersburg.[50]

From their mentors and professors, the Russian students learned how to view the frontier through Western European eyes. For the naturalists, this meant viewing the unknown lands as nature waiting to be identified, described, and named: to classify was to possess.[51] Gmelin and Steller were passionate and ambitious in cataloguing Siberia-specific plant and animal species. All three Europeans and Krashenninkov published their personal expedition journals as multi-volume travelogues—in multiple European languages. There was a market for adventures in Siberia. Krasheninnikov appeared in five languages and ran to seven printings.[52] The Russian graduate students also learned how to study the Indigenous peoples already inhabiting the regions. Müller prepared exhaustive instructions for his graduate students and assistants, entitled "Instruction on how one should describe peoples, particularly Siberian," written in 1740. He detailed the objectives and methods of their study in 923 articles.[53] The academic detachment acquired much knowledge, but met with tragedies that became infamous. Owing to presumed abductions, shipwreck, and malnutrition, the fatality rate for the final two crews surpassed 40 per cent.[54]

Science takes root and flowers

Müller's herculean efforts to record oral histories and collect archival documents earned him the title of the "father of Siberian history." His detailed descriptions of Indigenous peoples, languages, spiritual practices, family structures, and political systems became the first proto-ethnography of its kind. He took pains (and risked his then good standing with the state) to collect non-canonical versions of the history of contact between Indigenous peoples and representatives of the Russian state.[55] The material collected by Gmelin became the foundation of botanical knowledge of Siberia. He described over 1,178 plants with professional drawings, which were eventually published in his four-volume *Flora Sibirica*.[56] He was held in high regard by Linnaeus, who named 60 species of plants after him.[57] In 1750, Linnaeus wrote to Krasheninnikov with the highest praise for the botanists of Russia, "[M]ore unknown plants have been found in ten years in the Russian Empire, than in the whole world in half a century."[58]

Müller, Gmelin, and Steller also drafted plans for more effective management of Siberia, including suggestions for increasing Russian settlement and improving its economic development. Their backgrounds in medicine and chemistry were evident in suggestions to improve the organisation of medical care, to utilise certain native plant species as resources for medicinal preparations, and to convince would-be settlers that a locally sourced diet, though absent bread, could be healthy and affordably maintained in the Far East. The discovery of coal and iron ore deposits in the Kuznetsk district along the Tom River led Müller and Gmelin to recommend developing metallurgical industry on the site. Regarding restive borderlands in Western Siberia, Müller suggested incentivising settlement by offering enserfed peasants land and freedom in return for cultivating "uninhabited" lands.[59]

By the mid eighteenth century, Peter's initiatives in the sciences were flowering. In 1745, Mikhail Lomonosov became the first ethnic Russian professor at the Academy of Sciences, marking a significant turning point in its history.[60] In addition to his contributions in the fields of chemistry, physics, and metallurgy, Lomonosov was a formidable advocate for the sciences and education. He lobbied to establish Moscow University in 1755 and subsequently wrote its first charter.[61] His humble origins in a northern fishing town made the ideal raw material for a heroic narrative of upward mobility not through patronage but owing to his intelligence, work ethic, and fierce patriotism. Lomonosov, like his contemporaries during this era at the Academy of Sciences, was conscious of the global-level achievements they were seeking that would raise Russia's status to that of its Western European role models. In 1747, he published an ode in which he praised the achievements of the Great Northern Expedition, and lauded Bering as "the Russian Columbus."[62] His intent in the comparison and in constructing a myth around Bering is clear: finally, though late to enter, the Russian Empire had its own heroes of the age of exploration and discovery. Lomonosov's colleague, the same Stepan Krasheninnikov, was promoted to professor of botany and natural history in 1750.[63] Peter's plans to plant Western science in Russia and to raise Russian achievements to compete with European ones were succeeding.

By the mid eighteenth century, Peter's institutional reforms aimed at expanding the number of medical practitioners and physicians had firmly taken root as well. The early decades of importing physicians and scientists produced two phenomena that led to Russification of the medical field and the Academy of Sciences over the course of the eighteenth century. Firstly, Peter and Leibniz's plans for a three-tiered educational system gradually paid off, and domestic pathways to the professions grew in Russia. Secondly, some of the doctors, professors, and educated servitors who came from Western Europe ended up settling in Russia. Their progeny tended to follow their fathers' example, and made careers in medicine, sciences, and the arts, while assimilating to Russian culture and society. The list goes on.

In the second half of the eighteenth century, Catherine the Great (1762–96), herself a German, would extend Peter's ideas of civilising the Russian metropole and establishing a Europeanised, educated elite. While she continued to import Western European intellectuals, like Peter's master gardeners, she also fostered the growth of Russian-born agents of cultural colonisation, especially in the areas of medicine and midwifery.[64] Informed by the classifying and rationalising spirit of the age, she sought to send these Europeanised agents of science and medicine throughout the empire to study the diversity of peoples of the empire, and to spread Enlightenment. Thus, re-inscribing the circle of Peter's symbiotic colonisation process.

Notes

1 Quoted in Michael Gordin, "The Importation of Being Earnest: The Early St. Petersburg Academy of Sciences," *Isis* 91, no. 1 (2000): 1–31, here p. 7; A.M. Feofanov, "Obrazovatel'nye strategii elity rossiiskogo dvorianstva vo vtoroi polovine XVIII v.," *Voprosy istorii* 1 (2020): 164–75.

2 The best general history of academic science in Russia up to the mid nineteenth century is still Alexander Vucinich, *Science in Russian Culture: A History to 1860* (Stanford, CA: Stanford University Press, 1963). On the history of Western medicine in early modern Russia, see Ya. Chistovich, *Ocherki iz istorii russkikh meditsinskikh uchrezhdenii XVIII stoletiia* (St. Petersburg: Tip. Ya. Troya, 1870) and B.N. Palkin, *K istorii russkoi meditsiny XVIII veka* (Alma-Ata: Kazakhstanskii gosudarstvennyi meditsinskii institute, 1953).

3 I am borrowing the phrase "cultivation of science" ("*vyrashcheivania nauka*") to describe the process of implanting the sciences in eighteenth-century Russia from S.I. Vavilov by way of L.N. Pushkarev, "K Iubileiu Akademii Nauk SSSR: akademiia nauk i russkaia kul'tura xviii veka," *Voprosy istorii* 5 (1974): 32.

4 On the conceptualisation of European Russia as the metropole and the eastern or "Asiatic" regions as the colonies, see Mark Bassin, "Expansion and Colonialism on the Eastern Frontier: Views of Siberia and the Far East in Pre-Petrine Russia," *Journal of Historical Geography* 14, no. 1 (1988): 5. The history

of Russian conquest and expansion is long and complicated, and the scholarship is deep as well. The European territories of Russia (then called Muscovy) were fully consolidated in the fifteenth century under Ivan III. In the sixteenth century, Ivan the IV (the "terrible") conquered the khanates of Kazan, Astrakhan, and Sibir' (1552–80). Expansion continued in the sixteenth century, with the designation "Russian Empire" (*Rossisskaia imperiia*) adopted in 1721 during the reign of Peter the Great. On the history of Russian imperialism and colonisation, and the state's dynamic views of its incorporated territories and peoples, see Michael Khodarkovsky, *Russia's Steppe Frontier: The Making of a Colonial Empire, 1500–1800* (Bloomington: Indiana University Press, 2002); Yuri Slezkine, *Arctic Mirrors: Russia and the Small Peoples of the North* (Ithaca, NY: Cornell University Press, 1994).

5　Sverker Sörlin, "Ordering the World for Europe: Science as Intelligence and Information as Seen from the Northern Periphery," *Osiris* 15 (2000): 51–69. This special issue, "Nature and Empire: Science and the Colonial Empire," includes 16 articles, none addressing the Russian context.

6　The scholarship on the history of science in eighteenth-century Russia has yet to elaborate fully the connections between imperial policy and the development of academic and practical science. One exception is Sunderland on geography as a science becoming an imperial lens and tool. William Sunderland, "Imperial Space: Territorial Thought and Practice in the Eighteenth Century," in *Russian Empire: Space, People, Power, 1700–1930*, ed. Jane Burbank, Mark Von Hagen, and Anatolyi Remnev (Bloomington: Indiana University Press, 2007), 33–66. Recent work by Marina Loskutova on the nineteenth century treats scientific knowledge as critical to the administration of the empire; see her "'Svedeniia o klimate, pochve, obraze khoziaistva i gospodstvuiushchikh rasteniiakh dolzhny byt' sobrany...': prosveshchennaia biurokratiia, gumbol'dtovskaia nauka i mestnoe znanie v Rossiiskoi Imperii vtoroi chetverti XIX v.," *Ab Imperio: Studies of New Imperial History and Nationalism in the Post-Soviet Space* 4 (2012): 111–56.

7　For Boerhaave's influence in eighteenth-century Russia, see Iu. V. Natochin, "Stanovlenie fiziologii v rossii: xviii vek," *Istoriko-biologicheskie issledoveniia* 8, no. 2 (2016): 9–24, here p. 10. See also Nicholas Hans, "Russian Students at Leyden in the 18th Century," *The Slavonic and East European Review* 35, no. 85 (1957): 551–62.

8　Harold J. Cook, "Boerhaave and the Flight from Reason in Medicine," *Bulletin of the History of Medicine* 74, no. 2 (2000): 221–40.

9　Dániel Margócsy, "'Refer to Folio and Number': Encyclopedia, the Exchange of Curiosities, and Practices of Identification before Linnaeus," *Journal of the History of Ideas* 71, no. 1 (2010): 63–89.

10　Margery Rowell, "Linnaeus and Botanists in Eighteenth-Century Russia," *Taxon* 29, no. 1 (1980): 15–26; Sörlin, "Ordering the World for Europe," 60. Lisbet Koerner's acclaimed monograph about Linnaeus only mentions Russia in passing, as an example of the linguistic challenges encountered by botanists when cataloguing plant species according to their vernacular names. Lisbet Koerner, *Linnaeus: Nation and Nature* (Cambridge, MA: Harvard University Press, 1999), 46.

11　Georg Steller, *Journal of a Voyage with Bering*, ed. O.W. Frost (Stanford, CA: Stanford University Press, 1993), 47–57; Vucinich, *Science in Russian Culture*, 58–63, 99–105; Evgenii V. Anisimov, *The Reforms of Peter the Great: Progress through Coercion*, trans. John T. Alexander (Armonk, NY: M.E. Sharpe, 1993), 87–169, 244–63.

12　For an overview of Peter's reign and his extensive reforms, see Lindsey Hughes, *Russia in the Age of Peter the Great* (New Haven, CT: Yale University Press, 1998).

13　Inge F. Hendriks, James G. Bovill, Dmitrii A. Zhuravlev, Dmitrii A. Zhuravlev, Fredrik Boer, and Pancras C.W. Hogendoorn, "The Development of Russian Medicine in the Petrine Era and the Role of Dutch Doctors in This Process," *Vestnik sankt-peterburgskogo universiteta- meditsina* 14, no. 2 (2019): 159, doi.org/10.21638/spbu11.2019.208; John T. Alexander, "Medical Developments in Petrine Russia," *Canadian-American Slavic Studies* 8, no. 2 (1974): 199–217, reprinted in *Peter the Great Transforms Russia*, ed. James Cracraft (Lexington, MA: D.C. Heath and Company, 1991), 193–208.

14　Alexander, "Medical Developments," 194–6.

15　Hendriks et al., "The Development of Russian Medicine," 161.

16　The lack of secondary institutions to prepare Russian students was a significant hindrance. As a consequence, seminaries served as the only domestic route to eventually entering the profession of medicine. The state's desperate need for trained practitioners led to a tradition of recruiting and funding the higher education of Russian former seminary students abroad. f. 730, op. 2, d. 1, ll. 15–17 ob., Rossiiskii gosudarstvennyii istoricheskii arkhiv (Russian State Historical Archive, hereafter RGIA), St. Petersburg; Hendriks et al., "The Development of Russian Medicine," 162–3.

17　Hans, "Russian Students at Leyden," 551–62.

18 Peter drafted the rules for the hospitals, stipulating the desired quotas of graduates to meet the needs of the military. Hendriks et al., "The Development of Russian Medicine," 165; V.M. Richter, *Istoriia meditsnyi v rossii* (Moscow, 1814), 3: 240–1.

19 Quoted in A.N. Pishchita and N.G. Gocharova, *Evoliutsiia pravovogo regulirovaniia zdravookhraneniia v Rossii: Istoiko-pravovye aspekty* (St. Petersburg: Rossiskaia akademiia nauk, 2007), 10.

20 Pishchita and Goncharova, *Evoliutsiia*, 3–11. See Chapter 2, "The State and Midwifery," in Anna Kuxhausen, *From the Womb to the Body Politic: Raising the Nation in Enlightenment Russia* (Madison: University of Wisconsin Press, 2013) and John T. Alexander, *Bubonic Plague in Early Modern Russia: Public Health and Urban Disaster* (Oxford: Oxford University Press, 2003), 40–50. Owing to the role of the state as chief employer of physicians, the profession of medicine evolved in Russia such that doctors remained relatively low-status, underpaid, and reported to non-physicians within the state bureaucracy, especially in the provinces. Nancy Mandelker Frieden, *Russian Physicians in an Era of Reform and Revolution, 1856–1905* (Princeton, NJ: Princeton University Press, 1982), 30–48.

21 The one exception concerned the hospital school in Moscow, which remained under the purview of the Holy Synod. For a quick overview of the state medical administration in this era, see Alexander, *Bubonic Plague*, 38–41.

22 f. 1295, o. 1, RGIA, St. Petersburg, includes reports from physicians posted to provincial offices of the Medical Chancellery (later Medical Collegium) around the empire.

23 Hendriks et al., "The Development of Russian Medicine," 159.

24 Alexander Lipski, "The Foundation of the Russian Academy of Sciences," *Isis* 44, no. 4 (1953): 349–54. Gordin, seeking to revise the view of Peter's reforms as mainly utilitarian, argues that the Russian Academy of Sciences was intended to function not only as a research centre but also as a site for promoting European etiquette and manners. Gordin, "The Importation of Being Earnest."

25 Quoted in Yuri Slezkine, "Naturalists Versus Nations: Eighteenth-Century Russian Scholars Confront Ethnic Diversity," *Representations* 47 (1994): 171, doi.org/10.2307/2928790.

26 Vucinich, *Science in Russian Russian Culture*, 45–48.

27 Peter built his new capital on territory acquired from the defeated Swedes. Lipski, "Russian Academy of Sciences," 350–1; Vucinich, *Science in Russian Culture*, 47; Ludmilla Schulze, "The Russification of the St. Petersburg Academy of Sciences and Arts in the Eighteenth Century," *The British Journal for the History of Science* 18, no. 3 (1985): 305–35.

28 Quoted in Lipski, "Russian Academy of Sciences," 351.

29 Quoted in Lipski, "Russian Academy of Sciences," 351.

30 Gordin, "The Importation of Being Earnest," 4; Lipski, "Russian Academy of Sciences," 350–3.

31 Ronald Calinger, *Leonhard Euler: Mathematical Genius in the Enlightenment* (Princeton, NJ: Princeton University Press, 2016), 38.

32 Lipski, "Russian Academy of Sciences," 353.

33 Lipski, "Russian Academy of Sciences," 349–54.

34 A.I. Rusanov, "Beginnings of Chemical Science in Russia," *Zhurnal obshchei khimii* 81, no. 1 (2011): 4, doi.org/10.1134/S1070363211010026. Euler married an expat and raised children in Russia, returned to Prussia for 25 years, and then came back to Russia where he conducted research into old age, even after his eyesight failed. Nicolas Fuss, *Eloge de Monsieur Léonard Euler, lu à l'Académie impériale des sciences de S.- Pétersbourg dans son assemblée du 23 octobre 1783 par M. Nicolas Fuss*, 1783, facsimile reprint. On Euler's many productive years in Russia with lively details about the collegial life of the Academy, see Calinger, *Leonhard Euler*, chs. 2, 3, and 13.

35 Margócsy, "'Refer to Folio,'" 63–5. Amman maintained participation in a wide and growing network of botanists and natural historians around Europe. Through this correspondence network, these collectors and classifiers of plant life actively exchanged seeds and plants. One of the most active and successful solicitors of botanical material was Linnaeus.

36 Lipski, "Russian Academy of Sciences," 353.

37 Vucinich, *Science in Russian Culture*, 58, 76; T.V. Stankiuvich, *Kunstkamera peterburgskoi akademii nauk* (Moscow and Leningrad, 1953), 28–40; Lipski, "Russian Academy of Sciences," 349–53; Schulze, "Russification," 305–7.

38 Quoted in Calinger, *Leonhard Euler*, 38.

39 This dream of another passage from Europe to the Orient was at least as old as the sixteenth century. Anisimov, *Reforms of Peter the Great*, 255–63.

40 Anisimov, *Reforms of Peter the Great*, 259–60; Vucinich, *Science in Russian Culture*, 58–60; Sunderland, "Imperial Space," 33–66.

41 "Istoriia kunstkamery" on the official website for the Museum of the Kunstkamera. www.kunstkamera. ru/exposition/kunst_hist.

42 Vucinich, *Science in Russian Culture*, 59–60. Messerschmidt passed away in 1735. Steller married his widow, through whom Messerschmidt's unpublished notebooks and maps passed into Steller's possession. His journals finally found a publisher in the twentieth century. D.G. Messerschmidt, *Forschungsreise durch Sibirien, 1720–1727*, eds Eduard Winter and N.A. Figurovskii, 5 vols. (Berlin: Akademie-Verlag, 1962–77).

43 On the shift that began with Peter to viewing territories as resources to be exploited, see Sunderland, "Imperial Space."

44 One such project submitted to Peter the Great set out elaborate plans for Russian colonization in the Americas. "Proekt o kolonizatsii ameriki, podannyi petry pervomy," in I. Zabelin, *Moskvityanin*, vol. 1 (1851), 121–4.

45 These ambitions included massive population resettlements of the sort only attempted by force during the twentieth century. Anisimov, *Reforms of Peter the Great*, 259–60.

46 A.Kh. Elert, "Great Northern Expedition: In the Wake of the Academic Detachment," *Science First Hand* 18, no. 6 (2007): 6–11; Vucinich, *Science in Russian Culture*, 59–62; Lipski, "Russian Academy of Sciences," 353.

47 Elert, "Great Northern Expedition," 6–11.

48 Sometimes the orders for supplies were so extensive and the regional office so remote, it took many months to fulfil them. Steller spent many months in Irkutsk and studied Lake Baikal environs while waiting for officials to collect the resources they needed. Frank N. Egerton, "A History of the Ecological Sciences, Part 27: Naturalists Explore Russia and the North Pacific During the 1700s," *Bulletin of the Ecological Society of America* (2008): 43, doi.org/10.1890/0012-9623(2008)89[39:AHOT ES]2.0.CO;2.

49 This early-modern version of intellectual property rights was spelled out in the scholars' contracts.

50 Elert, "Great Northern Expedition," 6–11. Many of these Russian graduate students had been recruited from the Slavic–Greek–Latin Academy of the Holy Synod to enrol in the Academy of Sciences' university and prepare to accompany the expedition in these roles. Vucinich, *Science in Russian Culture*, 99–105.

51 Mary Louise Pratt, *Imperial Eyes: Travel Writing and Transculturation* (New York: Routledge, 2007), 7. Pratt writes that scientific exploration created "new ways of encoding Europe's imperial ambitions," 23–4. On the problem of "Russian" identity in a multi-ethnic empire in which those classifying and naming were themselves foreigners (or felt themselves to be), see Slezkine, "Nations and Naturalists," 170–95. On "cultural cartography" and Western European construction of Eastern Europe as the periphery of the Enlightenment, see Larry Wolff, *Inventing Eastern Europe: The Map of Civilization on the Mind of the Enlightenment* (Stanford, CA: Stanford University Press, 1994), 17–48, 144–93.

52 Pushkarev, "K iubileiu Akademii Nauk," 32.

53 Note that Müller went by Miller in Russia. For a few translated excerpts from some of Müller's hard-to-access or unpublished works, by a well-known Russian historian, see A.Kh. Elert, "A Description of Siberian Peoples: The Knights of the Taiga," *Science First Hand* 18, no. 6 (2007): 54–7; "G.F. Müller: On Peoples' Spiritual Qualities," *Science First Hand* 26, no. 2 (2010): 1–7; A.Kh. Elert, *Narody sibiri v trudakh G.F. Millera* (Novosibirsk: Izd-vo Instituta arkheologii i ėtnografii, 1999).

54 Steller, *Journal of a Voyage*, 109–41, 165–9. Other accounts pale in quality compared to Steller's first-hand reports.

55 A.Kh. Elert, " 'It is Hard to Handle the Koryaks Using Firing Arms': On the History of Russian–Koryak Relations," *Science First Hand* 52, no. 2 (2019): 52–67. Some documents from the Great Northern Expedition have been rescued from archives and published. See A.I. Alekseev and R.V. Makarova, eds., *Russkie ekspeditsii po izucheniiu severnoi chasti tikhogo okeana v pervoi polovine XVIII v.* (Moscow: Nauka, 1989); N. Okhotina-Lind and P. Ul'f Meller, eds., *Vtoraia Kamchatskaia ekspeditsiia: Dokumenty 1734–1736* (Saint Petersburg: Nestor-Istoriia, 2009). According to Elert, thousands more lie waiting in Rossiiskii gosudarstvennyi arkhiv drevnykh aktov, Moscow (Russian State Archive of Early Acts).

56 A.Kh. Elert, "A Journey across Siberia," *Science First Hand* 18, no. 6 (2007): 12–17.

57 Rusanov, "Chemical Science in Russia," 4–5.

58 Rowell, "Linnaeus and Botanists," 21.

59 Some of the plans were far ahead of their time, and they were not implemented until the nineteenth and twentieth centuries. For example, the metallurgical project was developed in the 1930s. A.Kh. Elert, "Great Northern Expedition," 6–11.

60 The status of Lomonosov in the history of Russian science, education, and culture cannot be overstated. Pushkarev, "K iubileiu akademii nauk," 28–38. For a reconsideration, see Steven A. Usitalo, *The Invention of Mikhail Lomonosov: A Russian National Myth* (Brighton, MA: Academic Studies Press, 2013), 129–66. For another revision that supports Usitalo's emphasis on Lomonosov's expert self-promotion, see Simon Werret, "Green Is the Colour: St. Petersburg's Chemical Laboratories and Competing Visions of Chemistry in the Eighteenth Century," *AMBIX* 60, no. 2 (2013): 122–38, doi. org/10.1179/0002698013Z.00000000027; Rusanov, "Chemical Science in Russia," 5–9.

61 Pushkarev, "K iubileiu Akademii Nauk," 30.

62 Yuri Sokolov, "Columbus, the Discovery of America, and Russia," *GeoJournal* 26, no. 4 (1992): 501.

63 Vucinich, *Science in Russian Culture*,105. By the end of the eighteenth century, the majority of the faculty were Russian. Schulze, "The Russification of the St. Petersburg Academy of Sciences," 305–35.

64 See, for example, Anna Kuxhausen, "The Modern Miracles of Mother's Milk: The New Science of Maternity in Enlightenment Russia," *Ab Imperio* 3 (2009): 94–118; and Chapters 2, 5, and 6 in Kuxhausen, *From the Womb to the Body Politic*. On the educated elite during Catherine's era, see Andreas Schönle, Andrei Zorin, and Aleksei Evstratov, eds., *The Europeanized Elite in Russia, 1762–1825* (Dekalb, IL: Northern Illinois University Press, 2016). For scientific expeditions initiated by Catherine and her ambitions regarding cultural colonization, see Andreas Schönle and Andrei Zorin, "The Europeanized Self Colonizing the Provinces," in *On the Periphery of Europe, 1762–1825* (Dekalb, IL: Northern Illinois University Press, 2018), 171–98.

18

SCIENTIFIC KNOWLEDGE IN THE QING EMPIRE

Engaging with the world, 1644–1911

James Flowers

The Manchu-led expansionist Qing Empire (1644–1911) presided over systems of administration more effective than its predecessors, achieving a level of material productivity and prosperity far beyond that of any earlier Chinese dynasty.[1] Although many scholars accept this characterisation of the Qing, it is still not the established view of all scholars, including in China itself.[2] In the second half of the twentieth century, the father of modern Chinese history in the United States and influential Harvard scholar, John Fairbank, for instance, established the widespread view of the Qing Empire as stagnant and weak in comparison to the modern West.[3] The Fairbank historical model of a hapless Qing dynasty correlated with the view of many early-twentieth-century Chinese thinkers who viewed it as an ossified regime that stifled science and intellectual inquiry.[4] In contrast to this ahistorical failure narrative of a weak Qing, examining the evidence shows a Qing empire with many accomplishments, with flourishing fields of knowledge, including mathematics, astronomy, medicine, geography, cartography, engineering, and the natural sciences. Nonetheless, many historians have ignored or marginalised the sciences in their analyses of China, which the historian of science Nathan Sivin refers to as the "awesome taboo."[5] In fact the sciences were central to the remarkable Qing achievements in statecraft and in governing the empire. In short, the idea which some historians have promoted, of the Qing exhibiting little interest in science, is one which was invented to serve the political interests of twentieth-century ruling elites hoping to demean the merits of their predecessors.[6]

Historians have analysed the Qing Empire using diverse methodologies, probing different dimensions of the past, whether they be political, economic, or social history. A growing number of historians of science and medicine in China, however, have argued that such historical understanding is limited when the central role of what we call science—or as the Chinese actors understood it, the investigation of things (*gewu*)—is not integrated into the broader history. The early Qing emperors and many of the imperial relatives and their courtiers focused on science as central to Qing imperial formation and governance. For example, mathematics, as well as medicine and other disciplines, was central to the Kangxi emperor's (1654–1722; r. 1661–1722) study schedule. Whereas Joseph Needham, the pioneering scholar of the history of science in China, did most of his research on the periods preceding the Qing, his point that China's history was rich in science laid a foundation for the next generation of scholars to analyse the processes of knowledge production in the Qing period.[7] The historian

William Rowe has made a convincing case for Qing dynamism and successes, with Benjamin Elman's 2005 book *On Their Own Terms: Science in China, 1550–1900* arguing that science was central to such vibrancy.[8] Hence, rather than adopting the teleological view of an empire heading towards weakness and victimhood, since the early 2000s, most Qing scholars have subsequently confirmed, challenged, complicated, and built on the interpretation of the Qing as sophisticated and multidimensional.

And that the imperial rulers were Manchu, and not Han Chinese, has emerged as an important factor in new analyses of the Qing Empire.[9] In this regard, historians of science, while not always directly focusing on the question of Manchu identity, have nevertheless demonstrated that the Manchu rulers often acted as intermediaries between European missionary scholars and Han Chinese scholars. Some of these scholars have also recently argued that such complexity complicates the previous assumptions of a simplistic East–West binary, with Western science dominant over a China devoid of science.[10] Instead scholars are moving towards the idea that there were more commonalities shared among a complex array of participants in knowledge making than previously thought.[11]

Historians of science and medicine have also shown that scholar-officials outside the imperial court were involved in the daily practice of the investigation of things.[12] At the local level, and across the empire, Qing officials used technical knowledge to govern provinces, counties, and districts.[13] In conjunction with local villagers and according to local conditions, officials busied themselves with applying knowledge in the service, for example, of agriculture, forestry, land management and conservation, irrigation and water conservancy, and construction engineering.[14]

Historians are trained to recover and analyse the past, and many scholars are reluctant to opine on connections with the present. Moreover, scholars of Chinese history have often confined their focus to a certain historical time period. Volker Scheid is an exception in that he draws a thread of continuity in medicine in the Qing to modern-day China and even to modern-day United States.[15] Although Scheid is less interested in drawing connections between medicine and other fields of knowledge making, he demonstrates not only how medical knowledge developed rapidly during the Qing dynasty but also how those medical practices remain important today. Qing medicine continues in the twenty-first century as a major presence in people's lives in China, generally in East Asia, and increasingly on a global scale. The question remains, though, is the Qing contribution to the global scientific community limited to medicine, and only a recent phenomenon?

For the most part, scholars have analysed the Qing as a site of knowledge exchange in which Chinese and Manchu scholars drew on local textual and practical knowledge, both native to the Empire and brought in from outside. Only recently, however, have scholars begun to show that the Qing Empire not only learnt from missionaries, but also contributed scientific knowledge to the global community of scholars. In addition, the Qing was the first empire to operate high-volume industrial production for export, especially ceramics.[16] In response to the famous European thirst for Chinese tea, the British were fascinated with Qing tea cultivation methods.[17] The history of Qing export to Europe of luxury items—ceramics, tea, and, of course, high-quality silk—is well known and the subject of numerous studies.[18] Less well known is the Jesuits' transmission of Chinese medical knowledge to a significant number of physicians in Europe during the seventeenth century.[19] Yet another example of knowledge transmission, closely related to Qing imperial power, was the Kangxi cartographic project. Mario Cams convincingly shows that what scholars had hitherto thought of as a Jesuit mapmaking project was actually characterised by significant Qing input. The Qing court accepted French expertise by adding it to their own existing store of knowledge, thus producing new and improved cartographic techniques. In the eighteenth century, this project was the largest and arguably most significant geographical surveying project in the world to date.[20] Thus the Qing imperial capital Beijing, and the Qing imperial court, in particular, acted as an important node of knowledge transmission not only within the empire but also on the Eurasian continent as a whole.

This chapter first focuses on what we call science in the Qing court, then discusses some examples of science practised outside the court, and finally, concludes on the importance of medicine during the Qing dynasty.

Science in the imperial court

Several scholars have accepted the trajectory of long-standing Qing interests in the natural world. Notably, Benjamin Elman traces a thread of intense study of the sciences in China, from the late Ming until the end of the Qing.[21] Close engagement with European missionaries was not simply a process of unidirectional learning from Europe. Rather, the attention to mastery of European natural learning over several centuries during the Qing meant learning "on its own terms." Scholars have examined the significance of the intimate relationship between the earlier Qing emperors, most notably the Kangxi emperor, and the royal princes and Jesuit missionary-scholars who worked in the imperial palace as advisors and teachers.[22] Most scholars agree on the extraordinary dynamism of these emperors personally engaging in investigation of knowledge. This was not a case of emperors relying on experts' advice and then making decisions accordingly. Rather, the emperors and the princes embarked on extensive study programmes, most notably in mathematics. For example, the Kangxi emperor studied on a one-to-one basis with the Flemish Jesuit scholar Ferdinand Verbiest (1623–88). The Kangxi emperor even famously adjudicated a dispute, in 1660, on calendrical astronomy, between the Chinese senior official, Yang Guangxian (1597–1669), who was critical of the Western calendar, and the Jesuits, in particular, Johann Adam Schall von Bell (1591–1666), who presided over the astronomical bureau.[23] In order to understand the issues, Kangxi had embarked on his own study of astronomy and the calendrical sciences.[24]

The fact of personal imperial involvement meant there was an unprecedented emphasis on science in the Qing. For 50 years, Kangxi and the princes established various networks of learning, connecting Manchus, Chinese, French, Portuguese, and others, spanning the Eurasian continent. Catherine Jami foregrounds her work on science in the Qing court by reminding us that many of the most pertinent works were written in Manchu, not Chinese language. Since most present-day scholars have only read the Chinese materials, we have yet to uncover the depth and breadth of the court's involvement in scientific investigation.[25] Nevertheless, even what has been uncovered so far has demonstrated a dynamic Qing court engaging with science. Mathematics, often regarded as the touchstone of both modernity and universality in the sciences, came to be appropriated as part of the construction of rulership by the Qing dynasty, to which the historiography had long denied participation in either modernity or universality. One of the reasons for the success of the Jesuits' teaching was the perceived relevance of their mathematical knowledge to statecraft. Calendrical methods were linked to essential affairs of state. Astronomy as an imperial monopoly symbolised the emperor's role as an intermediary between heaven and earth. For this, and other reasons, Kangxi understood the sciences as important for ruling, perhaps even parallel with the classical literary texts. As the earlier Jesuit missionary Matteo Ricci (1552–1610) had argued, mathematics was essential in holding imperial power through warfare. It was necessary in logistical planning, surveying distances, battle formations, and producing weaponry.[26]

Not only did Manchu princes and some of their royal relatives study and practise the sciences, but also did Chinese literati scholars close to the court. For example, the famous mathematician Mei Wending (1633–1721) challenged the view that the Jesuit sciences were new.[27] He pointed to earlier Chinese mathematical texts to argue for the unity of Chinese and foreign mathematics. This view came to predominate among the literati and also the Manchu royals. The central goal here was syncretism, combining the Jesuit ideas with existing and new Chinese ideas to produce improved universal knowledge.

The effort to integrate scientific knowledge and practices arguably saw its clearest expression in cartography.[28] The Kangxi emperor commissioned the Jesuits to survey the entire empire between 1707 and 1718.[29] For the Qing, better maps were tied to imperial expansion, with new surveying and mapmaking techniques essential. Until a generation ago, this massive project was understood as a French Jesuit one. Revising our understanding, Cams has shown that the project was as much Qing as it was French.[30] The largest mapping project the world had seen to date, the Qing surveyors married new techniques to existing practices, a phenomenon common to the development of the sciences in general. Cams thus defines the process as interculturally hybrid, and argues that we need to move beyond categorising knowledge as either European or Asian.

A similar argument of the importance of intercultural exchange may be applied to the production of clocks, porcelain, high-quality glassware, and jade artefacts, among other products.[31] For example, in 1693, Kangxi established a number of workshops that manufactured luxury items. These workshops were staffed by Jesuits and Chinese workmen. A number of porcelain kilns from the seventeenth to the nineteenth centuries constituted an industry unrivalled in the world for its quality and its export power. Kangxi established control of the most well-known kiln, Jingdezhen. Chinese porcelain was known across the Eurasian continent, characterised from the eighteenth century by a hybrid style influenced by European painting techniques.[32]

According to much of the historiography, the Qing had its heyday in the Kangxi, Yongzheng (1678–1735; r. 1709–35), and Qianlong (1711–99; r. 1735–96) reign periods. In this established view, after this Golden Age the Qing went into gradual but serious decline in strength and capacity. Well known is the precarity of the Jesuits in the eighteenth century when they were expelled in stages from China, reflecting an inward turn among literati. John Fairbank established the view that the Opium War (1839–42) marked a turning point after which a failing Qing China was compelled to haltingly find a path to modernity in response to an encroaching British imperialism. However, science and knowledge production during the Qing period may be better understood if we acknowledge a process of continuity, rather than over-emphasising a hard break in the mid nineteenth century. Despite an interim period of a few decades from about 1800 to 1850, when there was minimal Qing engagement with the world, Elman argues against the view of decline by insisting on continuity in the Qing's use of science and expert knowledge for the sake of financial income and imperial governance.[33] Moreover, there was a deep engagement in exchange of knowledge with Protestant—and mainly British—missionaries, starting in the mid nineteenth century. However, Chinese thinkers, scholars, and officials did not suddenly turn to science due to a rupture caused by British incursions. Rather, they were working from a solid base of knowledge built by their forebears since the seventeenth century and earlier.

In short, rather than Qing failure in the nineteenth century, the substantial Qing-led arsenals and industrial enterprises demonstrated strength and resourcefulness. The pursuit of Western munitions technology began in earnest when the senior official Zeng Guofan (1811–72) established an arsenal in Anhui in 1862. In the late nineteenth century the arsenals and plants were, for example, well ahead of Meiji Japan in scale and sophistication. Due to the persistence of the Qing failure narrative, the array of arsenals, shipyards, factories, and translation bureaus

Figure 18.1 Vase, China, Qing dynasty, Qianlong period (1736–96), porcelain with underglaze cobalt oxide decoration. Honolulu Museum of Art. Hiart, CC0, via Wikimedia Commons.

has been undervalued. Rather than a Western intervention or imposition upon a Chinese way of life, these examples of enterprises demonstrated a continuing Chinese fascination with the application of science and technology.

The significant attention that the Qing court paid to science belies the stereotype of an isolationist Qing China hostile to Western knowledge, and therefore stagnant and weak. The Qing Empire built its immense power on its ability to rule a vast area, including increasing the population through attention to agriculture and industry, based on scientific knowledge. Although subsequent emperors did not achieve the intensity of Kangxi's studies of science, on the whole, the evidence shows the opposite of a stagnant empire. More often than not, the Qing court was not only committed to science, but also made scientific investigation a key priority in ruling the empire. Furthermore, rather than thinking of science as Western, it is more useful to think of hybrid knowledge learnt from missionaries and others combined with existing forms of knowledge and experience in China. Rather than using terms such as stagnant and sclerotic, it would be more accurate to describe the Qing court as curious and keen to learn newly available scientific knowledge. The Qing–Jesuit project in the Imperial Palace may thus be best characterised as a major intercultural project that had been hitherto unprecedented in Chinese ruling circles.

Outside the palace

While much of the scholarship has concentrated on scholars connected to the palace studying astronomy and cartography, some historians have argued that a range of actors not directly associated with the court were also involved with areas of knowledge that we call science.

There were, to be sure, real limitations in the acquisition of new scientific knowledge during the Qing dynasty. The best-known provincial efforts were a group of scholars in the Lower Yangzi region, who through what they called "evidential studies" (*kaozheng*), claimed that the essential elements of learning could be found in Chinese classical texts. In this way, the imagined glories of Chinese antiquity could be revived. Famous mathematicians, such as Mei Wending and his grandson Mei Juecheng (d. 1763), along with well-known scholars such as Gu Yanwu (1613–82), Dai Zhen (1724–77), and Qian Daxin (1728–1804), made textual scholarship their priority.[34] Thus, for these scholars inquiries into natural phenomena were dependent on recovering classical texts, and not experimentation.

More recently there has been an effort to revisit the question of science outside Beijing, and to look beyond the "evidential studies" scholars, by taking seriously the writing and practices of local scholars and scholar-officials. For example, scholars such as Dagmar Schäfer,[35] Francesca Bray,[36] and Madeleine Zelin,[37] among others,[38] have also analysed specific technologies. Schäfer has focused on craft technologies, Bray on agriculture, and Zelin on salt production. One factor in the neglect of what we call science in the general history of the Qing is that science was not an actors' category for the people of that time and place. For example, in Schäfer's study of Song Yingxing (1587–1666?), a scholar-official whose life straddled the end of the Ming period (1368–1644) and the beginning of the Qing, he engaged in years of study and application of craft work. However, as with most scholars during both the Ming and Qing dynasties, he was careful to frame his work in the language of Confucian scholasticism. Song understood himself as working on the "investigation of things." He observed processes of nature and documented these in all their detail; his "scientific" career thus complicates the characterisation of Chinese scholarship as solely focused on metaphysical philosophical musings. Pertinent to Chinese scholars practising science "on their own terms,"[39] Song understood the things he observed and the objects he worked with in terms of *li* (pattern or principle), *qi*, yin yang, and five phases.

Francesca Bray defines science as knowledge about natural, material processes expressed in declarative, transmissible form. She also argues that many Chinese officials in the field practised science as a core activity in their work.[40] From this perspective, technology is the generation of material goods and social relationships that contribute to knowledge production defined as science. Thus the fact that large numbers of Chinese officials working for the state were deeply involved in agricultural production meant they were practising a "scientific" activity. Many officials produced agronomic scholarly treatises that were circulated and subsequently improved upon by other officials. The state actively facilitated these networks by recording and transmitting agricultural information. Agronomy was state science in late imperial China, in a way that had not occurred in Western states before the nineteenth century. Despite the role of a scientific state, scholars needed to draw on illiterate peasants to encode their new innovative agricultural knowledge. Thus agricultural science was not simply scholarly philology but was rather a process of applying knowledge involving both highly educated elite scholars and large numbers of peasants. The illustrious Qing official Chen Hongmou (1696–1771) exemplifies the scholar-official who was involved in practical technical knowledge, such as large water conservancy projects.[41] Although there is debate about the extent of technological innovation during the Qing, most agree that the proliferation of print culture reveals a high level of interest in, transmission of, and commodification of technology during this period.[42]

This chapter so far has discussed science in the imperial palace as well as scholar-officials' involvement in scientific activities but now we turn to examine scientific practice beyond state supervision. For example, in her study of the Furong salt production facilities, Madeleine Zelin examines in close detail an advanced industrial community in southern Sichuan from the eighteenth to the early twentieth centuries. This business history supports the larger argument that

Figure 18.2 Zigong salt wells, Qing dynasty. Photo from Zigong Salt Museum, 2004. Phreakster 1998 at en.wikipedia, Public domain, via Wikimedia Commons.

a benign and almost *laissez-faire* Qing state facilitated industry with a dense network of roads and watercourses throughout the empire. Although some historians have argued that a stagnant Qing operated with a cultural bias against business that precluded industrial development in China, the case of the Furong salt production facilities shows the scale and sophistication of the technology required to operate such large industries. As Zelin explains, "The percussion drill technology, as well as the craftsmanship behind the construction and maintenance of wells, was as advanced as any available in the nineteenth-century world."[43] Until the twentieth century, the Sichuan enterprise operated through Indigenous technology, with private Chinese entrepreneurial ownership and management.

The phenomenon of hybrid intercultural knowledge production was not only an imperial palace phenomenon in the Qing. In a study of British naturalists in Guangzhou from the nineteenth to the early twentieth centuries, Fa-ti Fan shows that natural scientific exploration was a widespread phenomenon.[44] Many British explorers and scholars first made their study of sinology and of China through the examination, recording, and categorisation of plants and animals. Although the British, for the most part, ridiculed the Chinese categorisations of the natural world, knowledge production was a shared project, with the British fundamentally reliant on the Chinese in their search for scientific knowledge.

As an increasing number of historians research science in the Qing, evidence is pushing the field to rethink the perception of the period as one of stagnation and anti-science metaphysical speculations. While there was a limitation in that the "evidential studies" scholars prioritised mining classical texts for wisdom from the past, this was only part of the story.[45] Scholars such as Song Yingxing and Chen Hongmou busied themselves with reading texts, but also with practical applications in the field.[46] Only a minority of scholars could afford to sit in libraries and read old texts. Agriculture was officially proclaimed as an ideal pursuit for bookish scholars, yet in real terms it required practical scientific knowledge and expertise.[47] Such knowledge was also applied to industrial development with aspects of late-Qing enterprise being equal to that in Europe.[48]

Medicine

Medicine is the field of knowledge that most fits the epithet of Chinese science "on its own terms."[49] Chinese ideologues in the twentieth century felt confident to coin the phrase "Chinese medicine," which continues to have purchase globally in the twenty-first century. No other field was similarly named nor comparably successful. We do not hear today of "Chinese astronomy" or "Chinese mathematics," even if some may claim geomancy and astrology as Chinese.[50] Some historians have emphasised the historical uniqueness of classical Chinese medical thinking and practice.[51] However, scholars are now beginning to show that there was more similarity with medicine in the West than previously acknowledged. Until the nineteenth century, much of medical practice in the West was based on humoral theory, with some similar features to medicine in China.[52]

As discussed above, the early Qing emperors committed themselves to shaping knowledge production. They famously commissioned publication of many volumes across a range of scholarly fields, including medicine. During the eighteenth century, three imperial editorial projects on medicine were completed. The two projects that the Qianlong emperor commissioned, *Golden Mirror of the Medical Lineage* (*Yizong Jinjian*, 1742) and *Complete Collection of the Four Treasuries* (*Siku Quanshu*, 1782) definitively augmented the authority of elite physicians of the Lower Yangzi region.[53] This powerful elite gained imperial patronage to define the official style of medicine in the empire. However, the official imperially patronised medicine was multifaceted. On the one hand, the Qing court commissioned Chinese medical texts, but on the other hand, the court generated a large medical literature in the Manchu language, much of it still unstudied by scholars.[54]

In thinking about medicine in the Qing, Florence Bretelle-Establet questions whether present-day scholars have ascribed too much importance to a state-led project in terms of what people in China as a whole actually practised.[55] Medical literature from the far south of the empire, specifically Yunnan, Guangxi, and Guangdong provinces, shows that there is a vast medical literature demonstrating a thriving and diverse field of medical practices. Just as the lower Yangzi region had an identifiable southern style in comparison to a northern style of medicine,[56] there was a greater plurality of styles of medicine moving out to the far southern frontiers.

The plurality of medical styles also saw physicians thinking more about the physicality of the body. For example, the medical fields of *fuke* (gynaecology) and *chanke* (obstetrics) from the seventeenth to the nineteenth centuries give a close-up view of medicine for women, specifically around the question of childbirth.[57] Despite an increasing subordination of female difference to a master narrative of bodily androgyny, Chinese doctors were also interested in the physical structure of the body, including the womb.

With regards to herbal medicine, the Qing period experienced a surge in new knowledge embodied in an ever-expanding pharmacy.[58] This vibrant growth in new knowledge also elevated the status of ordinary people's experiences, demonstrating that knowledge formation was not limited to the palace and elite scholars.

病者給錢數百名為門脈
者四吊四百文四吊八百不等如來到門首看
此中國醫道之圖也京中醫士有太醫御醫
之稱乃是在太醫院應差者如有人請馬錢

Figure 18.3   Doctor taking woman's pulse. Watercolour by Zhou Pei Qun, c. 1890, Wellcome Library, London. CC BY 4.0.

Another trajectory in understanding science and medicine in the Qing lies in its place in contributing to world knowledge. The Jesuits also played a role in transmitting Chinese medical knowledge to Europe.[59] There was considerable interest in Europe for information about Chinese herbs, acupuncture, moxibustion, and pulse diagnosis. Rather than incommensurability of knowledge systems in Europe and China, as often portrayed in the historiography,[60] there were elements of unity of science across East and West, as some scholars and physicians emphasised commonalities more than differences between China and Europe, and recognised common narrative forms in medical diagnosis.[61]

As with the scholar-official Song Yixing, physicians during the Qing practised medicine with a theoretical framework of qi, yin yang, and the five phases. While analysing physicians and their medical practice during the Qing, some scholars have, in different ways, connected Qing medicine with present-day China. While scholars have examined the meta-geographical concepts as applied to medicine and related nosology, specifically warm disease (*wenbing*),[62] other studies demonstrate the elite status of physicians in the Lower Yangzi region who built and strengthened lineages with essential social roles in late imperial society.[63] For example, one scholar demonstrated how the nosology category of warm diseases (*wenbing*) that physicians formulated in the aftermath of epidemics in the late Ming survived into the twenty-first century when it was used by physicians who found it useful for thinking about how to treat SARS patients.[64] In a similar trajectory, another scholar demonstrated the elite status of physicians in the Lower Yangzi region who built and strengthened lineages with essential social roles in late imperial society and showed that they continued to play influential roles in the state institutionalisation of medicine in the People's Republic of China in the twentieth century.[65]

The medical theories of qi, yin yang, and five phases continue to inform medical practice in China, but also in substantial numbers of communities across the world. The period of transmission began in the Qing, thus turning on its head the idea of the Qing only receiving ideas and having little impact on world knowledge. Chinese physicians operated in most towns in Australia, for example, and have served the health needs of many people by writing herbal prescriptions and practising acupuncture since the mid nineteenth century.[66]

Conclusion

From the seventeenth century to the twentieth century, the Qing Empire ruled a vast, mostly land-based continental empire, during a time when the population more than doubled. While scholars have debated the strengths and weaknesses of its ruling ideologies, such as interpretations of Confucianism and developments in philology, much of the statecraft required to govern the empire was based on pragmatic knowledge we would now call science. The view of a China that newly came to science in the twentieth century has been overstated. Native improvements in agriculture, innovations in medicine, and industrial developments were important factors in overall Qing strength. Furthermore, while Qing science was certainly an attribute of imperial rule, this official sanction of expert and research knowledge created the ability for local officials, scholars, and healers to pursue their own observations and scientific practices. And finally, recent scholarship has shown that Jesuit scholars not only translated Chinese textual knowledge such as Confucian texts into European languages and took them to Europe but also transmitted there a wide range of scientific knowledge, including cartography and medicine. The field has clearly moved past the idea of a binary between Western science and Chinese science to recognise the collaborative, hybrid, and intercultural nature of scientific knowledge making on the Eurasian continent.

Notes

1 William Rowe, *China's Last Empire: The Great Qing* (Cambridge, MA: Belknap Press, 2009), 1.
2 For analysis of Qing historiography in China, see Feng Erkang, "Studies of Qing History: Past, Present, and Problems," *Chinese Studies in History* 43, no. 2 (2009/2010): 20–32; Li Huaiyin, *Reinventing Modern China: Imagination and Authenticity in Chinese Historical Writing* (Honolulu: University of Hawaii Press, 2012).
  For a representative Chinese language Qing history, see Li Zhiting [李治亭], *Qing shi* [清史, Qing history] (Shanghai: Shanghai People's Publishing House, 2002).
3 John Fairbank, ed., *The Chinese World Order: Traditional China's Foreign Relations* (Cambridge, MA: Harvard University Press, 1968); Wang Xudong and Li Junxiang, "Modernization and the Study of Modern Chinese History: A Brief Thesis," *Chinese Studies in History*, 43, no. 1 (2009): 46–60.
4 For example, see Iwo Amelung's discussion in which he cites the intellectual Chen Duxiu (陳獨秀), Iwo Amelung, "Science and National Salvation in Early Twentieth Century China," in *Revisiting the "Sick Man of Asia": Discourses of Weakness in Late 19th and Early 20th Century China*, ed. Iwo Amelung and Sebastian Riebold (Frankfurt: Campus Verlag, 2018), esp. 22–3.
5 Nathan Sivin, "Drawing Insights from Chinese Medicine," *Journal of Chinese Philosophy* 34, no. 1 (2007): 43–55, here p. 45.
6 Feng, "Studies of Qing History." For representative scholarship on the claim of the absence of science in the Qing Empire, see H. Floris Cohen, "The Nonemergence of Early Modern Science outside Western Science," in *The Scientific Revolution: A Historiographical Inquiry* (Chicago: University of Chicago Press, 1994), chap. 6.
7 Joseph Needham, *Science and Civilisation in China*, vol. 1 (Cambridge: Cambridge University Press, 1954).
8 Rowe, *China's Last Empire*; Benjamin Elman, *On Their Own Terms: Science in China, 1550–1900* (Cambridge, MA: Harvard University Press, 2005).

9 Marta Hanson, "The Significance of Manchu Medical Sources in the Qing" (Proceedings of the First North American Conference on Manchu Studies, Portland, OR, May 9–10, 2003): 131–75; Wadley Stephen and Carsten Naeher with Keith Dede, eds., *Studies in Manchu Literature and History*, Series Tunguso Sibirica 15, vol. 1 (Wiesbaden: Harrassowitz Verlag, 2006); Catherine Jami, "Imperial Science Written in Manchu in Early Qing China: Does it Matter?," in *Looking at It From Asia: The Processes that Shaped the Sources of History of Science*, ed. Florence Bretelle-Establet (Milton Keynes: Springer, 2010), 371–92.

10 Elman, *On Their Own Terms*; Catherine Jami, *The Emperor's New Mathematics: Western Learning and Imperial Authority during the Kangxi Reign, 1662–1722* (Oxford: Oxford University Press, 2012).

11 Mario Cams, *Companions in Geography: East–West Collaboration in the Mapping of Qing China, c. 1685–1735* (Leiden: Brill, 2017).

12 Dagmar Schäfer, *The Crafting of the 10,000 Things: Knowledge and Technology in Seventeenth-Century China* (Chicago: The University of Chicago Press, 2011).

13 William Rowe, "Political, Social, and Economic Factors Affecting the Transmission of Technical Knowledge in Early Modern China," in *Cultures of Knowledge: Technology in Chinese History*, ed. Dagmar Schäfer (Leiden: Brill, 2012), 25–44; William Rowe, *Saving the World: Chen Hongmou and Elite Consciousness in Eighteenth-Century China* (Palo Alto, CA: Stanford University Press, 2002); Morris Low, "Beyond Joseph Needham: Science, Technology, and Medicine in East and Southeast Asia," in "Beyond Joseph Needham: Science, Technology, and Medicine in East and Southeast Asia," ed. Morris Low, special issue, *Osiris* 13 (1998): 1–6; Francesca Bray, "Technics and Civilization in Late Imperial China: An Essay in the Cultural History of Technology," *Osiris* 13 (1998): 11–33; Mark Elvin, "Who was Responsible for the Weather? Moral Meteorology in Late Imperial China," *Osiris* 13 (1998): 213–37.

14 Ian Miller, *Fir and Empire: The Transformation of Forests in Early Modern China* (Seattle: University of Washington Press, 2020); Bray, "Technics and Civilization"; Madeleine Zelin, *The Merchants of Zigong: The Industrial Entrepreneurship in Early Modern China* (New York: Columbia University Press, 2005).

15 Volker Scheid, *Currents of Tradition in Chinese Medicine, 1626–2006* (Seattle, WA: Eastland Press, 2007).

16 Elman, *On Their Own Terms*, 445; Michael Dillon, "Transport and Marketing in the Development of the Jingdezhen Porcelain Industry during the Ming and Qing Dynasties," *Journal of the Economic and Social History of the Orient* 35, no. 3 (1992): 278–90; Anne Gerritsen, *The City of Blue and White: Chinese Porcelain and the Early Modern World* (Cambridge: Cambridge University Press, 2020).

17 Fa-ti Fan, *British Naturalists in Qing China: Science, Empire, and Cultural Encounter* (Cambridge, MA: Harvard University Press, 2004); Andrew Liu, *Tea War: A History of Capitalism in China and India* (New Haven, CT: Yale University Press, 2020).

18 For silk, see Dagmar Schäfer, "Silken Strands: Making Technology Work in China," in Schäfer, *Cultures of Knowledge*, 45–74.

19 Marta Hanson and Gianna Pomata, "Medicinal Formulas and Experiential Knowledge in the Seventeenth-Century Epistemic Exchange between China and Europe," *Isis* 108, no. 1 (2017): 1–25; Marta Hanson and Gianna Pomata, "Travels of a Chinese Pulse Treatise: The Latin and French Translations of the *Tuzhu maijue bianzhen* 圖註脈訣辨真 (1650s–1730s)," in *Translation at Work: Chinese Medicine in the Global Age*, ed. Harold Cook (Leiden: Brill, 2020), 23–57; Linda Barnes, *Needles, Herbs, Gods, and Ghosts: China, Healing, and the West to 1848* (Cambridge, MA: Harvard University Press, 2005); Roberta Bivins, *Alternative Medicine: A History* (Oxford: Oxford University Press), 2007.

20 Mario Cams, "Converging Interests and Scientific Circulation between Paris and Beijing (1685–1735): The Path toward a New Qing Cartographic Practice," *Revue d'histoire des sciences* 70, no. 1 (2017): 47–78; Mario Cams, "Not Just a Jesuit Atlas of China: Qing Imperial Cartography and Its European Connections," *Imago Mundi* 69, no. 2 (2017): 188–201.

21 Elman, *On Their Own Terms*; Benjamin Elman, *A Cultural History of Modern Science in China* (Cambridge, MA: Harvard University Press, 2006).

22 Pingyi Chu, "Scientific Dispute in the Imperial Court: The 1664 Calendar Case," *Chinese Science* 14 (1997): 7–34; Pingyi Chu, "Trust, Instruments, and Cross-Cultural Scientific Exchanges: Chinese Debate over the Shape of the Earth, 1600–1800," *Science in Context* 12, no. 3 (1999): 385–411; Pingyu Chu, "Remembering our Grand Tradition: The Historical Memory of the Scientific Exchanges between China and Europe, 1600–1800," *History of Science* 41 (2003): 193–215; Qi Han, "Emperor, Prince, and Literati: Role of the Princes in the Organization of Scientific Activities in Early Qing

Period," in *Current Perspectives in the History of Science in East Asia*, ed. Yung Sik Kim and Francesca Bray (Seoul: Seoul National University Press, 1999), 209–16.

23 Jami, *Emperor's New Mathematics*; Han, "Emperor, Prince, and Literati."

24 Han, "Emperor, Prince, and Literati"; Chu, "Scientific Dispute"; Chu, "Remembering our Grand"; Elman, *On Their Own Terms*.

25 Jami, *Emperor's New Mathematics*. However, for an example of a scholar using some Manchu-language material related to science at the Qing court, see Sare Aricanli, "Diversifying the Center: Authority and Representation within the Context of Multiplicity in 18th Century Qing Imperial Medicine" (PhD diss., Princeton University, 2016), 165–7.

26 Jami, *Emperor's New Mathematics*.

27 Jami, *Emperor's New Mathematics*.

28 Cams, *Companions in Geography*. Also see Laura Hostetler, *Qing Colonial Enterprise: Ethnography* and *Cartography in Early Modern China* (Chicago: University of Chicago Press, 2001).

29 Elman, *On Their Own Terms*.

30 Cams, *Companions in Geography*.

31 Elman, *On Their Own Terms*; Dagmar Schäfer, "Knowledge by design-architecture and Jade models during the Qianlong 乾隆 reign (1735–1796)," in *The Structures of Practical Knowledge*, ed. Matteo Valleriani (Cham: Springer, 2017), 271–86.

32 Elman, *On Their Own Terms*, 212–16; Michel Beurdeley and Guy Raindre, *Qing Porcelain* (New York: Rizzoli, 1986).

33 Elman, *On Their Own Terms*.

34 Benjamin Elman, *From Philosophy to Philology: Intellectual and Social Aspects of Change in Late Imperial China* (Cambridge, MA: Harvard University Press, 1984).

35 Schäfer, *Crafting of the 10,000 Things*.

36 Francesca Bray, "Science, Technique, Technology: Passages between Matter and Knowledge in Imperial Chinese Agriculture," *The British Journal for the History of Science* 41, no. 3 (2008): 319–44.

37 Zelin, *Merchants of Zigong*.

38 Hans-Ulrich Vogel, "Copper Smelting and Fuel Consumption in Yunnan, Eighteenth to Nineteenth Centuries," in *Metals, Monies, and Markets in Early Modern Societies: East Asian and Global Perspectives*, ed. Thomas Hirzel and Nanny Kim (London: Transaction Publishers, 2008), 119–70; Werner Burger, "Coin Production during the Qianlong and Jiaqing Reigns (1736–1820): Issues in Cash and Silver Supply," in Hirzel and Kim, *Metals, Monies, and Markets*, 171–90; Thomas Mullaney, "Semiotic Sovereignty: The 1871 Chinese Telegraph Code in Historical Perspective," in *Science and Technology in Modern China, 1880s–1940s*, ed. Jing Tsu and Benjamin Elman (Leiden: Brill, 2014), 153–83; Andrea Bréard, *Nine Chapters on Mathematical Modernity: Essays on the Global Historical Entanglements of the Science of Numbers in China* (Heidelberg: Springer, 2019).

39 Elman, *On Their Own Terms*.

40 Bray, "Science, Technique, Technology."

41 Rowe, *Saving the World*.

42 Rowe, "Political, Social, and Economic Factors."

43 Zelin, *Merchants of Zigong*, 169.

44 Fan, *British Naturalists*; Erik Mueggler, *The Paper Road: Archive and Experience in the Botanical Exploration of West China and Tibet* (Berkeley: University of California Press, 2011).

45 Elman, *From Philosophy to Philology*.

46 Schäfer, *Crafting of the 10,000 Things*; Rowe, "Political, Social, and Economic Factors."

47 Bray, "Science, Technique, Technology."

48 Zelin, *Merchants of Zigong*.

49 Elman, *On Their Own Terms*.

50 T.J. Hinrichs and Linda Barnes, eds., *Chinese Medicine and Healing: An Illustrated History* (Cambridge, MA: Belknap Press, 2013), 171.

51 Shigehisa Kuriyama, *The Expressiveness of the Body and the Divergence of Greek and Chinese Medicine* (New York: Zone Books, 2002).

52 Nathan Sivin, "Science and Medicine in Chinese History," in *Heritage of China: Contemporary Perspectives on Chinese Civilization*, ed. Paul Ropp (Berkeley: University of California Press, 1990), 164–96, here p. 167; John Harley Warner, "From Specificity to Universalism in Medical Therapeutics: Transformation in the 19th-Century United States," in *Sickness and Health: Readings in the History of Medicine and Public*

Health, ed. Judith Leavitt and Ron Numbers, 3rd ed. (Madison: University of Wisconsin Press, 1997), 87–101.

53 Marta Hanson, "The 'Golden Mirror' in the Imperial Court of the Qianlong Emperor, 1739–1742," *Early Science and Medicine* 8, no. 2 (2003): 111–47.

54 Marta Hanson, "Northern Purgatives, Southern Restoratives: Ming Medical Regionalism," *Asian Medicine: Tradition and Modernity* 2, no. 2 (2006): 115–70; He Bian, "Of Wounded Bodies and the Old Manchu Archive: Documenting Personnel Management in the Early Manchu State," *Saksaha: A Journal of Manchu Studies* 16 (2019), doi.org/10.3998/saksaha.13401746.0016.001; Beatriz Puente-Ballesteros, "Jesuit Medicine in the Kangxi Court (1662–1722): Imperial Networks and Patronage," *East Asian Science, Technology, and Medicine* 34 (2011): 86–162.

55 Florence Bretelle-Establet, "Is the Lower Yangzi Region the Only Seat of Medical Knowledge in Late Imperial China? A Glance at the Far South Region and Its Medical Documents," in *Looking at It from Asia: The Processes that Shaped the Sources of History of Science,* ed. Florence Bretelle-Establet (Milton Keynes: Springer, 2010).

56 Hanson, "Northern Purgatives, Southern Restoratives"; Bretelle-Establet, "Lower Yangzi Region."

57 Yi-Li Wu, *Reproducing Women: Medicine, Metaphor, and Childbirth in Late Imperial China* (Berkeley: University of California Press, 2010).

58 He Bian, "An Ever-Expanding Pharmacy: Zhao Xuemin and the Conditions for New Knowledge in Eighteenth Century China," *Harvard Journal of Asiatic Studies* 77, no. 2 (2017): 287–319; He Bian, *Know Your Remedies: Pharmacy and Culture in Early Modern China* (Princeton, NJ: Princeton University Press, 2020).

59 Nicolas Standaert, *Handbook of Christianity in China,* vol. 1, *635–1800* (Leiden: Brill, 2001), 113–906; Hanson and Pomata, "Medicinal Formulas and Experiential Knowledge."

60 Robert Wardy, *Aristotle in China: Language, Categories and Translation* (Cambridge: Cambridge University Press, 2004).

61 Hanson and Pomata, "Travels of a Chinese Pulse Treatise."

62 Marta Hanson, *Speaking of Epidemics in Chinese Medicine: Disease and the Geographic Imagination in Late Imperial China* (New York: Routledge, 2011).

63 Scheid, *Currents of Tradition.*

64 Hanson, *Epidemics in Chinese Medicine.*

65 Scheid, *Currents of Tradition.*

66 Nadia Rhook, "'The Chinese Doctor James Lamsey': Performing Medical Sovereignty and Property in Settler Colonial Bendigo," *Postcolonial Studies* 23, no. 1 (2020): 58–78; Michelle Bootcov, "Dr George On Lee (葉七秀): Not Just a Medical Practitioner in Colonial Australia," *Chinese Southern Diaspora Studies* 8 (2019): 82–101.

19

EMPIRE, CULTIVATION, AND THE ENVIRONMENT IN SOUTHEAST ASIA SINCE 1500

Timothy P. Barnard

The initial foray of Western powers into Southeast Asia occurred through the Spanish and Portuguese naval fleets in the sixteenth century. While the Spanish eventually gained influence over the Philippines, the Portuguese were the more successful of the Iberian powers following their capture of the trade centre of Melaka in 1511 and the establishment of outposts throughout the Indonesian Archipelago as well as Burma and Siam over the subsequent decades. After several centuries of an imperial presence that was mainly based in fortified trade ports and the entry of trade companies, with the English East India Company (EIC) and the United (Dutch) East India Company (Vereenigde Oost-Indische Compagnie, hereafter VOC) growing in influence in the seventeenth and eighteenth centuries, power transferred to national governments in the mid twentieth century. Over several centuries, imperialism—in its many forms—became strongly established in Southeast Asia.

This process was rooted in the exploitation of the environment these external powers encountered. This environment was tropical; dense rainforests dominated the land, while the seas and rivers not only provided a wealth of resources but also acted as the conduits for the movements of peoples, ideas, and goods. Products from this environment drew these imperial powers to Southeast Asia. The holy trinity of spices—cloves, nutmeg, and mace—were among the most valuable items in early-modern European markets and, combined with other plants such as black pepper, cinnamon, and tea, were a key stimulus drawing explorers, merchants, and sailors. The environment and how it was understood was a key component of imperialism.

The role that understanding the environment would play in the imperial enterprise can be seen in Portuguese efforts to survey the wonders of the region in the decades following their arrival. One of the earliest accounts came from Tomé Pires, an apothecary, who gathered information on the landscapes, geographies, and histories of the region in his *Suma Oriental*, with the goal of furthering Portuguese efforts to harness the environmental riches of the region. In his account of Sumatra, for example, Pires described the island as a place that "has gold in great quantities, edible camphor of two kinds, pepper, silk, benzoin, apothecary's lignaloes; it has honey, wax, pitch, Sulphur, cotton, many rattans, which are canes from which they make mats."[1] Understanding and exploiting this environment was the foundation of imperial rule.

A counterpart to Tomé Pires during the era of trade companies, which subsequently replaced the Portuguese Empires in the seventeenth century, was Georg Everhard Rumpf, better known as Rumphius. A VOC employee posted to Maluku (commonly known as the Moluccas, or Spice Islands), Rumphius spent decades surveying the array of plants and animals that he encountered. Written in the vein of a pre-Linnaean classification system, these works combined the aesthetic and scientific, and their publication in Europe opened up the exotic world of Southeast Asia to scientific classification, pushing it beyond the exaggerated tale of previous adventurers.[2] Such surveys of the natural world ultimately had practical uses for economic botany, the identification and classification of plants in an effort to cultivate them for profit. Once this base was established, the natural world could be understood, and then exploited. Such efforts were not only at the core of imperial rule but also imperial enterprise.[3]

The trade that the VOC and EIC oversaw flowed to Europe throughout the early-modern era, and this foreign influence over the environment went beyond the documentary gaze to one that encouraged the cultivation of certain crops. The Dutch and Spanish initially attempted to control this trade from the Indonesian Archipelago and the Philippines through monopolies on specific products. The VOC was able to achieve this for clove, nutmeg, and mace, as they were only produced in an isolated region in the far eastern archipelago, but was unable to do so with other products, such as black pepper. The Spanish were able to operate a quasi-monopoly on some Asian goods through the Spanish Galleon Trade across the Pacific Ocean, made possible by the relatively short distances between the Philippines and other trade emporia in Southeast Asia.

These monopolies, and their focus on the cultivation of specific crops, had the power to alter environments, and are most clearly seen with regards to clove cultivation in Maluku. The VOC gained control over Ambon as well as Ternate-Tidore, key polities in the region in the early seventeenth century. Following a period of jockeying for influence in the region among the Portuguese, Spanish, and English, the VOC began to enforce a monopoly over the existing spice trade, which led to the extermination of much of the population of the Banda Islands, the main site for the production of nutmeg and mace, in the 1620s. In their place, a pliant group of planters that supplied cloves to the Dutch company arose, reinforcing the role that this landscape would play with regard to larger global empires and markets during the period.[4] While this is an extreme example of the influence that the European presence had on Southeast Asian environments and societies, it slowly began to replicate throughout the region in the development of plantation agriculture, usually focusing on the cultivation of cash crops that could be exported to world markets, a process that altered landscapes and societies.

One of the first sites in which this became apparent during the period of high imperialism from 1800 onwards was in Java. Following the Napoleonic Wars, the transfer of sovereignty over the island to the government of the Netherlands from the VOC, and a costly war of conquest in the central region of the island, Dutch imperial forces were concerned with how their presence could be financed and, hopefully, even become profitable. The solution was the development of the Cultivation System (*Cultuurstelsel*), through which the government required after 1830 that a portion of the agricultural production of every community be devoted to export crops, legally justified and enforced as a form of taxation. The cultivation of crops such as coffee, sugar, and indigo quickly led to massive profits, creating a surplus that made the utilisation of the Southeast Asian environment an important component of Dutch colonial rule.

While it appeared to not be overly onerous on paper, in reality the Cultivation System led to massive disruption of the social, economic, and natural landscape of Java as it essentially transformed the island into a huge plantation. Prior to direct imperial rule, most agricultural production was directed toward the cultivation of rice. As land and labour were shifted toward the production of export crops, which had to be processed and pass through a series of proxies, villages often fell into poverty and occasional famine while their traditional political systems became distorted to meet the demands of the colonial government. Much of this was overlooked at the time, however, due to the massive profits the system generated, which constituted between 20 and 30 per cent of Dutch national income between 1830 and 1870.[5]

The financial success of the Cultivation System in Java meant it became a model for imperial nations throughout the region. As other powers expanded their influence over Southeast Asian societies they implemented important aspects of the system. The only difference was that the crops grown on plantations varied, depending on the society as well as the geography of a specific place. This can be seen clearly in mainland Southeast Asia near the mouths of the Mekong and Irrawaddy Rivers, where vast deltas of rainforest and coastal swamplands came under the axe in the mid nineteenth century, when new lands were opened to rice cultivation, primarily for export. The access that global capital provided meant that these former wetlands were now cultivated for profit on international markets, and led to the rise of new urban centres, such as Rangoon and Saigon, anchoring the regional colonial economy. This transformation also resulted in demographic shifts, as various ethnic groups moved into these regions, a process that often altered the balance of power in the societies, whether they were under the direct control of European empires or not.[6]

The shift in agriculture from locally consumed food crops to export crops under colonial rule transformed the environment in Southeast Asia. While produce from the forests and seas had originally drawn the interest of Western powers, the ability to identify other riches and exploit them on a much larger scale soon became an obsession. The Cultivation System reflected how the region could produce unimaginable profits if the right crops could be identified. As imperialism entered a new era, the ability of outside powers to exert control over the environment also expanded. Much of this was based in science, the ability to test and determine what would most likely profit the imperial power, whether it was a trade company or a national government. Key to these developments was understanding the environment of the region, and the introduction of plants that were not native to Southeast Asia that could be cultivated for profit on global markets. These efforts were centred in botanic gardens, which arose as Western powers expanded their influence in the region. The cultivation of nature soon became vital in the imperial process.

Botanic gardens and the environment

Botanic gardens in Europe were first conceptualised as scientific institutions during the Renaissance, when they were closely linked to medical faculties, where scholars and physicians studied plants for their therapeutic value. Following the exponential growth of the known natural world after the discovery of the Americas and the exploration of Africa and Asia, botanical gardens soon expanded to serve the entire university, and naturalists forged connections with political institutions, on the basis of their botanical expertise, and used this to influence imperial and colonial policy. By the eighteenth century, governments were supporting the efforts of botanists in the hope that new (profitable) plants could be identified and adapted to new environments around the globe, with Kew Gardens in England becoming a model for the development of what came to be known as economic botany. Gardens established in the colonies would eventually serve an important role in this enterprise.[7]

Colonial botanic gardens were centres of research, making them important in the development of imperial societies and economies. Scattered throughout the world, these scientific institutions were key components of empire, becoming repositories for the gathering of information about new lands and new plants. Nature could be cultivated and shaped to serve the interests of medical schools, trading companies, and governments within their boundaries. While European botanic gardens played an important role at the centre of global networks, their expansion overseas led to new understandings of distant lands. In addition, although botanic gardens were created to serve the interests of a specific trade company or imperial government, they often cooperated and competed with each other as they developed within an expanding transnational understanding of science and botany. Behind such utopian ideals, nevertheless, was the goal of cultivating nature and science to serve the needs of empire. In this regard botanic gardens were one of the main tools of experimentation, colonisation, and exploitation. While scientists associated with these gardens explored the region and identified new plants, their main task was to test the ability of foreign plants that had potential for profit to acclimatise to new environments. They thus became important centres from which the power of imperialism radiated throughout Southeast Asia, ultimately transforming the environment.

The Dutch developed the first modern botanic garden in Southeast Asia in Buitenzorg in the outskirts of Batavia (modern-day Jakarta) in 1817.[8] For its first two decades the institution was a problematic presence within the imperial administration, with officials unsure of the role the garden was to play. It was allowed to exist, however, as long as it did not place any demands on the colonial budget or profits. Once an herbarium was constructed in the 1840s, the Buitenzorg Botanic Garden came into its own. It quickly became a centre for research in tropical botany, and proved its worth to the imperial enterprise through the development of a number of plantation crops, usually imported from other parts of the globe. The most influential plant developed by Dutch colonial scholars was *Cinchona*, a genus of plants native to the Andes Mountains, from which the bark is the source of quinine, the main ingredient used in the treatment of malaria. Little was known about *Cinchona* in the early nineteenth century, but European expansion would have been difficult without it, as malaria was a disease that decimated early colonial settlements. The ability to produce quinine on a large scale would not only be profitable, but also ensure the health of imperial officials and troops as they expanded their influence in tropical environments.[9]

The development of *Cinchona* cultivation on the island of Java was unexpected, as most of the early research on the plant took place in the network of British botanical gardens. By the mid-1850s botanists at Kew Gardens had overseen efforts to collect *Cinchona* seeds from South America and their transfer to British-controlled botanic gardens throughout the world, where early methods of cultivation for mass production and processing took place. These efforts were successful, and by 1865 the India Office arranged for *Cinchona* to be grown in the Calcutta Botanic Gardens, thus providing the model for acclimatisation of economically useful plants in tropical botanic gardens for the needs of the Empire.[10]

The amount of quinine produced from these acclimatised trees, however, was highly variable, and many mature trees had low quantities of the compound. Dutch officials continued to scour the globe in an attempt to improve their own yields, as they had been conducting their own experiments in Java during the same period. This eventually led the Dutch Consul-General in London to purchase 20,000 seeds of a species that Kew Gardens had refused to buy. These seeds were sent to Buitenzorg and the seedlings were grafted on to existing stock. Dutch botanists associated with the Buitenzorg Botanic Garden began experimenting with these plants, and this eventually led to the development of much greater quinine yields. The results were spectacular, and the Netherlands East Indies quickly came to dominate the market in quinine production by the early 1870s, eventually producing over 90 per cent of the global supply of this antimalarial between 1890 and 1940, and enhancing the ability of the coloniser to expand their influence.[11]

Figure 19.1 Women on Java at work on a *Cinchona* plantation, c. 1930. KITLV collection, Leiden University Library. CC BY 4.0.

Dutch scientists—under the rubric of imperial or economic botany—at Buitenzorg had not only acclimatised a foreign plant and increased its yield, but also strengthened the presence of other imperial powers in the tropical world, as they sought to duplicate Dutch scientific methods. This achievement became a model for botanical gardens globally, and motivated the British to seek a similar success. Hundreds of plants underwent experimentation in an attempt to unlock their potential to serve imperial governments. As colonial botanic gardens developed, and experimentation continued, networks of cooperation and science grew between them. This is most clearly seen in efforts to develop rubber, which was becoming increasingly important for industrialised economies as a material with a multitude of useful applications ranging from tyres to gaskets.

Rubber, however, does not simply come from one species. Numerous plants from trees to woody climbers produce latex, a gum resin. The quality and properties of this gum resin vary depending on the plant of origin, and botanists began working on a range of possibilities to identify the most productive, with each major garden concentrating on a specific genus. Botanists at the Peradeniya Gardens in Ceylon, for example, spent much of their time focusing on the *Ceara* genus, while the Dutch in Buitenzorg fixated on *Ceara* and *Castilloa* and the French dedicated their efforts on *Ficus*, as did the Dutch in an attempt to cover the numerous possibilities. Singaporean botanists, in the meantime, focused on gutta-percha (*Palaquium gutta*) as well as the *Hevea* plant.[12]

It was the efforts at the Singapore Botanic Gardens that soon became the cornerstone for an industry while also encapsulating the amalgamation of the forces of imperialism, science, and the environment. *Hevea brasiliensis* originated from trees that roaming labourers harvested in the

Amazonian jungle and had been sent to Kew in the mid-1870s. By 1877 Kew Gardens sent 22 rubber seedlings to Singapore, and over the next two decades botanists—mainly under the supervision of H.N. Ridley—in Singapore identified the ideal conditions for its growth, such as cultivating *Hevea brasiliensis* on moist, well-drained grounds, as well as harvesting and processing techniques that produced the maximum amount of latex.

Ridley and his associates laid the foundation for a transformative industry through these experiments. The effect was staggering, and statistics reflect the influence that such knowledge held for Southeast Asia and its environment as it expanded to encompass and change landscapes throughout the region. To establish this industry the Singapore Botanic Gardens distributed over seven million rubber seeds in the first decade of the twentieth century. Between 1897 and 1922 the amount of land under rubber cultivation in Malaya expanded from 345 acres (139 ha) to 2,304,231 acres (900,000 ha, or 900 square kilometres). Throughout South and Southeast Asia, a similar expansion occurred, with another 900,000 hectares planted in Ceylon, the Netherlands East Indies, Burma, and French Indochina, while Africa also became a site of cultivation. The result was the displacement of South America, and particularly Brazil, as the world supplier of rubber. In 1900, "wild rubber" from South America and Africa made up 98 percent of global exports. Within two decades, they controlled less than 6 percent of the total, with more than 80 percent of rubber exports originating from Malaya and the Netherlands East Indies.[13]

The impact of rubber, and the general plantation economy, on the environment was massive. Long-standing forests were cut, transportation networks created, and communities developed. New migrants—from within and without colonies—flooded the rubber estates. Javanese migrated to Sumatra, Indians to Malaya, and North Vietnamese to southern Vietnam, altering the demographic makeup of societies. These new migrants radically simplified a landscape of tremendous biodiversity, all for the cultivation of a single crop that produced profits for—mainly—Western capitalist firms, making the continuing theme of exploiting the produce of the Southeast Asian environment for imperial desires a theme across centuries.

The exploitation of the Southeast Asian environment continued into the twentieth century. A transition occurred, however, as departments of agriculture soon displaced botanic gardens with the primary remit of cultivation for profit. Although it occurred in all colonies, it was most clearly seen in the British Empire. Under this new approach, Joseph Chamberlain, who was Secretary of State for the Colonies from 1895 until 1905, promoted the exploitation of the environment for the economic security of Britain. This resulted in funds and personnel being channelled toward research in tropical medicine, agriculture, and forestry, as well as a range of new disciplines that would integrate the various British possessions further into the global economy. As Joseph Hodge has argued, this not only required new categories of colonial officials, rooted in scientific and academic expertise, it also led to new departments that further solidified imperial control over the landscape, all in service to the production of profits through the manipulation and cultivation of foreign plants.[14] Such efforts were not limited to the British Empire. It was a policy that was also mirrored in the strategies that American, French, and Dutch colonial governments employed in other polities in Southeast Asia.

While cloves, nutmeg, and mace had originally attracted imperialist forces to Southeast Asia, and the cultivation of cotton, indigo, and eventually rubber generated huge profits for the imperial enterprise, plantation agriculture eventually shifted to new products to fulfil new needs following independence for the various nations in the decades after World War II. Among these new products cultivated for export and profit was African oil palm (*Elaeis guineensis*), which produces the highest yield per hectare of any oil crop. Its acclimatisation to Southeast Asia began in botanic gardens, with H.N. Ridley identifying its potential as early as 1907.[15] The

Figure 19.2 Hevea brasiliensis specimen sheet. Ridley collected this one in 1897 when he was conducting tests to optimise its yield and the techniques to do so. Courtesy of SING Herbarium, Singapore Botanic Gardens.

growth of this export crop led to further deforestation throughout Indonesia, Malaysia, and even southern Thailand in the late twentieth century, following a basic pattern that had already been established during the colonial era. The fundamental lesson of all of these changes was that the environment and its exploitation in Southeast Asia would continue, and it was usually for the benefit of external, imperial, forces that profited from it. This transformation was based in a scientific understanding of how to achieve this, and in the cultivation of non-native plants on a massive scale.

Notes

1 Tomé Pires, *The Suma Oriental of Tomé Pires, an Account of the East, from the Red Sea to Japan, written in Malacca and India, 1512–1515* (London: The Hakluyt Society, 1944), 136.

2 Genie Yoo, "Wars and Wonders: The Inter-Island Information Networks of Georg Everhard Rumphius," *The British Journal for the History of Science* 51, no. 4 (2018): 559–84; E.M. Beekman, *The Poison Tree: Selected Writings of Rumphius on the Natural History of the Indies* (Kuala Lumpur: Oxford University Press, 1993), 1–40.

3 Bernard Cohn, *Colonialism and Its Form of Knowledge: The British in India* (Princeton, NJ: Princeton University Press, 1996); Richard Drayton, *Nature's Government: Science, Imperial Britain, and the "Improvement" of the World* (New Haven, CT: Yale University Press, 2000); Timothy P. Barnard, "The Rafflesia in the Natural and Imperial Imagination of the East India Company in Southeast Asia," in *The East India Company and the Natural World*, ed. Vinita Damoradaran, Anna Winterbottom, and Alan Lester (London: Palgrave Macmillan, 2015), 147–66.

4 Leonard Y. Andaya, *The World of Maluku: Eastern Indonesia in the Early Modern Period* (Honolulu: University of Hawaii Press, 1993).

5 C. Fasseur, "The Cultivation System and Its Impact on the Dutch Colonial Economy and the Indigenous Society in Nineteenth-Century Java," in *Two Colonial Empires: Comparative Essays on the History of India and Indonesia in the Nineteenth Century*, ed. C.A. Bayly and D.H.A. Kolff (Dordrecht: Martinus Nijhoff, 1986), 138.

6 Michael Adas, *The Burma Delta: Economic Development and Social Change on an Asian Rice Frontier, 1852–1941* (Madison: The University of Wisconsin Press, 1974); Michitake Aso, *Rubber and the Making of Vietnam: An Ecological History, 1897–1975* (Chapel Hill: The University of North Carolina Press, 2018).

7 Drayton, *Nature's Government*; Richard Grove, *Green Imperialism: Colonial Expansion, Tropical Edens, and the Origins of Environmentalism, 1600–1860* (New York: Cambridge University Press, 1995).

8 Timothy P. Barnard, *Nature's Colony: Empire, Nation and Environment in the Singapore Botanic Gardens* (Singapore: NUS Press, 2016); Andrew Goss, *The Floracrats: State-Sponsored Science and the Failure of the Enlightenment in Indonesia* (Madison: University of Wisconsin Press, 2011).

9 Daniel R. Headrick, *The Tools of Empire: Technology and European Imperialism in the Nineteenth Century* (Oxford: Oxford University Press, 1982), 58–79; P.D. Curtin, "The White Man's Grave: Image and Reality, 1780–1850," *The Journal of British Studies* 1, no. 1 (1961): 107.

10 Goss, *The Floracrats*, 35–47; Drayton, *Nature's Government*, 206–11.

11 Andrew Goss, "Building the World's Supply of Quinine: Dutch Colonialism and the Origins of the Global Pharmaceutical Industry," *Endeavour* 38, no. 1 (2014): 8–18; Goss, *The Floracrats*, 48–56; Drayton, *Nature's Government*, 210.

12 Barnard, *Nature's Colony*, 134–52.

13 J.H. Drabble, *Rubber in Malaya, 1876–1922: The Genesis of an Industry* (Kuala Lumpur: Oxford University Press, 1973), 212–30; Barnard, *Nature's Colony*, 134–52.

14 Joseph Morgan Hodge, *Triumph of the Expert: Agrarian Doctrines of Development and the Legacies of British Colonialism* (Athens: Ohio University Press, 2007), 8–9; Drayton, *Nature's Government*, 255–60.

15 H.N. Ridley, "The Oil Palm," *Agricultural Bulletin of the Straits and Federated Malay States* 6, no. 2 (1907): 37–40; Richard H.V. Corley and P.B. Tinker, *The Oil Palm* (Oxford: Blackwell Science, 2003), 1–6.

20

SCIENCE AND ITS PUBLICS IN BRITISH INDIA

Charu Singh

This essay surveys the changing relationship between science and society in British India through the analytic of publics, conceptualised here as epistemic and political communities which are constituted by discursive and material practices of communication. Over the past three decades, an extensive body of scholarship has built an effective critique of science as colonial knowledge and as a tool of imperial domination and epistemic violence.[1] A first wave of macro-surveys since the 1990s has given way to detailed, empirical case studies of specific scientific disciplines.[2] Following the anti-diffusionist, constructionist, and global turns in the discipline more widely, recent studies have moved past the study of science as a set of ideas and Western knowledge towards an analysis of the sciences as practices and bodies of knowledge, embedded in particular spatial contexts, and practised by local actors in locally relevant political and cultural agendas.[3] Foregrounding the concept of publics, this essay surveys the history of colonial science in South Asia thematically to chart a shift in scholarly focus from the imperial sciences to the colonised subject—not as the passive recipient of "Western science," but as practitioner and participant in global scientific, technological, and medical modernity.[4]

While sociologists of scientific knowledge have used the concept of publics to study the construction of social and cultural boundaries between the practitioners of science and lay communities, the concept has been useful to scholars of South Asian history and politics for querying the nature of the relationship between communities, individuals, and the state under colonialism.[5] Since the late 1990s, historians of science in South Asia have used the concept to study the acceptance, contestation, and appropriation of "alien" scientific knowledge by colonial actors, by drawing quite explicitly on the British history of the communication of science to multiple audiences.[6] Foundational accounts proposed a chronology and typology of the acculturation of science by local actors in colonial society. In this model, such acculturation was enacted in three successive phases: first by the autodidact rooted in local intellectual traditions, then by the scientist-cum-renaissance mind, and finally by the professional scientist of the modern scientific research system.[7] Recent scholarship has elaborated the sociohistorical processes by which such acculturation occurred and demonstrated that scientific inquiry was taken up by several actors and constituted heterogeneous publics in various regions. At the same time, different social groups in colonial society had starkly different access to scientific, medical, and technical education and expertise, and this differential access continues to inflect the public lives of science in South Asia.

This essay connects earlier models with recent historical case studies to reinterpret the relations between imperial sciences and colonial society. It examines the heterogeneous and multivalent knowledge practices of local actors—subjects of colonial governance and conscripts of scientific modernity but not passive targets of "science popularisation"—by raising the following questions: Who could participate in the worlds of colonial science, and on what terms? Who laboured for science and was waged as its expert or its proletariat? And finally, how were mass publics for science forged under colonial conditions? In the following sections, the essay explores these questions in relation to three kinds of local actors: the pandit; the *daktar* and the engineer; and the everyday reader and consumer of science in the colony.

Pandits as teachers, translators, and pupils

If the learned society and its scholar-administrators, exemplified by the Asiatic Society and William Jones (1746–94) and other orientalists provide the foundational starting point for histories of the colonial episteme, pandits—Hindu scholars of the Brahman caste trained in the Sanskrit legal, moral, and philosophical sciences (*śāstra*)—are seen as their primary local interlocutors.[8] Philology, the classical science of textual interpretation practised in knowledge cultures across the world, was the meeting ground for Europeans in the employ of the East India Company (hereafter the Company) and South Asia's precolonial elites.[9] In this highly asymmetrical cross-cultural epistemic exchange, Hindu and Muslim scholars taught the new political elites the languages, legal codes, land systems, and religious and social mores of the colony.[10] Serving as "go-betweens" and brokers, teachers, and translators, these traditional custodians of knowledge provided the practical knowledge that enabled early colonial rule.[11]

The early Company state's dependence on such actors is well illustrated in the case of legal administration. In 1772, when Warren Hastings, Governor General of Bengal, decided to administer the Company's Muslim and Hindu subjects according to their own laws, a team of 11 pandits was asked to create a reliable digest of Hindu law, called the *Vivādārṇavasetu* ("The sea of litigations"). This text was then translated into Persian, from which Nathaniel Brassey Halhed translated it into English, as the Code of Gentoo Laws (1776).[12] Halhed, Jones, and other orientalists such as Henry Thomas Colebrooke, Horace Hayman Wilson, and James Prinsep in Calcutta, employed pandits and munshis to learn South Asian languages, often paying them out of their own salaries.[13] By 1800, the Company also employed scholars of Sanskrit, Persian, and Arabic at the College of Fort William, established to train young officials in Indian languages. Pandits also played an important role in the parallel structure of knowledge that emerged in south India, at institutions such as the Madras Literary Society (1817) and the College of Fort St. George (1812), where they served as interlocutors to such figures as Colin Mackenzie and Francis Whyte Ellis.[14] Contrary to the stereotype of the hide-bound pandit, they assisted European astronomers at observatories, facilitated the scientific collections amassed by European surveyors and military commanders, and themselves surveyed regions where Europeans were barred entry.[15]

Several South Asian languages had long-standing traditions of language analysis and philological practices in the subcontinent and the philological expertise of Indian interlocutors was a crucial element in the so-called "discoveries" of the early colonial period. While European scholars depended on this expertise, they maintained a suspicion of the trustworthiness and credibility of their native collaborators. The significant "discoveries," such as the idea of the Indo-European language family and the idea that languages of south India were historically related to each other and not derived from Sanskrit, emerged out of these interactions between Europeans and Indians and their historically distinct traditions of language analysis.[16] These discoveries and their attendant civilisational hierarchies have left enduring legacies for cultural identity and belonging in South Asia.

In the past two decades, scholars have traced the historical transformations in the relationship between pandits and orientalists from the late eighteenth century into the nineteenth century. Attending closely to the social, political, and cultural contexts in which orientalist scholarship was produced, they have demonstrated that new European knowledge was presented to the pandits—seen as India's "traditional intelligentsia"—as the first stage of its introduction and acceptance by the rest of Indian society. At educational institutions of Sanskrit learning sponsored by the colonial state, the pandits were transformed into the state's pupils.[17] Under James Robert Ballantyne, the Scottish orientalist who was in charge of the Benares Sanskrit College, principles of European astronomy were introduced to the pandits. Convinced that his pupils would only be persuaded of European scientific truths if they were presented to them in Sanskrit—South Asia's classical language of science—Ballantyne invested state resources in an entire infrastructure of pedagogy and knowledge distribution to meet those ends. With the aid of his pupils, he translated European scientific knowledge into Sanskrit, including the creation of a Sanskrit terminology to render the technical terms of the sciences.[18]

The "new pandit," equipped with a number of new knowledge practices as a result of such exchanges with orientalists, emerged at institutions such as Sanskrit colleges and the colonial bureaucracy.[19] These new skills included the adoption of new norms of historical study, standards of textual criticism, and procedures for editing, translating, and publishing traditional literature.[20] By the mid-nineteenth century, pandits intervened in the emerging colonial public sphere by adopting print, becoming publishers, joining literary societies, and bringing *śāstra* to bear upon contemporary concerns such as social reform.[21]

Education and work: science as an Indian vocation

Indian engineers may today appear ubiquitous in the global information economy, but in the assessments of nineteenth-century British administrators, the Hindus were not a "mechanical race."[22] Contrary to such orientalist stereotypes of the metaphysical east, science, and engineering and medicine in particular, emerged as distinctly Indian vocations in the twentieth century. These historic transformations in professional and social identity in South Asia were linked to the state's need for an underclass of lowly professionals as well as a labouring proletariat to staff army hospitals and surveys. Consequently, a number of institutions to provide scientific, medical, and technical education emerged in the nineteenth century. Rarely acknowledged in the archives, the Indian workers of science in the colony comprised a large number of local actors: from *patwaris* (village officials) responsible for maintaining accurate crop and field records in the countryside; petty officials responsible for recording readings of meteorological instruments dispersed deep in the countryside; foremen and overseers responsible for the everyday work at botanical gardens and experimental farms; low-level engineers to be employed on the construction of large public works like railways; to menial revenue officials responsible for releasing water from irrigation canals.

Crucially, the capacity of Indian subjects to access specialist education and their aspirations for scientific futures was defined by their existing socioeconomic location. The recognition of the central role of Indian labour within colonial scientific practice makes possible an analysis of this differential access, its postcolonial legacy, and the making of heterogeneous Indian publics of science.

The early Company state's military expansion and political consolidation over extremely variable environments and socially diverse populations required intensive information gathering. Geographical and revenue surveying, the collection of botanical specimens, natural historical inquiries into flora and fauna, forests and agricultural practices, the description and classification of populations were ongoing activities of colonial state-formation. The information gathering essential for imperial administration provided the "stuff of science" to scientific publics located in Europe. Specimens collected and reports produced in the course of military campaigns or administrative surveys in the colony were rendered into the collections of scientific institutions such as the Company's India Museum established at its Leadenhall Street headquarters in 1801.[23] Scottish surgeons trained in distinct traditions of natural history, such as the conjectural history of the Scottish enlightenment, employed new methods of survey and a new "literary technology"—the survey report—to communicate their findings to the military state and to the metropolitan public.[24]

Colonial surveying was critically dependent upon Indian labour. The proletariat of imperial science included those employed on the surveying teams of the Great Trigonometrical Survey as well as those on the more regular revenue survey and settlement operations. A Revenue Survey School had been attached to the Madras Observatory from the 1790s, and even youths of mixed European–Asian parentage from the Madras Male Asylum were trained for survey work.[25] Imperial policy recognised the colonial state's dependence upon this proletariat through the creation of institutions such as the engineering colleges established at Roorkee, Madras, Poona, and Shibpur over the course of the nineteenth century. These colleges had an engineering class, and also trained local youths to assist British engineers as subordinates in building the infrastructure—roads, bridges, buildings, and irrigation works—which underlay the colonial economy under the aegis of the Public Works Department (PWD) (1854).

By the end of the nineteenth century, a large proportion of labouring masses belonging to lower castes had become the object of the colonial state's technical education policy. The state aimed to provide minimal training and useful skills to maintain a steady flow of "technically trained men" to the railways, mining, and other industrial enterprises. For example, in 1892, the Lucknow Industrial School was established to provide apprentices at railway workshops, and train smiths and carpenters.[26] The Indian National Congress (1885) supported the state's drive for the establishment of technical schools and colleges and the introduction of modern arts and industry to reform the "traditional" Indian artisan and labourer.[27]

Indians from different regions and with varying institutional resources, socioeconomic power, and caste locations accessed scientific education in distinct ways, and recent scholarship has detailed some of these trajectories. For instance, in the early nineteenth century, the Hindu elites of Calcutta came together to appeal British authorities for the foundation of a college to introduce into India "the literature and science of Europe." At Hindu College (1817), the institution towards which these urban elites contributed a significant sum, the curriculum included instruction in "history, geography, chronology, astronomy, mathematics, chemistry and other sciences."[28] When the London-based British India Society (1821) donated some scientific instruments to the college in 1824, the students were also exposed to the material culture of experimental natural philosophy. The Bengali *bhadralok*, who formed the main constituency of modern science education at the college, showed a distinct preference for the theoretical aspects of the curriculum over the experimental lectures.[29]

European physicians and surgeons employed by the Company depended on Indian medical assistants to ensure the health of its troops, and of European populations in Company factories and urban settlements. In 1822, the Company established the Native Medical Institution, later replaced by the Calcutta Medical College (1835), to train medical assistants drawn from the higher castes of Hindu society.[30] Such "sub-assistant surgeons," it was hoped, would further disseminate Western medical knowledge into Indian society by training apprentices, compounders, dressers, and other subordinates. Even as Indians were employed in the colonial medical services, they also participated in the medical market through private medical practice.

Daktari, a vernacular neologism (calqued off "doctor"), was the term used to describe the complex medical practices of the Bengali *daktars*, a large social group that included Hindus and Muslims. *Daktars'* training ranged from apprenticeship in the army or a dispensary and college education, to higher medical degrees acquired in London or Edinburgh, and their practice varied across allopathy, homeopathy, and ayurvedic medicine. Deeply embedded in the local social and political context of their spheres of practice, *daktars* availed of the nineteenth-century publishing boom and medical market to carve a distinctive social identity for themselves. *Daktari* conceptions of the diseased body and its causes asserted that Bengali constitutions differed from those of Europeans and articulated culturally specific projects of national physical development under colonialism.[31]

A distinct social and professional identity also emerged among Indian engineers belonging to the PWD of the colonial state by the early twentieth century, a period that saw the expansion of engineering activity beyond public works, military, and railways towards large-scale industry, and the growing employment of natives. A small number of Indian engineers began to join metropolitan professional bodies, and by the 1910s, more local professional institutions emerged in this period, such as the Punjab PWD Congress (1912) and the Institution of Engineers (1920). By the 1910s, when Indian nationalists protested against the colonial administration of Indian railways and called for the greater employment of Indians on the railways, they made these claims in the name of an Indian public.[32]

Since Max Weber, the sociology of science has examined science as a vocation. Studies of the scientific life, the value attached to scientific knowledge, and the ethical implications of choosing a career in science have been a mainstay of such analysis. The examples above point to two distinctively Indian stories, which are part of a wider trend in the public lives of science in other colonial and non-Western contexts. Over the course of the twentieth century, the figures of the doctor and the engineer have reorganised social aspiration, professional mobility, and cultural status in Indian society. The growth of the Indian middle class was tied to these professions, especially after 1947. They shaped the ideals of work and upward mobility and defined what it meant to have a "good profession." These colonial histories variously enable and constrain the abilities of South Asians to craft scientific selves along the lines of caste. Emerging critiques of "merit" in this context point to new directions in the historical sociology and anthropology of science, education, and caste.[33]

Readers and consumers: everyday colonial publics

By the late nineteenth century, Indian subjects lived conscripted to a world increasingly reshaped under the sign and influence of science. Colonial subjectivity was defined by: large-scale geopolitical and biopolitical projects such as cartographic surveys, the decennial census, and the control of epidemics; the creation of new educational institutions and the rise of new professionals such as graduates, lawyers, doctors, and engineers; and the growth of large infrastructures of transport and agrarian capitalism, such as railways and irrigation canals.[34] From the 1880s, following the expansion of world trade and the integration of multiple communications networks (steamships, railroads, postal services, and newspapers), a large number of new industrially produced commodities from the west began to find mass markets in South Asia. The import and consumption of small-scale technologies such as sewing machines, typewriters, bicycles, cameras, and clocks marked a new technological awareness amongst their Indian users.[35] Alongside, the growth of a medical market offered a range of cures, introducing homeopathy and allopathy alongside the subcontinent's more traditional remedies.[36] The acculturation of these new sciences, technologies, and therapies by Indian subjects took place in multiple settings—museums, classrooms, and offices as much as markets, medical dispensaries, and post offices. As science-learners at schools and universities and as readers of literary periodicals and newspapers filled with advertisements for many of these products, Indians became consumers of technoscientific modernity and put it to work within their own local agendas.

The domain of reproductive politics provides an example of the complex ways in which the availability of new small technologies, their adoption by social groups with disposable incomes, and their deployment in local agendas created Indian consumers of technoscientific modernity. In the 1920s and 30s, modern, mass-produced forms of contraception became commercially available in urban markets. These included mechanical contraceptives such as condoms as well as chemical preparations such as "Contraceptin."[37] Contraceptive products, medical manuals, and advisory literature on sex, birth control, and reproductive health were widely advertised in English and vernacular print media. The books advertised included the works of global eugenics campaigners such as Marie Stopes's *Married Love* and *Wise Parenthood*, written as birth control manuals for lay audiences. Indian writers translated such works by foreign authors and also wrote their own manuals, in English as well as the vernaculars.[38]

The availability of technologies of contraception intersected with the uptake of eugenics in British India. Recent scholarship shows that the global eugenics movement had a large Indian constituency. Urban, middle-class supporters of birth control formed themselves into voluntary organisations, such as the Madras Neo-Malthusian League (MNML, 1928). The MNML held regular public meetings, ran a small library, published pamphlets, and even brought out a bi-monthly journal, the *Madras Birth Control Bulletin*. The membership of the League was comprised largely of Brahmin men, elite members of the colonial bureaucracy. The movement did not represent a critique of the colonial state's lack of support for contraception, nor was its main aim to control the reproductive practices of urban masses and the poor. Rather, the MNML's support for contraception was more focused on the changing conjugality and sexual practices among urban Brahmins themselves. The language of science reinforced their agenda of caste consolidation and self-regulation.[39]

Small technological devices and instruments found a wide range of contexts in which they were deployed and given local meaning. For example, the pocket watch and the thermometer, the microscope and the stethoscope, were embraced by practitioners of Ayurveda, who mobilised these unspectacular, everyday objects to visualise the body in new ways. These technologies literally materialised patients' bodies through new metaphors or "physiograms" for representing, understanding, and treating the Ayurvedic body.[40]

As these mediations between traditional and modern therapeutics by new technologies suggest, Indian intellectual engagement with the globalising European sciences was often worked out within existing cosmologies, through available religious metaphors and moral vocabularies. The significance of language, translation, and other forms of discursive practice—essential tools for reconstituting older knowledge communities and calling new publics into being—is increasingly being recognised in the scholarship. As language itself became a contested site of reform and agitation, reformers and language activists in several regions took on the work of transforming South Asian vernaculars into effective media of administration, literary expression, and mass education.[41] A significant aspect of this linguistic modernisation was the desire to transform the vernaculars into languages of science education and the media for creating knowledgeable, science-literate national publics.[42] In the early-twentieth-century context of mass politics and growing conflict between religions, scientific and medical knowledge was deployed in novel ways to imagine new political collectives and to actualise the body politic of new national imaginaries.[43]

These issues of translation, knowledge economies, and linguistic identity became ever more significant in South Asia, a region with a higher population and far greater linguistic diversity than Europe. Here, colonial education policy institutionalised a linguistic division of labour between English and the vernaculars, at a time when languages of science in Europe were being reordered by significantly different logics.[44] English was privileged as the language of higher scientific education and scientific research in South Asia, while the vernaculars were regarded as the media of "translated science." Ironically, even as vernacular languages remain the languages of primary and secondary education in postcolonial South Asia, the function of science writing in the vernacular has been understood as making available scientific knowledge to "the lower educational levels and to the general public."[45] Although scientific and medical knowledge in translation began to circulate widely in the regional literary spheres from the second half of the nineteenth century, its languages had a significant impact on the nature of knowledge made available as science to vernacular publics. Histories of imperial and postcolonial science are increasingly reckoning with the fraught legacy of this bifurcation and now attending to the very media of the public lives of science in the subcontinent.[46]

Conclusion

This essay reframes questions of imperial science as questions about its practitioners, patrons, and publics. Imperial science in South Asia altered the norms of traditional authority, defined the image of authoritative knowledge in colonial society, and came to ineluctably shape desirable futures for individuals and collectives. The promise of a science education reorganised colonial subjectivity and aspirations and still continues to have a hold on middle-class desires of a "good vocation." This essay deployed the analytic of publics in its synoptic and partial account of the sciences in British India to highlight a significant shift underway in the historiography. This shift centres not the imperial sciences, nor its learned societies or surveys, its institutions and paradigms; instead, it follows the colonial subject, as a maker of her own scientific lives. Scientific lives in postcolonial South Asia continue to bear the traces of differential paths to education, expertise, and social power.

Notes

1 The pioneering surveys include Deepak Kumar, *Science and the Raj, 1857–1905* (New Delhi: Oxford University Press, 1995); Partha Chatterjee, ed., *Texts of Power: Emerging Disciplines in Colonial Bengal* (Minneapolis: University of Minnesota Press, 1995); Zaheer Baber, *The Science of Empire: Scientific Knowledge, Civilisation, and Colonial Rule in India* (Albany: State University of New York Press, 1996); Christopher A. Bayly, *Empire and Information: Intelligence Gathering and Social Communication in India, 1780–1870* (Cambridge: Cambridge University Press, 1996); Bernard Cohn, *Colonialism and its Forms of Knowledge: The British in India* (Princeton, NJ: Princeton University Press, 1996); Gyan Prakash, *Another Reason: Science and the Imagination of Modern India* (Princeton, NJ: Princeton University Press, 1999); David Arnold, *The New Cambridge History of India*, vol. 3, pt. 5, *Science, Technology, and Medicine in Colonial India* (Cambridge: Cambridge University Press, 2000); Pratik Chakrabarti, *Western Science in Modern India: Metropolitan Methods, Colonial Practices* (Delhi: Permanent Black, 2004); Dhruv Raina and S. Irfan Habib, *Domesticating Modern Science: A Social History of Science and Culture in Colonial India* (New Delhi: Tulika Books, 2004).

2 For example, Matthew H. Edney, *Mapping an Empire: The Geographical Construction of British India, 1765–1843* (Chicago: University of Chicago Press, 1997); Patricia Uberoi, Nandini Sundar, and Satish Deshpande, *Anthropology in the East: Founders of Indian Sociology and Anthropology* (Delhi: Permanent

Black, 2007); Pratik Chakrabarti, *Bacteriology in British India: Laboratory Medicine and the Tropics* (Rochester, NY: University of Rochester Press, 2012); Joydeep Sen, *Astronomy in India, 1784–1876* (London: Pickering and Chatto, 2014).

3 On anti-diffusion, co-constitution, and circulation, see Kapil Raj, *Relocating Modern Science: Circulation and the Construction of Knowledge in South Asia and Europe, 1650–1900* (Basingstoke: Palgrave Macmillan, 2007). On global connections in the history of science, see Sujit Sivasundaram, "Sciences and the Global: On Methods, Questions, and Theory," *Isis* 101, no. 1 (2010): 146–58. On reading South Asian histories of science against Eurocentric narratives and within an Asia-centred framework, see Prakash Kumar, Projit B. Mukharji, and Amit Prasad, "Decolonizing Science in Asia," *Verge* 4, no. 1 (2018): 24–43.

4 For key explorations of the refiguration of European disciplines in South Asia, see Projit B. Mukharji, *Nationalizing the Body: The Medical Market, Print and Daktari Medicine* (London: Anthem Press, 2009); Jahnavi Phalkey, *Atomic State: Big Science in Twentieth-Century India* (Ranikhet: Permanent Black, 2013); Shinjini Das, *Vernacular Medicine in Colonial India: Family, Market and Homeopathy* (Cambridge: Cambridge University Press, 2019). Recent special issues provide the best guides to this new scholarship; see Pratik Chakrabarti, ed., "States of Healing," special section, *South Asian History and Culture* 4, no. 1 (2013); Jahnavi Phalkey, ed., "Science, History and Modern India," focus section, *Isis* 104, no. 2 (2013); Asif A. Siddiqi, ed., "Technology in the South Asian Imaginary," special issue, *History and Technology* 31, no. 4 (2015); Prakash Kumar, ed., "New Histories of Technology in South Asia," special issue, *Technology and Culture* 60, no. 4 (October 2019); and Minakshi Menon, ed., "Indigenous Knowledges and Colonial Sciences in South Asia," special issue, *South Asian History and Culture*, forthcoming. See also Rohan Deb Roy and Guy Attewell, eds., *Locating the Medical: Explorations in South Asian History* (New Delhi: Oxford University Press, 2018).

5 Steven Shapin, "Science and the Public," in *Companion to the History of Modern Science*, ed. Robert Olby et al. (London: Routledge, 1990), 990–1007; Sandra Freitag, "Introduction: 'The Public' and its Meanings in South Asia," *South Asia: Journal of South Asian Studies* 14, no. 1 (1991): 1–13; J. Barton Scott and Brannon Ingram, "What is a Public? Notes from South Asia," *South Asia: Journal of South Asian Studies* 38, no. 3 (2015): 357–70.

6 Satpal Sangwan, "Science and its Public in British India: Problematic of Diffusion and Social Appropriation," in *Uncharted Terrains: Essays on Science Popularisation in Pre-Independence India*, ed. Narender K. Sehgal, Satpal Sangwan, and Subodh Mahanti (New Delhi: Vigyan Prasar, 2000), 13–53.

7 Dhruv Raina, "Lamenting the Past, Anticipating the Future: A Chronology of Popular Science Writing in India (1850–1914)," in Sehgal, Sangwan, and Mahanti, *Uncharted Terrains*, 54–66.

8 Early scholarship on the Asiatic society, such as O.P. Kejariwal, *The Asiatic Society of Bengal and the Discovery of India's Past, 1784–1838* (Delhi: Oxford University Press, 1988), has been substantially revised; see Raj, *Relocating Modern Science*. The Company's early interlocutors also included highly literate Muslim elites. This section's focus on pandits reflects a slant in the scholarship and by no means sidelines the equally important epistemic engagement between orientalists and *maulavis* and munshis.

9 For philology's precolonial histories, see Rajeev Kinra, "Cultures of Comparative Philology in the Early Modern Indo-Persian World," *Philological Encounters* 1 (2016): 225–87; Whitney Cox, *Modes of Philology in Medieval South India* (Leiden: Brill, 2017). See also Sheldon Pollock, ed., *Forms of Knowledge in Early Modern Asia: Explorations in the Intellectual History of India and Tibet, 1500–1800* (Durham, NC: Duke University Press, 2011).

10 Rosane Rocher, "British Orientalism in the Eighteenth Century: The Dialectics of Knowledge and Government," in *Orientalism and the Postcolonial Predicament: Perspectives on South Asia*, ed. Carol Breckenridge and Peter van der Veer (Philadelphia: University of Pennsylvania Press, 1993).

11 See Simon Schaffer, "The Asiatic Enlightenments of British Astronomy," in *The Brokered World: Go-Betweens and Global Intelligence, 1770–1820*, ed. Simon Schaffer, Lissa Roberts, Kapil Raj, and James Delbourgo (Sagamore Beach, MA: Science History Publications, 2009); Kapil Raj, "Mapping Knowledge Go-Betweens in Calcutta, 1770–1820," in Schaffer et al, *Brokered World*.

12 Raj, *Relocating Modern Science*, 95–138.

13 Sisir Kumar Das, *Sahibs and Munshis: An Account of the College of Fort William* (New Delhi: Orion Publications, 1960).

14 Thomas Trautmann, ed., *The Madras School of Orientalism: Producing Knowledge in Colonial South India* (New Delhi: Oxford University Press, 2009); Rama Sundari Mantena, *The Origins of Modern Historiography in India: Antiquarianism and Philology, 1780–1880* (Basingstoke: Palgrave Macmillan, 2012).

15 Simon Schaffer, "Instruments and Ingenuity between India and Britain," *Bulletin of the Scientific Instrument Society* 140 (2019): 1–13; Raj, *Relocating Modern Science*, 181–222.

16 Thomas Trautmann, *Aryans and British India* (Berkeley: University of California Press, 1997); Thomas Trautmann, *Languages and Nations: The Dravidian Proof in Colonial Madras* (Berkeley: University of California Press, 2006); see also Kinra, "Cultures of Comparative Philology."

17 Michael S. Dodson, *Orientalism, Empire and National Culture: India, 1770–1880* (Basingstoke: Palgrave Macmillan, 2007).

18 Michael S. Dodson, "Translating Science, Translating Empire: The Power of Language in Colonial North India," *Comparative Studies in Society and History* 47, no. 4 (2005): 809–35.

19 Madhav M. Deshpande, "Pandit and Professor: Transformations in Nineteenth-Century Maharashtra," in *The Pandit: Traditional Scholarship in India*, ed. Axel Michaels (New Delhi: Manohar, 2001).

20 Brian A. Hatcher, "Sastric Modernity: Mediating Sanskrit Knowledge in Colonial Bengal," in *Modernities in Asian Perspective*, ed. Kausik Bandyopadhyay (Kolkata: Setu Prakashani, 2010).

21 Francesca Orsini, "Pandits, Printers and Others: Publishing in Nineteenth-Century Benares," in *Print Areas: Book History in India*, ed. Abhijit Gupta and Swapan Chakravorty (New Delhi: Permanent Black, 2004), 103–38.

22 Richard Temple, Governor of Bombay, 1881, quoted in Ross Bassett, *The Technological Indian* (Cambridge, MA: Harvard University Press, 2016), 2.

23 Jessica Ratcliff, "The East India Company, the Company's Museum, and the Political Economy of Natural History in the Early Nineteenth Century," *Isis*, 107, no. 3 (2016): 495–517.

24 Minakshi Menon, "Transferable Surveys: Natural History from the Hebrides to South India," *Journal of Scottish Historical Studies*, 38, no. 1 (2018): 143–59. See also Jessica Ratcliff, "Hand-in-Hand with the Survey: Surveying and the Accumulation of Knowledge Capital at India House during the Napoleonic Wars," *Notes and Records of the Royal Society* 73 (2019): 149–66.

25 Edney, *Mapping an Empire*, 172.

26 Arun Kumar, "Skilling and Its Histories: Labour Market, Technical Knowledge and the Making of Skilled Workers in Colonial India (1880–1910)," *Journal of South Asian Development*, 13, no. 3 (2018): 249–71. See also Shahana Bhattacharya, "Transforming Skin, Changing Caste: Technical Education in Leather Production in India, 1900–1950," *Indian Economic and Social History Review* 55, no. 3 (2018): 307–43.

27 Bidisha Dhar, "Technical Education Discourse in India: State and the Artisans, 1880s–1914," *Studies in History*, 33, no. 2 (2017): 213–33.

28 Raj, *Relocating Modern Science*, 159–80.

29 Scholarly focus on colonial Calcutta has made it possible to track the evolving nature of the Bengali urban elites' attitude to science over the colonial period, but far less is known about other urban centres such as Lahore, Bombay, Madras, and Benares. See Dhruv Raina and S. Irfan Habib, "Bhadralok Perceptions of Science, Technology and Cultural Nationalism," *Indian Economic and Social History Review* 32, no. 1 (1995): 95–117; John Bosco Lourdusamy, *Science and National Consciousness in Bengal: 1870–1930* (New Delhi: Orient Longman, 2004).

30 Ishita Pande, *Medicine, Race and Liberalism in British Bengal: Symptoms of Empire* (London: Routledge, 2009).

31 Mukharji, *Nationalizing the Body*.

32 Aparajith Ramnath, *The Birth of an Indian Profession: Engineers, Industry and the State, 1900–47* (New Delhi: Oxford University Press, 2017); Stefan Tetzlaff, "Engineers and Social Change in Colonial and Postcolonial India: Considerations between Recent Literature and Future Research Possibilities," *Südasien-Chronik/South Asia Chronicle* 7 (2017): 449–61.

33 Sunandan K.N., "From Acharam to Knowledge: Claims of Caste Dominance in Twentieth-century Malabar," *History and Sociology of South Asia* 9, no. 2 (2015): 174–92; Ajantha Subramanian, *The Caste of Merit: Engineering Education in India* (Cambridge, MA: Harvard University Press, 2019); Renny Thomas, "Brahmins as Scientists and Science as Brahmins' Calling: Caste in an Indian Scientific Research Institute," *Public Understanding of Science* 29, no. 3 (2020): 306–18.

34 For the transformations in subjectivity wrought by these infrastructures of mass transport, see Ritika Prasad, " 'Time-Sense': Railways and Temporality in Colonial India," *Modern Asian Studies* 47, no. 4 (2013): 1252–82; Aparajita Mukhopadhyay, *Imperial Technology and "Native" Agency: A Social History of Railways in Colonial India, 1850–1920* (London: Routledge, 2018).

35 David Arnold, *Everyday Technology: Machines and the Making of India's Modernity* (Chicago: University of Chicago Press, 2013).

36 On the role of advertising in producing the new medical consumer, see Madhuri Sharma, "Creating a Consumer: Exploring Medical Advertisements in Colonial India," in *The Social History of Health and Medicine in Colonial India*, ed. Biswamoy Pati and Mark Harrison (London: Routledge, 2009), 213–28; Douglas E. Haynes, "Selling Masculinity: Advertisements for Sex Tonics and the Making of Modern Conjugality in Western India, 1900–1945," *South Asia: Journal of South Asian Studies* 35, no. 4 (2012): 787–831.

37 Sarah Hodges, *Contraception, Colonialism and Commerce: Birth Control in South India, 1920–1940* (Aldershot: Ashgate, 2008).

38 For the growing literature on Indian interpreters of global sexual science, see Luzia Savary, "Vernacular Eugenics? *Santati-Śāstra* in Popular Hindi Advisory Literature (1900–1940)," *South Asia: Journal of South Asian Studies* 37, no. 3 (2014): 381–97; Shrikant Botre and Douglas E. Haynes, "Understanding R. D. Karve: Brahmacharya, Modernity and the Appropriation of Global Sexual Science in Western India, 1927–1953," in *A Global History of Sexual Science, 1880–1960*, ed. Veronika Fuechtner, Douglas E. Haynes, and Ryan M. Jones (Berkeley: University of California Press, 2017), 163–185; Ishita Pande, "Time for Sex: The Education of Desire and the Conduct of Childhood in Hindu Sexology," in Fuechtner, Hayes, and Jones, *Global History of Sexual Science*, 279–301; Durba Mitra, *Indian Sex Life: Sexuality and the Colonial Origins of Modern Social Thought* (Princeton, NJ: Princeton University Press, 2020).

39 Hodges, *Contraception, Colonialism and Commerce*.

40 Projit B. Mukharji, *Doctoring Traditions: Ayurveda, Small Technologies, and Braided Sciences* (Chicago: University of Chicago Press, 2016); Projit B. Mukharji, "Akarnan: The Stethoscope and Making of Modern Ayurveda, Bengal, c. 1894–1952," *Technology and Culture* 60, no. 4 (2019): 953–78.

41 There is a vast scholarship on each region; see Lisa Mitchell, *Language, Emotion, and Politics in South India: The Making of a Mother Tongue* (Bloomington: Indiana University Press, 2009) for a conceptual analysis of the new emotional commitment to language by the end of the nineteenth century.

42 Dodson, "Translating Science, Translating Empire"; A.R. Venkatachalapathy, "Coining Words: Language and Politics in Late Colonial Tamil Nadu," reprinted in *In Those Days There Was No Coffee: Essays in Cultural History* (New Delhi: Yoda Press, 2006); Kavita S. Datla, "A Worldly Vernacular: Urdu at Osmania University," *Modern Asian Studies* 43, no. 5 (2009): 1117–48; Gautham Reddy, "The Andhra Sahitya Parishat: Language, Nation and Empire in Colonial South India (1911–15)," *Indian Economic and Social History Review* 56, no. 3 (2019): 283–310; Charu Singh, "Science in the Vernacular? Translation, Terminology and Lexicography in the *Hindi Scientific Glossary* (1906)," *South Asian History and Culture*, forthcoming.

43 Kavita Sivaramakrishnan, *Old Potions, New Bottles: Recasting Indigenous Medicine in Colonial Punjab (1850–1945)* (Hyderabad: Orient Longman, 2006); Guy Attewell, *Refiguring Unani Tibb: Plural Healing in Late Colonial India* (Hyderabad: Orient Longman, 2007); Richard S. Weiss, *Recipes for Immortality Medicine, Religion, and Community in South India* (Oxford: Oxford University Press, 2009); Mukharji, *Nationalizing the Body*; Rachel Berger, *Ayurveda Made Modern: Political Histories of Indigenous Medicine in North India, 1900–1955* (New York: Palgrave Macmillan, 2013).

44 Michael D. Gordin, *Scientific Babel: How Science Was Done Before and After Global English* (Chicago: University of Chicago Press, 2015.

45 Partha Chatterjee, "The Disciplines in Colonial Bengal," in Chatterjee, *Texts of Power*, 17.

46 Das, *Vernacular Medicine in Colonial India*; Mitra, *Indian Sex Life*.

21

FROM HISTORY OF SCIENCE TO HISTORY OF KNOWLEDGE?

Themes and perspectives in colonial Australasia

James Beattie and Ruth A. Morgan

On his return to Scotland in 1862, physician and botanist William Launder Lindsay (1829–80) reflected on his recent tour of New Zealand.[1] In the pamphlet *Place and Power of Natural History in Colonisation*, he observed that "the systematic, economical, and complete development of ... [Otago's] resources can be effectually accomplished only by the aid of scientific observations and deductions."[2] Such views typified his generation's approach to science in Australasia as being primarily concerned with matters of practical use to colonial development. For historians of science too, Lindsay's outlook reflects the predominant focus of the field in Australasia on "science in the service of empire" during the long nineteenth century.[3]

In this short review chapter, we map such approaches to the history of science in Australasia, and make a case for new directions in the field. We argue that, rather than the frameworks that the history of science has traditionally offered, the application of a history of knowledge lens would better serve historians seeking to make sense of the ways in which Indigenous and non-Indigenous peoples interpreted and understood the natural and human worlds of settler-colonial Australasia. Such an approach, we show, is already underway in some of the most exciting scholarship on Australia and Aotearoa-New Zealand, in which historians are exploring opportunities for the study of the nature and legacies of cross-cultural encounters in colonial knowledge production.

Shifting the historical inquiry from the history of science to histories of "knowledge" and "systems of knowledge," we argue, more accurately reflects the nature and reality of colonial encounter in Australasia. Across these settler colonies, interactions between Aboriginal Australians, Māori, Chinese, South Asians, Europeans, and others shaped, elided, and contested understandings of the natural world, both locally and in the metropole. After all, in the nineteenth century, the "authority" of Western knowledge was far from established, especially as regards to health, materia medica, climate, and the like, which faced considerable epistemic challenges.[4]

Rethinking the history of science in Australasia reflects a much wider disciplinary conversation. As Lorraine Daston has argued,

> [t]he original disciplinary narrative of the history of science is simply untenable on scholarly grounds, undermined by the careful historicism and aversion to anachronism

and teleology that has characterized the most rigorous and imaginative work in the field for the last forty years.[5]

The history of knowledge is a broader and more capacious framing that acknowledges the existence of a variety of different knowledge systems, while allowing for no system to take analytical precedence over another. Further still, the history of knowledge releases historians from the intellectual straitjacket of determining everything in relation to whether or not it is indeed science, that is, Western and modern science. In advocating for history of knowledge, we are not, we wish to make clear, arguing that scholars abandon their examination of the history of science. Far from it. What we advocate instead is that science be regarded as just one system among several in the colonial context of Australasia, albeit one that was contested and privileged from the late eighteenth to the early twentieth centuries.

Before we examine these arguments in favour of history of knowledge, we overview some of the dominant themes and approaches to the field known as history of science in Australasia. We note that social history approaches still dominate Australasian science historiography. A common focus remains biography, particularly of heroic men of science (and sometimes women), as well as institutional histories of the formal associations and patronage networks of scientific endeavour.[6] This holds across a broad range of topics, whether the study relates to aspects of medical history, conservation, or botanical collecting.[7] Some notable exceptions include Pamela Wood's cultural history of dirt, which explores contested lay and professional ideas about disease transmission, cleanliness, and the state, or Charlotte Macdonald's comparative study of gender and physical fitness, which ranges widely over cultural history, health history, and nationalism.[8]

Across Australasian history of science, some areas are much better covered than others. For example, there are sophisticated studies of the professionalisation and cultural history of ideas of climate, and its relationship to agricultural science, meteorology, and medicine. As well, there are contextualised studies on debates over what is and is not science.[9] Scholarship on Darwin, led mainly by John Stenhouse in New Zealand and Iain McCalman in Australia, has provided nuanced understandings of the manner in which Darwin's ideas were received, debated, and interpreted in colonial Australasia and the wider Pacific.[10] Stenhouse and others have also provided subtle examinations of the relationship between science and religion, or more accurately, sciences and religions.[11] Given the historic dominance of settler agriculture in Australasia, there is a core of studies on this topic: notably within the last decade, this includes Cameron Muir's monograph on agriculture in Australia, and Tom Brooking and Eric Pawson's works on New Zealand's grasslands revolution.[12] There are many histories of medicine, particularly that of psychiatry and mental health institutions, as well as cultural histories of health and race, ranging from phrenology to eugenics.[13] Given their role in imperial science, histories of conservation, botanical science, and acclimatisation continue to figure prominently in Australasia's science historiography.[14] There are also a number of works on the region's scientific and technical education,[15] as well as ethnography and collecting.[16] As we show below, these seemingly well-traversed areas of study are becoming subject to fresh interpretative frameworks that emphasise a more dynamic, agential, and networked set of processes of knowledge production than earlier approaches have allowed.[17]

Over the past two decades, the field has become increasingly attuned to the significance of gender, race, and class to histories of science and knowledge production in colonial Australasia.[18] Guided by the insights of gender and social history, scholars have moved beyond a compensatory approach to more nuanced and sophisticated studies that analyse the construction and breaking down of gender categories within science in the home, in the garden, and in other spheres.[19] For example, Rebecca Rice's work on female botanical illustrators demonstrates the manner in which a number of white women in Aotearoa were able to extend women's perceived roles in botanical art to produce scientific works on botany.[20] In Australia, Caroline Jordan argues that middle-class women were able to extend their perceived traditional role as nurturers and artists into areas such as professional botanical science.[21]

Detailed case studies have the potential to provide evidence of the circumvention of masculine stereotypes within the practice of colonial science. For example, the diary of teenage Laura Taylor contains pressed petals, leaves, feathers, and natural history sketches. On her 14th birthday, in 1847, her father gave her a microscope so she could examine God's works more closely.[22] The microscope would, her father explained, "reveal some of the wonders of the minute portion of creation."[23] Similar work by Rosi Crane on the institutional membership of the Otago branch of the New Zealand Institute indicates a desire amongst its members to encourage female participation in science, yet at the same time also highlights the circumscriptions around those opportunities.[24] Another refreshing approach to gender studies and science comes from the realm of medical topography, which has examined gendered and racialised ideas about the impact of different climates on human constitutions. Deployed in the Australasian colonies, these ideas justified white health migration to cooler, more temperate climes as well as exploitative labour regimes of non-whites long into the twentieth century.[25]

Drawing on the insights of subaltern studies, Indigenous studies, and science and technology studies, historians of science and knowledge production have sought to dismantle heroic narratives of imperial and colonial exploration, commerce, and scientific enterprise across Australasia. Revisiting these accounts recognises the many Indigenous and local knowledge brokers, guides, and intermediaries, on whom non-Indigenous people depended, on land and at sea.[26] Although such gendered and racialised encounters were vital on the frontier, they were rarely acknowledged in metropolitan contexts.[27] Interdisciplinary studies using records from the Australian colonies show the extent to which collectors of all manner of specimens, whether mineral, biological, ornithological, or zoological, relied on Indigenous peoples and their knowledge of the so-called natural world.[28] Analysing the specimens themselves and their contested taxonomy points to the potential for studies of "ethnographic natural history" to reveal the nature and structures of knowledge-production, and their legacies for the present.[29]

Much recent work in Aotearoa, often conducted by Māori scholars, focuses on the retrieval of lost knowledge and practices, often using *whakapapa* (genealogy) and other traditional means, commonly triangulated with work found in historical manuscripts.[30] In Australia too, Indigenous researchers are using the colonial record to show the extent and diversity of Aboriginal landscape management at contact, not least as a political manoeuvre to defy persistent racist attitudes.[31] Others too are re-examining oral accounts and ethnographies from the colonial period to interpret Indigenous experiences and knowledge of geological and astronomical events.[32] Although there are obvious problems with the reliance on such manuscripts, as historians we encourage others to ask different questions of these sources, to approach them not (only) as a source of lost traditions, but rather to help answer questions as to how knowledge transfer worked in practice, which knowledge traditions survived and why, as well as how that knowledge was redeployed in the light of changed circumstances and different knowledge traditions, such as Western science. In doing so, non-Indigenous historians need to be attentive to Indigenous ways of approaching the past, which means language learning, consultation from a project's beginning and, ideally, cooperation on projects of this nature.[33]

With these ethical approaches in mind, we encourage further studies of the medium of knowledge transfer (including science), whether through oral traditions, educational systems, written fora, or material culture.[34] How did different groups translate (linguistically and in the Latourian sense), or reinterpret, the others' knowledge? Were there particular realms of knowledge—such as over medicinal properties of plants or observed weather patterns—that encouraged interaction and discussion, or others which discouraged points of interaction? Such an approach might include analysis of the reception and interpretation of Western science within Indigenous and non-Western groups. Complex and unequal as these relationships were, Indigenous people had agency to interpret and use selectively for their own purposes aspects of Western knowledge systems, including science.[35] Consider, for instance, the engagement and use of Western ethnography for the purposes, say, of cultural renaissance, in the case of Aotearoa-New Zealand,[36] or the role of science in attempting to arrest the decline of the Māori population.[37] Indeed, a close reading of colonial texts reveals that debates between tradition and modernity, religion and science, were not restricted to Australasia's white settler population.[38]

Another particularly fruitful avenue of study concerns interactions between different knowledge systems. For example, many leading Māori chiefs sent their sons to study agriculture at the Church Missionary Society School in Parramatta, New South Wales, in the early nineteenth century. What, also, of interactions between Indigenous peoples in Australia and Māori? Both peoples travelled widely, and encountered each other in Australia and Aotearoa—in the latter, several Indigenous Australians became incorporated into Māori tribal groups, especially during the early years of contact from the late eighteenth century and early nineteenth century. Likewise, there are countless examples of Europeans learning from Māori, whether through the adoption of Māori methods of growing potatoes (itself based on their experience of raising kumara in mounds), the learning of weather systems based on navigation, or exchanges of botanical knowledge and the properties of plants.[39] An example of the last includes the French-born nursing nun, Mother Mary Aubert (1835–1926). She enthusiastically explored and exploited some of the medical potential of New Zealand's plants by preparing and selling herbal remedies commercially and using them for the care of many Māori to whom she ministered.[40]

In colonial Australasia, another particularly fascinating area concerns the role of Western and Chinese practices of health care on the goldfields of Australasia. New research is revealing that European miners had access to Chinese medical practitioners and Chinese men used European hospitals in New Zealand.[41] In Australia, material culture offers insights into the practice of traditional Chinese medicine in Melbourne and Ballarat at the turn of the twentieth century, while colonial newspapers show the racialised contestation of medical authority in Bendigo.[42] South Asian doctors navigated the equally complex terrain of white corporeal and affective vulnerability, as Nadia Rhook has shown in colonial Victoria.[43] Similarly, studying particular sites of interaction—or zones of cultural contact—can offer an especially rich area of research into cross-cultural interactions.[44] Following the work of Fa-ti Fan on botanical and artistic interactions between Europeans and Chinese in Treaty Port Canton, research in New Zealand is revealing the role of Chinese-run market gardens as sites of technological and plant transfer between settlers and Cantonese.[45]

Although feminist and postcolonial approaches are infusing the most dynamic scholarship of colonial knowledge production, we observe that the field of history of science in Australasia has been relatively averse to explicit engagements with theory over the past two decades. Indeed, Australasian historians of science have long taken as given the polycentrism of colonial (Western) science for which Roy MacLeod argued in the 1980s.[46] Their attention has since turned to the very processes and means of (cross-cultural) knowledge production, circulation, and dissemination. This includes examining dynamic interactions between particular peoples, places, and institutions, the use of differing and contested technologies, and the impact of historical contingencies on those histories. These approaches build on conceptual innovations such as Actor–Network Theory and "boundary objects" which offer analytical tools to map relationships and movements of knowledge, material objects, and meanings.[47] Libby Robin has demonstrated the relationship between science and First People's knowledge in the collection and identification of the platypus. She shows that, in 1887, Cambridge University's William Caldwell relied exclusively on local Aboriginal knowledge of the natural history of the platypus for its collection, in turn stimulating an economy in collecting the animal.[48] Jodi Frawley has also shown how mangoes underwent "re-articulation" as they were represented and realised in different botanical settings across the British Empire.[49] Similarly, for colonial New Zealand, these approaches help to demonstrate how the collection of birds (kiwi: *Apteryx* spp.) and plants (native to New Zealand) set in motion a set of relationships drawing in Māori knowledge and collection practices, British museums and naturalists, as well as botanical illustrators and scientific journals within a web of patronage networks.[50]

With these theoretical insights in mind, we also argue for the importance of bringing together material and cultural analyses of knowledge production. This means paying due attention to the material dimensions and impacts of practices of science, whether of the role of natural history collection on extinction or the ecological effects of experimental acclimatisation.[51] In some cases these concerns could impact on the practice of science and scientific collecting. For example, concerns by the Canterbury Museum Board in the 1880s around the impact of scientific collecting on ferns led to the board pressuring its director to stop collecting birds, something he only reluctantly did.[52] Similarly, we argue that it is impossible to examine aspects of hybridisation or genetics in a botanical garden without considering the networks of trade, commerce, and culture which encouraged such experimentation or the potential ecological impacts of hybridised plants as weeds in a new environment.[53] More recently, historical reconstructions and analyses of the global trade in natural history specimens during the nineteenth century are enriching understandings of museum collections.[54] These studies recognise a vital need to broaden history of science to include consideration of the environmental conditions in which newly introduced organisms lived, whether in the laboratory, the garden, the field, the atmosphere, or the body. This means, quite literally, a greater consideration of the physical or material place of science in Australasia.[55] One way of achieving this is to encourage greater collaboration between environmental historians and historians of science, especially given the enduring significance of introduced organisms in Australasia. Katrina Ford's work on agriculture, germ theories, and public health in colonial New Zealand is just one of many such approaches which could be taken.[56]

Our final points relate to broadening the topics of study of knowledge systems and professions to include those of cartography and surveying, sailing and navigation, mining and irrigation engineering, sanitary inspection, *tohunga* (experts), and other groups, and also to examine the role of non-Europeans in such professions.[57] We are particularly interested in contextualising the making of disciplines, such as the contestation and definition of specialised knowledge systems, whereby contemporaries made claims to knowledge and expertise in the face of a rapidly changing technological world in the colonial period.[58] Finally, we call for further work on histories of disease in colonial Australasia. Examining the ways in which diseases, their vectors, and their impacts were understood across different cultures and periods offers historians opportunities to consider the ways in which particular disease regimes were shaped and to what effect.

Conclusion

Although more traditional approaches to the history of science linger in Australasia, we applaud the rise of innovative and dynamic approaches that interpret colonial science as just one of many knowledge systems at work during the long nineteenth century. Understanding interactions of different knowledge systems in the settler colonies of Australasia draws these encounters into much wider scholarly debates that recognise the contested and plural nature of knowledge and knowledge production, and the relationships that structure them. Such approaches warrant more interdisciplinary and collaborative methods across cultures and institutions that acknowledge and attend to the ongoing legacies of colonial science.

Notes

1 We wish to acknowledge helpful conversations with Catherine Abu-Nemeh, Warwick Brunton, and Pauline Harris (Rongomaiwahine, Ngāti Rakaipaka, and Ngāti Kahungunu), in formulating the ideas for this chapter.
2 *Otago Colonist*, 24 January 1862, 4.
3 John Gascoigne, *Science in the Service of Empire: Joseph Banks, the British State and the Uses of Science in the Age of Revolution* (Cambridge: Cambridge University Press, 1998).
4 Joanna Bishop, "The Role of Medicinal Plants in New Zealand's Settler Medical Culture, 1850s–1920s" (PhD diss., University of Waikato, 2005); Michael Belgrave, "Medicine and the Rise of Health Professionals in New Zealand, 1860–1939," in *A Healthy Country: Essays on the Social History of Medicine in New Zealand*, ed. Linda Bryder (Wellington: Bridget Williams Books, 1991), 7–24.
5 Lorraine Daston "The History of Science and the History of Knowledge," *Know* 1, no. 2 (2017): 144. See also Sven Dupré and Geert Somsen, "The History of Knowledge and the Future of Knowledge Societies," *Berichte zur Wissenschafts-Geschichte* 42, nos. 2–3 (2019): 186–99.
6 Alessandro Antonello and Ruth A. Morgan, "Making and Unmaking Bodies: Embodying Knowledge and Place in Environmental History," *International Review of Environmental History* 4, no.1 (2018): 1–14; Ruth Barton, "'Not *Merely* a Scientific Society': The New Zealand Institute and its Affiliates, c.1868–1900," *Journal of the Royal Society of New Zealand* 47, no. 1 (2017): 33–40, doi.org/10.1080/03036758.2016.1207680; Rebecca Priestley, "A Survey of the History of Science in New Zealand 1769–1992," *History Compass* 8, no. 6 (2010): 474–90. Note also Gascoigne, *Science in the Service of Empire*.
7 Note, for example, Derek Dow, *Maori Health and Government Policy, 1840–1940* (Wellington: Victoria University Press, 1999); Paul Star, *Thomas Potts of Canterbury: Colonist and Conservationist* (Dunedin: Otago University Press, 2020); Ross Galbreath, *Scholars and Gentlemen Both: G.M. Thomson and Allan Thomson in New Zealand. Science and Education* (Wellington: Royal Society of New Zealand, 2002).
8 Pamela Wood, *Dirt: Filth and Decay in a New World Arcadia* (Dunedin: University of Otago Press, 2005); Charlotte Macdonald, *Strong, Beautiful and Modern: National Fitness in Britain, New Zealand, Australia and Canada, 1935–1960* (Wellington: BWB Books, 2011).
9 Ruth Morgan, *Running Out? Water in Western Australia* (Perth: UWA Publishing; 2015); James Beattie, Emily O'Gorman, and Matt Henry, eds., *Climate, Science, and Colonization: Histories from Australia and New Zealand* (New York: Palgrave Macmillan, 2014); James Beattie, "Environmental Anxiety in New Zealand, 1840–1941: Climate Change, Soil Erosion, Sand Drift, Flooding and Forest Conservation," *Environment and History* 9, no. 4 (2003): 379–92; Emily O'Gorman, *Flood Country: An Environmental History of the Murray–Darling Basin* (Collingwood, Victoria: CSIRO Publishing, 2012); Don Garden, *Floods, Droughts and Cyclones: El Niños that Shaped our Colonial Past* (Melbourne: Australian Scholarly Publishing, 2009). For a summary, see E. O'Gorman, J. Beattie, and M. Henry, "Histories of Climate, Science, and Colonization in Australia and New Zealand, 1800–1945," *WIRES: Climate Change*, 2016, doi.org/10.1002/wcc.426. And a recent article: Ciaran Doolin, "Norway Comes to New Zealand: Edward Kidson, Jørgen Holmboe and the Modernisation of Australasian Meteorology," *Bulletin of the American Meteorological Society* (August 2020): 1–46, doi.org/10.1175/BAMS-D-20-0058.1.
10 John Stenhouse, "Darwinism in New Zealand, 1859–1900," in *Disseminating Darwin: The Role of Place, Race, Religion and Gender*, ed. R.L. Numbers and J. Stenhouse (Cambridge: Cambridge University Press, 1999), 61–89; D. B. Paul, J. Stenhouse, and H.G. Spencer, eds., *Eugenics at the Edges of Empire: New Zealand, Australia, Canada and South Africa* (Cham, Switzerland: Palgrave MacMillan, 2018), doi.org/10.1007/978-3-319-64686-2; Iain McCalman and Nigel Erskine, eds., *In the Wake of the Beagle: Science in the Southern Oceans from the Age of Darwin* (Sydney: UNSW Press, 2009); Iain McCalman, *Darwin's Armada: How Four Voyagers to Australasia Won the Battle for Evolution and Changed the World* (Melbourne: Penguin, 2009).
11 Ron Numbers and John Stenhouse, "Anti-Evolutionism in the Antipodes: From Protesting Evolution to Promoting Creationism in New Zealand," in *The Cultures of Creationism: Anti-Evolutionism in*

English-Speaking Countries, ed. S. Coleman and L. Carlin (Aldershot, England: Ashgate, 2018), 125–44; James Beattie and John Stenhouse, "Empire, Environment and Religion: God and Nature in Nineteenth-Century New Zealand," *Environment and History* 13, no. 4 (2007): 413–46; James Beattie, "Science, Religion, and Drought: Rainmaking Experiments and Prayers in North Otago, 1889–1911," in Beattie, O'Gorman, and Henry, *Climate, Science, Colonization*, 137–55, https://link.springer.com/chapter/10.1057/9781137333933_8.

12 Cameron Muir, *The Broken Promise of Agricultural Progress: An Environmental History* (New York: Routledge/Earthscan, 2014); Tom Brooking and Eric Pawson, *Seeds of Empire: The Environmental Transformation of New Zealand* (London: I.B. Tauris, 2010); R. Vaughan Wood, "Soil Fertility Management in Nineteenth Century New Zealand Agriculture" (PhD diss., University of Otago, 2003).

13 See, for example, Catharine Coleborne, *Reading Madness: Gender and Difference in the Colonial Asylum in Victoria, Australia, 1848–1888* (Perth: Network Books, 2007); Catharine Coleborne and Dolly MacKinnon, *Exhibiting Madness in Museums: Remembering Psychiatry through Collections and Display* (New York: Routledge 2011); Paul, Stenhouse, and Spencer, *Eugenics at the Edges*; Warwick Brunton, *The Medicine of the Future: A history of the Department of Preventive and Social Medicine, University of Otago, 1886–2011* (Dunedin: Department of Preventive and Social Medicine, University of Otago, 2011); Geoffrey Rice, *Black November: The 1918 Influenza Pandemic in New Zealand*, with the assistance of Linda Bryder, 2nd ed. (Christchurch: Canterbury University Press, 2005); Linda Bryder, *A Voice for Mothers; The Plunket Society and Infant Welfare 1907–2000* (Auckland: Auckland University Press, 2003); Linda Bryder, *A History of the "Unfortunate Experiment" at National Women's Hospital* (Auckland: Auckland University Press, 2009); Rebecca Rice, "Conversazione in the Colonial Museum," *Journal of the Royal Society of New Zealand* 47, no. 1 (2017): 41–7, doi.org/10.1080/03036758.2016.1189439; James Dunk, *Bedlam at Botany Bay* (Sydney: NewSouth Publishing, 2019); Effie Karageorgos, "The Bushman at War: Gendered Medical Responses to Combat Breakdown in South Africa, 1899–1902," *Journal of Australian Studies* 44 (2020): 18–32. For cultural histories of health, eugenics, and phrenology, see Warwick Anderson, *The Cultivation of Whiteness: Science, Health and Racial Destiny in Australia* (Melbourne: Melbourne University Press, 2002); Alison Bashford, "'Is White Australia Possible?' Race, Colonialism and Tropical Medicine," *Ethnic and Racial Studies* 23, no. 2 (2000): 248–71; Paul Turnbull, "British Anthropological Thought in Colonial Practice: The Appropriation of Indigenous Australian Bodies, 1860–1880," in *Foreign Bodies: Oceania and the Science of Race 1750–1940*, ed. B. Douglas and C. Ballard (Canberra: ANU Press, 2008), 205–28; Stephen Garton, "Eugenics in Australia and New Zealand: Laboratories of Racial Science," in *The Oxford Handbook of the History of Eugenics*, ed. A. Bashford and P. Levine (New York: Oxford University Press, 2010), 243–57; Bronwen Douglas, *Science, Voyages and Encounters in Oceania, 1511–1850* (Basingstoke: Palgrave Macmillan, 2014); Alexandra Roginski, "Talking Heads on a Murray River Mission: Phrenological Lecturers and Their Aboriginal Receptions Decoded," *History Australia* 16, no. 4 (2019): 714–32.

14 Thomas Dunlap, *Nature and the English Diaspora: Environment and History in the United States, Canada, Australia, and New Zealand* (Cambridge: Cambridge University Press, 1999); Sara Maroske, "Science by Correspondence: Ferdinand Mueller and Botany in Nineteenth Century Australia" (PhD diss., University of Melbourne, 2005); Sara Maroske, "Ferdinand von Mueller and the Shape of Nature: Nineteenth Century Systems of Plant Classification," *Historical Records of Australian Science* 17 (2006): 147–68; Jodi Frawley, "Botanical Knowledges, Settling Australia: Sydney Botanic Gardens, 1896–1924" (PhD diss., University of Sydney, 2009); Peter H. Hoffenberg, "Nineteenth-Century Australian Scientists and the Unholy Australian Trinity: Overcoming Distance, Exile and Wandering at Exhibitions," *British Scholar* 2 (2010): 227–53; A.M. Lucas, "Specimens and the Currency of Honour: The Museum Trade of Ferdinand von Mueller," *Historical Records of Australian Science* 24 (2013): 15–39; A.M. Lucas, "Evolving Contexts of Collecting: the Australian Experience," in *Naturalists in the Field: Collecting, Recording and Preserving the Natural World from the Fifteenth to the Twenty-First Century*, ed. A. McGregor (Leiden: Brill, 2018), 806–62; Pete Minard, *All Things Harmless, Useful, and Ornamental. Environmental Transformation through Species Acclimatization, from Colonial Australia to the World* (Chapel Hill: University of North Carolina Press, 2019); Carolyn M. King, *Invasive Predators in New Zealand: Disaster on Four Small Paws* (Basingstoke: Palgrave Macmillan, 2020).

15 Francis Reid, "'The Democratic Politician Does Not Concern Himself with Science': Class and Professionalization in the New Zealand Institute, 1867–1903," *Tuhinga* 16 (2005): 21–31; A.M. Lucas, Sara Marsoke, and Andrew Brown-May, "Bringing Science to the Public: Ferdinand von Mueller and Botanical Education in Victorian Victoria," *Annals of Science* 63 (2006): 25–57; Linden Gilbank, "University Botany in Colonial Victoria: Frederick McCoy's Botanical Classes and Collections at

the University of Melbourne," *Historical Records of Australian Science* 19 (2008): 53–82; James Beattie, "Natural History, Conservation and Health: Scottish-Trained Doctors in New Zealand, 1790–1920s," *Immigrants & Minorities: Historical Studies in Ethnicity, Migration and Diaspora* 29, no. 3 (2011): 281–307; Lindy A. Orthia, "'Laudably Communicating to the World': Science in Sydney's Public Culture, 1788–1821," *Historical Records of Australian Science* 27 (2016): 1–12.

16 Conal McCarthy, *Exhibiting Māori: A History of Colonial Cultures of Display* (London: Bloomsbury, 2007); Rosi Crane, Bronwyn Labrum, and Angela Wanhalla, "Introduction: Museum Histories in Aotearoa New Zealand: Intersections of the Local and the Global," *Museum History Journal* 13, no. 1 (2020): 1–7, doi.org/10.1080/19369816.2020.1759004; Tom Griffiths, *Hunters and Collectors: The Antiquarian Imagination in Australia* (Cambridge: Cambridge University Press, 1996); Amiria Henare, *Museums, Anthropology and Imperial Exchange* (Cambridge: Cambridge University Press, 2005); Alexandra Roginski, *The Hanged Man and the Body Thief: Finding Lives in a Museum Mystery* (Melbourne: Monash University Publishing, 2015); Rosi Crane, "Show and Tell: TJ Parker and Late Nineteenth-Century Science in Dunedin," *Journal of the Royal Society of New Zealand* 47, no. 1 (2017): 61–6; Maria Nugent and Gaye Sculthorpe, "A Shield Loaded with History: Encounters, Objects and Exhibitions," *Australian Historical Studies* 49, no. 1 (2018): 28–43.

17 See, for example, Brett M. Bennett and Joseph Hodge, eds., *Science and Empire: Knowledge and Networks of Science across the British Empire, 1800–1970* (Basingstoke: Palgrave Macmillan, 2011); James Beattie, *Empire and Environmental Anxiety, 1800–1920: Health, Science, Art and Conservation in South Asia and Australasia* (Basingstoke: Palgrave Macmillan, 2011).

18 Antonello and Morgan, "Making and Unmaking Bodies," 1–14.

19 There is a strong need to move beyond compensatory history approaches. See Kate Hannah, "Finding Matilda: Deconstructing Women's Invisibility in Finding New Zealand's Scientific Heritage," *Journal of the Royal Society of New Zealand* 47, no. 2 (2017): 148–55, doi.org/10.1080/03036758. 2017.1305975.

20 Rebecca Rice, "'My Dear Hooker': The Botanical Landscape in Colonial New Zealand," *Museum History Journal* 13, no. 1 (2020): 20–41.

21 Caroline Jordan, *Picturesque Pursuits: Colonial Women Artists and the Amateur Tradition* (Carlton: Melbourne University Press, 2005).

22 Kathryn G. Mercer, "The Missionary Naturalist: A Case Study of New Zealand Scientific Communication to 1870" (master's thesis, Victoria University of Wellington, 2000), 35.

23 Richard Taylor to Laura Taylor, 26 March 1847, Mission House (Whanganui), Taylor Family Papers, MS-Papers-6817-01, Alexander Turnbull Library, Wellington, New Zealand.

24 Rosi Crane, "Otago Museum: Women, Visitors and Social Purpose" (lecture for SCIS301 "Historical Issues in Science in Society," Victoria University of Wellington, June 2020).

25 Anderson, *Cultivation of Whiteness*; Alexander Cameron-Smith, "Australian Imperialism and International Health in the Pacific Islands," *Australian Historical Studies* 41 (2010): 5774; Ruth A. Morgan, "Health, Hearth and Empire: Climate, Race and Reproduction in British India and Western Australia," *Environment and History* (forthcoming 2021).

26 For overviews of this literature in the Australasian context, see Shino Konishi, Maria Nugent, and Tiffany Shellam, "Exploration Archives and Indigenous Histories: An Introduction," in *Indigenous Intermediaries: New Perspectives on Exploration Archives*, ed. S. Konishi, M. Nugent, and T. Shellam (Canberra: ANU Press, 2015); Tiffany Shellam, Maria Nugent, Shino Konishi, and Allison Cadzow, "Brokering in Colonial Exploration: Biographies, Geographies, and Histories," in Konshi, Nugent, and Shellam, *Brokers and Boundaries*. Scholarship is not nearly as developed in Aotearoa, but see Tony Ballantyne, "Paper, Pen, and Print: The Transformation of the Kai Tahu Knowledge Order," *Comparative Studies in Society & History* 53, no. 2 (2011): 232–60.

27 See for example Harriet Mercer, "'White Men Would Consult Them': Aboriginal Women and Atmospheric Knowledge in Colonial Tasmania, c.1790s–1840s," *Environment and History* (forthcoming 2021).

28 See for example Fred Cahir, Ian D. Clark, and Philip A. Clarke, *Aboriginal Biocultural Knowledge in South Eastern Australia: Perspectives of Early Colonists* (Melbourne: CSIRO Publishing, 2018); Penny Olsen and Lynette Russell, *Australia's First Naturalists: Indigenous Peoples' Contributions to Early Zoology* (Canberra: NLA Publishing, 2019).

29 Luke Keogh, "Duboisia Pituri: A Natural History," *Historical Records of Australian Science* 22 (2011): 199–214; Anna Haebich, "Biological Colonization in the Land of Flowers," in *Ecocritical Concerns and the Australian Continent*, ed. B. Neumeier and H. Tiffin (London: Lexington Books, 2020), 75–90.

30 See for example *Te Taiao: Māori and the Natural World* (Auckland: David Bateman, 2010); Priscilla M. Wehi, Murray P. Cox, Tom Roa, and Hēmi Whaanga, "Human Perceptions of Megafaunal Extinction Events Revealed by Linguistic Analysis of Indigenous Oral Traditions," *Human Ecology* 46 (2018): doi.org/10.1007/s10745-018-0004-0; Khyla J. Russell, "Landscape: Perceptions of Kai Tahu I Mua, Āianei, Ā Muri Ake" (PhD diss., University of Otago, 2000); Michael J. Stevens, "Kāi Tahu me te Hopu Tītī ki Rakiura: An Exception to 'Colonial Rule'?," *Journal of Pacific History* 41, no. 3 (2006): 273–91.

31 Bruce Pascoe, *Dark Emu: Black Seeds – Agriculture or Accident?* (Broome: Magabala Books, 2014). For a critique of this approach, see Billy Griffiths and Lynette Russell, "What We Were Told: Responses to 65,000 Years of Aboriginal History," *Aboriginal History* 42 (2018): 31–54.

32 Patrick Nunn and Nicholas J. Reid, "Aboriginal Memories of Inundation of the Australian Coast Dating from more than 7000 Years Ago," *Australian Geographer* 47 (2016): 11–47; Duane Hamacher, "Observations of Red-Giant Variable Stars by Aboriginal Australians," *Australian Journal of Anthropology* 29 (2018): 89–107; Pauline Harris, Rangi Matamua, Takirirangi Smith, Hoturoa Kerr, and Toa Waaka, "A Review of Māori Astronomy in Aotearoa-New Zealand', *Journal of Astronomical History and Heritage* 13 (2013): 325–66.

33 Te Maire Tau, "Mātauranga Māori as an Epistemology," in *Te Pouhere Kōrero* (Palmerston North: Pouhere Kōrero, 1999); see also Laura Rademaker, "Why Historians Need Linguists (and Linguists Need Historians)," in *Language, Land and Song: Studies in Honour of Luise Hercus*, ed. P.K. Austin, H. Koch, and J. Simpson (London: EL Publishing, 2016), 480–93. An exciting step in this direction is the conference organised by Hugh Slotten, "Encounters and Exchanges: Exploring the History of Science, Technology and Mātauranga (Indigenous Knowledge)" (University of Otago and the Tōtaranui 250 Trust, Blenheim, New Zealand, 1–3 December 2020).

34 An excellent exemplar is Ballantyne, "Paper, Pen, and Print."

35 Megan Pōtiki, "Te Hū o Moho: The Call of the Extinct Moho. The Death of the Māori Language at Ōtākou," *Te Pouhere Kōrero* 9 (2020): 27–49.

36 Henare, *Museums.*

37 Derek A. Dow, "'Pruned of Its Dangers': The Tohunga Suppression Act 1907," *Health and History* 3, no. 1 (2001): 41–64.

38 Henare, *Museum*; J. Heine, "Colonial Anxieties and the Construction of Identities: The Employment of Maori Women in Chinese Market Gardens, Auckland, 1929" (master's thesis, University of Waikato, 2006).

39 R.J.M. Fuller, "Ethnobotany: Major Developments of a Discipline Abroad, Reflected in New Zealand," *New Zealand Journal of Botany* 51, no. 2 (2013): 116–38, doi.org/10.1080/0028825X.2013.778298; Jonathan West, *The Face of Nature: An Environmental History of the Otago Peninsula* (Dunedin: Otago University Press, 2017); Ray Hargreaves, "Changing Māori Agriculture in Pre-Waitangi New Zealand," *Journal of the Polynesian Society* 72, no. 2 (1963): 101–17; Peter Holland and Jim Williams, "Pioneer Settlers Recognizing and Responding to the Climatic Challenges of Southern New Zealand," in Beattie, O'Gorman, and Henry, *Climate, Science, and Colonization*, 81–98.

40 Cited in James Beattie, "Colonial Geographies of Settlement: Vegetation, Towns, Disease and Well-Being in Aotearoa/New Zealand, 1830s–1930s," *Environment and History* 14, no. 4 (2008): 583–610.

41 James Beattie and Joanna Boileau, "'Cultivated with Great Carefulness': Chinese Market Gardening, Urban Food Supplies and Public Health in Australasia, 1860s–1950s," *New Zealand Journal of History* 54, no. 2 (2020): 100–28; James Beattie, "Chinese Migrant Landscapes: Environmental Exchanges between South China and New Zealand," manuscript.

42 Rey Tiquia, "'Bottling' and Australian Medical Tradition: Chinese Medicine in Australia in the Early 1900s," *Otherland* 9 (2004): 203–15; Nadia Rhook, "'The Chinese Doctor James Lamsey': Performing Medical Sovereignty and Property in Settler Colonial Bendigo," *Postcolonial Studies* 23 (2020): 58–78.

43 Nadia Rhook, "Affective Counter Networks: Healing, Trade, and Indian Strategies of In/dependence in Early 'White Melbourne,'" *Journal of Colonialism and History* 19 (2018), doi.org/10.1353/cch.2018.0012; Nadia Rhook, "The Balms of White Grief: Indian Doctors, Vulnerability and Pride in Victoria, 1890–1912," *Itinerario* 42 (2018): 33–49.

44 On this approach, see James Beattie, ed., *Gardens at the Frontier: New Methodological Perspectives on Garden History and Designed Landscapes* (London: Routledge, 2018).

45 Fa-ti Fan, *British Naturalists in Qing China: Science, Empire, and Cultural Encounter* (Cambridge, MA: Harvard University Press, 2004); Joanna Boileau, *Chinese Market Gardening in Australia and New Zealand: Gardens of Prosperity* (London: Palgrave Macmillan, 2017); Beattie and Boileau, "'Cultivated

with Great Carefulness'"; James Beattie, "Chinese Farming, Rural Enterprise and Environmental Change in Aotearoa New Zealand, 1870s–1950s" (23rd Biennal NZASIA Conference, Victoria University of Wellington, 25–27 November 2019).

46 This is well summarised and critiqued in Tony Ballantyne, "Empire, Knowledge and Culture: From Proto-Globalization to Modern Globalization," in *Globalization in World History*, ed. A.G. Hopkins (London: Pimlico, 2002), 115–40. For the history of science specifically, see Warwick Anderson, "Remembering the Spread of Western Science," *Historical Records of Australian Science* 29 (2018): 73–81.

47 For example, Bruno Latour, *Science in Action: How to Follow Scientists and Engineers through Society* (Cambridge, MA: Harvard University Press, 1987); Susan Leigh Star and James R. Griesemer, "Institutional Ecology, 'Translations' and Boundary Objects: Amateurs and Professionals in Berkeley's Museum of Vertebrate Zoology, 1907–39," *Social Studies of Science* 19, no. 3 (1989): 387–420.

48 Libby Robin, *How a Continent Created a Nation* (Sydney: UNSW Press, 2007).

49 Jodi Frawley, "Making Mangoes Move," *Transforming Cultures eJournal* 3 (2008): 165–84.

50 James Beattie, "Thomas McDonnell's Opium: Circulating, Plants, Patronage, and Power in Britain, China and New Zealand, 1830s–1850s," in *The Botany of Empire in the Long Eighteenth Century*, ed. Sarah Burke Cahalan and Yota Basaki (Washington, DC: Dumbarton Oaks/Harvard University Press, 2017), 163–88.

51 Eric Pawson and Tom Brooking, eds., *Making a New Land: Environmental Histories of New Zealand* (Dunedin: Otago University Press, 2013); Ross Galbreath, *Walter Buller: The Reluctant Conservationist* (Wellington: GP Books, 1989).

52 Paul Star, *Thomas Potts of Canterbury: Colonist and Conservationist* (Dunedin: Otago University Press, 2020), 88.

53 Minard, *All Things*; King, *Invasive Predators*.

54 Simon Ville, Claire Wright, and Jude Philp, "Macleay's Choice: Transacting the Natural History Trade in the Nineteenth Century," *Journal of the History of Biology* 53 (2020): 345–75.

55 David Livingstone, *Putting Science in its Place: Geographies of Scientific Knowledge* (Chicago: Chicago University Press, 2003).

56 Katrina Ford, "The Very Life-Blood of the Country: Germs, Dairying and Public Health in New Zealand, 1890–1910," *New Zealand Journal of History* 47 (2013): 157–84; see also Tony Nightingale, *White Collars and Gumboots: A History of the Ministry of Agriculture and Fisheries, 1892–1992* (Palmerston North: Ministry for Culture and Heritage, 1992).

57 Brad Patterson, "Reading between the Lines: People, Politics and the Conduct of Surveys in the Southern North Island, New Zealand 1840–1876" (PhD diss., Victoria University of Wellington, 1984); Giselle Byrnes, *Boundary Markers: Land Surveying and the Colonisation of New Zealand* (Wellington: Bridget Williams Books, 2001); Paul Carter, *Lie of the Land* (London: Faber and Faber, 1996).

58 Thomas F. Gieryn, *Cultural Boundaries of Science: Credibility on the Line* (Chicago: University of Chicago Press, 1999).

22

EMPIRES AND SCIENCE

The case of the sixteenth-century Iberian Empire

Antonio Barrera-Osorio

The Iberian Empire created modern science, and modern science created the empire. Navigation, to give but one example, made possible the establishment of empires in America and the empire pushed medieval navigational practices in the direction of modern science.[1] In the Christian-European context, empire and science were asymmetrically co-created at the time of expansion.[2] As the Iberian Empire linked different regions and socio-political communities, knowing these regions and communities became integral to the formation of the empire. This chapter focuses on the Spanish American empire as a case study of the co-creation of empire and science in the sixteenth century.

By the beginning of the sixteenth century, the Castilian crown had already incorporated a few kingdoms and territories under its jurisdiction and political influence. In 1479, Portugal and Castile signed a treaty in which Portugal accepted Castilian's jurisdiction over the Canary Islands—and Castile accepted Portuguese jurisdiction over Madeira, the Azores, and Cape Verde Islands.[3] Early in 1492, the Sultan of Granada, Abu Abdallah Muhammad XII, surrendered the Emirate of Granada to Isabella I (1451–1504) and Ferdinand II (1452–1516). The Muslim kingdom of Granada and its Andalusi Arabic-speaking Muslim population became part of Castile. Both the Canary Islands and Granada reconfigured the political ambitions of the Castilian crown. The crown had now extended its political, economic, religious, and legal reach into territories and people with different religious beliefs and practices, languages, and socio-political realities. Castile was becoming both a composite monarchy and a composite empire.[4]

Acknowledging this reality, on 3 May 1493, Pope Alexander VI (1431–1503) issued a series of bills that granted and "assigned forever" the Castilian kings possession of all

> the remote and unknown mainlands and islands lying towards the western parts and the ocean sea, that have been discovered or hereafter may be discovered by you or your envoys, whom you have equipped therefor not without great hardships, dangers and expense—and with them all their lordships, cities, castles, places, villages, rights and jurisdictions; provided however these countries have not been in the actual temporal possession of any Christian lords.[5]

These papal bills, together with the invasion of the Canary Islands and Granada and the Portuguese precedents in Africa, constituted the early components in the formation of the Spanish Empire.[6]

A few months later, in October, Cristóbal Colón (Christopher Columbus), on his own initiative, incorporated a few Taino-inhabited Caribbean islands into the Castilian crown. In 1495 the Canaries' Guanches, who had resisted the Christian invasion since the early fifteenth century, surrendered to the invaders: those who resisted were captured and sold into slavery.[7] In 1497 Castilian forces took over Melilla, and in 1509, they did the same over Oran. After the invasion of Granada and the Caribbean Islands, "the empire of the kings of Spain" began, slowly and accidentally, "spreading across the western lands."[8]

This internal and external process of expansion and consolidation of royal power continued throughout the century.[9] In 1512, Castilian forces took over the kingdom of Navarra and placed it under the Castilian crown. In 1519, Charles I (1500–58), who had inherited the crowns of Castile and Aragon by design and accident, became Charles V, the Holy Roman Emperor, after using financial incentives linked to the American lands to obtain the support of the German electors. Charles V's dominions included the Holy Roman Empire, the crowns of Castile and Aragon, the Low Countries, Naples, Sicily, Sardinia, Oran, and Santo Domingo, Puerto Rico, and Cuba in the Caribbean. In 1521, Spanish and Tlaxcalan forces, and, unknown to the actors at the time, lethal pathogens, defeated the Aztec empire and incorporated it into the Castilian crown. In 1532 Spaniards, again with the support of pathogens, took over an unstable Inca empire after a civil war and incorporated it into the Castilian crown.

In a matter of about 50 years, accidents, pathogens, political and military alliances, marriages, surviving children, ships, maps, instruments, reports, plants, animals, and violence transformed Castile's small kingdom into a composite, large, fragmented, and multilingual empire that linked, among other sites, Milan, Oran, Seville, the Canary Islands, Santo Domingo, Tenochtitlan, and Cuzco: an empire characterised by profound demographic and ecological transformations and by the establishment of structures for the circulation of things, reports, and expertise.[10] For example, throughout the sixteenth century, sugar—as in the plant, and knowledge, expertise, labour practices, tools, and mills associated with the cultivation of sugar—travelled from the Canary Islands to Santo Domingo, and from there to Puerto Rico, Jamaica, Veracruz, Puebla, Tuxtla and Cuernavaca in Mexico, and then to the audiencia of Guatemala. Sugar reached Caracas, Maracaibo, Cartagena, and Santa Marta in Tierra Firme. As sugar travelled, its cultivation transformed the land, produced knowledge about these regions, transformed labour relations, and linked these regions of the empire to European ports via mules, canoes, and ships.[11]

This multilingual, religiously and ethnically diverse empire challenged the Castilian crown's political organisation and its expansion—how to incorporate and control multilingual and religiously diverse people and regions into the crown? Lurking behind the religious and political challenges was the epistemological task of knowing the sites and peoples incorporated, even nominally, and absorbing that knowledge into this empire. The act of communicating with people who became subjects to the Castilian crown was an aspect of this epistemological challenge. This became apparent early. One day in the late fifteenth century, as the scholar Elio Antonio de Nebrija (1444–1522) was working on a new Castilian grammar treatise, Queen Isabella I (1474–1504), curious about the utility of a book on Castilian grammar, since Castilians knew Castilian, asked him "*que podia aprovechar*" (what would be the use of) such a book? Hernando de Talavera (1428–1507), the bishop of Avila and confessor of the queen, responded for Nebrija: this book was useful because "after your highness" has placed "under her rule many barbarous people and nations with strange languages—*lenguas peregrinas*," and imposed your laws, they need to learn our language. Nebrija corroborated: with his book, the defeated can learn the Castilian language. The reference to *lenguas peregrinas* points toward the difficulties of incorporating different communities into this *Castilian*-speaking empire and to a central problem of understanding, communicating, and producing knowledge within this multilingual and diverse empire.

When Christian kings and queens subjugated non-Christian people and their territory, they subsumed them under their crown and laws as subjects. The queens and kings were then the guardians of the law in the acquired kingdoms, too.[12] In 1492, Nebrija succinctly summarised this idea on the first page of his grammar: "*siempre la lengua fue compañera del imperio*" (language has always been a companion of the empire).[13]

A long-distance and multilingual empire created political, legal, and administrative challenges. The Castilian crown sought the advice of Spanish scholars on the legitimacy of this accidental empire. Spanish scholars agreed that the papacy, as the papal bills mentioned above claimed to do, did not have the power to grant the Castilian crown possession of the American nations and lands. Still, by the 1530s, the Spanish presence was an accomplished fact. The Salamanca scholars argued that the Castilian empire was about *dominium* (private property and lordship), *ius gentium* (international law), and *bellum iustum* (just war).[14] In their view, *ius gentium* allowed all nations to "live with one another in harmony," own private property, divide territories, and praise God.[15] Those who resisted or opposed this political and legal order became enemies of the empire in a just war that validated the Christian invasion and colonisation of their territories, and the selling of prisoners of war into slavery.

This legal framework allowed the Castilian crown to bring together regions with their own distinctive religious, political, economic, and social practices under a single crown. Over time, the crown attempted to establish a single yet limited administrative structure and uniform systems of taxation, commerce, labour and justice, and religion over different regions. Thus the empire emerged as an administrative and legal structure that bound together peoples and regions engaged in commerce, colonisation, slavery, religious conversion, exploitation of natural resources, and control of local populations.[16]

Drawing on the Aragonese empire and Castilian traditions, the Castilian kings established an administrative structure in the American territories consisting of viceroyalties, *audiencias* (highest administrative and judicial courts), and city councils. Drawing on the Portuguese Empire, the Castilian kings established a house of trade—the Casa de la Contratación (1508), to organise commercial and navigational activities related to the American territories.[17] As these territories' administration became increasingly complex, the Castilian crown established the Council of Indies (1524) to organise the kingdoms' administration. This designation was critical: the Castilian administration used the term "kingdom" to refer to the American territories under Castilian jurisdiction.

The term "kingdom" implied a territorial jurisdiction with its own socio-political traditions and institutions: peoples and nations belonging to an empire different from each other and with different knowledge communities and experts. For example, Muslims in Andalucia (1492), Tainos in Santo Domingo (1493), Aztecs in Mexico (1521), Incas in Peru (1534), Muisca in Colombia (1537), Mapuche in Chile (1544), and Ais in Florida (1566), to name just a few of the nations and territories brought under different negotiated levels of imperial administrative and legal control—after violent encounters and the establishment of alliances.[18] By the early 1500s, both the Castilian crown and its composite empire were emerging projects of colonisation. As an emerging empire, Castile reorganised peoples, practices, sites, and ideas in new socio-political configurations and ways of knowing.

This empire was organised around several interconnected centres—different ones at different moments throughout the years. For instance, in the first half of the sixteenth century, some of these centres were Seville, Barcelona, Santo Domingo, and Brussels; in the late sixteenth century, some of them were Mexico City, Madrid, Seville, and Lisbon. The sixteenth- and seventeenth-century Castilian empire was a long-distance empire structured around a web of geographically, politically, and legally connected sites, each with its decision-making processes, authorities, histories, and commercial and communication infrastructures (in all cases, these structures predated the presence of the Spaniards). Local authorities exercised imperial power in their areas of jurisdiction and authorities, merchants, farmers, and artisans produced local knowledge about people and things.[19]

Christianity, Castilian, and a standard legal system brought them together into a single political unity: the Iberian Empire. However, neither Catholicism nor Castilian was the only religion or language in the empire. Muslims, Guanches, Tainos, Mayas, Mexica, Inkas, Muiscas, Mapuches negotiated, in a process of transculturation, their practices and languages as they became subjects of the empire.[20]

They also had and produced knowledge, of course, about themselves and the land, about the animals, plants, insects, minerals, stars, gods, and spirits that lived with and around them. Like their territories and labour, this local knowledge was integrated into the empire—with different levels of success. The history of empires as a political structure that seeks to control territories, resources, and peoples is also a history about the control of local knowledge.[21] This history of the empire to control local knowledge is part of the history of modern science and colonialism.[22]

By 1570, there was a Castilian empire across the Atlantic ocean with two viceroyalties, one in Mexico and the other in Peru; with an economic system based on international and regional commerce, mining, and agricultural production; with a labour system extending throughout the American territories based on Indigenous and slave labour; and with an infrastructure of roads and maritime routes linking American villages, towns, cities, ports, and markets with ports and urban sites in Castile and Europe. Some of those who became subordinated to Euro-Christian institutions and authorities learned Nebrija's Castilian and adopted Christian gods into their religious practices. Yet those who sought to impose their institutions and practices over local communities had to learn too about the languages, religions, history, political and economic systems, and natural history of these communities. The production of knowledge on, and the gathering of information about, the non-Castilian-speaking people and territories under Castilian jurisdiction was central to establishing the Spanish Empire. In all areas of the empire, local experts had the knowledge the Christians needed to operate in those areas. In some areas, new ways of knowing and a new type of knowledge emerged, as was the case of the Caribbean region.[23] As these different communities learned from each other, new practices and ways of knowing emerged in the empire.

Thus, in the context of a composite empire such as the Iberian Empire, empirical methods emerged as a tool for Europeans to verify the knowledge of local people. Generating information was central to the long-distance, multilingual, and regionally diverse Castilian empire. In America, the Christian invaders established tools for gathering information about the people and regions under control and colonisation. These tools emerged from the interactions between private invaders, Native Americans, and royal officials as they needed to know about each other and the regions under occupation.

As Columbus gained some information about the islands, their natural products, and inhabitants from the Tainos, in a mutual process in which the Tainos learned about him and his companions and things, Columbus reported the information he learned from them back to the kings. As the kings read and talked about Columbus' reports, they understood "how great this business of yours [Columbus] has been," and they asked for more, and more specific, information.[24] So in 1494, Columbus sent a report to the Catholic kings in which he provided some information about the Caribbean islands. This report prompted the kings to ask for more information: "in particular," they wanted to know "the number of islands he had discovered with their names, and the names he had given them, and their distances from each other, and what he had found in each one," and a report on their seasons, and their crops. The kings also asked him to send "all the hawks possible, and different birds."[25] In 1502, the kings asked Columbus to "provide a complete report ... of the size of the islands and to report on all of the islands, and their people, and their quality."[26]

These instructions established the practice of gathering information in the form of reports about the natural and human worlds of America.[27] The gathering of information based on both locals' knowledge and direct experience, and its articulation in reports, emerged in the context of interactions between Christians, locals, and crown officials. These reports constituted a central tool of the empire for the production and dissemination of knowledge. This knowledge, gathered from locals, served commercial, political, and scientific purposes. The Iberian Empire created modern science; modern science, in turn, fostered the empire and its expansion.

The knowledge empire: private initiatives and knowledge gathering

As more Spaniards arrived at the Caribbean islands, and as they began moving into Central America and Mexico, they understood that they needed to gather information about these territories and to establish an infrastructure for its circulation.[28] They were not the only ones gathering information and circulating it. As African slaves arrived in the Caribbean—to work, for instance, in sugar plantations, mines, and urban settings, they also learned from Indigenous people, and generated their own knowledge, and circulated it. African slaves, Indigenous people, and Spaniards came into contact with each other in the Caribbean islands, and all learned from each other and created new knowledge too. The soldier Pedro de Osma (fl. 1568), who sent a report on medicinal things from Peru to doctor Nicolás Monardes (c. 1512–88) in Seville, after he had read Monardes' book composed of reports from travellers to Mexico and Florida, travelled with a slave woman who was cured by an Indigenous doctor in Margarita Island. Thus Osma learned from the reports published by Monardes; Monardes incorporated Osma's report in a new edition of his first book; the slave woman learned from the Indigenous doctor, and both Osma and the slave woman took that knowledge with them to Peru. The empire established connections between Margarita Island, the Andes, and Seville where Spanish soldiers, doctors, slaves, and Indigenous practitioners came in contact to share and recreate natural knowledge and practices.[29]

In Spain, merchants, friars, royal officials, and entrepreneurs needed information too. The emerging, long-distance Spanish Empire of the sixteenth century posed two immediate problems: establishing a reliable communication system and learning about the territories and people incorporated into the crown. To establish the communication infrastructure, the Castilian crown had by the early years of the sixteenth century established a fleet system linking Europe and America and a network of maps, instruments, and training workshops to navigate the Atlantic Ocean. Ships, maps, instruments, and training workshops, already established for navigating the Atlantic Ocean, now embedded different types of knowledge and reconfigured relationships between sailors, cosmographers, merchants, and royal officials. They were tasked to produce knowledge of shared meaning and importance. This network of humans and things organised the infrastructure, and practices, of communication between Europe and America.[30] By 1525, the Castilian crown established a postal service linking the American territories to

Castile's sites of the imperial government.[31] This postal service supported the circulation of reports and news—knowledge and information—between Europe and America.

Central to the establishment of this long-distance empire and its infrastructure of communication was the production of knowledge. Everywhere the Christians arrived in the American territories, they learned about the land, its resources, foods, dangers, stars, and infrastructure from Indigenous people: their knowledge made possible the establishment of the Euro-Christian empires in America. To cite a few examples: in 1492 Columbus set the pattern of gathering information from Native Americans. He travelled in a particular direction in the Caribbean "because all the Indians I bring with me, and others, made signs to this southern quarter, as the direction of the island they call Samoet, where the gold is."[32] When the pilot Sebastian Cabot arrived at the Río de la Plata region, he asked Indigenous people about natural commodities.[33] The merchant Antonio de Villasante sent a report on medicines to the Council of Indies in 1528 based on his Taino wife's knowledge and her relatives. The Franciscan Bernardino de Sahagún (ca. 1499–1590) collected information on medicinal plants from Indigenous physicians of Tlatelolco: Gaspar Matias, Pedro de Santiago, Francisco Simon, Pedro de Raquena, Miguel Garcia, and Miguel Motolinia.[34] In 1552, an Indigenous physician from Mexico wrote a book on medicinal plants known as the *Códice de la Cruz-Badiano*; the same year, the viceroy Francisco de Mendoza offered it to Philip II in Spain. Monardes understood that many of the reports he received from travellers returning from America originated in Indigenous accounts.[35]

In the Iberian peninsula, the production of knowledge about the colonies reflected the American context. As the invasion of the American lands began, the crown received information from those travelling to the Caribbean without formally requesting it. Yet soon, as the example of Columbus mentioned above shows, the queen and king ordered explorers to send reports about the people and regions they were trying to place under Castilian control. However, this reporting practice was not the only way of gathering information in Spain about the American world. The chronicler Peter Martyr (1457–1526) became an information-gathering officer as soon as he learned of Columbus' voyage. He invited those coming from America—sailors, captains, royal officials—to his lodgings and asked them questions about the Tainos, chocolate, and the Gulf Stream; he also gathered information from those who came to the court to report their voyages and experiences. He collected news from explorers, pilots, officers, merchants, and Indigenous people, and wrote a series of report letters about America. He compiled these reports in *De Orbe Novo*—published in 1511 and expanded in 1516 and 1530.[36] Parallel to these developments, the crown continued to receive information from those travelling to the Caribbean. In 1508 the royal official, Alonso de Zuazo (fl. 1515), for instance, wrote a report on natural things from Hispaniola Island to Charles V.[37] Up to this point, merchants and entrepreneurs sent their accounts on their own; the crown had also been requesting reports from explorers and merchants as well as from its officials.

The different written and oral reports for gathering information fostered the institutionalisation of these practices in imperial sites. In 1508, the crown established the position of the Chief Pilot at the Casa de Contratación—a navigational office—to collect information for making charts (and also for the training of pilots), and ordered pilots returning from America to provide navigational information to the cosmographers at the Casa, and to the cosmographers to organise this information into charts and reports. This association of pilots, sailors, oral and written reports, maps, navigational instruments, ships, cargo, oceanic routes and winds, and cosmographers produced new knowledge about navigation, geography, and cosmography. As the empire incorporated more regions, peoples, and resources, with different socioeconomic organisations, languages, and religious practices, the Castilian established other associations of people and things for understanding and learning about these territories. This was particularly true with the formal additions of some of the largest territories in the first half of the sixteenth century, including the lands of the Mexica in the 1520s, the Incas in the 1530s, and the Muiscas in the 1540s.

In 1532, the crown added a new mechanism for collecting information. It appointed a *Cronista de Indias* (Chronicler of Indies) to collect and organise information about the New World. The royal official, humanist, and natural historian Gonzalo Fernández de Oviedo (1478–1557), who in 1526 had published a short natural history entitled *De la natural hystoria de las Indias*, proposed an expedition to collect information from areas he had not visited yet to complete a more extensive natural history. The Council of Indies supported the expedition, but it seemed too expensive, and they decided on a different proposal. He was to stay in Spain, and the crown sent decrees to its officials in America requesting information for Fernández de Oviedo. Once again, reports based on local expertise became central in the production of knowledge and the making of the empire.

In the mid-1550s, just after the ascension of Philip II (r. 1556–98), the cosmographer Alonso de Santa Cruz (c. 1500–72) proposed asking specific questions of explorers and colonists, rather than asking, as had been the case, for general information about the land and its natural things. He proposed questions about the latitude and longitude of places and ports; the geographical characteristics and healthiness of the land; descriptions of rivers, mountains, lakes, and fountains; and information about mines, minerals, stones, pearls, animals, monsters, trees, fruits, spices, drugs, and herbs. There were questions concerning the Indigenous people—their kingdoms and provinces, borders, towns and cities, costumes, rites, types of knowledge, books, arms, general trade arrangements, and items they traded. Santa Cruz's proposal constituted an expansion of practices already institutionalised at the navigational office in the House of Trade and deployed by such royal officials as Peter Martyr and Gonzalo Fernández de Oviedo.[38]

In the 1560s, a broad knowledge culture based on reports and established practices and institutions emerged across the Spanish American empire. The practices and institutions for gathering information from Native Americans and those who had travelled in the empire shaped practices on both sides of the Atlantic. Simultaneous with the navigational work at the House of Trade and Santa Cruz's proposal, in Mexico Bernardino de Sahagún elaborated a questionnaire to write his *Historia general de las cosas de la Nueva España*. Sahagún sent questionnaires (reconstructed from his book) to Indian villages, asking for information about aspects of their culture, society, and land. On natural history, the questionnaire included questions, for instance, about names of animals, the history of the names, physical description of the animals, their respective environments, activities, food, their customs, ways of hunting or catching them if so, popular histories and sayings about them.[39] In 1567, the physician Pedro Arias de Benavides (fl. 1566) published his book on New World medicines in Spain. In his *Secretos de Chirurgia*, he described his experiences as a surgeon in Guatemala and Mexico, the herbs and medicines used in the New World, and their uses by the Indigenous people.[40] In the mid-1560s, the royal official Tomás López Medel (c. 1520–c. 1582) began writing a treatise on America's natural world. His work was the first attempt to situate America's nature within the classical framework of the elements and their properties.[41]

Similarly, in the 1560s, the physician Nicolás Monardes, mentioned above, whose practice was based in Seville, began publishing a series of reports on medicinal plants from the American territories.[42] Like Martyr, Santa Cruz, and the cosmographers at the House, he gathered information by asking people arriving at Seville about medicinal plants they had used in the colonies. Monardes asked soldiers, merchants, Franciscans, royal officials, and women about new medicines and plants. They brought American medicines in their bags. The doctor asked about the medicines' names, uses, and characteristics; and their experiences using them. Most of his informers' experiences referred back to Indigenous uses of those herbs (as well as Indigenous names).[43] At some point, he started receiving samples of medicines with accompanying reports. He established a botanical garden—and he was not the only one to do so in Seville.[44] He fully trusted the reports accompanying the samples and the accounts he gathered from travellers arriving at Seville. In 1565, 1569, and 1574 Monardes published three reports on these medicines. This knowledge culture extended through the empire. This is the book that prompted the soldier Pedro de Osma to search for medicines in Peru. As mentioned above, he sent his report, and samples, to Monardes who included it in his second book on medicines.[45]

The production of knowledge was not limited to navigational, cosmographical, and medical domains, but it also covered mining, agricultural, and hydrographical activities too.[46] In 1553 Bartolome Medina (c. 1497–1585), a tailor in Seville, arrived in Mexico with the project of using mercury for the exploitation of silver—a technique he claimed to have learned in his trade from a German who moved within Charles V's possessions from the Holy Roman Empire (Germany) to Seville (Castile). In Mexico, Medina worked and experimented at the Pachuca region, a market centre of obsidian production under the Aztec empire. In 1554, Medina found a method for using mercury amalgamation to extract silver from ore on an industrial scale. This method was called the patio method or the amalgamation process. It became widely used in Mexico in the 1550s. The system was exported to Peru in the 1570s; and later to Europe. The amalgamation process transformed silver production and labour organisation in both Mexico and Peru.[47] Similarly to the examples discussed above, the production of this mining knowledge emerged from imperial networks, in this case, of German, Spanish, and Inca miners, artisans, and officers working and living in different power positions within the empire.[48]

The empire connected diverse regions and peoples and allowed for the establishment of tools, institutions, and systems for the circulation of people, things, and information. Franciscans in Mexico, soldiers in Peru, miners in Mexico and Peru, and physicians and cosmographers in Seville developed standard practices for knowledge production in the context of the empire, that is, in the context of networks of, and interactions between, people with different backgrounds and information: pilots, royal officials, and cosmographers in Spanish's ports and Indigenous people, merchants, artisans, and Franciscans in the streets of Mexico. These standard practices consisted of personal observations and collaborative, if hierarchically organised, approaches to knowledge production. Personal observations were collected in the form of reports, in many cases accompanied by samples and drawings. Then these reports, in turn, were evaluated by experts (the natural historian, the cosmographer, and the physician) who transformed them into knowledge (books about natural history, medicine, and cosmography). This transformation occurred at particular institutions: the Casa de la Contratación, the Council of Indies, botanical gardens, hospitals, and viceroyal courts.

These practices were institutionalised in the 1570s. This institutionalisation constituted a significant event in the history of science, as it established reporting activities based on empirical practices and the circulation of those reports within imperial sites of knowledge production. At the centre of these reporting activities was the Indigenous knowledge of the inhabitants of the American territories. The history of science is entangled with the knowledge of Native Americans.

Royal initiatives and knowledge gathering

The Council of Indies was formally established in 1524 to deal with the administrative, ecclesiastical, commercial, military, and judicial matters of the empire. Initially it was composed of a president and three counsellors plus a supporting administrative staff. The counsellors advised the crown on all matters concerning the American kingdoms. Those matters reached the council in reports and documents generated in different parts of the empire: paper and ink connected these sites and provided the tools for the circulation of information, knowledge, legislation, decisions, questions, and stories. The circulation of these reports linking imperial sites was central for the creation, establishment, and preservation of the empire.[49]

Juan de Ovando, the president of the Council of Indies from 1571 to 1575, had been tasked earlier, in 1569, to inspect the Council of Indies. As a result of this inspection, he reformed the council's legislative and administrative activities and established systematic information-gathering practices based on the practices described in the previous section. Years later, the chief chronicler, Antonio de Herrera (appointed 1596; d. 1624), explained that Ovando ordered the collecting of "the most accurate reports found in Spain and the Indies about the events of the discovery of the Indies," given the lack and uncertainty at the Council regarding that matter.[50] Ovando's visit to the Council resulted in the statutes of 1571—the *Ordenanzas Reales del Consejo de Indias*—and in the formalisation of a programme for collecting information about the natural world and geography of the Indies.

These statutes created the chief cosmographer-chronicler's office, and Juan López de Velasco (1530–98) was appointed to this office. His functions consisted of writing the history of the Indies, censuring proposed accounts about the Indies, and collecting geographical and natural information about America for the government of the territories.[51] More generally, the statutes of 1571 established the following provision:

> Since nothing can be understood nor appropriately dealt with if its subject is not first known by the people that would have to know and decide about it, We order and command that those [officials] of Our Council of Indies endeavour with particular study and care always to have a complete and certain description and inquiry about all the things concerning the condition of the Indies, from land to sea, natural and moral [things], eternal and temporal, ecclesiastical and secular, past and present and those [things] that in time could fall under government or legal jurisdiction.[52]

The statutes formalised two methods for gathering information: the expeditions and the questionnaires. In 1571 the crown sent Dr Francisco Hernández to collect information about medicinal plants and samples of them. In the same year, the Council circulated a questionnaire of 200 chapters that would eventually evolve into a 55-chapter questionnaire by 1577. In the next section, I discuss these information-gathering tools.

In 1571, Dr Francisco Hernández left Spain and went to Mexico and Peru. Philip II ordered him to travel to the two largest kingdoms of the empire to gather both samples and "information generally about herbs, trees, and medicinal plants;"[53] to write down their descriptions, uses, and manner of cultivation; to consult with "physicians, surgeons, herbalists, Indians, and other curious people" knowledgeable in "herbs, trees, and medicinal plants;" to experience and test their properties; and to send to Spain samples of medicines and herbs noteworthy.[54] This expedition was the result of the empire consisting of a collection of regions under a single ruler, focused on expanding its political and economic power; the circulation of reports with information on the regions and peoples in the empire; and established infrastructures and communities of knowledge in the territories of the empire discussed above. This Mexican expedition was not only an expedition to access Indigenous knowledge about medicine and to gather medicinal plants, it was also about the emergence of epistemological practices within the empire to study nature. In Hernández's Mexican expedition, science and empire co-created each other.

By 1573, in a letter to the king, Hernández explained the magnitude of the Mexican expedition:

> What I may now advise Your Majesty is that four volumes of paintings of plants have been completed recently, in which there are 1,100 paintings, and another in which there are 200 animals, all exotic and native to this region, and scripts in draft and almost half of the descriptions in fair copy, of the nature, climate of the places to which they are native, the sounds they make, and their characteristics, according to the Indians, whose experience stretches over hundreds of years here. I have relied both on the evidence of other curious persons and of the doctors of this land and my own experiences, beyond what can be deduced using the rules of medicine. In all this, great care has been taken that no plant is painted unless I have seen it ten or more times in different seasons, smelled and tasted all its parts and asked more than twenty Indian doctors, each one individually, and considered how they agree and differ, and unless I have subjected it to the rigorous methods of identification and examination that I have developed here for this project.[55]

Hernández used the passive voice in reference to these four volumes—"*que están hechos*"— that "have been completed recently."[56] By whom? Hernández refers to the "Indians, whose experience stretches over hundreds of years," "other curious people," "the doctors of this land," "my own experience," and "more than twenty Indian doctors." Yet the paintings, writing, and circulation of samples also involved Indigenous people, and he did not mention them. The empire created, through violence and alliances, the conditions, first, for this asymmetrical exchange and circulation of knowledge—the Indigenous people's work and expertise is sort of acknowledged, sort of pushed to the side; and, second, for the legitimation of an epistemological method that validated Indigenous people's empirical knowledge. Hernández "asked more than twenty Indian doctors, each one individually, and considered how they agree and differ" as he saw, smelled, and tested all the plants contained in these volumes.[57] Hernández stayed in Mexico for six years; in February 1577 he returned to Spain.[58]

This was not the only expedition of this period. In 1583 López de Velasco proposed to the Council that a cosmographical expedition to the New World be carried out by Jaime Juan, "from Valencia, a man expert in mathematics and calculations in astronomy." Velasco defined the tasks of the expedition as follows: first, to "take the altitude or elevation of the places where he would go"; second, "to ascertain the deviation of the compass in relation to the pole in the said places"; third, "once this deviation is found, he would come to know the longitude and the east–west navigation by means of the instrument of longitudes that he carries with him"; fourth, to "observe the eclipses of the moon that would happen in order to find out the longitudes and distances from province to province"; fifth, to "make an inquiry into and report of the times and hours of the high and low tides of the sea in the coasts and seas where he would go"; and sixth, to organise the observation of moon eclipses in Spain and America. Finally, López de Velasco explained that:

> the truth and precision of the instruments and the intelligence of this said Jaime Juan in the use of them, cannot be judged without looking at and examining them. But if those instruments were those made by Juan de Herrera, chief master of architecture of His Majesty, they can be well taken as certain and well designed.[59]

The Council found Jaime Juan appropriate for the expedition. They ordered him to go first to New Spain (Mexico) and then to the Philippines. In New Spain, he was ordered to meet the cosmographer Francisco Domínguez and in the Philippines to collect the papers of Fray Martín de Rada. The Council expected the expedition to last between six and eight years and assigned Juan an annual salary.[60] Besides the calculations and observations, the Council also requested that Juan "attempt to make maps, in particular of those lands and provinces where he would go... and make a separate report about the noteworthy things of the said provinces."[61]

By 1584 Juan was already in New Spain making observations. He prepared a report on the eclipse of 17 November 1584. Several people helped him: Gabriel Gudiel, Francisco Domínguez (who had helped the royal physician Francisco Hernández during his expedition), and the physician Agustín Farfán. When Jaime Juan died, perhaps that year or the next, the expedition came to a halt. Years later, the Council would use this precedent to convince the king to support yet another expedition to the New World.[62]

From Oviedo's proposal of a natural history expedition in 1525 to Hernández's medical expedition of 1570, and including Juan's cosmographical expedition of 1583, expeditions became one more tool in the state-sponsored programme to collect empirical information about the American kingdoms. The use of questionnaires also continued. Juan de Ovando himself had already begun the process of systematically collecting information about America. In 1569, and extending practices already established within different sites of the empire (House of Trade, viceroyal courts, religious houses), Ovando circulated in the Indies a questionnaire of 37 chapters. Later, in 1571, upon his advice, the Council of Indies circulated a question-naire of 200 chapters to those coming from America—as the Casa pilots and cosmographers had been doing for the past 60 years. In 1573 there was yet another questionnaire of 135 chapters. The 200-chapter questionnaire was the basis for the questionnaire of 1577 (reduced to 50 chapters) by Juan López de Velasco.[63] This final questionnaire was reprinted with minor changes in 1584.[64] The report's political purpose was laid down in the title of the question-naire, "Instruction and memory for the reports that have to be done for the description of the Indies, commanded by His Majesty, for their good government and their ennoblement." The questionnaires reconfigured the Spanish Empire's web of knowledge production by linking experts and royal officials in both America and Europe. The answers provided not only infor-mation about the American territories but also a representation of the empire that, in turn, shaped the empire itself.

The questionnaire explained the mode of circulation: it was to be sent, via Viceroys or Audiences (who had to answer the questionnaire as well), to governors, magistrates (*corregidores*), and mayors. These officials were instructed to draw up a list of all Spanish and Indian towns under their jurisdiction and send the "Instruction and Memory" to these towns. In each town, the questionnaire was to be sent to the municipal council members. If there were no council, it would be sent to a priest (or to a friar if there were no priests). The "Instruction and Memory" had to be sent back with the reports to the official who had sent it, for further distribution, even in manuscript form, among towns that had not received the "Instruction and Memory."

The government officials who received the "Instruction and Memory" could either answer it by themselves (the actual questions were the second part, the memorial) or find "people intelligent of the things of the land" (*personas ynteligentes de las cossas de la tierra*) to answer it. In general, more than one person answered the questions. A notary testified to the informants' identity, who signed their report, as requested in Chapter 50 of the instruction. The notary also testified to the authenticity of the report thus elaborated.[65] Parallel to the 1577 Questionnaire, the Cosmographer-Chronicler López de Velasco elaborated a set of instructions to collect information on lunar eclipses and to determine, with the information collected, latitudes and longitudes of American places. Once again, his instructions were the result of the statutes of 1571.

Expeditions and questionnaires emerged in the context of the sixteenth-century long-distance Spanish American empire. The empire connected diverse territories and peoples, each with its own socioeconomic infrastructures and communities of knowledge and expertise. The empire created a framework for the circulation of existing knowledge and a method to verify that information. In the process, the empire pushed certain types of knowledge to the margins and centred empirical knowledge as a valid form of knowledge.

Conclusions

The Iberian Empire established the institutions and practices of modern science as an empirical activity based on the articulation of experience in reports and the circulation of these reports. The empire connected regions with different peoples and resources, people with their own knowledge and expertise. As Christians entered in contact with these communities of knowledge, they established, and eventually institutionalised, systems for verifying local knowledge: testing and reporting methods. Native American knowledge underlay the production of knowledge in the Iberian Empire: they provided the Spaniards with the knowledge about "all the things concerning the condition of the Indies, from land to sea." The Spaniards were the intermediaries between Native Americans and other Spaniards. In the process, they erased Native Americans' contributions to knowledge production in the early-modern empires. Finally, communities in the different regions of the Iberian Empire established their own system and practices of knowledge: the Spaniards, with their information-gathering institutions at the House of Trade, Council of Indies, and viceroyal courts, established but one system of knowledge-making practices. In the Caribbean, for instance, African practitioners established their own system of knowledge with similar empirical methods as the one the Spaniards had established in other sites of the empire. These empirical systems coexisted in the empire and created the empire.[66]

Notes

1 David N. Livingstone, "Region: Cultures of Science," in *Putting Science in Its Place: Geographies of Scientific Knowledge* (Chicago: University of Chicago Press, 2003), 97; Malyn Newitt, *A History of Portuguese Overseas Expansion, 1400–1668* (London: Routledge, 2005), 248.
2 Mignolo claims that the expansion of Western capitalism implied the expansion of Western epistemology, whereas I argue that in the sixteenth and seventeenth centuries, empire and science were co-created at the time of expansion. Mignolo's argument applies to a latter moment of development since he seems to assume that capitalism and science were already established at the time of expansion; see Walter Mignolo, "The Geopolitics of Knowledge and the Colonial Difference," *The South Atlantic Quarterly* 101, no. 1 (2002): 57–96, here p. 59.
3 Newitt, *Portuguese Overseas Expansion*, 38.
4 Jane Burbank and Frederick Cooper, *Empires in World History: Power and the Politics of Difference* (Princeton, NJ: Princeton University Press, 2010), 117.
5 Frances Gardiner Davenport, *European Treaties Bearing on the History of the United States and its Dependencies to 1648* (Washington, DC: The Carnegie Institute of Washington, 1917), 68.
6 Nicholas Canny, ed., *The Oxford History of the British Empire*, vol. 1, *The Origins of Empire: British Overseas Enterprise to the Close of the Seventeenth Century* (Oxford: Oxford University Press, 1988), 39; Anthony Pagden, *Lords of all the World: Ideologies of Empire in Spain, Britain and France c. 1500–c. 1800* (New Haven, CT: Yale University Press, 1995), 32; Leslie Bethell, ed., *Historia de América Latina*, vol. 1, *América Latina colonial: la América Precolombina y la conquista* (Barcelona: Editorial Crítica, 1984), 131.

7 Stefan Halikowski Smith, "The Mid-Atlantic Islands: A Theatre of Early Modern Ecocide?" *International Review of Social History* 55 (2010): 51–77, here p. 75. On Canarian slaves, see Alberto Vieira, "Sugar Islands The Sugar Economy of Madeira and the Canaries, 1450–1650," in *Tropical Babylons*, ed. Stuart B. Schwartz (Chapel Hill: The University of North Carolina Press, 2004), 57.

8 Jerónimo Zurita y Castro, *Historia del rey Don Hernando el Católico. De las empresas, y ligas de Italia* (Zaragoza, 1580), f. 17r.

9 John H. Elliott, *Imperial Spain, 1469–1716* (London: Penguin Books, 2002), 43.

10 These transformations were particularly clear in the islands; see R.H. Grove, *Green Imperialism: Colonial Expansion, Tropical Island Edens and the Origins of Environmentalism, 1600–1860* (Cambridge: Cambridge University Press, 1995); Jason W. Moore, "Madeira, Sugar, and the Conquest of Nature in the 'First' Sixteenth Century, Part II: From Regional Crisis to Commodity Frontier, 1506–1530," *Review: Fernand Braudel Center* 33, no. 1 (2010): 1–24.

11 Justo L. del Río Moreno y Lorenzo E. López y Sebastián, "El comercio azucarero de la Española en el siglo XVI. Presión monopolística y alternativas locales," *Revista Complutense de Historia de América* 17 (1991): 39–78, here p. 42. For a study of the profound transformations introduced with the cultivation of sugar, see the case of Madeira: Jason W. Moore, "Sugar and the Expansion of the Early Modern World-Economy: Commodity Frontiers, Ecological Transformation, and Industrialization" *Review: Fernand Braudel Center* 23, no. 3 (2000): 409–33.

12 Mario Góngora, *El estado en el derecho indiano: época de fundación (1492–1570)* (Santiago de Chile: Instituto de Investigaciones Histórico-Culturales, 1951), 16.

13 Antonio de Nebrija, *A la mui alta e assi esclarecida princesa doña Isabel la tercera deste nombre Reina i señora natural de españa e las islas de nuestro mar. Comienza la gramatica que nuevamente hizo el maestro Antonio de Lebrixa sobre la lengua castellana. e pone primero el prologo* (Salamanca, 1492), a.ii.

14 Martti Koskenniemi, "Empire and International Law: The Real Spanish Contribution," *The University of Toronto Law Journal* 61, no. 1 (2011): 1–36, here p. 11; Pagden, *Lords*, 47.

15 *Las Siete Partidas del Sabio Rey don Alonso el Nono, nuevamente Glosadas, por el Licenciado Gregorio Lopez, del Consejo Real de Indias de su Magestad* (Salamanca, 1576), 1: f. 5v.

16 Burbank and Cooper, *Empires*, 126.

17 I explore this topic and the themes of this chapter in Antonio Barrera-Osorio, *Experiencing Nature: The Spanish American Empire and the Early Scientific Revolution* (Austin: University of Texas Press, 2006).

18 Henry Kamen, *Empire: How Spain Became a World Power, 1492–1763* (New York: HarperCollins Publishers, 2003), 112–13.

19 On the model center–periphery, see Miles Ogborn, *Indian Ink: Script and Print in the Making of the English East India Company* (Chicago: The University of Chicago Press, 2007), 2–3. On the different networks organising the empire, see Pablo F. Gómez, *The Experiential Caribbean: Creating Knowledge and Healing in the Early Modern Atlantic* (Chapel Hill: University of North Carolina Press, 2017), 23.

20 Transculturation describes a process of sociocultural transformation and creation as different groups come in contact in different power positions. The Cuban sociologist Fernando Ortiz proposed this term in his *Contrapunteo Cubano del Tabaco y el Azúcar* (Caracas: Biblioteca Ayacucho, 1987), 96; Mary Louise Pratt, "Arts of the Contact Zone," *Profession* (1991): 33–40, here p. 36.

21 On empires, resources, territories, and people, see Burbank and Cooper, *Empires*. On the connection of empire and knowledge, see Barrera-Osorio, *Experiencing*; Arndt Brendecke, *The Empirical Empire: Spanish Colonial Rule and the Politics of Knowledge* (Berlin: Walter de Gruyter, 2016).

22 Linda Tuhiwal Smith, *Decolonizing Methodologies: Research and Indigenous Peoples* (London: Zed Books and University of Otago Press, 1999), 64. Walter Mignolo discusses the connections between colonisation and knowledge in his work; see, for instance Mignolo, "Geopolitics of Knowledge." My work aims to look at the historical process by which modern science emerged in a Christian-European imperial and American colonial context. José Bustamante García, "La empresa naturalista de Felipe II y la primera expedición científica en suelo americano: la creación del modelo expedicionario renacentista" in *Felipe II (1527–1598): Europa y la monarquía católica: Congreso Internacional*, vol. 4, *Europa dividida, la monarquía católica de Felipe II* (Madrid: Universidad Autónoma de Madrid, 1998), 40.

23 Gómez, *The Experiential Caribbean*, 186.

24 This letter is published in Martín Fernández de Navarrete, ed., *Colección de los viages y descubrimientos que hicieron por mar los españoles desde fines del siglo XV* (Madrid, 1825), 2: 109.

25 The letter from the kings to Columbus is printed in Fernández de Navarrete, ed., *Colección de los viages*, 2: 154; the quotes come from Antonio de Herrera, *Historia General de los Hechos de los Castellanos en*

las Islas i Tierra Firme del Mar Oceano. Escrita por Antonio de Herrera Coronista Major de Su Magestad de las Indias y su Coronista de Castilla (Madrid, 1601), 1: 76.

26 For the 1502 instructions to Columbus, see Navarrete, ed., *Colección*, 1: 277–81.

27 This idea comes from Marcos Jiménez de la Espada, who explored it in his introductory study, the "Antecedentes," *Relaciones Geográficas de Indias* (Madrid: Biblioteca de Autores Españoles, 1965), 1: xxi.

28 On empires and distance, see Fernand Braudel, *The Mediterranean and the Mediterranean World in the Age of Philip II* (Berkeley: University of California Press, 1995), 1: 371.

29 Nicolás Monardes, *Primera y segunda y tercera partes de la historia medicinal de las cosas que se traen de nuestras Indias Occidentales que sirven en medicina. Tratado de la piedra bezaar, y dela yerva escuerçonera. Dialogo de las grandezas del hierro y de sus virtudes medicinales. Tratado de la nieve y del bever frio* (Seville, 1574), f. 75v.

30 On networks of humans and things, see Steven J. Harris, "Long-Distance Corporations, Big Science, and the Geography of Knowledge," *Configurations* 6, no. 2 (1998): 269–304; John Law, "On the Methods of Long Distance Control: Vessels, Navigation, and the Portuguese Route to India," The Centre for Science Studies, Lancaster University, 2001, www.comp.lancs.ac.uk/sociology/papers/Law-Methods-of-Long-Distance-Control.pdf.

31 On the postal service or *correo mayor,* see *Cedulario Indiano: Recopilado por Diego de Encinas* (Madrid: Real Academia de la Historia, 2018), Libro Segundo, 301.

32 Clements R. Markham, ed., *The Journal of Christopher Columbus (During his First Voyage, 1492–93) and Documents Relating the Voyages of John Cabot and Gaspar Corte Real* (London: Hakluyt Society, 1893), 48.

33 On Cabot's informants, see Sebastian Cabot's 1530 deposition in Henry Harrisse, *John Cabot the Discoverer of North-America and Sebastian Cabot, a Chapter of the Maritime History of England under the Tudors, 1496–1557* (London: Benjamin Franklin Steves, 1896), 423.

34 Bernardino de Sahagún, *Historia general de las cosas de Nueva España* (Mexico City: Fomento Cultural Banamex, 1982), 2: 603; María Luisa Rodríguez-Sala, *Los médicos en la Nueva España: Roles profesionales, organizacionales y sociales: etapa de formación y asimilación (1553–1621)* (Mexico City: Universidad Nacional Autónoma de México, 2014), 126.

35 Monardes, *Primera*, f. 5v–9v.

36 Peter Martyr d'Anghiera, *De Orbe Novo: The Eight Decades of Peter Martyr D'Anghera,* 2 vols. (New York: G.P. Putnam's Sons, 1912), see examples in 1: 347 and 1: 411.

37 Letter from Alonso de Zuazo to Charles V in Marcos Jiménez de la Espada, *Relaciones Geográficas de Indias: Perú* (Madrid: Biblioteca de Autores Españoles, 1965), I: xvii.

38 Barrera-Osorio, *Experiencing*; Raquel Alvarez Peláez, *La Conquista de la Naturaleza Americana* (Madrid: Consejo Superior de Investigaciones Científicas, 1993).

39 See Josefina García Quintana and Alfredo López Austin, "Prólogo," in *Historia general de las cosas de Nueva España,* ed. Fray Bernardo de Sahagún (Mexico City: Fomento Cultural Banamex, 1982), 1: xiii.

40 Pedro Arias de Benavides, *Secretos de Chirurgia, especial de las enfermedades de Morbo galico y Lamparones y Mirrarchia, y asimismo la manera como se curan los Indios de llagas y heridas y otras passiones en las Indias, muy util y provechoso para en España y otros muchos secretos de chirurgia hasta agora no escritos* (Valladolid: F. Fernandez de Cordoua, 1567).

41 See Medel's treatise in Berta Ares Queijia, *Tomás López Medel: Trayectoria de un clérigo-oidor ante el nuevo mundo* (Guadalajara: Instituto Provincial de Cultura Marqués de Santillana, 1993), 409–593.

42 These reports were published together in the 1574 edition; see Monardes, *Primera*.

43 See some examples in Monardes, *Primera*, f. 18v, 19 r, 23r, 27r, 29r, 121r.

44 On Monardes's garden, see Monardes, *Primera*, f. 67v. On gardens and collections, see also Barrera-Osorio, *Experiencing,* 120ff.

45 Monardes, *Primera*, f. 74v.

46 On hydrographical activities, the case of the construction of the *desagüe* en Mexico City is relevant. See Antonio Barrera-Osorio, "Experts, Nature, and the Making of Atlantic Empiricism," *Osiris* 25, no. 1 (2010): 129–48; Vera Candiani, *Dreaming of Dry Land: Environmental Transformation in Colonial Mexico City* (Stanford, CA: Stanford University Press, 2014).

47 On Medina and the amalgamation method, see Luis Muro, "Bartolomé de Medina, introductor del beneficio de patio en Nueva España," *Historia Mexicana* 13, no. 4 (1964): 517–31; Barrera-Osorio, *Experiencing,* 69. On the spread of the amalgamation method, see P.J. Bakewell, *Silver Mining and Society in Colonial Mexico: Zacatecas, 1546–1700* (Cambridge: Cambridge University Press, 1971), 138.

48 Alison Bigelow, "Incorporating Indigenous Knowledge into Extractive Economies: The Science of Colonial Silver," *The Extractive Industries and Society* 3, no. 1 (2016): 117–23, doi.org/10.1016/

j.exis.2015.11.001; Modesto Bargalló, *La minería y la metalurgia en la América Española durante la época Colonial* (Mexico City, 1955), 522. The Italian Gemelli Careri visited the mines in Pachuco and discussed them, with illustrations, in his book of voyages; see Byron Ellsworth Hamann, "The Mirror of Las Meninas: Cochineal, Silver, and Clay," *The Art Bulletin* 92, no. 1/2 (2010): 6–35, here p. 24.

49 On the Council of Indies and the use of reports, see, for instance, Ernesto Schäfer, *El Consejo Real y Supremo de las Indias: Su Historia, Organización y Labor Administrativa hasta la Terminación de la Casa de Austria* (Seville: Escuela de Estudio Hispano-Americanos, 1947), 2: 330 and 2: 406.

50 Antonio de Herrera, *Historia General de los Hechos de los Castellanos en las Islas i Tierra Firme del Mar Oceano. Escrita por Antonio de Herrera Coronista (sic) Major de Su Magestad de las Indias y su Coronisata de Castilla,* vol. 1, *Al licenciado Paulo de Laguna Presidente del Real y Supremo Consejo de las Indias* (Madrid, 1601–15); David Goodman, *Power and Penury: Government, Technology, and Science in Philip II's Spain* (Cambridge: Cambridge University Press, 1988), 68ff; Peláez, *Conquista de la Naturaleza,* 131ff.; Manuel Jiménez de la Espada, *Relaciones Geográficas de Indias: Perú* (Madrid: Biblioteca de Autores Españoles, 1965), 1: 59. The Casa had its own system of information.

51 *Colección de documentos inéditos relativos al descubrimiento, conquista y colonización de las antiguas posesiones españolas en América y Oceanía, sacados de los Archivos del Reino, y muy especialmente del de Indias. Competentemente Autorizada* (Madrid: Imprenta de José María Perez, 1871), 16: 457–59.

52 Archivo General de Indias, Seville (hereafter cited as AGI), Indiferente 856, no. 3. See also *Recopilación de leyes de los reinos de las Indias: mandadas imprimir y publicar por la magestad católica del rey Don Carlos II, nuestro señor* (Madrid, 1841), vol. 1, book II, tit. II, law VI, 154.

53 Simon Varey, ed., *The Mexican Treasury: The Writings of Dr. Francisco Hernández* (Stanford, CA: Stanford University Press, 2000), 46; Instrucciones al Dr. Francisco Hernández, 1 November 1570, Madrid, AGI, Indiferente 1228.

54 AGI, Indiferente, 1228; Varey, *Mexican Treasury,* 46.

55 Varey, *Mexican Treasury,* 53.

56 For the letter in Spanish, José Toribio Medina, *Biblioteca Hispano-Americana (1493–1810)* (Santiago de Chile: printed and engraved at the author's house, 1900), 2:275.

57 Varey, *Mexican Treasury,* 53.

58 On Hernández the best, and still the one single source used by all the scholars who study the work of Hernández, is Germán Somolinos D'Ardois, *Vida y Obra de Francisco Hernández,* vol. 1, *Obras Completas de Francisco Hernández* (Mexico City: Universidad Autónoma de México, 1960).

59 AGI, Indiferente 740, no. 103. On Jaime Juan's expedition, see Barrera-Osorio, *Experiencing Nature,* 98; Maria Portuondo, *Secret Science: Spanish Cosmography and the New World* (Chicago: University of Chicago Press, 2009), 86.

60 Consulta del Consejo in AGI, Indiferente 740, no. 103.

61 Instrucciones a Jaime Juan in AGI, Indiferente 740, no. 103.

62 AGI, Indiferente 868.

63 On these questionnaires, see Jiménez de la Espada, *Relaciones,* 1: 50 and 1: 59.

64 See the "Instruction and Memory" of 1584 in Rene Acuña, "Instrucción y Memoria de las relaciones que se han de hacer para la descripción de las Indias que su majestad manda hacer, para el buen gobierno y ennoblecimiento dellas," in *Relaciones Geográficas del Siglo XVI* (Mexico City: Universidad Nacional Autónoma de México, 1982), vol. 1: 73–8.

65 See examples of this in Rene Acuña, *Relaciones,* 1: 75 and 1: 92.

66 On the Caribbean practitioners and their system of knowledge and practices, see Gómez, *The Experiential Caribbean.*

23

SCIENCE IN EARLY NORTH AMERICA

Cameron B. Strang

In 1778, as Patriots in 13 of Britain's North American colonies fought for independence, the Spanish cartographer Bernardo de Miera y Pacheco was completing a map of the Colorado plateau and eastern Great Basin. Miera had been part of an exploring expedition led by two Franciscan missionaries, Atanasio Domínguez and Silvestre Vélez de Escalante, that sought an overland route from New Mexico to Monterey to supply Spain's new outposts in California. Spain's expansion into California, the Domínguez–Escalante expedition, and dozens of other eighteenth-century Spanish scientific expeditions were fuelled by fears of Russian, British, and French incursions into North America.

Part of the story of Miera's map is predictable. As histories of cartography have told us again and again, maps promote imperialism, and Miera included a stunning image of the Pope speeding into the Great Basin on a chariot pulled by the lions of Castile. But science and imperialism could also intersect in surprising ways in North America, and Europeans were hardly the only people who dominated neighbours or generated knowledge. Far from projecting Spanish imperialism, Miera's map reflected Indigenous power. It depicted Native nations with clearly bounded territories, roofless "habitations of Spaniards ruined by enemies," and a long note on how Comanches had mastered "horses and iron weapons" to become "lords and masters of all the buffalo country."[1] These Comanche imperialists even sought knowledge in ways that were all too similar to the worst practices of Euro-American men of science. In 1850, Comanches scraped off the flesh of two Black Seminole girls to determine if "beneath the cuticle the flesh was black like the colour of the exterior." They then burned the girls "to ascertain whether fire produced the same sensations of pain as with their own people, and tried various other experiments, which were attended with the most acute torture."[2] Euro-American men of science did not have a monopoly on studying human difference through violence.

Despite its focus on Native power, Miera's map would eventually be repurposed to animate expansionist dreams. Alexander von Humboldt copied the map from Mexican archives and, in 1804, shared it with President Thomas Jefferson, who consulted it while planning journeys of exploration and intrigue into the Spanish Southwest. Anglo-American cartographers continued plagiarising Miera's chart into the 1840s, eagerly perpetuating its most enticing errors (especially the rivers that supposedly ran from the Great Basin to the Pacific) while effacing the boundaries of sovereign Native nations to portray the West as empty space awaiting white settlement.[3]

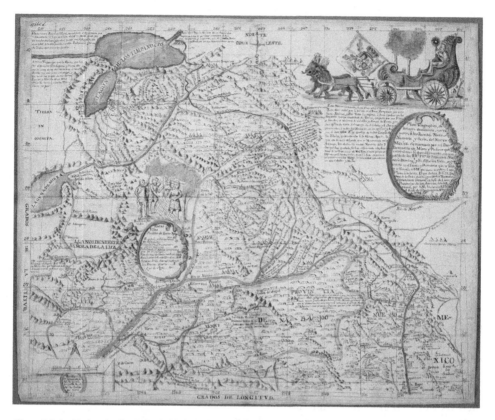

Figure 23.1 Bernardo de Miera y Pacheco, "*Plano geographico, de la tierra descubierta, nuevamente, a los rumbos norte, noroeste y oeste, del Nuevo Mexico*," Chihuahua, 1778. © The British Library Board. Add. 17661.d.

Aggressive US expansion—as well as ongoing rivalries among European empires—ensured that knowledge of North America's geography, nature, and peoples remained as contextualised by imperialism after 1776 as it had been since the 1500s.

Histories of science and imperialism in North America have usually fallen into four broadly geographic categories. Atlantic histories have emphasised connections between eastern colonies and European metropoles from the 1500s to around 1800.[4] Histories of the continental interior have looked mostly at exploration and mapping from the 1500s to the late 1800s.[5] Pacific histories have concentrated disproportionately on famous scientific expeditions from the mid-1700s to the early 1800s.[6] Arctic histories, lastly, have leaned toward exploration and ethnography from the 1500s into the 1900s.[7] Beyond these regional fields, scholarship has usually been limited to the ideas, networks, and practices of individual empires, colonies, or nations.[8] This is, of course, a practical approach when it comes to completing in-depth archival research, but such intra-national histories have also had the unfortunate effect of reifying boundaries that were often protean in North America itself.

A continental perspective can tie these regional and imperial histories together. This is because, for one, collisions and exchanges with Indigenous peoples were fundamental to colonial projects across North America, and a continental lens—far more than Atlantic or Pacific ones—helps ensure that Native nations are not marginalised in Eurocentric stories of imperial expansion and scientific progress. It is also because exploration and research in all parts of North America took place within a broader continental (and, really, global) setting of imperial competition.[9] Contests for land, resources, souls, prestige, and influence among European powers and, eventually, the United States conditioned the pursuit of science—by which I mean the study of the natural world—from Alaska to Florida. In addition, concentrating on the era of imperial rivalry offers a useful way to define the temporal limits of "early" (post-European encounter) North America since the period lasted from the incursions of the 1500s through to the 1860s, when Russia gave up its American claims and the boundaries between a newly confederated Canada, the just-reunited United States, and a recently diminished Mexico were finally settled.

This essay seeks to identify some of the historical and historiographical themes that could scaffold a continental history of science and empire in early North America. To be sure, scientific practices and the material conditions undergirding them varied considerably across place and time. But encounters with Native groups and inter-imperial rivalries—and all of the violence, imaginings, and entanglements that went with these—remained constants that contextualised the production of knowledge throughout Spanish, French, British, Dutch, Russian, and US territories from the 1500s to the mid-1800s. More specifically, scientific practitioners in all of these regions had comparable concerns like exploring and charting lands, justifying imperialism through science, interpreting human difference, and building and manipulating knowledge networks. My hope is that identifying such themes will help specialists embedded in North America's various regional and imperial historiographies recognise connections, commonalities, and, perhaps, exceptional differences across an uncomfortably shared continent.

Exploration and cartography

European exploration of New World nature began around the year 1000 as multiple expeditions of Norse mariners scouted the coast from Baffin Island to Vínland. According to Kirsten A. Seaver, Norse voyagers searched for farmlands and resources like furs and timber amid ongoing conflicts with Native peoples and, it seems, remembered how to access these resources for generations after Leif Eirikson first sighted America's shores.[10] These excursions set a long-standing precedent of combining information gathering with violence. Spanish *entradas* during the 1500s were essentially armies of conquest that included cartographers, naturalists, and ethnographic observers. Hernando de Soto, who ranged from Florida to Louisiana from 1539 to 1542, even favoured torture as a way of extracting information from Native people because it seemed to guarantee their credibility.[11] This potent combination of exploration and violence persisted in North America throughout the era of imperial competition. As the United States invaded the Southwest in the 1840s and 50s, the federal government and the Smithsonian Institution depended on the US Army to chart the region and collect specimens while conquering Indigenous and Mexican territories.[12]

Exploration unfolded amid tensions between what travellers expected to find and actual circumstances and encounters on the ground. This was particularly evident in searches to find a navigable waterway through the continent, an enterprise that brought together Atlantic, interior, Pacific, and Arctic historiographies. From the 1500s to 1800s, officials and entrepreneurs in each of the powers competing for North America looked for a passage because contemporary geographic theories and delusions of grandeur convinced them of its reality. Such expectations then led explorers to interpret what they saw as proof that a waterway did exist, just over the horizon. As historian William Goetzmann first argued over 50 years ago, cultural assumptions fundamentally affected explorers' findings: for example, Thomas Jefferson's 1803 instructions "programmed" Lewis and Clark's search for a transcontinental water route.[13]

Preconceptions and directions certainly mattered, but more recent scholarship has made it clear that exploration was never merely a performance of imperial visions. For instance, the instructions that the Académie des Sciences penned to order the comte de La Pérouse's search for a passage from the Pacific in the 1780s included collecting human skulls, but tense encounters with Tlingits ended with the French contributing gifts to Natives' graves instead of robbing them.[14] Indeed, encounters with Native peoples marked the information and itineraries of even the most carefully planned excursions. Eskimo cartographers laid out Alaska's north-eastern coast for Otto von Kotzebue as he sought an Arctic passage for the Russian Empire in 1816, a Hidatsa map and Shoshone guide got Lewis and Clark to the Pacific, and an Inuit woman named Iligliuk drew a dozen maps for William Parry in 1821 that he read as evidence that a passage existed north of Hudson's Bay.[15] While such anecdotes of Native people's contributions to European science are important, too little scholarship examines how Native Americans themselves created and circulated knowledge within this same milieu of imperial competition. Julie Cruikshank, for one, examined how James Cook's and La Pérouse's expeditions in search of a transcontinental passage met Tlingits and Athapaskans who were themselves exploring coastal regions recently laid bare by receding glaciers. But whereas the Europeans aimed to extract atomised scientific facts about these places, these Native peoples studied and remembered the natural and social world as inseparable.[16]

More so than any other science, cartography equipped officials and colonisers to see Native peoples' homelands as their own. As Daniel Clayton posited in his study of Vancouver Island, imperial mapping involved an ongoing process of abstraction through which geographic knowledge that was created amid social encounters became increasingly decontextualised as it travelled further from its point of origin, eventually making Indigenous knowledge invisible and turning Indigenous homelands into cartographic outlines fit to be disputed among rival powers.[17] But Indigenous peoples' geographic knowledge and political decisions nevertheless remained visible on imperial maps. David Bernstein has suggested that, through actions like drawing maps, fighting wars, and even selectively ceding lands, Native nations on the Great Plains shaped the outlines and content of the maps that guided Anglo-Americans as they envisioned expansion.[18] Still, the relationship between Indigenous knowledge and imperial maps was often deeply ambiguous. Paul Mapp has explored how lack of trust, communication problems, and even Native peoples' ignorance of neighbouring regions ensured that, even though Spanish, French, and British cartographers worked assiduously to co-opt Indigenous peoples' geographic information, the charts they made did more to perpetuate uncertainty about the continental interior than foster dominion over it.[19]

North American cartography was, moreover, thoroughly rooted in imperial rivalry. In the Dutch colony of New Netherland, for instance, charting, naming, and illustrating the mid-Atlantic coast upheld Dutch claims over far more space than its scanty colonial population could sustain.[20] Colonial mapping enforced and exaggerated boundaries, but it could also aim to blur them: Dutch merchant Augustine Herrman's map of the Chesapeake emphasised fluid borders and commercial ties between Dutch and English America but, after its publication in 1673, English officials transformed the map into a tool of English empire building.[21] The ways officials used maps did not necessarily correspond with the intent of their creators and, sometimes, the goals of mapmakers themselves changed over time. In 1768, Benjamin Franklin drafted a hydrographic chart of the Gulf Stream as evidence that Britain could, with colonists' help, dominate eastern North America from the Mississippi to the Arctic. But during the Revolutionary War, Franklin repurposed this chart to assert that Anglo-Americans could, with French help, rule these territories themselves.[22] Perhaps the most large-scale example of how maps created to buttress existing empires could also prop up new ones occurred in the aftermath of the Seven

Years' War (1754–63), when Britain's Board of Trade ordered a comprehensive survey of North America to consolidate its new acquisitions and stave off ongoing French and Spanish intrigue. As Max Edelson has revealed, this massive cartographic project represented a shift in imperialism itself: the British Empire drawn on these maps would be centralised, allowing officials in London to replace white settlers as the main agents of colonisation. British colonists, however, saw such centralisation as a step toward tyranny and a denial of their own hunger for western lands. These grievances launched a revolution and, after achieving independence, US citizens repurposed the Board of Trade's maps "to see an expansive continental empire in which the restrained energies of settler colonialism would be unleashed and redirected toward the west."[23] In short, scholars have shown that imperialism could inspire maps, and maps could inspire imperialism.

Perceptions and peoples

It was not only maps that promoted particular visions of empire; perceptions of the natural world and schemes to alter it could be central to justifying imperialism. As Christopher Parsons has demonstrated, French colonists in seventeenth-century Canada believed that "cultivating" local plants and people—both of which seemed to be underdeveloped, "*sauvage*" varieties of European ones—made their colonial project just and good.[24] Anglo naturalists in eighteenth-century Nova Scotia held a similar ideology they termed "improvement." These men sought to reform agronomy and industry in ways that would make the region's very climate more temperate, a move that both legitimised colonisation and encouraged more of it.[25] Shannon Lee Dawdy has argued that the colonisation of New Orleans in the early 1700s was in and of itself "an intellectual experiment." Efforts to make science inextricable from Louisiana's settlement manifested in projects—collecting specimens, testing crops, building observatories, writing natural histories—that promised to validate the colony's existence despite its sordid reputation and failure to turn a profit.[26]

Producing and compiling knowledge about nature could also pave the way for colonies to transition into independent (but nevertheless imperial) nations. In the first half of the nineteenth century, "inventory science"—which entailed collecting data in fields like geology, meteorology, and natural history—seemed to prove that Canada's prosperity depended on both uniting far-flung provinces and extending control over Indigenous homelands.[27] In the United States, according to Michael Adas, the intellectual underpinnings of expansion depended less on synthesising natural knowledge than "ideological imperatives that nature was to be mastered, resources exploited to the fullest, and technologies invented as these enterprises required." This set of beliefs satisfied Anglo-Americans that the displacement of Native people and, eventually, overseas colonialism were part of a benevolent process of human progress.[28]

Nowhere was science more fully woven into the fabric of imperialism than in the ideas and practices meant to differentiate some humans from others. Most of the harshest notions of race located human variation in inheritable bodily differences. Joyce Chaplin has reconstructed an early iteration of this perspective among seventeenth-century English colonists who compared Native people's high mortality rate from imported diseases like smallpox to Europeans' relatively low rate. Since the English believed that these diseases were endemic to North America, they reasoned that Indigenous bodies were less well adapted to life in America than English ones. This line of thought fuelled other beliefs about English superiority, such as the notion that Native Americans would always lag behind in technological advancement because they had to dedicate a disproportionate amount of their creative genius to maintaining their health. Ultimately for Chaplin, Anglo-Americans' belief in their exceptional right to dominate North

America was rooted more in their bodies than their culture.[29] Racial distinctions were similarly strict in early-eighteenth-century Louisiana, where French colonists "consistently emphasized their belief in the inherent superiority of whites over blacks by invoking quasi-biological conceptualizations" of race adapted from *ancien régime* hierarchies of nobles and commoners. In both the Old and New Worlds, elites feared miscegenation because the mixing of two "species" would lead to a "bad race."[30] From New England to New Orleans, then, idiosyncratic but nonetheless biology-based notions of race justified white power well before the rise of so-called scientific racism in the nineteenth century.[31] Research into the nature of human difference after the late 1700s remained thoroughly rooted in the death and enslavement of non-whites. As Ann Fabian has shown, Anglo craniologists studied Native skulls collected amid conquest and African-American skulls collected amid slavery to contend that these races were, and always would be, inferior. In sum, imperialism facilitated scientific research that made North America's racial hierarchies seem natural.[32]

Yet European and Anglo-American colonisers did not all subscribe to theories of biological racial difference, and many remained committed to the idea that variations in colour and culture did not make some peoples permanently better than others. Sharon Block has observed that Anglos in Britain's Atlantic colonies may have focused less on race than "complexion," a set of traits based in humoral theory that characterised individuals according to place and temperament rather than fixed racial categories.[33] Spanish and Russian ethnographers in California, moreover, studied Native cultures to better understand the origins and development of human societies on the whole. David Weber has found that the men of science in Alejandro Malaspina's 1791 venture to the Pacific coast "tended to portray Indians' behaviour as rationally adaptive and often virtuous, no matter how much their societies differed from European norms." These ethnographic observations took place amid imperial contests for land and prestige: not only did Spanish officials order this expedition to the Pacific coast to help stave off Russian and British incursions, but Malaspina himself was eager to employ a humane approach to ethnography to reverse the lingering "black legend" of Spanish cruelty and, instead, contrast Spanish enlightenment with English brutality.[34] Claims to coastal lands and national pride similarly inspired St. Petersburg's Academy of Sciences to send Ilya Gavrilovich Voznesenky to study the plants, minerals, and people of Russian Alaska and California. During his sojourn in northern California from 1840 to 1841 (when he gathered what remains the world's most extensive ethnographic collection from pre-gold-rush California), Voznesenky both rejected biological racism and leaned on the black legend to posit that the condition of Indigenous Californians had less to do with inherent difference than cultural adaptations to escape being "enslaved by the Spanish."[35] And while some nineteenth-century Euro-American linguists considered Native languages to be evidence of inherent mental inferiority, others, like Peter Stephen Du Ponceau, took advantage of the United States' sprawling network of military personnel to collect evidence that Native languages were sophisticated and, thus, that Native people shared Anglos' potential for genius. Such linguists nevertheless used their work to argue for ongoing US expansion, albeit to create what they considered a benevolent empire that would uplift non-whites rather than oppress them.[36] Views of human difference based in culture, environment, and biology all took shape in a context of imperialism and could be harnessed to encourage further expansion.

Networks

The powers competing for North America relied on a strikingly similar set of investigators and patrons to generate natural knowledge. This is especially clear in the case of fur traders.

French, Russian, Anglo-American, and Spanish fur traders were among the first to explore much of North America, and their expeditions were often thoroughly multinational.[37] In 1795, for example, Spain sent the Scottish fur trader James Mackay and the Welsh nationalist John Evans (both of whom had opted to become Spanish subjects) to explore the upper Missouri River in search of a transcontinental waterway and establish trade relations with the Mandans. Evans did eventually reach the Mandan towns and seize British trading posts but failed to reach the Pacific or achieve his personal goal of finding the mythical Welsh Indians.[38] In Russian Alaska, *promyshlenniki* (fur traders) collected an enormous wealth of information about geography and natural history, much of which was violently extracted from the same Native families they held hostage to ensure a steady supply of furs. Alexey Postnikov and Marvin Falk have uncovered how the combined cartographic work of *promyshlenniki* during the eighteenth and nineteenth centuries—collected and compiled by hydrographers at the Russian Admiralty—constituted a far more extensive mapping project than the famed British, Spanish, and French expeditions to the north Pacific coast.[39] Further east, agents of Britain's Hudson's Bay Company (HBC) engaged in all sorts of scientific work in coordination with the Royal Society, including observing the 1769 transit of Venus, collecting plants and animals, studying Native languages, and—in the case of Thomas Hutchins's experiments with mercury coagulation—testing the transferability of European technology into the Arctic.[40]

Fur traders did not just perform scientific work. They supported it. The Russian-American Company, which won a monopoly on all fur trading in Russian America in 1799, funded scientific expeditions and distributed circulars to traders from the Aleutian Islands to California with detailed instructions on how to gather specimens and geographic data for scholars in St. Petersburg.[41] The Russian-American Company supported these projects as part of its competition for Arctic furs with the HBC, itself the patron of a wide range of scientific activities. According to Ted Binnema, the HBC—which at its height claimed one-quarter of North America—promoted science primarily to counteract a reputation for selfishness and secrecy in hopes of ensuring ongoing official support for the Company's monopoly. Thus, while some of the scientific work supported by the HBC did lead to economic and territorial gains, much of it actually hurt the Company's bottom line but bolstered its "brand," an important reminder that both collectors and patrons operated in a world in which public opinion, the whims of officials, and reputation all had to be taken into consideration.[42] Power, like knowledge, depended on convoluted and sometimes hidden networks.

The notion that colonists in North America were (as Raymond Phineas Stearns put it in 1970) "relatively minor contributors of information, feeding American scientific data … to English and occasionally European patrons" has collapsed: historians now see colonial scientific networks as complex webs of collectors, patrons, and interpreters connected through multiple layers of social relationships.[43] Susan Scott Parrish has unpacked how closeness to American nature and a collective faith in empirical observation gave men and women in North America considerable authority in transatlantic knowledge networks that operated through reciprocal relationships of deference and gift giving.[44] Colonists' desire for recognition and ongoing patronage could even influence the theories and generalisations of European natural philosophers. In eighteenth-century French Canada, the collections that naturalists in Québec sent to the Académie des Sciences increasingly emphasised plants that were unlike those of Europe, a move that vitalised French intellectuals' shift from viewing the nature of New France as merely uncultivated to essentially alien.[45] Knowledge networks, moreover, offered multiple paths for individuals throughout North America to fashion identities as experts. In the early 1800s, western geographic observers like William Clark depicted the continental interior for eastern audiences primarily to advance their own political careers while naturalists in Philadelphia, especially at the American Philosophical Society, tried to cast themselves as the heirs to the London *savants* who had ordered American nature before the Revolution by seeking data and specimens from collectors throughout the United States.[46]

Lastly, information networks did not simply connect so-called centres and peripheries. There were multiple smaller nodes throughout the continent that organised research and synthesised its findings. These hubs began forming almost immediately after colonisation and lasted throughout the era of imperial competition. In 1566, just one year after founding Saint Augustine, the first European colony in North America, the governor of Spanish Florida wrote up scientific instructions for Juan Pardo and his party (which included soldiers and an alchemist) to explore the continental interior, an expedition that reached into present-day Tennessee.[47] Some 250 years later, the chief manager of the Russian-American Company in New Archangel wrote scientific instructions for fur traders, built an observatory to collect astronomical data, and kept a scientific library. And around this same time, the Natural History Society of Montreal distributed questionnaires to fur traders throughout Canada to gather information about Native peoples and valuable resources.[48] Such continental nodes were hubs that connected colonists, Indigenous peoples, and European natural philosophers, all of whom were embedded in wider webs of exploitation, exchange, and competition.

Conclusion

The age of imperial rivalry for the continent came to an end during the 1860s, but both the United States and Canada continued to colonise lands within their territorial claims and, by the 1890s, the United States had embarked on its own era of overseas expansion. Indeed, during the second half of the twentieth century, a new age of imperial competition between the United States and the Soviet Union would make the ties between science and government power in North America tighter than ever. Imperial rivalry, then, may offer one way to frame the long history of science in North America since 1500.

Although rethinking the temporal limits of the story of North American science could be a promising path forward, the larger challenge facing scholars is to reimagine its main protagonists. If a continental perspective proves to be an appealing way to move the field away from national and imperial histories, then we may take the additional step of re-centring Native American epistemologies (despite enormous variations among nations and across time) as the most persistent and widespread paths to knowledge on the continent. This means going beyond examples of how Indigenous informants added to European science and, instead, unpacking Native peoples' own methods, goals, and networks with the same care that we have given to those of Euro-Americans. Such histories could trace continuities from the pre-contact era but could also reveal how the ways Native Americans produced and applied natural knowledge after the 1500s were part of the same context of imperial competition that mattered so much to Europeans' scientific work. Once we have uncovered the histories of how and why Indigenous peoples from Hudson's Bay to San Diego pursued natural knowledge amid continental and global geopolitical rivalries, we might then start to consider whether "science" and "empire" were the most consequential aspects of the larger history of knowledge and power in North America.

Notes

1 Bernardo de Miera y Pacheco, "Plano Geographico, de la tierra descubierta, nuevamente," Chihuahua, 1778, The British Library, London, Add. Ms. 17.66t.D, reproduced in John L. Kessell, *Whither the*

Waters: Mapping the Great Basin from Bernardo de Miera to John C. Frémont (Albuquerque: University of New Mexico Press, 2017), xii; Carl I. Wheat, *Mapping the Transmississippi West, 1540–1861*, vol. 1, *The Spanish Entrada to the Louisiana Purchase, 1540–1804* (San Francisco, CA: Institute of Historical Cartography, 1957), 94–116.

2 Randolph B. Marcy, *Thirty Years of Army Life on the Border* (New York: Harper & Brothers, 1866), 55. Pekka Hämäläinen, *The Comanche Empire* (New Haven, CT: Yale University Press, 2008).

3 On the Anglo-American afterlives of Miera's map see Claudio Saunt, *West of the Revolution: An Uncommon History of 1776* (New York: W.W. Norton & Company, 2014), 113–14; Kessell, *Whither the Waters*, 55–61.

4 Raymond Phineas Stearns, *Science in the British Colonies of America* (Urbana: University of Illinois Press, 1970); Joyce E. Chaplin, *Subject Matter: Technology, the Body, and Science on the Anglo-American Frontier, 1500–1676* (Cambridge, MA: Harvard University Press, 2001); Susan Scott Parrish, *American Curiosity: Cultures of Natural History in the Colonial British Atlantic World* (Chapel Hill: University of North Carolina Press, 2006); Christopher M. Parsons, *A Not-So-New World: Empire and Environment in French Colonial North America* (Philadelphia: University of Pennsylvania Press, 2018).

5 Alfred Barnaby Thomas, *After Coronado: Spanish Exploration Northeast of New Mexico, 1696–1727* (Norman: University of Oklahoma Press, 1966); William H. Goetzmann, *Exploration and Empire: The Explorer and the Scientist In the Winning of the American West* (New York: Knopf, 1966); James P. Ronda, *Lewis and Clark among the Indians* (Lincoln: University of Nebraska Press, 1984); Paul W. Mapp, *The Elusive West and the Contest for Empire, 1713–1763* (Chapel Hill: University of North Carolina Press, 2011).

6 Donald C. Cutter, *Malaspina & Galiano: Spanish Voyages to the Northwest Coast, 1791 & 1792* (Seattle: University of Washington Press, 1991); Stephen W. Haycox et al., eds., *Enlightenment and Exploration in the North Pacific, 1741–1805* (Seattle: University of Washington Press, 1997); Alexey Postnikov and Marvin Falk, *Exploring and Mapping Alaska: The Russian America Era, 1741–1867*, trans. Lydia Black (Fairbanks: University of Alaska Press, 2015); John McAleer and Nigel Rigby, *Captain Cook and the Pacific: Art, Exploration & Empire* (New Haven, CT: Yale University Press, 2017).

7 Trevor Harvey Levere, *Science and the Canadian Arctic: A Century of Exploration, 1818–1918* (Cambridge: Cambridge University Press, 1993); Glyndwr Williams, *Arctic Labyrinth: The Quest for the Northwest Passage* (Berkeley: University of California Press, 2010); Theodore Binnema, *Enlightened Zeal: The Hudson's Bay Company and Scientific Networks, 1670–1870* (Toronto: University of Toronto Press, 2014); Philip J. Hatfield, *Lines in the Ice: Exploring the Roof of the World* (Montreal: McGill-Queen's University Press, 2016).

8 There are exceptions, including Mapp, *The Elusive West*; Cameron B. Strang, *Frontiers of Science: Imperialism and Natural Knowledge in the Gulf South Borderlands, 1500–1850* (Chapel Hill: University of North Carolina Press, 2018).

9 For the purposes of this volume, North America is defined as the region north of the present-day boundaries of Mexico. On the scope and potential of continental history, a perspective that prioritises the dizzying array of encounters throughout North America rather than deterministic national histories, see Alan Taylor, *American Colonies* (New York: Viking, 2001); Elizabeth A. Fenn, "Whither the Rest of the Continent?," *Journal of the Early Republic* 24, no. 2 (2004), 167–75; Michael Witgen, "Rethinking Colonial History as Continental History," *The William and Mary Quarterly* 69, no. 3 (2012), 527–30; Juliana Barr and Edward Countryman, eds., *Contested Spaces of Early America* (Philadelphia: University of Pennsylvania Press, 2014).

10 Like later Europeans, powerful Norse families tried to keep this valuable information secret, but Seaver speculates that Norse Greenlanders may have lent their expertise to English navigators in the late 1400s. Kirsten A. Seaver, *The Frozen Echo: Greenland and the Exploration of North America, ca. A.D. 1000–1500* (Stanford, CA: Stanford University Press, 1996), 1–2, 23–32, 255, 311.

11 Wheat, *Mapping the Transmississippi West, 1540–1861*, 1: 18–21, 1: 31–2; Strang, *Frontiers of Science*, 26–31.

12 Robert V. Bruce, *The Launching of Modern American Science, 1846–1876* (Ithaca, NY: Cornell University Press, 1988), 201–14.

13 Goetzmann, *Exploration and Empire*, 5. See also Domenic Vitiello, "Reading the Corp of Discovery Backwards: The Metropolitan Context of Lewis and Clark's Expedition," in *The Shortest and Most Convenient Route: Lewis and Clark in Context*, ed. Robert S. Cox (Philadelphia, PA: American Philosophical Society, 2004), 12–51; James P. Ronda, "Dreams and Discoveries: Exploring the American West, 1760–1815," *The William and Mary Quarterly* 46, no. 1 (1989), 145–62.

14 Julie Cruikshank, *Do Glaciers Listen?: Local Knowledge, Colonial Encounters, and Social Imagination* (Vancouver: UBC Press, 2005), 144–5.

15 Postnikov and Falk, *Exploring and Mapping Alaska*, 247–52; Ronda, *Lewis and Clark among the Indians*, 127–8, 252–3; Williams, *Arctic Labyrinth*, 217–18.

16 Cruikshank, *Do Glaciers Listen?*, 2, 4, 31, 127–47.

17 Daniel Wright Clayton, *Islands of Truth: The Imperial Fashioning of Vancouver Island* (Vancouver: UBC Press, 2000).

18 David Bernstein, *How the West Was Drawn: Mapping, Indians, and the Construction of the Trans-Mississippi West* (Lincoln: University of Nebraska Press, 2018).

19 Mapp, *The Elusive West*.

20 Benjamin Schmidt, "Mapping an Empire: Cartographic and Colonial Rivalry in Seventeenth-Century Dutch and English North America," *The William and Mary Quarterly* 54, no. 3 (1997), 549–78; Elizabeth A. Sutton, *Capitalism and Cartography in the Dutch Golden Age* (Chicago: University of Chicago Press, 2015).

21 Christian J. Koot, *A Biography of a Map in Motion: Augustine Herrman's Chesapeake* (New York: New York University Press, 2018).

22 Joyce E. Chaplin, *The First Scientific American: Benjamin Franklin and The Pursuit of Genius* (New York: Basic Books, 2006), 177–200, 289.

23 S. Max Edelson, *The New Map of Empire: How Britain Imagined America before Independence* (Cambridge, MA: Harvard University Press, 2017), 4. Indeed, the ways US citizens read maps and other geographic productions could both inspire and alter their visions of expansion. See Martin Brückner, *The Geographic Revolution in Early America: Maps, Literacy, and National Identity* (Chapel Hill: University of North Carolina Press, 2006), 204–63; Michele Currie Navakas, *Liquid Landscape: Geography and Settlement at the Edge of Early America* (Philadelphia: University of Pennsylvania Press, 2017).

24 Parsons, *A Not-So-New World*, 1–96.

25 Anya Zilberstein, *A Temperate Empire: Making Climate Change in Early America* (New York: Oxford University Press, 2016), 148–73.

26 Shannon Lee Dawdy, *Building the Devil's Empire: French Colonial New Orleans* (Chicago: University of Chicago Press, 2008), 27.

27 Suzanne Elizabeth Zeller, *Inventing Canada: Early Victorian Science and the Idea of a Transcontinental Nation* (Toronto: University of Toronto Press, 1987).

28 Michael Adas, *Dominance by Design: Technological Imperatives and America's Civilizing Mission* (Cambridge, MA: Belknap Press of Harvard University Press, 2006), 74.

29 Chaplin, *Subject Matter*.

30 Guillaume Aubert, "'The Blood of France': Race and Purity of Blood in the French Atlantic World," *The William and Mary Quarterly* 61, no. 3 (2004), 448, 472, 476.

31 The classic study of this shift in perception among Anglo-Americans is Winthrop D. Jordan, *White over Black: American Attitudes toward the Negro, 1550–1812* (Chapel Hill: University of North Carolina Press, 1968).

32 Ann Fabian, *The Skull Collectors: Race, Science, and America's Unburied Dead* (Chicago: The University of Chicago Press, 2010).

33 Sharon Block, *Colonial Complexions: Race and Bodies in Eighteenth-Century America* (Philadelphia: University of Pennsylvania Press, 2018).

34 David J. Weber, *Bárbaros: Spaniards and Their Savages in the Age of Enlightenment* (New Haven, CT: Yale University Press, 2005), 38, 49. On the broader scientific work of Malaspina's expedition to Alaska and California, see Cutter, *Malaspina & Galiano*.

35 Quoted in Travis Hudson and Craig D. Bates, *Treasures from Native California: The Legacy of Russian Exploration*, ed. Thomas C. Blackburn and John R. Johnson (Walnut Creek, CA: Left Coast Press, 2015), 12; A.I. Alekseev, *The Odyssey of a Russian Scientist: I.G. Voznesenskii in Alaska, California and Siberia, 1839–1849*, ed. Richard Austin Pierce, trans. Wilma C. Follette (Kingston, Ontario: Limestone Press, 1987), esp. 1–61; Morgan B. Sherwood, "Science in Russian America, 1741 to 1865," *The Pacific Northwest Quarterly* 58, no. 1 (1967), 35. Similarly, the Russian missionary and ethnographer Ioann Veniaminov consistently praised the intellectual abilities of Native peoples during the 1820s and 30s. Postnikov and Falk, *Exploring and Mapping Alaska*, 223–8.

36 Sean P. Harvey, *Native Tongues: Colonialism and Race from Encounter to the Reservation* (Cambridge, MA: Harvard University Press, 2015); Cameron Strang, "Scientific Instructions and Native American

Linguistics in the Imperial United States: The Department of War's 1826 Vocabulary," *Journal of the Early Republic* 37, no. 3 (2017), 399–427.

37 Frank Norall, *Bourgmont, Explorer of the Missouri, 1698–1725* (Lincoln: University of Nebraska Press, 1988); A.P. Nasatir, *Before Lewis and Clark: Documents Illustrating the History of Missouri, 1785–1804* (Lincoln, NE: Bison Book, 1990), 1: 1–115; Goetzmann, *Exploration and Empire*, 3–180. Similarly, French, Spanish, Russian, and Anglo missionaries engaged in many of these same scientific pursuits, including exploration, ethnography, and natural history. See Louis Nicolas et al., *The Codex Canadensis and the Writings of Louis Nicolas: The Natural History of the New World* (Montreal: McGill-Queen's University Press, 2011); Postnikov and Falk, *Exploring and Mapping Alaska*, 212–28; Harvey, *Native Tongues*, 28, 113, 119, 122, 220; Robert A. Kittle, *Franciscan Frontiersmen: How Three Adventurers Charted the West* (Norman: University of Oklahoma Press, 2017).

38 W. Raymond Wood, *Prologue to Lewis and Clark: The Mackay and Evans Expedition* (Norman: University of Oklahoma Press, 2003).

39 Postnikov and Falk, *Exploring and Mapping Alaska*, 77–93, 189–90.

40 Wheat, *Mapping the Transmississippi West, 1540–1861*, 1: 178–80; Stearns, *Science in the British Colonies of America*, 247–57; Levere, *Science and the Canadian Arctic*, 98–120; Stuart Houston, Tim Ball, and Mary Houston, *Eighteenth-Century Naturalists of Hudson Bay* (Montreal: McGill-Queen's University Press, 2003).

41 Sherwood, "Science in Russian America," 4, 36, 38; Postnikov and Falk, *Exploring and Mapping Alaska*, 211–12, 294–9, 460.

42 Binnema, *Enlightened Zeal*, 13.

43 Stearns, *Science in the British Colonies of America*, 4.

44 Parrish, *American Curiosity*. See also James Delbourgo, *A Most Amazing Scene of Wonders: Electricity and Enlightenment in Early America* (Cambridge, MA: Harvard University Press, 2006), 20–1, 143; Kathleen S. Murphy, "To Make Florida Answer to Its Name: John Ellis, Bernard Romans and the Atlantic Science of British West Florida," *The British Journal for the History of Science* 47, no. 1 (2014), 43–65.

45 Parsons, *A Not-So-New World*, 125–51.

46 Peter J. Kastor, *William Clark's World: Describing America in an Age of Unknowns* (New Haven, CT: Yale University Press, 2011); Andrew J Lewis, *A Democracy of Facts: Natural History in the Early Republic* (Philadelphia: University of Pennsylvania Press, 2011).

47 Strang, *Frontiers of Science*, 34.

48 Alekseev, *The Odyssey of a Russian Scientist*, 4–5, 31; Postnikov and Falk, *Exploring and Mapping Alaska*, 296–309; Zeller, *Inventing Canada*, 5.

24

SCIENCE, THE UNITED STATES, AND LATIN AMERICA

Megan Raby

Can historians productively approach US–Latin American scientific relations through the lens of science and empire? This relationship holds an ambiguous position within the broader historiography of science, imperialism, and colonialism, where the overwhelming focus has been on European empires. While the historiography of Iberian colonial science in Latin America is robust, the relationship between science and US imperialism in the region has long remained at the margins. For many years, historians of US science rarely strayed beyond the bounds of the continental United States. US imperialism has been invisible, or at least peripheral, to the interests of most scholars working on the history of "American" science. In contrast, historians of Latin American science more readily confronted the historical influence of US hegemony on the development of national scientific communities. Yet, insights have long been slow to cross the boundaries between these national and regional historiographies.

Emerging from the interstices of these fields, a growing body of scholarship is beginning to explore the relationships among US and Latin American scientific communities and their connections to projections of political and economic power. More than simply filling a historiographic gap, histories of science in the context of US–Latin American relations can contribute to a deeper theorisation of science and empire. A closer examination of the shared and contested ground between US and Latin American science illuminates the dynamics of power in scientific practice and the production of knowledge across a range of hegemonic and imperial forms. New works highlight imperial transitions, transnational networks and identities, and the imbrication of nationalism, internationalism, and imperialism. In this way, a "science and empire" framing has the potential to connect US and Latin American historiographies of science, while in the process also raising fundamental questions of relevance to historians of science, imperialism, and colonialism in other regions of the globe.

US science and empire in Latin America

The historiography of US–Latin American scientific relations is not a single coherent entity. Distinct strands have developed in response to the concerns of differing national and regional historiographies—a situation that mirrors the persistent gap between the fields of US diplomatic history and Latin American history.[1] While new works have begun to weave these strands together, a closer look at their origins can clarify the central questions that still animate this heterogeneous field. One major strand, the historiography of US science and empire, has situated US–Latin American scientific relations within the broad historical sweep of US imperialism.

The historiography of US science and empire began, however, with an orientation toward Anglo-American scientific relations. Most works of the 1960s and 1970s focused on US scientists' historical role as colonial junior partners to European science, rather than as participants in the projection of imperial power across the continent and hemisphere. This scholarship both influenced and was influenced by George Basalla's diffusionist model, and tended to approach US science as a case study of a successful transition from a state of peripheral and dependent "colonial science" to one of "independent science."[2] The question was: how had US scientists outgrown their institutional and intellectual reliance on a European metropole (a stage lasting long after political independence), to finally achieve an equal footing within the international scientific community? Implicitly and explicitly, this literature suggested there could be lessons for contemporary postcolonial nations in a closer examination of the successful transcendence of US science from its dependent status. Yet it was also immersed in Cold War and exceptionalist discourses—by eschewing analysis of US imperialism and emphasising the relationship between scientific development and democracy, capitalist enterprise, and US cultural values. This orientation left little room for engagement with Latin America's shared history of colonisation and independence.

Overwhelmingly, this early literature positioned the scientific activities of US settler colonists in relation to their cultural subordination to Europe, yet a few significant studies examined the role of science in the appropriation of land from Native American and Spanish-speaking peoples. In particular, William Goetzmann's work on US explorers, scientists, and army engineers and "the winning of the American West" opened up scholarly examination of how the production of knowledge had enabled the US to produce an empire.[3] Although writing in an unapologetically exceptionalist and celebratory mode, Goetzmann argued that scientific activities were not only integral to the consolidation of western territory, but also performed significant ideological and cultural work. A key example was the Mexican Boundary Survey, which not only demarcated the US–Mexico border following the Mexican–American War, but also produced a vast quantity of botanical, zoological, geological, and anthropological collections and publications. These shaped both the US popular imagination and public policy toward the West. As the Smithsonian and newly formed US Geological Survey inventoried the productions of lands wrested from Mexico, the US federal government also became a major sponsor of science for the first time.[4] While Goetzmann's account failed to consider Mexican perspectives and scientific activities—including significant cooperation and conflict between the Mexican and US survey commissions—US–Mexican relations nevertheless played a pivotal role.[5] This work positioned imperial expansion as central to the narrative of US science, a catalyst for the rise of national scientific institutions and culture.

By the 1980s, historians began to turn away from exceptionalism by placing US science in comparative context with settler colonial societies around the world. This could have opened dialogue with Latin American history, but the focus remained on science in Anglo settler colonies.[6] Scholars debated whether the emerging field should be called the history of "American science"—emphasising the development of a unique national scientific style—or, more neutrally, "science in America."[7] But both framings elided Latin America and the potential for historical attention to North–South connections and comparisons.

Not until the 1990s and early 2000s did new perspectives on US science, imperialism, and culture begin to emerge. These increasingly engaged with the history of US colonialism beyond continental North America. In his explorations of the place of the biological sciences within

US national culture, Philip J. Pauly took particular care to situate the intellectual and policy debates of life scientists within broader imperialist and anti-imperialist discourse at the turn of the twentieth century.[8] For example, Pauly examined conflict within the US Department of Agriculture as scientists grappled with the consequences of the 1898 Spanish–American War and annexation of Hawai'i. On one hand, ecological "cosmopolitans" sought to increase the importation of useful plants and animals from around the globe, including to and from the newly acquired former Spanish colonies of Puerto Rico, Cuba, the Philippines, and Guam in the Caribbean and Pacific. Ecological "nativists," on the other hand, sought restrictions on the circulation of species, fearing an invasion of exotics within the United States and its ecologically vulnerable "island dependencies."[9] US biologists not only responded to very real environmental changes—such as the devastating spread of mongoose in the Caribbean—but also expressed concerns about non-native species that reflected broader US public anxieties about empire and the incorporation of populations of non-white and Spanish-speaking people into the national body politic.

Historians of science also began to expand the view beyond the US federal government, and in the process they drew closer analytic connections to the broader, growing field of science and empire. New work explored how scientists' private institutional "empire building" and political empire intersected after 1898 in ways that could be compared and contrasted with the British case. With only limited federal support for science in the new colonies, scientific institutions in competing US metropolitan centres often took the lead—as in the New York Academy of Sciences' extensive research activities throughout Puerto Rico.[10] The international agricultural, medical, and scientific activities of US-based philanthropic foundations, especially the Rockefeller Foundation, also became the focus of more careful historical analysis. This began to reveal the complexity of local engagements. Most significantly, the contributors to *Missionaries of Science: The Rockefeller Foundation and Latin America*, edited by Marcos Cueto, challenged a view that framed US philanthropic efforts in international development as simply an informal arm of US imperialism. Deborah Fitzgerald demonstrated, for example, how the foundation's attempts to export US agricultural models to Mexico fizzled out when they failed to enrol local support.[11]

More nuanced and critical in tone than earlier works, this scholarship as a whole still centred on the US scientific community and used US-based archives. With the notable and significant exception of Cueto's volume, cross-fertilisation with Latin American scholarship remained limited. Historians of US science approached the question of empire armed largely with the bread-and-butter concerns of the history of science field—such as professionalisation, institution building, patronage, and the relationship between expert knowledge and public culture.

A quite different approach to science, empire, and US–Latin American relations was brewing. In the wake of the First Gulf War, a new cultural historiography of US empire emerged—as signalled by the publication of two now-classic volumes, Amy Kaplan and Donald E. Pease's *Cultures of United States Imperialism* and Gilbert M. Joseph, Catherine LeGrand, and Ricardo Donato Salvatore's *Close Encounters of Empire: Writing the Cultural History of US–Latin American Relations*.[12] In these volumes and works that followed, the production of scientific knowledge stood as just one of many cultural practices of US imperialism. Salvatore, for example, outlined how "businessmen, teachers, social reformers, scientists, missionaries, and diplomats" all contributed to a collective "enterprise of knowledge" about South America that functioned to justify the expansion of US engagement and intervention beyond territories of formal political control.[13] Others engaged with science and other forms of knowledge production as related to public health, food politics, reproduction and population control, and racial ideologies.[14] Informed by postcolonial studies, this new scholarship emphasised the ways in which knowledge practices inscribed Latin American difference and reinforced gendered and racialised hierarchies of power.

By the early 2000s, then, scholarship coming both from within the discipline of the history of science and from the cultural history of US empire had thoroughly challenged exceptionalist views of US science. Whether in a range of colonial and neocolonial contexts or within the sphere of domestic imperialist discourse, this scholarship proved that US science could be productively analysed in much the same way that scholars approached science in the European empires. A literature was emerging that revealed not only that imperialism was a significant shaping factor in the development of US science, but also that knowledge production played a fundamental role in the projection of US power in Latin America.

While this literature centred on US scientists and institutions, scholars paid increasing attention to the historical agency of Latin Americans—as scientists and politicians involved in international development, as patients and research subjects contesting US hegemony, and as participants in discourse about knowledge and sovereignty. It is not by accident that two of the most influential volumes were edited or co-edited by Latin American historians (Cueto, from Peru, and Salvatore, from Argentina). This new turn came with the infusion of scholarship based in Latin America.

Latin American science and the limits of US hegemony

Not surprisingly, the history of US–Latin American relations looked quite different when approached from the South. For scholars from Latin American countries and whose research centred on archives within the region, new sets of historical actors emerged from the margins of dominant narratives, ultimately to challenge and reframe them. Perhaps even more fundamentally, this scholarship was motivated by a different set of concerns, deeply connected to questions of the relationship between science and technology, economic dependency, and national development. With traditionally much closer ties to the social sciences and policy circles, Latin American scholarship has often been propelled by an interest in understanding how histories of science might inform efforts to develop national scientific capacities.[15] Studies of US–Latin American encounters thus emerged as scholars pursued the construction of broader narratives about the formation of national scientific communities and institutions. Yet, while the US historiography explored the causes and consequences of the rise of national independence and international dominance in science, Latin American authors often grappled with the persistence or re-emergence of dependent and colonial relationships.

Despite these major differences, in many ways the two historiographies also developed in parallel as they responded to broader intellectual trends and Cold War geopolitics. As in the US historiography, early Latin American histories of science sought a place within existing Eurocentric narratives.[16] Beginning in the 1950s and 1960s, scholars sought to understand how "Western" science had diffused to Latin America and worked to identify the—seemingly all too rare—contributions of Latin Americans to global science. This work likewise drew heavily on post-World War II theories of economic and technological development, including Basalla's effort to model the international spread of science. In this mode, US models and institutions at times appeared alongside those from Europe as a scientific centre that set the standards by which progress in the periphery was measured.

By the 1970s, however, the ascendance of world systems and dependency theory in the social sciences in Latin America opened up ways to undermine such models—even as these frameworks continued to position the region as peripheral in the global arena of knowledge production. Scientists and policy makers began to articulate critique of what they identified as modern scientific colonialism or imperialism, where foreign, and especially US, priorities took precedence over local and national interests. Such critique took many forms. For example, Gerardo Budowski, a Venezuelan ecologist and Director General of the International Union for the Conservation of Nature, spoke out in a widely circulated piece against the way "Scientists from developed countries descend upon developing countries to collect, 'protect' or capture and take home flora, fauna and professional prestige," without acknowledging local assistance or sharing the benefits of their research. He argued, "every country has the right to utilize and present to the world its own scientific resources."[17] In contrast, the Argentine mathematician and social theorist Oscar Varsavsky focused on the cultural hegemony of Northern science, expressing frustration that Latin American students were drawn to pursue training abroad in high-prestige research areas, neglecting research problems he considered to be of more direct social utility at home. From this point of view, acquiescence to the financial and cultural power of Northern scientific institutions had left Latin American countries without the infrastructure to produce expertise relevant to their own needs.[18] Until they developed approaches appropriate to their specific national contexts, science and technology would remain an import product.[19]

For their part, historians responded by working to uncover the particular conditions that had either fostered or impeded the growth of scientific institutions and attitudes within Latin America's diverse national contexts.[20] At times, this literature could still be quite negative in orientation, focusing on questions of why national scientific institutions had failed to emerge or persist, or why, when successful, they rarely fulfilled expectations for bringing technological and material progress to society as a whole. Yet, beginning in the 1980s, as the historiography of science broadly shifted toward cultural approaches, historians of science in Latin America increasingly began to explore how "peripheral science" might instead be understood within its own specific context and evaluated on its own terms.[21]

A more complex view of the international power dynamics of science was possible from this point of view, one in which both US and Latin American participants exerted their agency to shape historical events and scientific research priorities. A pathbreaking work in this respect was Marcos Cueto's study of the emergence and institutionalisation of the field of high-altitude physiology in Peru, *Excelencia científica en la periferia*.[22] Cueto demonstrated how Peruvian biologists were able to successfully develop their own original biomedical research programme by marshalling the unique conditions of Andean nature and the support of diverse local and foreign actors—including US-based funders like the Rockefeller Foundation, National Institutes of Health, and Air Force. In a similar vein, Nancy Leys Stepan examined how Latin Americans engaged the international eugenics movement. Eugenics, Stepan noted, "was more than a set of national programs embedded in national debates; it was also a part of international relations."[23] While US eugenicists attempted to use international organisations like the Pan American Union to push an extreme and racialised eugenic agenda, Latin American scientists and physicians forcefully rejected their ideas and policies, opting instead for a more moderate view of eugenics—one compatible with social medicine and based on less rigid ideas about race. While still attentive to disparities of power and resources, this new scholarship delved into complex cases that could not be described as mere "imported science" or top-down cultural imperialism.

In the 2000s, historians built on this work and developed new theoretical tools to interrogate the dynamics of power and knowledge in a wider range of encounters. Moving beyond the dichotomy of local versus Western science, Stuart McCook proposed the concept of "creole science" to support a more nuanced examination of how knowledge is transformed as it takes root in new places.[24] Studying the role of agricultural science in nation building in a variety of national and colonial contexts, McCook showed how a diverse array of scientists selectively adopted and adapted institutional models from the United States to suit the needs of agriculture in the Caribbean—often amid rapidly shifting economic, political, and natural environments. Other scholars analysed the tensions inherent in US–Latin American scientific exchanges and collaborations. Anne-Emanuelle Birn, for example, revealed an unlikely "marriage of con-venience" in the three-decades-long relationship between the Rockefeller Foundation and Revolutionary Mexico.[25] Despite mutual suspicion, Mexican politicians and the foundation found common cause in hookworm eradication. Similarly, Camilo Quintero Toro explored a case of "imperialism by invitation," as Colombian naturalists welcomed US scientific interest in Colombia's diverse bird species.[26] While US ornithologists benefited from their privileged ability to study Colombian birds, Colombian scientists leveraged US involvement to their own ends in order to garner a larger place for nature in Colombia's national identity. In each of these cases, both sides found ways to make the interaction serve their own interests, even amid imbalances of power and differing priorities.

Such studies helped to illuminate not only the specific routes by which science, medicine, and agriculture travelled across national borders, but also demonstrated the multiple ways that nationalism and imperialism could be entangled in US–Latin American encounters. These works were also frequently multi-archival and intent on recovering the role of a broader range of actors—from across national borders and beyond the realm of elite scientists and politicians. The door was finally open for far more substantial integration of Latin American and US scholarship.

Crossing borders: science and power

Over the past decade, scholars have shown that US–Latin American scientific relations were never a one-way street. The historiographies of both US and Latin American science that emerged in the mid twentieth century were once wedded to linear models of scientific diffusion and development, but new work at the intersection of these fields has moved well beyond core–periphery models. At the same time, this scholarship has not lost sight of power. Indeed, because it has had to confront the complexities of sovereignty, hegemony, and identity across such a diverse region and field of encounters, this scholarship has much to offer historians of science more generally in theorising power, knowledge, and connection.[27]

First, US–Latin American scientific encounters shaped both sides. Alfred McCoy and Francisco Scarano's landmark *Colonial Crucible* made this clear by examining the ways that US imperialism transformed not only the territories over which the US ruled, but also the US state itself.[28] Significantly, this volume integrated projects of colonial knowledge produc-tion and expert management—including in medicine, botany, forestry, education, and racial categorisation—alongside work on legal regimes, the military, and other more traditional topics in diplomatic history. These could be closely intertwined, as Mariola Espinosa has shown, in the ways that public health and humanitarian concerns served to justify the US invasion of Cuba in 1898.[29] At the same time, administering an "archipelago" of colonies not only contributed to the emergence of the national security state, but also the environmental man-agement state. Paul Sutter explored how US entomologists and sanitarians' efforts to classify and control mosquitos and people in the landscape of the Panama Canal Zone during the early twentieth century "reverberated throughout the United States."[30] Methods developed in a colonial context were imported back home as public health became increasingly federalised. While McCoy and Scarano's volume focused on territories of US rule and occupation (in Latin America these included Cuba, Puerto Rico, Haiti, the Dominican Republic, Virgin Islands, and Panama Canal Zone), it also problematised placing sharp divisions between sites of formal and informal colonialism. Building on this approach, more recent work has examined how earth and environmental scientists both took advantage of and worked to extend US hegemony across a range of territorial contexts, in the process reshaping not only environments and ideas but also national policy.[31] Science also played a key role in US cultural diplomacy in Latin America during the Cold War.[32] Yet, as a wealth of new scholarship has shown for twentieth-century Argentina, Brazil, and Chile, US expertise and development models could be appropriated to suit the agenda of neoliberal states and modernising elites.[33] To bring the instrumental and ideological power of science into focus, future scholarship must continue to blur the foreign and domestic.

Second, relationships of power have been far more complex and dynamic than once presumed. Power and knowledge might reside in unexpected places or take unexpected forms. In her work on wild yams and synthetic hormone production in Mexico, for example, Gabriela Soto Laveaga has shown how *campesinos'* appropriation of chemical language and involvement in the pharmaceutical industry challenged the power of US and transnational corporations and celebrated Mexican science and chemists.[34] Networks of power and knowledge are also "lumpy" and uneven; discontinuities are as important as continuities and flow.[35] In some disciplinary contexts, US scientists working in Latin America and the Caribbean insulated themselves from local scientific communities. Field ecologists often worked at research stations that, both by circumstance and design, largely excluded Latin American scientists. Only as the 1960s and 1970s brought a tide of nationalism and anticolonialism did cooperation become a political necessity.[36] In the context of agricultural research, however, US and Latin American scientists more readily formed shared networks.[37] At times, cooperative programmes might simply be the new guise in which old "layers of colonialism" were reconstituted, as Nicolás Cuvi has argued in the case of the Cinchona Program during World War II. While nominally engaged in a project of scientific collaboration, US participants withheld key information and ensured that Andean countries remained sites of extraction rather than full partners in the research and manufacture of quinine. The United States thereby joined a long line of imperial powers in exploiting *Cinchona* and Andean knowledge.[38] Yet, in other contexts, agricultural researchers and reformers could meet on relatively equal footing, as Tore Olsson has argued in examining exchanges between the US and Mexico during the 1930s and 1940s. The shared agrarian ideals of the Mexican Revolution and US New Deal provided common ground, at least for a time.[39] And of course, science and technology has also been involved in some Latin American countries' efforts to exert their own imperial ambitions in the Pacific and at their frontiers—contexts in which US actors might not be the central players, yet may still appear in a supporting role.[40]

Finally, although "US–Latin American scientific relations" suggests two-way encounters, scholarship has increasingly moved beyond a bilateral framing. Historians have begun to recognise the interaction of multiple state and non-state actors, including exchanges in which the US might be just one among many players.[41] Such a perspective has revealed South–North connections from the earliest days of the American "sister republics," providing an important reminder that US actors were not always the most dominant brokers of scientific knowledge.[42] Cameron Strang, for example, has explored the role of knowledge in US expansion into the Gulf South—a region in which the US vied for power not only with the Spanish Empire, but also the French, British, and multiple Indigenous groups. Nor did science itself have supremacy in the field of natural knowledge, where naturalists, healers, slaves, surveyors, and sages alike gathered observations in borderlands that were both epistemologically and politically unstable.[43] Identities, too, have been shown to be less stable than once presumed. Recent scholarship examines how individual scientists crafted transnational identities or "hybrid nationalities," navigating entangled allegiances not only to "their countries of birth and the nations where they practiced but also the international communities that shaped their professional identities."[44] Future work can push these insights even further by interrogating the dynamics of power and knowledge as mediated by global social movements, non-state actors, and transnational institutions, such as NGOs and multinational corporations.

Taken together, it is clear that the history of US–Latin American scientific relations has begun to outgrow the confines of long-standing national and regional historiographic traditions. Ultimately, this field can problematise the once taken-for-granted categories of nation, empire, region, and even science itself. Such work is crucial for the history of science as a whole because it provokes us to rethink the forms, directionality, and loci of knowledge and power, and to envision their myriad entanglements.

Notes

1 Max Paul Friedman, "Retiring the Puppets, Bringing Latin America Back In: Recent Scholarship on United States–Latin American Relations," *Diplomatic History* 27, no. 5 (2003): 621–36.

2 George Basalla, "The Spread of Western Science," *Science* 156, no. 3775 (1967): 611–22. For a critique, see Roy MacLeod, "On Visiting the 'Moving Metropolis': Reflections on the Architecture of Imperial Science," in *Scientific Colonialism: A Cross-Cultural Comparison*, ed. Nathan Reingold, and Marc Rothenberg (Washington, DC: Smithsonian Institution Press, 1987).

3 William H. Goetzmann, *Army Exploration in the American West, 1803–1863* (New Haven, CT: Yale University Press, 1959); William H. Goetzmann, *Exploration and Empire: The Explorer and the Scientist in the Winning of the American West* (New York: Knopf, 1966).

4 See also A. Hunter Dupree, *Science in the Federal Government: A History of Policies and Activities to 1940* (Cambridge, MA: Belknap Press of Harvard University Press, 1957).

5 Paula Rebert, *La Gran Línea: Mapping the United States–Mexico Boundary, 1849–1857* (Austin: University of Texas Press, 2001); Raymond B. Craib, *Cartographic Mexico: A History of State Fixations and Fugitive Landscapes* (Durham, NC: Duke University Press, 2004); Cameron B. Strang, *Frontiers of Science: Imperialism and Natural Knowledge in the Gulf South Borderlands, 1500–1850* (Chapel Hill: Omohundro Institute of Early American History and Culture and the University of North Carolina Press, 2018), 330–8.

6 For example, Nathan Reingold, *The Sciences in the American Context* (Washington, DC: Smithsonian Institution Press, 1979); Nathan Reingold and Marc Rothenberg, *Scientific Colonialism: A Cross-Cultural Comparison* (Washington, DC: Smithsonian Institution Press, 1987).

7 Clark A. Elliott, "Forum for the History of Science in America: Identity and Organization," *Isis* 90 (1999): 238–9.

8 Philip J. Pauly, "The Beauty and Menace of the Japanese Cherry Trees: Conflicting Visions of American Ecological Independence," *Isis* 87, no. 1 (1996): 51–73; Philip J. Pauly, *Biologists and the Promise of American Life: From Meriwether Lewis to Alfred Kinsey* (Princeton, NJ: Princeton University Press, 2000); Philip J. Pauly, *Fruits and Plains: The Horticultural Transformation of America* (Cambridge, MA: Harvard University Press, 2007).

9 Pauly, *Biologists and the Promise of American Life*, 79.

10 Simon Baatz, "Imperial Science and Metropolitan Ambition: The Scientific Survey of Puerto Rico, 1913–1934," *Annals of the New York Academy of Sciences* 776, no. 1 (1996): 1–16; Peter Philip Mickulas, *Britton's Botanical Empire: The New York Botanical Garden and American Botany, 1888–1929* (Bronx: New York Botanical Garden, 2007); Sharon Kingsland, *The Evolution of American Ecology, 1890–2000* (Baltimore, MD: Johns Hopkins University Press, 2005).

11 Marcos Cueto, ed. *Missionaries of Science: The Rockefeller Foundation and Latin America* (Bloomington: Indiana University Press, 1994); Deborah Fitzgerald, "Exporting American Agriculture: The Rockefeller Foundation in Mexico, 1943–53," *Social Studies of Science* 16, no. 3 (1986): 457–83. See also Anne-Emanuelle Birn, *Marriage of Convenience: Rockefeller International Health and Revolutionary Mexico* (Rochester, NY: University of Rochester Press, 2006); Steven Paul Palmer, *Launching Global Health: The Caribbean Odyssey of the Rockefeller Foundation* (Ann Arbor: University of Michigan Press, 2010).

12 Amy Kaplan and Donald E. Pease, eds. *Cultures of United States Imperialism* (Durham, NC: Duke University Press, 1993); Gilbert M. Joseph, Catherine LeGrand, and Ricardo Donato Salvatore,

eds., *Close Encounters of Empire: Writing the Cultural History of US–Latin American Relations* (Durham, NC: Duke University Press, 1998).

13 Ricardo D. Salvatore, "The Enterprise of Knowledge: Representational Machines of Informal Empire," in Joseph, LeGrand, and Salvatore, *Close Encounters of Empire*, 94.

14 Steven Palmer, "Central American Encounters With Rockefeller Public Health, 1914–1921," in Joseph, LeGrand, and Salvatore, *Close Encounters of Empire*, 311–32; Lauren Derby, "Gringo Chickens With Worms: Food and Nationalism in the Dominican Republic," in Joseph, LeGrand, and Salvatore, *Close Encounters of Empire*, 451–96; Laura Briggs, *Reproducing Empire: Race, Sex, Science, and US Imperialism in Puerto Rico* (Berkeley: University of California Press, 2002). Outside of Latin America, some of the most analytically powerful work addressed US rule in the Philippines as it intersected with racial ideology and medicine: Paul A. Kramer, *The Blood of Government: Race, Empire, the United States, & the Philippines* (Chapel Hill: University of North Carolina Press, 2006); Warwick Anderson, *Colonial Pathologies: American Tropical Medicine, Race, and Hygiene in the Philippines* (Durham, NC: Duke University Press, 2006).

15 For more on this, including the emergence of institutions supporting research in science and technology studies and policy, see Antonio Arellano Hernández, Rigas Arvanitis, and Dominique Vinck, "Circulación y Vinculación Mundial de Conocimientos. Elementos de la Antropología de los Conocimientos en y Sobre América Latina," *Redes* 18, no. 34 (2012): 15–23; Pablo Kreimer and Hebe Vessuri, "Latin American Science, Technology, and Society: A Historical and Reflexive Approach," *Tapuya: Latin American Science, Technology and Society* 1 (2018): 21–2.

16 Fernando de Azevedo, *As Ciências No Brasil* (São Paulo: Edições Melhoramentos, 1955); Elí de Gortari, *La Ciencia en la Historia de México* (Mexico: Fondo de Cultura Económica, 1963); José López Sánchez, *Tomás Romay y el Origen de la Ciencia en Cuba* (Havana: Academia de Ciencias, 1964). For a more detailed critique of this literature, see the introductory chapter of Juan José Saldaña, ed. *Historia Social de las Ciencias en América Latina* (Mexico: Coordinación de Humanidades, 1996). In translation, see Juan José Saldaña, ed. *Science in Latin America: A History* (Austin: University of Texas Press, 2006).

17 Gerardo Budowski, "Scientific Imperialism," *Science and Public Policy* 2, no. 8 (1975): 354. On tensions between US and Latin American biologists, see Chapter 5 of Megan Raby, *American Tropics: The Caribbean Roots of Biodiversity Science* (Chapel Hill: University of North Carolina Press, 2017).

18 See for example Oscar Varsavsky, "Scientific Colonialism in the Hard Sciences," *The American Behavioral Scientist* 10 (1967): 22; Oscar Varsavsky, *Ciencia, Política y Cientificismo* (Buenos Aires: Centro Editor de América Latina, 1969). On how such critiques could be used to cut funding for basic science, see Alexis De Greiff A. and Mauricio Nieto Olarte, "What We Still Do Not Know About South–North Technoscientific Exchange: North-Centrism, Scientific Diffusion, and the Social Studies of Science," in *The Historiography of Contemporary Science, Technology, and Medicine: Writing Recent Science*, ed. Ronald Edmund Doel and Thomas Söderqvist (New York: Routledge, 2006), 247.

19 For a broader historiographic overview, see Eden Medina et al., *Beyond Imported Magic: Essays on Science, Technology, and Society in Latin America* (Cambridge, MA: MIT Press, 2014). On dependency theory and technology policy in Chile, see Eden Medina, *Cybernetic Revolutionaries: Technology and Politics in Allende's Chile* (Cambridge, MA: MIT Press, 2011).

20 Nancy Stepan, *Beginnings of Brazilian Science: Oswaldo Cruz, Medical Research and Policy, 1890–1920* (New York: Science History Publications, 1976); Simon Schwartzman, *Formação da Comunidade Científica No Brasil* (Rio de Janeiro: Financiadora de Estudos e Projetos, 1979); Elena Díaz, Yolanda Texera, and Hebe M.C. Vessuri, *La Ciencia Periférica: Ciencia y Sociedad en Venezuela* (Caracas, Venezuela: Monte Avila Editores, 1983); Emanuel Adler, *The Power of Ideology: The Quest for Technological Autonomy in Argentina and Brazil* (Berkeley: University of California Press, 1987); Hebe M.C. Vessuri ed. *Las Instituciones Científicas en la Historia de la Ciencia en Venezuela* (Caracas, Venezuela: Fondo Editorial Acta Científica Venezolana, 1987).

21 A key work setting forth this perspective is Saldaña, *Historia Social de las Ciencias en América Latina*; Saldaña, *Science in Latin America*.

22 Marcos Cueto, *Excelencia Científica en la Periferia: Actividades Científicas e Investigación Biomédica en el Perú 1890–1950* (Lima: GRADE, 1989). See also Marcos Cueto, "Andean Biology in Peru: Scientific Styles on the Periphery," *Isis* 80, no. 4 (1989): 640–58.

23 Nancy Leys Stepan, *"The Hour of Eugenics": Race, Gender, and Nation in Latin America* (Ithaca, NY: Cornell University Press, 1991), 171.

24 Stuart McCook, *States of Nature: Science, Agriculture, and Environment in the Spanish Caribbean, 1760–1940* (Austin: University of Texas Press, 2002).

25 Birn, *Marriage of Convenience*.

26 Camilo Quintero Toro, *Birds of Empire, Birds of Nation: A History of Science, Economy, and Conservation in United States–Colombia Relations* (Bogota: Universidad de los Andes, 2012).

27 In framing imperial histories in terms of power and connection, I draw on Paul A. Kramer, "Power and Connection: Imperial Histories of the United States in the World," *The American Historical Review* 116, no. 5 (2011): 1348–91.

28 Alfred W. McCoy and Francisco A. Scarano, eds. *Colonial Crucible: Empire in the Making of the Modern American State* (Madison: University of Wisconsin Press, 2009).

29 Mariola Espinosa, "A Fever for Empire: US Disease Eradication in Cuba as Colonial Public Health," in McCoy and Scarano, *Colonial Crucible*, 288–96; Mariola Espinosa, *Epidemic Invasions: Yellow Fever and the Limits of Cuban Independence, 1878–1930* (Chicago: University of Chicago Press, 2009).

30 Paul S. Sutter, "Tropical Conquest and the Rise of the Environmental Management State: The Case of US Sanitary Efforts in Panama," in McCoy and Scarano, *Colonial Crucible*, 326. On environmental management, science, and empire in the Canal Zone, see also Paul Sutter, "Nature's Agents or Agents of Empire? Entomological Workers and Environmental Change During the Construction of the Panama Canal," *Isis* 98, no. 4 (2007): 724–54; Ashley Carse et al., "Panama Canal Forum: From the Conquest of Nature to the Construction of New Ecologies," *Environmental History* 21, no. 2 (2016): 206–87; Ashley Carse, *Beyond the Big Ditch: Politics, Ecology, and Infrastructure at the Panama Canal* (Cambridge, MA: MIT Press, 2014); Megan Raby, "Ark and Archive: Making a Place for Long-Term Research on Barro Colorado Island, Panama," *Isis* 106, no. 4 (2015): 798–824; Marixa Lasso, *Erased: The Untold Story of the Panama Canal* (Cambridge, MA: Harvard University Press, 2019); Christine Keiner, *Deep Cut: Science, Power, and the Unbuilt Interoceanic Canal* (Athens: University of Georgia Press, 2020).

31 Megan Black, *The Global Interior: Mineral Frontiers and American Power* (Cambridge, MA: Harvard University Press, 2018); Raby, *American Tropics*.

32 Audra J. Wolfe, *Freedom's Laboratory: The Cold War Struggle for the Soul of Science* (Baltimore, MD: Johns Hopkins University Press, 2018); Patrick Iber, *Neither Peace Nor Freedom* (Cambridge, MA: Harvard University Press, 2015).

33 Julia Rodríguez, *Civilizing Argentina: Science, Medicine, and the Modern State* (Chapel Hill: University of North Carolina Press, 2006); Medina, *Cybernetic Revolutionaries*; Eve E. Buckley, *Technocrats and the Politics of Drought and Development in Twentieth-Century Brazil* (Chapel Hill: University of North Carolina Press, 2017); Javiera Barandiarán, *Science and Environment in Chile: The Politics of Expert Advice in a Neoliberal Democracy* (Cambridge, MA: MIT Press, 2018).

34 Gabriela Soto Laveaga, "The Conquest of Molecules: Wild Yams and American Scientists in Mexican Jungles," in McCoy and Scarano, *Colonial Crucible*, 297–308; Gabriela Soto Laveaga, *Jungle Laboratories: Mexican Peasants, National Projects, and the Making of the Pill* (Durham, NC: Duke University Press, 2009).

35 John Krige, ed. *How Knowledge Moves: Writing the Transnational History of Science and Technology* (Chicago: University of Chicago Press, 2019), 9.

36 Raby, *American Tropics*. US biologists' early cooperation in the Caribbean with British colonial scientists—mediated by scientific internationalism and a shared sense of "Anglo Saxon" identity—is the exception to local cooperation that proves the rule. Megan Raby, "A Laboratory for Tropical Ecology: Colonial Models and American Science at Cinchona, Jamaica," in *Spatializing the History of Ecology: Sites, Journeys, Mappings*, ed. Raf de Bont and Jens Lachmund (New York: Routledge, 2017).

37 Leida Fernández Prieto, "Islands of Knowledge: Science and Agriculture in the History of Latin America and the Caribbean," *Isis* 104, no. 4 (2013): 788–97; Leida Fernández Prieto, "Saberes Híbridos: Las Sugar Companys y la Moderna Plantación Azucarera en Cuba," *Asclepio* 67, no. 1 (2015): 1–15; Prakash Kumar et al., "Roundtable: New Narratives of the Green Revolution," *Agricultural History* 91, no. 3 (2017): 397–422; Timothy W. Lorek, "The Puerto Rican Connection: Recovering the 'Cultural Triangle' in Global Histories of Agricultural Development," *Agricultural History* 94, no. 1 (2020): 108–40.

38 Nicolás Cuvi, "The Cinchona Program (1940–1945): Science and Imperialism in the Exploitation of a Medicinal Plant," *Dynamis* 31 (2011): 183–206; Nicolás Cuvi, "Tecnociencia y Colonialismo en la Historia de las Cinchona," *Asclepio* 70, no. 1 (2018), doi.org/10.3989/asclepio.2018.08.

39 Tore C. Olsson, *Agrarian Crossings: Reformers and the Remaking of the US and Mexican Countryside* (Princeton, NJ: Princeton University Press, 2017). A comparable case in the history of conservation is Emily Wakild, "Border Chasm: International Boundary Parks and Mexican Conservation, 1935–1945," *Environmental History* 14, no. 3 (2009): 453–75.

40 Gregory T. Cushman, *Guano and the Opening of the Pacific World: A Global Ecological History* (Cambridge: Cambridge University Press, 2013); Felipe Fernandes Cruz, "Flight of the Steel Toucans: Aeronautics and Nation-Building in Brazil's Frontiers" (PhD diss., University of Texas at Austin, 2016); Elizabeth Hennessy, *On the Backs of Tortoises: Conserving Evolution in the Galápagos Islands* (New Haven, CT: Yale University Press, 2019); Felipe Fernandes Cruz, "Alberto Santos-Dumont and Brazilian Aviation," in *Oxford Research Encyclopedia of Latin American History*, 2020, doi.org/10.1093/acrefore/9780199366439.013.864.

41 Many excellent examples appear in Andra Chastain and Timothy Lorek, eds., *Itineraries of Expertise: Science, Technology, and the Environment in Latin America* (Pittsburgh, PA: University of Pittsburgh Press, 2020). Likewise, see Gisela Mateos and Edna Suárez-Díaz, "Technical Assistance in Movement: Nuclear Knowledge Crosses Latin American Borders," in Krige, *How Knowledge Moves*, 345–67.

42 James Delbourgo and Nicholas Dew, eds. *Science and Empire in the Atlantic World* (New York: Routledge, 2008); Lina Del Castillo, "Entangled Fates: French-Trained Naturalists, the First Colombian Republic, and the Materiality of Geopolitical Practice, 1819–1830," *The Hispanic American Historical Review* 98, no. 3 (2018): 407.

43 Strang, *Frontiers of Science.*

44 Chastain and Lorek, *Itineraries of Expertise*, 26; Adriana Minor, "Manuel Sandoval Vallarta: The Rise and Fall of a Transnational Actor at the Crossroad of World War II Science Mobilization," in Krige, *How Knowledge Moves*, 227–53; Michael J. Barany, "The Officer's Three Names: The Formal, Familiar, and Bureaucratic in the Transnational History of Scientific Fellowships," in Krige, *How Knowledge Moves*, 254–80; Olival Freire, Jr., "Scientific Exchanges Between the United States and Brazil in the Twentieth Century: Cultural Diplomacy and Transnational Movements," in Krige, *How Knowledge Moves*, 281–307.

25

ARCTIC SCIENCE

Nanna Katrine Lüders Kaalund

The Arctic has long captured the imagination of both scholarly and popular audiences, with numerous books, articles, exhibitions, and artworks devoted to the so-called achievements of European and Euro-American explorers. The perception of the Arctic as a *terra nullius,* a place waiting to be discovered by Europeans, has long influenced scientific research in and about the Arctic, and in numerous and complicated ways. As Michael Bravo and Sverker Sörlin write,

> The images of the region as a desolate place, a pristine natural laboratory for the field sciences, or alternatively, as a place of evolutionary survival for hunting societies have a longevity, if not an accuracy, that is as persistent as those so aptly identified by Edward Said for the Orient nearly 25 years ago.[1]

Though the Arctic has historically been presented as a natural laboratory, as an empty space for geographical and scientific discovery, recent literature has highlighted how the erasure of Indigenous peoples has functioned as part of the research methodologies of European and Euro-American scientists—both within and outside of the Arctic.[2]

In the first instance, it is instructive to consider what we mean by "The Arctic." Typically it is conceived as a large polar region, which currently spreads across Canada, the USA, Russia, Greenland, Sweden, Norway, Finland, Iceland, and the Arctic Ocean. Another way of describing the Arctic is with reference to the Arctic Circle, the tree line, or the 10 degree Celsius iso-thermal line—all of which are focused around geographical or cartographical features. The Arctic is also home to peoples whose identities do not square with the modern nation-state or geographical definitions.[3] Many inhabitants of the Arctic regions are Inuit (particularly in what is now Canada and Greenland) or Yupik (particular in Alaska and Russia) or Sámi (particu-larly in Finland and Sweden).[4] As with other European and North American imperial ventures throughout the world, British explorers in the Arctic claimed to discover areas that were already inhabited. In the European and Euro-American accounts of the Arctic, this plurality of cultures, languages, and histories of Arctic Indigenous peoples has often been erased.[5] Recent literature has highlighted how many discoveries in the Arctic were only new to European and Euro-American travellers, and that embedded within the rhetoric of scientific and geographical dis-covery is an imperialist agenda, which disregards the fact that the explored areas were already someone's home.[6]

The aim of this chapter is to introduce some of the approaches taken in Arctic studies to dismantle the emphasis on heroism and discovery in order to reconsider the practices and ideologies of nineteenth-century scientific research in the Arctic, and its legacies today. The paper is structured around four interlinked themes: "narratives of discovery," "cartography and geographical surveying," "the First International Polar Year of 1882," and "looking beyond the white explorer." The essay addresses the nineteenth-century British imperial context of Arctic research, and also examines the history of the other major European and North American stakeholders in the region, including Russia, Canada, the USA, Germany, Denmark, Norway, and Sweden, to name only the most prominent examples. For non-Arctic states, imaginations of the Polar regions are deeply culturally specific, and the approach taken to Arctic scientific research differs temporally and geographically.

Narratives of discovery, or discovering the narratives

Throughout the nineteenth century, Arctic travellers were instructed to research and collect anything of potential interest to naturalists and politicians in the metropole. This broad research programme included diverse fields such as geology, magnetism, oceanography, glaciology, meteorology, biology, botany, archaeology, and ethnography. It also included the collection and preservation of natural history specimens. The primary geographical aims, in addition to surveying and documenting lands, was to locate a Northwest Passage and reach the geographical North Pole. What united these diverse research activities was the desire of non-Arctic states, institutions, and individuals to catalogue and chart the Arctic regions, and to exercise dominance over land, routes, and resources through scientific research and geographical discovery. This dual focus on scientific and geographical discovery has had an enduring influence on European and Euro-American understandings of the Arctic, and the modes of communication through which these discovery narratives were constructed in texts and imagery.

Written accounts and visual depictions in all forms were central in shaping metropolitan conceptions of the Arctic, and these representations were often deliberately embedded within tropes of heroic masculinity and European supremacy.[7] Through print, the Arctic was represented and reimagined through such diverse channels as travel narratives, articles in elite scientific journals, poems, plays, and children's books. Travel narratives were a particularly popular form of literature in the nineteenth century, and were an important mode through which British audiences came to know the Arctic. Although billed as accounts of voyages and framed in a language of first-hand observation, travel narratives were highly curated portrayals of the Arctic, the explorers, and the nations that sponsored these expeditions. As Innes Keighren, Charles Withers, and Bill Bell write in *Travels into Print*, "The move into print was far from a straightforward recounting of events or the cataloguing of simple travel facts."[8] Drawing on insights from book history and periodical studies, scholars such as Keighren, Withers, and Bell, as well as Adriana Craciun, Heidi Hansson, Shane McCorristine, Anka Ryall, Efram Sera-Shriar, Robert David, Jen Hill, and Janice Cavell have shown how European conceptions of the Arctic and the explorers were written into existence.[9]

While there is much to be gained from studying individual explorers and the imperial policies of single nations, and the way scientific practice and narratives of discovery co-constructed specific visions of the Arctic, this can also obscure the international nature of Arctic exploration and scientific practice. Imperial expansionism in the Arctic took many different forms. As Michael Bravo and Sverker Sörlin have observed, national differences (sometimes very subtle) could have a big impact on the expeditions and the knowledge they produced, as well as the trajectories of imperial policies in the Arctic.[10] For example, scientific and geographical exploration of Greenland was part of the Danish imperial project, but Danish colonialism in Greenland took a different form to that of the British Empire.[11] Similarly, as Trevor Levere, Michael Robinson, and others have shown, this is also the case when looking at the Canadian and American contexts.[12] Discovery narratives and associated visions of the explorer were culturally and temporally specific, and should not be taken as universal, or as reflecting the lived experience of Arctic travel and research.

Tropes of exploration, discovery, and (white male) heroism were visual as well as literary. Visual representations of the Arctic took place through formats such as panoramas, the illustrated periodical press, woodcuts, and paintings. In *Arctic Spectacles*, Russell Potter shows how nineteenth-century developments in print and visual technologies, including the magic lantern and the cheap illustrated press, were mobilised to create and capitalise on an increasing fascination with the Arctic regions.[13] In the British context, *ice* was a particularly important trope, as the Arctic was overwhelmingly portrayed as a frozen wasteland. At the same time, ice also came to stand for the sublime. Beauty and danger, the thrills of the unknown, and the honour of overcoming hardships, those were some of the key visual and narratorial tropes through which British audiences encountered the Arctic. In addition to images in books, periodicals, and lectures, another venue through which Arctic visual culture was consumed was through the large and popular panoramas. The first Arctic panorama was opened in Leicester Square in 1819, and the format remained incredibly popular throughout the nineteenth century.[14] As the audience walked through the rotunda panorama, the large colourful paintings seemed to bring the Arctic to Britain, or transfer the audience to the Arctic. As Robert David observes in *The Arctic in the British Imagination*, in contrast with the black-and-white illustrations reprinted in periodicals, the panoramas were a colourful *tour de force* of the Arctic.[15]

In addition to such culturally significant symbolisms, images were part of scientific practices, in both explicit and subtle ways. For example, drawings of Arctic flora and fauna were, of course, directly used to categorise the Arctic world. However, representations of the natural landscape, of explorers researching in their meteorological observatory, of the aurora, and of encountered animals and plants, also functioned as part of the explorers' repertoire to portray themselves as trustworthy observers of natural phenomena.[16] As with the narrative accounts of the Arctic, the visual representations were also highly curated and were not simply a snapshot of the voyages. As Eavan O'Dochartaigh has shown, the visual representations of Arctic explorations made during the expeditions portrayed a more complex experience than the often highly sensationalised versions displayed in the metropole.[17] Visual and textual representations of Arctic Indigenous peoples were similarly mediated through European preconceptions, and the changing practices of ethnography and anthropology. This was further complicated with the development of photographic technologies, an insight reached by scholars such as Ingeborg Høvik.[18] Taken together, narratives and visual imagery played a key role in making the foreign tangible, in underwriting the credibility of the explorers, and in constructing specific portrayals of the Arctic.[19] This is an important point for Arctic exploration and science, as the credibility of the resulting scientific data depended on the perception of the explorer as an authoritative observer of Arctic phenomena. This included cartographical practices, and maps representing the natural worlds through a combination of visuality, text, and scientific theories.[20]

Cartography and geographical surveying

The concept of exploration is inseparable from the practices and ideologies of imperial expansionism. This includes cartography and geographical surveying. In the now classic work, *The*

New Nature of Maps, J.B. Harley challenged the perception of maps as mirrors of nature, and highlighted how maps instead reflect a socially constructed world. Drawing on Foucault, Harvey argued that maps are an "instrument of knowledge," which is "impregnated with power," as much a weapon of imperialism as guns and warships.[21] The power of maps was also epistemic, as mapping practices erased Indigenous cultures and histories through toponymic colonialism, understood as the colonial practice of naming and renaming places.[22] What does the history of maps, surveying, and exploration reveal when viewed within its larger context of colonial expansionism, and not just as a product of individual white male explorers? One approach has centred maps as the locus of scientific and imperial competition, to highlight the interplay between the theories and techniques of mapmaking, personal ambitions, and empire building.[23] Another approach focuses on recovering cross-cultural encounters, to investigate how the so-called discoveries of colonial explorers and settlers were co-produced by multiple actors, including Indigenous peoples.[24] There is frequently overlap between these two central ways of approaching surveying and mapping practices in the Arctic context, and both ways approach maps as inherently political. As Carl-Gösta Ojala and Jonas Monié Nordin write in relation to Sámi history, "land and the understanding of land have been, and still are, of central importance in the colonial confrontations and negotiations … And as such, they are always embedded in power relations."[25]

One of the main drivers behind European and Euro-American Arctic expeditions was the search for a Northwest Passage. The dream of locating and demarcating the Northwest Passage was an imperialistic one. Who owned the Arctic, and who had the right to its resources and potential trading routes, was a key motivating factor for the organisation of many European and Euro-American Arctic ventures. The Northwest Passage was envisaged as a new trading route between the Atlantic and Pacific oceans, and as an alternative to the route that sailors took around the southern horn of South America via the Straits of Magellan. The first British expeditions to the Arctic following the Napoleonic Wars were the 1818 twin expeditions in search of the Northwest Passage and geographical North Pole. Both geographical aims were predicated on the assumption that there were navigable water routes across the northern shores of what is now Canada. While the 1818 twin expeditions came nowhere close to completing their geographical aims, new expeditions soon followed. In the first half of the nineteenth century, one of the key proponents of British Arctic exploration was the second secretary to the Admiralty John Barrow.[26] As historian Christopher Lloyd has shown, Barrow worked tirelessly to direct British naval policies toward prioritising Arctic surveying in order to find the Northwest Passage.[27] By the early 1840s, there was an increasing sense that, even if a Northwest Passage could be found, it was impracticable as a trade route. Yet, Barrow petitioned for one more British expedition to the Arctic, which resulted in the 1845 disastrous venture led by John Franklin in the *Erebus* and *Terror*.[28]

Several theoretical frameworks were set out to predict what the uncharted regions of the Arctic might look like, and we see the results of this theorising in the form of maps and in the instructions provided to the various expeditions. The search for Franklin's lost expedition became linked with a central theoretical idea about the Arctic: the existence of an Open Polar Sea, also referred to as a Polynia. Already in 1595 the map produced by the Flemish cartographer Gerardus Mercator (namesake for the Mercator projection) visualised the North Pole as an open-water basin, through which the North Pole could be reached. It re-emerged with an increased popularity and sense of urgency in the early 1850s, as cartographic theories and practices were central to the organisation of the many missions sent out in search of Franklin's lost expedition. Organisers sought to predict not only what was in the unmapped areas of the Arctic, but also where the lost crew could have ended up. In the early 1850s, the American philanthropist Henry Grinnell co-financed two searching ventures to find Franklin and his lost crew, known as the First and Second Grinnell Expeditions. The theory of the Open Polar Sea shaped the organisation of both expeditions, as the organisers argued Franklin's lost expedition could have reached this open water.[29]

Though explorers drew on the rhetoric of direct observation to establish their Arctic regional expertise and the associated veracity of their data, the expeditions and their data were clearly shaped by theory. The historiography on the relationship between travel and armchair theory, between fieldwork and the laboratory, is well developed. For example, Vanessa Heggie has shown in *Higher and Colder* how the multifaceted practices of Polar researchers seamlessly moved between what could be categorised as the field and the laboratory.[30] In this way, Arctic researchers drew on complex strategies in seeking to control their research environment, in the Polar regions and in the European centres of learning. In turn, the location of the field and the laboratory was also changeable, as the ships that took explorers to the Arctic also functioned as laboratories, and permanent or semi-permanent research stations were built in the field.[31]

The First International Polar Year and beyond

The First International Polar Year (IPY) took place between 1882 and 1883, and it was, in many ways, the start of a new era for approaching Polar research—encompassing both the Arctic and Antarctic. Up until the IPY geographical exploration and scientific research in the Arctic were intertwined—a dual dependency where both shaped the other. The IPY set out an alternative framework, one that decoupled science from exploration. The idea for what would become the IPY was first proposed by the Austro-Hungarian naval officer Karl Weyprecht in 1875, but it took several years to get the project under way. One of the big stumbling blocks was the need for deliberate and open international collaboration at an official, rather than informal, level. While the so-called "magnetic crusade" of the 1830s through 1850s involved a high degree of collaboration and data sharing, the IPY was the first grand-scale internationally structured cooperative scientific project.[32]

In 1880, two years before the IPY, Denmark, Russia, Norway, and Austro-Hungary committed to securing the finances required for participation at the Second International Polar Conference. The following year saw the addition of the Netherlands, the USA, and France, with the later additions of Germany, Finland, Sweden, and, at the last moment, Canada and Britain. The composition of nations contributing to the venture is also a good indicator of the perceived importance of non-Arctic states in securing some level of geopolitical power in the Arctic. For example, although Alaska remained sparsely settled by Euro-Americans following the US purchase of the territory from Russia in 1867 until the end of the nineteenth century, it can be seen as part of the wider imperial project to expand American presence in the Arctic. As shown by Susan Barr, Cornelia Lüdecke, and David Rothenberg, nationalistic and imperialistic concerns influenced the scientific research undertaken in the Arctic both during the planning of the First IPY and after.[33]

The Second IPY took place between 1932 and 1933. The Third IPY is also known as the International Geophysical Year and took place between 1957 and 1958. The Fourth IPY took place between 2007 and 2008. The longevity of the format of the IPY does not, however, mean that the First IPY established a unified Polar research programme with international cooperation. Nor did the IPY's emphasis on decoupling travel and science necessarily take hold. This enduring and continuously renegotiated relationship between Arctic fieldwork, exploration, and scientific authority has been well highlighted in the work of scholars such as Klaus Dodd and Richard Powell.[34] Who had political and economic priority in the Arctic remained closely linked with scientific and geographical investigations in the Arctic, and was as contested after the First IPY as it had been previously. In the modern day, these debates play out, for example, in the composition of the Arctic Council and in the establishment of posts for trade and military purposes. With the shifts brought on by climate change, these are debates that surely will intensify. The story of Arctic science has been, and still is, one of imperialistic expansionism and epistemic warfare, even when framed as an endeavour in international cooperation.

Looking beyond the white explorer

Looking beyond the traditional concept of the white male Arctic explorer not only offers a way to break down how this category was constructed and functioned historically and in the present, but also allows us to reconsider who was undertaking scientific research in the Arctic, and how they were doing it. Religious missionaries and colonial settlers were, for example, also important producers of knowledge about the Arctic, with their own problematic past.[35] Gender is another historiographical focus that has opened up the categories of Arctic knowledge makers and travellers.[36] In North America, the Hudson's Bay Company (HBC) was a significant producer of scientific knowledge. Founded in 1670, the HBC held an absolute monopoly on trade in large regions of the Arctic for extended periods. Scholars such as Debra Lindsay and Ted Binnema have shown the complex role of this form of economic imperialism in the development of scientific societies in Canada and the USA.[37] In the Danish imperial context, the *Kongelige Grønlandske Handel* (f. 1774) had a monopoly on trade in Greenland. Scientific research and geographical discovery for the purpose of resource extracting were central to both companies. Finally, the way Arctic Indigenous peoples contributed to Western scientific understandings of the Arctic has also been the focus of recent literature.

While European and Euro-American discovery and mapping of the Arctic were portrayed as grand scientific and human achievements, these were also physical and epistemic acts of violence against Arctic Indigenous peoples. It is well known that when the English privateer-turned-knight Martin Frobisher surveyed the Arctic in the service of Queen Elizabeth I in 1577, he abducted three Inuit from Qikiqtaaluk (Baffin Island) and forced them to return with him to England. European and Euro-American explorers continued this practice of abducting or deceiving Arctic peoples into travelling with them back to the metropole in the centuries that followed Frobisher's expedition. For example, in 1897, the American explorer Robert Peary made six Inughuit travel back with him from North Greenland to the USA, where they were displayed and exploited under the guise of scientific research.[38] In this way, Northern Indigenous peoples were treated as specimens for scientific research, in particular for the social, human, and life sciences, and as objects of curiosity displayed for the amusement of others.[39]

European explorers often sought to hide the fact that they drew on Indigenous knowledge and labour to fulfil the aims of their expeditions. What is more, in the British context this erasure of Indigenous expertise formed part of the toolkit for explorers, and was a central way that they asserted regional and scientific expertise in the region. This had consequences for surveying practices in unexpected ways. Consider again the lost Franklin expedition. Its location remained a mystery for European and Euro-Americans until the *Erebus* was located in 2014. The *Terror* was found in 2016. However, in 1854 the British surgeon-explorer John Rae learned that many Inuit knew the direction taken by Franklin's expedition, and that the crew members had succumbed to starvation. Yet, when Rae reported this to the British Admiralty, he was admonished for accepting the testimony of Inuit.[40] In the British context, as in that of other non-Arctic nations, the knowledge, history, and experiences of Arctic Indigenous peoples have been erased through cultural and scientific practices that were inherently racist, elitist, and imperialistic. European racial and cultural biases influenced all areas of Arctic exploration, and the focus on discovery narratives, of so-called heroic explorers and their trials and tribulations in the Arctic, has obscured a much more complex story.[41]

More often than not, explorers relied on Inuit and other Northern Indigenous groups to survive in the Arctic, and to complete their scientific research. That is, the entire project of Arctic exploration was contingent on accessing the labour and knowledge of extra-European peoples. As scholars such as Karen Routledge and Sheila Nickerson have shown, the history of Arctic exploration looks remarkably different to the standard European explorer-focused accounts when we look beyond the European explorers.[42] Furthermore, there is a growing body of work that not only considers Inuit versions of encounters, but also takes seriously Inuit knowledge and scientific practice as significant in their own right.[43]

Mirroring other developments in the history of science, technology, and medicine in recent years, Arctic studies scholars are (slowly) beginning to incorporate research from decolonial, postcolonial, and global studies scholars into their work. For those engaged in contemporary research, one response has been to reassess fieldwork practices, and to move towards frameworks of knowledge making that prioritise co-development and co-production of research questions and formats between foreign scientists and local communities.[44] However well intentioned these efforts are, problems remain. As Alethea Anaquq-Baril's documentary film *Angry Inuk* demonstrates, there needs to be a willingness on behalf of non-Indigenous researchers, politicians, and activists to learn from Indigenous communities, and to accept traditional lifestyles and practices.[45] As Anaquq-Baril's film shows, non-Indigenous peoples still approach the Arctic and its Indigenous communities from a Eurocentric and paternalistic standpoint. Recovering experiences and achievements of Arctic Indigenous peoples working as part of European and Euro-American expeditions, and the structures through which their work and knowledge were erased, is ongoing, and involves reconsiderations of biases in the archives.[46] Finally, there has been a growth in research that does not centre on colonial encounters at all, but instead works to tell Indigenous-focused histories as the primary goal.[47] This includes recovering Indigenous knowledge about the natural world, not merely for academic purposes but as part of broader efforts to reclaim what colonialisation has taken.

Notes

1 Michael Bravo and Sverker Sörlin, eds., *Narrating the Arctic: A Cultural History of Nordic Scientific Practices* (Canton, MA: Science History Publications, 2002), vii.

2 Notable examples include Andrew Stuhl, *Unfreezing the Arctic: Science, Colonialism, and the Transformation of Inuit Lands* (Chicago: University of Chicago Press, 2016); Adriana Craciun and Mary Terrall, eds. *Curious Encounters: Voyaging, Collecting, and Making Knowledge in the Long Eighteenth Century* (Toronto: University of Toronto Press, 2019); Julie Cruikshank, *Do Glaciers Listen?: Local Knowledge, Colonial Encounters, and Social Imagination* (Vancouver: UBC Press, 2010); H.G. Jones, "Teaching the Explorers: Contributions of One Baffin Family to History and Geography," *Terrae Incognitae* 34, no. 1 (2002): 73–81.

3 I draw in particular on the insights of Aqqaluk Lynge, *Inuit Culture and International Policy*, Canadian Papers in Peace Studies 3 (Toronto, ON: Science for Peace/Samuel Stevens, 1992); Mia M. Bennett et al., "Articulating the Arctic: Contrasting State and Inuit Maps of the Canadian North," *Polar Record* 52, no. 6 (2016): 630–44; Yvon Csonka, "The Yupik People and Its Neighbours in Chukotka: Eight Decades of Rapid Changes," *Études/Inuit/Studies* 31, no. 1/2 (2007): 23–37.

4 These definitions are in themselves fraught; see also Claudio Aporta, "The Trail as Home: Inuit and Their Pan-Arctic Network of Routes," *Human Ecology* 37, no. 2 (2009): 131–46; Yvon Csonka, "Changing Inuit Historicities in West Greenland and Nunavut," *History and Anthropology* 16, no. 3 (2005): 321–34; Karen Langgård, "An Examination of Greenlandic Awareness of Ethnicity and National Self-Consciousness through Texts Produced by Greenlanders 1860s–1920s," *Études/Inuit/*

Studies 22, no. 1 (1998): 83–107; Gary N. Wilson and Heather A. Smith, "The Inuit Circumpolar Council in an Era of Global and Local Change," *International Journal* 66, no. 4 (2011): 909–21.

5 Robert G. David, *The Arctic in the British Imagination 1818–1914* (Manchester: Manchester University Press, 2000); Jen Hill, *White Horizon: The Arctic in the Nineteenth-Century British Imagination* (Albany: State University of New York Press, 2009).

6 Although some expeditions went north of permanent human inhabitation, they still relied on Indigenous knowledge and labour, and used the northernmost towns as bases. See Adriana Craciun, *Writing Arctic Disaster: Authorship and Exploration* (Cambridge: Cambridge University Press, 2016); Michael Trevor Bravo, "The Postcolonial Arctic," *Moving Worlds: A Journal of Transcultural Writings* 15 (2015): 93–111; Janice Cavell, "Representing Akaitcho: European Vision and Revision in the Writing of John Franklin's Narrative of a Journey to the Shores of the Polar Sea," *Polar Record* 44, no. 1 (2008): 25–34. See also Miguel A. Cabañas et al., eds., *Politics, Identity, and Mobility in Travel Writing* (New York: Routledge, 2015).

7 On the practices of creating tropes of heroism and masculinity in the Arctic, see Laurie Garrison, "Virtual Reality and Subjective Responses: Narrating the Search for the Franklin Expedition through Robert Burford's Panorama," *Early Popular Visual Culture* 10, no. 1 (2012): 7–22; Huw W.G. Lewis-Jones, "'Heroism Displayed': Revisiting the Franklin Gallery at the Royal Naval Exhibition, 1891," *Polar Record* 41, no. 3 (July 2005): 185–203; Huw Lewis-Jones, *Imagining the Arctic: Heroism, Spectacle and Polar Exploration* (London: Bloomsbury Publishing, 2017); Ingeborg Høvik, "Heroism and Imperialism in the Arctic: Edwin Landseer's Man Proposes – God Disposes," *Nordlit*, no. 23 (2008): 183–94; Johanne M. Bruun, "Invading the Whiteness: Science, (Sub)Terrain, and US Militarisation of the Greenland Ice Sheet," *Geopolitics* 25, no. 1 (2020): 167–88.

8 Innes M. Keighren, Charles W.J. Withers, and Bill Bell, *Travels Into Print: Exploration, Writing, and Publishing with John Murray, 1773–1859* (Chicago: University of Chicago Press, 2015), 2.

9 Adriana Craciun, *Writing Arctic Disaster* (Cambridge: Cambridge University Press, 2016); Heidi Hansson, "Henrietta Kent and the Feminised North," *Nordlit: Tidsskrift i Litteratur Og Kultur* 11, no. 2 (2007): 71–96; Shane McCorristine, *Spectral Arctic: A History of Dreams and Ghosts in Polar Exploration* (London: UCL Press, 2018); Anka Ryall, Johan Schimanski, and Henning Howlid Wærp, *Arctic Discourses* (Cambridge: Cambridge Scholars Publishing, 2010); Heidi Hansson, Maria Lindgren Leavenworth, and Anka Ryall, *The Arctic in Literature for Children and Young Adults* (London: Routledge, 2020); Efram Sera-Shriar, "Civilizing the Natives: Richard King and his Ethnographic Writings on Indigenous Northerners," in *Made Modern: Science and Technology in Canadian History*, ed. Edward Jones-Imhotep and Tina Adcock (Vancouver: University of British Columbia Press, 2018), 39–59; Robert G. David, *The Arctic in the British Imagination 1818–1914* (Manchester: Manchester University Press, 2000); Jen Hill, *White Horizon: The Arctic in the Nineteenth-Century British Imagination* (Albany: State University of New York Press, 2009); Janice Cavell, *Tracing the Connected Narrative: Arctic Exploration in British Print Culture, 1818–1860* (Toronto: University of Toronto Press, 2008).

10 Bravo and Sverker Sörlin, *Narrating the Arctic*, 19.

11 Søren Rud, "Erobringen af Grønland: Opdagelsesrejser, etnologi og forstanderskab i attenhundredetallet," *Historisk Tidsskrift* 106, no. 2 (2013): 488–520; Søren Rud, *Colonialism in Greenland: Tradition, Governance and Legacy* (Basingstoke: Palgrave Macmillan, 2017); Kirsten Hastrup, "Ultima Thule: Anthropology and the Call of the Unknown," *The Journal of the Royal Anthropological Institute* 13, no. 4 (2007): 789–804. For an example of the complex legacies of Danish colonialism in Greenland and the construction of identities see Lill Rastad Bjørst, "Stories, Emotions, Partnerships and the Quest for Stable Relationships in the Greenlandic Mining Sector," *Polar Record* 56 (2020): 1–13.

12 Trevor H. Levere, *Science and the Canadian Arctic: A Century of Exploration, 1818–1918* (Cambridge: Cambridge University Press, 2004). See also Edward Jones-Imhotep and Tina Adcock, eds., *Made Modern: Science and Technology in Canadian History* (Vancouver: UBC Press, 2018); Michael Robinson, *The Coldest Crucible: Arctic Exploration and American Culture* (Chicago: University of Chicago Press, 2006).

13 Russell A. Potter, *Arctic Spectacles: The Frozen North in Visual Culture, 1818–1875* (Seattle: University of Washington Press, 2007).

14 For more on Arctic panoramas, see Hill, *White Horizon*, 130–84; Russell A. Potter and Douglas W. Wamsley, "The Sublime yet Awful Grandeur: The Arctic Panoramas of Elisha Kent Kane," *Polar Record* 35, no. 194 (1999): 193–206; Ralph O'Connor, *The Earth on Show: Fossils and the Poetics of Popular Science, 1802–1856* (Chicago: University of Chicago Press, 2008). For more on science and

photography, see Geoffrey Belknap, *From a Photograph: Authenticity, Science and the Periodical Press, 1870–1890* (London: Bloomsbury Publishing, 2016).

15 David, *The Arctic in the British Imagination*, 150–6.

16 See for example Katharine Anderson, *Predicting the Weather: Victorians and the Science of Meteorology* (Chicago: University of Chicago Press, 2005); Henrika Kuklick, "Personal Equations: Reflections on the History of Fieldwork, with Special Reference to Sociocultural Anthropology," *Isis* 102, no. 1 (2011): 1–33; Daniela Bleichmar, *Visible Empire: Botanical Expeditions and Visual Culture in the Hispanic Enlightenment* (Chicago: University of Chicago Press, 2012).

17 Eavan O'Dochartaigh, "From Science to Sensation: A Study of Visual and Literary Representation in Arctic Exploration in the Mid-19th Century," National University of Ireland, Galway, 2018; Eavan O'Dochartaigh, "Exceedingly Good Friends: The Representation of Indigenous People during the Franklin Search Expeditions to the Arctic, 1847–59," *Victorian Studies* 61, no. 2 (2019): 255–67.

18 Ingeborg Høvik, "Reproducing the Indigenous: John Møller's Studio Portraits of Greenlanders in Context," *Acta Borealia* 33, no. 2 (2016): 166–88; Ingeborg Høvik, "Framing the Arctic: Reconsidering Roald Amundsen's Gjøa Expedition Imagery," *Nordlit*, no. 35 (2015): 137–60.

19 I.S. MacLaren, "The Aesthetic Map of the North, 1845–1859," *Arctic* 38, no. 2 (1985): 89–103; Ingeborg Høvik, "Heroism and Imperialism in the Arctic; Edwin Landseer's Man Proposes – God Disposes," *Nordlit*, no. 23 (2008): 183–94.

20 Material culture, including museum displays, has also played a highly significant role in shaping understandings of the Arctic. Charlotte Connelly and Claire Warrior, "Survey Stories in the History of British Polar Exploration: Museums, Objects and People," *Notes and Records: The Royal Society Journal of the History of Science* 73, no. 2 (2019): 259–74.

21 J.B. Harley, *The New Nature of Maps: Essays in the History of Cartography*, ed. Paul Laxton (Baltimore, MD: Johns Hopkins University Press, 2001).

22 See also Charles W.J. Withers, "Authorizing Landscape: 'Authority', Naming and the Ordnance Survey's Mapping of the Scottish Highlands in the Nineteenth Century," *Journal of Historical Geography* 26, no. 4 (2000): 532–54; Sarah Cogos, Marie Roué, and Samuel Roturier, "Sami Place Names and Maps: Transmitting Knowledge of a Cultural Landscape in Contemporary Contexts," *Arctic, Antarctic, and Alpine Research* 49, no. 1 (2017): 43–51; Béatrice Collignon, "Les Toponymes Inuit, Mémoire du Territoire: Étude de l'Histoire des Inuinnait," *Anthropologie et Sociétés* 26, no. 2–3 (2002): 45–69.

23 Miles Ogborn and Charles W.J. Withers, *Georgian Geographies: Essays on Space, Place and Landscape in the Eighteenth Century* (Manchester: Manchester University Press, 2004); E. Tammiksaar, N.G. Sukhova, and I.R. Stone, "Hypothesis versus Fact: August Petermann and Polar Research," *Arctic* 52, no. 3 (1999): 237–43; Janice Cavell, "'As Far as 90 North': Joseph Elzéar Bernier's 1907 and 1909 Sovereignty Claims," *Polar Record* 46, no. 4 (2010): 372–3; Brandon Luedtke, "An Ice-Free Arctic Ocean: History, Science, and Scepticism," *Polar Record* 51, no. 2 (2015): 130–9; Philip J. Hatfield, *Lines in the Ice: Exploring the Roof of the World* (Montreal: McGill-Queen's Press, 2016).

24 Kelli Lyon Johnson, "Writing Deeper Maps: Mapmaking, Local Indigenous Knowledges, and Literary Nationalism in Native Women's Writing," *Studies in American Indian Literatures* 19, no. 4 (2007): 103–20.

25 Carl-Gösta Ojala and Jonas Monié Nordin, "Mapping Land and People in the North: Early Modern Colonial Expansion, Exploitation, and Knowledge," *Scandinavian Studies* 91, no. 1–2 (2019): 100.

26 Maurice James Ross, *Polar Pioneers: John Ross and James Clark Ross* (Montreal: McGill-Queen's Press, 1994).

27 Christopher Lloyd, *Mr. Barrow of the Admiralty: A Life of Sir John Barrow* (London: Irvington Publishers, 1970).

28 There are numerous studies on John Franklin and his lost expedition. Good starting points are Anthony Brandt, *The Man Who Ate His Boots: Sir John Franklin and the Tragic History of the Northwest Passage* (New York: Random House, 2011); Adriana Craciun, "Writing the Disaster: Franklin and Frankenstein," *Nineteenth-Century Literature* 65, no. 4 (2011): 433–80.

29 Mark Metzler Sawin, "Raising Kane: Elisha Kent Kane and the Culture of Fame in Antebellum America," *Transactions of the American Philosophical Society* 98, no. 3 (2008): i–368; Liz Cruwys, "Henry Grinnell and the American Franklin Searches," *Polar Record* 26, no. 158 (1990): 211–16; John Woitkowitz, "Science, Networks, and Knowledge Communities: August Petermann and the Construction of the Open Polar Sea" (paper presented at the Annual Meeting of the Canadian Historical Association, Vancouver, 2019).

30 Vanessa Heggie, *Higher and Colder: A History of Extreme Physiology and Exploration* (Chicago: University of Chicago Press, 2019); See also Henrika Kuklick and Robert E. Kohler, "Introduction," *Osiris* 11

(1996): 1–14; Klaus Dodds and Richard Powell, "Polar Geopolitics: New Researchers on the Polar Regions," *The Polar Journal* 3, no. 1 (2013): 1–8; Richard C. Powell, *Studying Arctic Fields: Cultures, Practices, and Environmental Sciences* (Montreal: McGill-Queen's University Press, 2017).

31 Antony Adler, "The Ship as Laboratory: Making Space for Field Science at Sea," *Journal of the History of Biology* 47, no. 3 (2014): 333–62.

32 Christopher Carter, "Magnetic Fever: Global Imperialism and Empiricism in the Nineteenth Century," *Transactions of the American Philosophical Society* 99, no. 4 (2009): i–168; John Cawood, "The Magnetic Crusade: Science and Politics in Early Victorian Britain," *Isis* 70, no. 4 (1979): 493–518; Edward J. Larson, "Public Science for a Global Empire: The British Quest for the South Magnetic Pole," *Isis* 102, no. 1 (2011): 34–59; Levere, *Science and the Canadian Arctic.*

33 Susan Barr, "The Expeditions of the First International Polar Year 1882–83," *Polar Research* 28, no. 2 (2009): 311–12; Susan Barr and Cornelia Lüdecke, eds., *The History of the International Polar Years (IPYs)* (Berlin: Springer Science & Business Media, 2010); Rip Bulkeley, "The First Three Polar Years – A General Overview," in Barr and Lüdecke, *History of the International Polar Years,* 1–6; David Rothenberg, "Making Science Global? Coordinated Enterprises in Nineteenth-Century Science," in *Globalizing Polar Science: Reconsidering the International Polar and Geophysical Years,* ed. Roger D. Launius, James Rodger Fleming, and David H. DeVorkin (New York: Palgrave Macmillan, 2010), 23–35; For more on international collaborations and the IPY see also Colin P. Summerhayes, "International Collaboration in Antarctica: The International Polar Years, the International Geophysical Year, and the Scientific Committee on Antarctic Research," *Polar Record* 44, no. 4 (2008): 321–34; Philip N. Cronenwett, "Publishing Arctic Science in the Nineteenth Century: The Case of the First International Polar Year," in Launius, Fleming, and DeVorkin, *Globalizing Polar Science,* 37–46; Christopher Carter, "Going Global in Polar Exploration: Nineteenth-Century American and British Nationalism and Peacetime Science," in Launius, Fleming, and DeVorkin, *Globalizing Polar Science,* 85–106.

34 Klaus Dodds and Richard Powell, "Polar Geopolitics: New Researchers on the Polar Regions," *The Polar Journal: Polar Geopolitics: New Researchers on the Polar Regions* 3, no. 1 (2013): 1–8; Richard C. Powell, *Studying Arctic Fields: Cultures, Practices, and Environmental Sciences* (Montreal: McGill-Queen's University Press, 2017). See also Nanna Katrine Lüders Kaalund, *Explorations in the Icy North: How Travel Narratives Shaped Arctic Science in the Nineteenth Century,* Science and Culture in the Nineteenth Century (Pittsburgh, PA: University of Pittsburgh Press, forthcoming).

35 Einar Lund Jensen, Hans Christian Gulløv, and Kristine Raahauge, *Cultural Encounters at Cape Farewell: The East Greenlandic Immigrants and the German Moravian Mission in the 19th Century* (Chicago: Museum Tusculanum Press, 2011); J.C.S. Mason, *The Moravian Church and the Missionary Awakening in England, 1760–1800* (Martlesham: Boydell & Brewer, 2001); Felicity Jensz, *German Moravian Missionaries in the British Colony of Victoria, Australia, 1848–1908: Influential Strangers* (Leiden: Brill, 2010).

36 Silke Reeploeg, "Women in the Arctic: Gendering Coloniality in Travel Narratives from the Far North, 1907–1930," *Scandinavian Studies* 91, no. 1–2 (2019): 182–204; Linda S. Bergmann, "Woman against a Background of White: The Representation of Self and Nature in Women's Arctic Narratives," *American Studies* 34, no. 2 (1993): 53–68; Jeannine Atkins, *How High Can We Climb?: The Story of Women Explorers* (New York: Farrar, Straus and Giroux, 2005).

37 Debra Lindsay, *Science in the Subarctic: Trappers, Traders and the Smithsonian Institution* (Washington, DC: Smithsonian Institution Press, 1993); Ted Binnema, *Enlightened Zeal: The Hudson's Bay Company and Scientific Networks, 1670–1870* (Toronto: University of Toronto Press, 2014).

38 Kenn Harper, "The Minik Affair: The Role of the American Museum of Natural History," *Polar Geography* 26, no. 1 (2002): 39–52; Kenn Harper, *Give Me My Father's Body: The Life of Minik, the New York Eskimo* (New York: Simon and Schuster, 2001).

39 Sadiah Qureshi, *Peoples on Parade: Exhibitions, Empire, and Anthropology in Nineteenth-Century Britain* (Chicago: University of Chicago Press, 2011); Rikke Andreassen and Anne Folke Henningsen, *Menneskeudstilling: fremvisninger af eksotiske mennesker i Zoologisk Have og Tivoli* (Copenhagen: Tiderne Skifter, 2011).

40 When the wrecks of the *Erebus* and *Terror* were discovered in the 2010s, they were found where Inuit had told Rae to search. For more see Gillian Hutchinson, *Sir John Franklin's Erebus and Terror Expedition: Lost and Found* (London: Bloomsbury Publishing, 2017); Nanna Katrine Lüders Kaalund, "What Happened to John Franklin? Danish and British Perspectives from Francis McClintock's Arctic Expedition, 1857–59," *Journal of Victorian Culture* 25, no. 2 (2020): 300–14; Janice Cavell, "Publishing

Sir John Franklin's Fate: Cannibalism, Journalism, and the 1881 Edition of Leopold McClintock's The Voyage of the 'Fox' in the Arctic Seas," *Book History* 16, no. 1 (2013): 155–84.

41 See for example Karen Routledge, *Do You See Ice? Inuit and Americans at Home and Away* (Chicago: The University of Chicago Press, 2018); Valerie Henitiuk, " 'Memory Is so Different Now': The Translation and Circulation of Inuit-Canadian Literature in English and French," *Perspectives* 25, no. 2 (2017): 245–59; Kirsten Thisted, "On Narrative Expectations: Greenlandic Oral Traditions about the Cultural Encounter between Inuit and Norsemen," *Scandinavian Studies* 73, no. 3 (2001): 253–96.

42 Sheila B. Nickerson, *Midnight to the North: The Untold Story of the Woman Who Saved the Polaris Expedition* (New York: Tarcher / Putnam, 2002); Karen Routledge, *Do You See Ice?* See also Jan Løve, *Hans Hendrik og Hans Ø: beretningen om Hans Hendrik og de to Hans Øer* (Charlottenlund: Det Grønlandske Selskab, 2016).

43 Notable examples include Pascale Laneuville, "Ontologie et Territorialité Inuit en Contexte d'exploitation Minière à Qamani'tuaq (Baker Lake) au Nunavut," *Études/Inuit/Studies* 38, no. 1/2 (2014): 197–216; Peter C. Dawson, "Seeing like an Inuit Family: The Relationship between House Form and Culture in Northern Canada," *Études/Inuit/Studies* 30, no. 2 (2006): 113–35; Meghan Walley, "Exploring Potential Archaeological Expressions of Nonbinary Gender in Pre-Contact Inuit Contexts," *Études/Inuit/Studies* 42, no. 1 (2018): 269–89: Edmund Searles, "Interpersonal Politics, Social Science Research and the Construction of Inuit Identity," *Études/Inuit/Studies* 25, no. 1/2 (2001): 101–19; Scot Nickels and Cathleen Knotsch, "Inuit Perspectives on Research Ethics: The Work of Inuit Nipingit," *Études/Inuit/Studies* 35, no. 1/2 (2011): 57–81; Anne Kendrick and Micheline Manseau, "Representing Traditional Knowledge: Resource Management and Inuit Knowledge of Barren-Ground Caribou," *Society & Natural Resources* 21, no. 5 (2008): 404–18; George W. Wenzel, "Traditional Ecological Knowledge and Inuit: Reflections on TEK Research and Ethics," *Arctic* 52, no. 2 (1999): 113–24; Francis Levesque, "Revisiting 'Inuit Qaujimajatuqangit': Inuit Knowledge, Culture, Language, and Values in Nunavut Institutions since 1999," *Etudes Inuit. Inuit Studies* 38, no. 1 (2015): 115–36; Heather Igloliorte, "Curating Inuit Qaujimajatuqangit: Inuit Knowledge in the Qallunaat Art Museum," *Art Journal* 76, no. 2 (2017): 100–13; Mia M. Bennett, "Unfreezing the Arctic: Science, Colonialism, and the Transformation of Inuit Lands," *Polar Geography* 41, no. 2 (2018): 139–42; Nancy Wachowich and John MacDonald, eds., *The Hands' Measure: Essays Honouring Leah Aksaajuq Otak's Contribution to Arctic Science* (Iqaluit, NU: Nunavut Arctic College Media, 2018).

44 See for example Scot Nickels and Cathleen Knotsch, "Inuit Perspectives on Research Ethics: The Work of Inuit Nipingit," *Études/Inuit/Studies* 35, no. 1/2 (2011): 57–81; Lawrence F. Felt and David Natcher, "Ethical Foundations and Principles for Collaborative Research with Inuit and Their Governments," *Études/Inuit/Studies* 35, no. 1/2 (2011): 107–26; Thomas F. Thornton and Adela Maciejewski Scheer, "Collaborative Engagement of Local and Traditional Knowledge and Science in Marine Environments: A Review," *Ecology and Society* 17, no. 3 (2012).

45 *Angry Inuk*, produced by Alethea Anaquq-Baril (The National Film Board of Canada/Unikkaat Studios and EyeSteelFilm, 2016).

46 Ann Laura Stoler, *Along the Archival Grain: Epistemic Anxieties and Colonial Common Sense* (Princeton, NJ: Princeton University Press, 2010).

47 Erin Keenan, Lucia M. Fanning, and Chris Milley, "Mobilizing Inuit Qaujimajatuqangit in Narwhal Management through Community Empowerment: A Case Study in Naujaat, Nunavut," *Arctic* 71, no. 1 (2018): 27–39; Heather Igloliorte, "Curating Inuit Qaujimajatuqangit: Inuit Knowledge in the Qallunaat Art Museum," *Art Journal* 76, no. 2 (2017): 100–13; Mary Wilman, *Governance through Inuit Qaujimajatuqangit: Changing the Paradigm for the Future of Inuit Society* (Quebec: International Arctic Social Sciences Association, 2002); Kaitlin Jessica Schwan and Ernie Lightman, "Fostering Resistance, Cultivating Decolonization: The Intersection of Canadian Colonial History and Contemporary Arts Programming With Inuit Youth," *Cultural Studies ↔ Critical Methodologies* 15, no. 1 (2013): 15–29.

26

SCIENCE AND DECOLONISATION IN UNESCO

Casper Andersen

When the United Nations was established in 1945, science was placed within the mandate of a new special agency, the United Nations Educational, Scientific, and Cultural Organization (UNESCO). UNESCO played a central role in international science during the post-war decades when formal colonial rule ended. There are widely different assessments of the nature of UNESCO's influence. In one line of argument, UNESCO is regarded as a decolonising force in science as the organisation promoted internationalism in science and provided support for the establishment of national science infrastructures in the newly independent countries joining the organisation during the first post-war decades. A different interpretation emphasises the continuity of imperial structures and discourses in UNESCO science, arguing in line with Mark Mazower's assertion that the UN "was a product of empire and, indeed, at least at the outset, regarded by those with colonies to keep as a more than adequate mechanism for its defense."[1]

A closer reading of the recent research literature on science in UNESCO—and of key primary sources—reveals a complex process captured neither in the UNESCO-the-decoloniser interpretation nor by a narrative of colonial continuity. Rather, understanding the place of science in UNESCO requires us to consider what Helen Tilley calls "the tripartite structure of knowledge production—national, imperial, and international."[2] The configuration of these institutional tiers shaped how and why scientific ideas and techniques travelled globally during the first decades after the Second World War. The structuring argument of this review essay is thus that to understand the role of UNESCO in the world of science during the endgame of empire, we need to take into account the national, international, and imperial layers in an evolving tripartite structure.[3]

In the first two sections of the essay, I review the literature on imperial legacies during UNESCO's founding moment and examine the organisation's endeavour to establish an international science infrastructure in a decolonising world. I then focus on the environmental sciences, which from small beginnings grew to become the largest area for UNESCO science in terms of budget. In the last section I discuss how decolonisation shaped UNESCO's engagements with race science—an area where historians have teased out thoroughly the layers in the tripartite structure of national, imperial, and international science in UNESCO.

Empire and UNESCO's founders

The founding of UNESCO constitutes a key moment in the history of scientific internationalism—the notion that science, given its alleged universal nature, can provide a shared platform for international collaboration and serve as a global unifier of humanity.[4] Founded at the close of the Second World War, UNESCO embodied an idealist belief that assigned a critical role to science in creating lasting peace in the war-torn world.[5] Scholars have unravelled the political and ideological underpinnings of UNESCO's scientific internationalism and identified its explicit and implicit imperial connotations and connections.

In this context the founders of UNESCO have provided an important inroad for historical research. The first Director-General of UNESCO, the British ecologist Julian Huxley, is a case in point. In the decades prior to his appointment to the Director-Generalship of UNESCO in 1945, Huxley had served as an expert on several science commissions for the British Colonial Office.[6] In his memoirs the Nigerian author Chinua Achebe recalls how, as a boarding school pupil in Eastern Nigeria in 1944, he encountered the famous biologist watching birds in the grounds while he inspected the school as member of the Elliot Commission for Higher Education in Africa.[7] Huxley combined a deep commitment to international organisations such as UNESCO with an outspoken support for the British Empire, which he regarded as an indispensable and progressive force, also in the post-war future.[8] The grandson of "Darwin's Bulldog" T.H. Huxley, Julian Huxley was born into the English scientific aristocracy. His views were rooted in British traditions of liberal imperialism that emphasised "partnerships" between colonisers and colonised with the British firmly in control. Based on a study of Huxley's worldview and position in UNESCO Glenda Sluga makes the broader point that "UNESCO's allegedly radical cosmopolitan purpose was beholden to the persistence not only of an Enlightenment-coddled trust in the universal power of knowledge and education but also of late-nineteenth century conceptions of empire and evolution."[9]

Huxley's own views on race were, indeed, composite. His interwar writings exposed race-based theories of human evolution as pseudoscience. They gained iconic status as authoritative anti-racist tracts. At the same time, he argued in favour of eugenics on a global scale in order to preserve and promote genetic diversity according to principles of population genetics.[10] These ideas were also included in his 1946 booklet, *UNESCO: Its Purpose and Its Philosophy*. In the short book he argued that a naturalist philosophy of evolutionary humanism should form the ideological bases for the organisation.[11] However, this foundational philosophy encountered political and religious opposition and did not play a significant role in the working programme of UNESCO's Science Sector. Huxley enjoyed greater success in securing UNESCO a place in the bourgeoning field of international nature conservation. During Huxley's short tenure as Director-General, UNESCO thus midwifed the establishment of the non-governmental organisation the International Union for the Preservation of Nature (IUPN) and he later mobilised his UNESCO connections to gain support for the World Wildlife Foundation (WWF), established in 1961.[12]

Scholars have also scrutinised closely the influence of the British biochemist and historian of science, Joseph Needham. Needham is credited as the main architect in the successful campaign to include science (and technology) in UNESCO's institutional mandate and he was head of UNESCO's Natural Sciences Sector from 1945 to 1948.[13] The historian Patrick Petitjean argues that Needham's influence in UNESCO was profound. A Marxist connected to the Social Relations of Science (SRS) movement in Britain, Needham subscribed to the idea that scientists had a deep responsibility to address directly political and social issues. This commitment structured the main priorities he set for the organisation.[14] Needham's wartime experiences as head of the Sino-British Science Cooperation Office in Chongqing in China were also key—not just for the instigation of his famous book series *Science and Civilization in China* but also for the formulation of the administrative philosophy behind his scientific programme in UNESCO.[15]

These influences condensed in what Needham in 1944 dubbed the "Periphery Principle." According to this principle, UNESCO should focus on supporting science and technology in the periphery, that is, outside what he called the scientifically advanced "bright zone" constituted by Western Europe and North America. Needham argued that scientists and engineers in the "dark areas" outside the "bright zone" needed the helping hand of international science the most.[16]

The periphery principle's dichotomy of bright and dark zones bears the mark of racialised colonial discourse. Yet Needham also expressed profound respect for scientists outside Europe and North America, and he would emphasise that there was room and need for two-way traffic between the zones. Indeed, Needham had come to regard science as an ecumenical human endeavour to which all peoples had contributed. As he saw it, the fundamental issue was not "racial" or "cultural" but economic and material. The problem was that international science cooperation had all but ignored the needs in the periphery. His explicitly non-Eurocentric vision for UNESCO was to turn international science into a truly ecumenical structure in which scientists from both zones could contribute together to solve the challenges of post-war reconstruction.[17]

Needham and Huxley thus brought different national, imperial, and international experiences and priorities into UNESCO's agenda for science. Needham was much more critical towards colonial structures than was Huxley, and he directly challenged ideas about so-called backward peoples and regions that seeped through many of Huxley's writings.[18] Their tenure at UNESCO was short and the nature and extent of their influence remain debated. Petitjean, the most prolific historian to analyse the early phase of science in UNESCO, emphasises the demise of the post-war idealism represented by Huxley, Needham, and other founding figures. He argues that the cosmopolitan ideas of world peace through science practically disappeared as the periphery principle was replaced by technical assistance programmes from the 1950s in which "[s]upport for science was reduced to a de-politicized instrumental support for economic development, the social aims (the well-fare of mankind) were forgotten, and less priority was given to the basic sciences in developing countries."[19]

According to some interpretations this shift represented a broader trend within UNESCO. Lynn Meskell thus identifies a related change in UNESCO's approach to archaeology and the international protection of heritage. She notes that programmes of technical assistance within a few years "edged out the particularly British imperial vision for UNESCO" and that "the largely European ideal of world government and civilizational uplift masterminded by classicists and scientists would be replaced by the post-Second World War technocrats, economists, and engineers who were setting the development agenda."[20]

Building science infrastructures in a decolonising world

During the first post-war decade, the main priority for UNESCO's Science Sector was to erect an infrastructure for science that would encompass the globe. By doing so in an era of formal decolonisation, UNESCO stirred up the layers in the tripartite structure of national, imperial, and international science in Europe, in the independent countries in Latin America, and in decolonising Asia and Africa.

Considering the scope of its global ambition, UNESCO had meagre financial resources and manpower at its disposal. UNESCO therefore sought to develop and work through international science organisations, such as the International Council of Scientific Unions (ICSU), which UNESCO helped to resurrect after the war.[21] When UNESCO tried to create a similar structure in engineering it came up against empire-based opposition. In 1946 the engineering institutions in the British Commonwealth bloc—at this time still an association of Britain and the "white" dominions—refused to join UNESCO's engineering organisation and chose instead to found the Commonwealth Engineers' Council (CEC). To the Commonwealth engineers, the recent war effort had proven not the demise but rather the vitality of the British Empire. The Commonwealth alternative to UNESCO's internationalism also held better prospects for retaining the influence in the remaining British colonies and protectorates. Building on long-lasting imperial connections between the engineering communities across the "British world" bound together by professional interests and a shared sense of culture, race, and history was therefore preferred over UNESCO's new Union for International Engineering Organizations (UATI).[22]

This case demonstrates how promoters of internationalism in science and technology at the post-war moment came up against powerful groups that regarded the connections and institutions of the empires as vibrant, future alternatives. Against this backdrop, institutional decolonisation in science and technology was a contested process, not a foregone conclusion. In the case of engineering the "British world" opposition to UNESCO's internationalism proved enduring. Only in 1968 did the CEC—by then a "multiracial" organisation—join UNESCO's UATI. By then newly established national engineering organisations in the countries that had gained independence from Britain had also joined UNESCO's international engineering platform. As the imperial alternative disappeared, scientific nationalism and internationalism could thus be reconciled.[23]

International scientific organisations provided one avenue for UNESCO's post-war institution building. Another was constituted by Field Science Co-operation Offices (FSCOs) which UNESCO—in line with the "periphery principle"—began to establish "in those regions of the world remote from the main centres of science and technology."[24] In 1947–8 UNESCO opened FSCOs in Rio de Janeiro (Latin America), Cairo (Middle East), Nanjing (Eastern Asia), and New Delhi (South Asia). In 1949 the Latin America office relocated to Montevideo while the Nanjing office was moved to Jakarta in 1951 after Indonesia had gained independence and had joined UNESCO the previous year.[25]

The FSCOs were small units staffed by a few UNESCO FSC officers who were to assist scientists based in the various regions. UNESCO discourse cautiously drew up a contrast between what was called "traditional colonial ways" and the so-called modern approaches that the UNESCO FSCOs were to embody. A UNESCO booklet published in 1950, for example, advised that expatriate FSC officers:

> should not be part of old-style colonial authority; should approach the people of underdeveloped areas with no assumption of superiority; should know that some of these people have produced great men in the past, and that all will be able to do so when once the elementary needs of civilized life have been secured; should be aware that peoples of underdeveloped areas have produced forms of art and culture as noble and beautiful as anything in that Euro-American technical civilization, which has dominated, and now unifies the world.[26]

The small FSCOs covered vast geographical spaces and were to serve liaison functions in science, medicine, engineering, and industrial development. The broadness left open an extremely wide gap between aspiration and what could be achieved on the ground. The FSCOs were later renamed Regional Offices for Science and Technology (ROSTs).[27]

UNESCO also made experiments to establish scientific institutions devoted to specific research areas. An important example of UNESCO-led institution building in post-war science was the International Institute for the Hylean Amazon (IIHA). Established on the initiative of UNESCO in the Amazon forest in 1946, the IIHA was the first flagship initiative of UNESCO's Science Sector.[28] The ambitious plan was to create a world-leading centre for the study of tropical environments. The IIHA, however, closed again in 1952 after a fierce debate during which Latin American scientists protested "against the so-called UNESCO colonialism" in international science at a time when the USA provided nearly 40 per cent of UNESCO's budgets.[29]

Thomas Mougey argues that the controversy over the IIHA was testament to competing post-war projects of scientific "worldmaking" based on rivalling conceptions of the social and natural world. For Joseph Needham and his supporters in UNESCO, the IIHA was to be an international research hub with connections to field stations throughout Asia, Africa, and South America. This model assigned key roles to imperial scientists and research institutions as the IIHA would collaborate with field stations administered by European powers and the USA in tropical regions in Asia, the Caribbean, and Africa in order to attain the projected global reach of the institute. The Latin American scientists constituted a second group who, under the leadership of the Brazilian chemist and UNESCO representative, Paulo Carneiro, protested against the global model for international science advocated by Needham. Instead this group pursued a future for the IIHA as a regional institute. The IIHA should help to emancipate the Amazon region and the nation-state be kept as the central unit of science development. Both models, however, shipwrecked as Brazil's technocratic elites rejected UNESCO's project completely. They closed down the IIHA and created instead a National Institute for Amazonian Research in pursuit of a high modernist project of "technocratic developmentalism."[30]

Mougey's analysis is noteworthy because it highlights amply the competing visions for science in a decolonising world. In doing so Mougey directs attention to the ways in which local dynamics played out in places where UNESCO tried to implement initiatives. This focus is in line with a growing research agenda to study the impact at the recipient end of UNESCO's many outgoing initiatives. Moving beyond the debates and expert committees inside UNESCO's Paris headquarters requires careful unravelling of the entangled international, imperial, and regional layers in UNESCO Science.[31]

In sub-Saharan Africa—where formal colonial rule endured longest—UNESCO's scientific internationalism for long got entangled with the infrastructures of imperial science. Indeed, after the war the French and British in particular revamped and expanded science activities in Africa as part of their developmental colonialism or "second colonial occupation."[32] Developmental colonialism prioritised expertise and initiatives in science, culture, and education—the areas covered by UNESCO. In the volatile late-colonial setting of the 1950s, UNESCO's Science Sector worked together with the colonial governments in sub-Saharan Africa, for example on the educational programme for the ill-fated British East Africa Groundnut Scheme in Tanganyika.[33] Moreover UNESCO, in spite of reservations among some UNESCO staff, supported financially the Commission for Technical Cooperation in Africa South of the Sahara (CCTA). The CCTA was a pan-colonial organisation established by Britain, France, Belgium, Portugal, South Africa, and Rhodesia to pool their science resources and expertise in the Africa colonies and—according to contemporary observers—prevent international interference in colonial affairs.[34]

This form of pan-colonial science collaboration would collapse along with the French and British colonial empires in Africa. Formal decolonisation, however, opened new opportunities for UNESCO and for other UN agencies with stakes in science. As newly independent African countries joined the organisation—more than 20 between 1959 and 1962 alone—UNESCO scrambled with other international organisations to obtain a position in the development agenda for independent Africa. New initiatives included the establishment in 1965 in Nairobi of UNESCO's Regional Office for Science and Technology in Africa (ROSTA) which supplemented the field offices set up in the late 1940s.[35]

The establishment of ROSTA may be said to mark the conclusion of the formal institutional decolonisation of UNESCO science. Indeed, by the early 1960s the colonial tides had decidedly shifted. UNESCO had become firmly committed to supporting "national science"—a mode of scientific knowledge production in which nation building was key and in which scientists in recently independent countries were influenced by a national ethos underpinned by their status as civil servants.[36] The 1960s and 1970s are remembered and regarded by many in the UN system as a golden age for science in UNESCO and the UN generally. UN development programmes during the 1960s prioritised science and technology and substantial funding was available for UNESCO. Priority areas included physics and engineering, the building of universities and other research institutions, and not least programmes for training a cadre of scientific personnel and instructors in the formerly colonised world.[37]

Conservation, environmental science, and decolonisation

It is well established that colonialism was an important context for the emergence of conservation, ecology, and environmental thought as arenas of scientific and political concern.[38] Conservation and environmental sciences were also central to UNESCO's science agenda from the founding and, as noted above, in 1948 UNESCO helped to establish the IUPN.[39] In the institutional context of UNESCO these fields of science are analysed most convincingly when we regard decolonisation as a lengthy process rather than a specific historical event.

Colonial authorities with little appetite for international interference in the management of wildlife and broader land issues in their colonies often viewed with wariness the post-war internationalisation of conservation through organisations such as UNESCO and the IUPN.[40] Yet, connections were often close and interests could overlap. The international UNESCO–IUPN network that was assembled at the post-war moment had a strong representation of people with backgrounds from colonial administration and projects—a trend that continued in UNESCO's Science Sector during the ensuing decades.[41]

The UNESCO connection was crucial for conservationists during the immediate post-war moment when conservation concerns moved from a relative marginal pre-war position into the realm of power politics in the bourgeoning UN system. In 1949 UNESCO co-organised with the UN Economic and Social Council (ECOSOC) a week-long International Technical Conference on the Protection of Nature.[42] This conference broadened the narrow focus on utilisation and resource scarcity endorsed by ECOSOC into a much wider concern for sustainable management of wildlife and environmental protection based on ecological understandings of the interdependence of human beings with other species.[43]

Indeed, from the time of Huxley, ecology featured prominently in UNESCO's Science Sector. It continued to do in subsequent decades, personified by people with long careers in UNESCO such as the French ecologist Alain Gille and the energetic Michel Batiste.[44] Several historians have explored the imperial links of ecology as a scientific discipline and its transformation in after-empire contexts.[45] In UNESCO continuities in this field were also notable beyond

Figure 26.1 UNESCO specialists in photo-elastic research at the Central Water and Power Research Station in Pune, India, 1954. UNESCO collection, Paris.

the end of colonial rule. For example, at a seminal UNESCO conference in Lagos in 1964 on the Organization of Research and Training in Africa in Relation to the Study, Conservation and Utilization of Natural Resources, the British ecologist E.B. Worthington gave the opening address on behalf of UNESCO. The ecological model he presented was practically identical to the one he employed in his magisterial book *Science in Africa* from 1938, published as part of Lord Hailey's Africa Survey conducted under the auspices of the British Colonial Office.[46] For Worthington this ecological model clearly worked for the scientific management of resources and environments in imperial, international, and national contexts during all phases of what his autobiography called "The Ecological Century."[47]

As Cold War tensions began to mount, the lofty ideas of world governance and world conscience slid to the background in UNESCO. However, the wave of independence offered UNESCO and the other UN agencies new opportunities as they took over the developmental missions from the imploding colonial empires. Environmental sciences became central to UNESCO also in this context. Initiatives launched included the Arid Zone Research Project begun in 1951. The project evolved into The Major Project on Arid Lands *(*1957–64) and later fed into UNESCO's Hydrological Decade from 1965 to 1974.[48] Ecology also featured prominently in UNESCO's Arid Zone research, as was also the case in the Man and the Biosphere programme, formally established in 1971.[49]

Such major undertakings were partly funded by the UN Extended Programmes of Technical Assistance (EPTA) and drew heavily on ideas and expertise developed in colonial settings. A significant project was the World Soil Map. This project was instigated in 1961 after the Russian soil scientist Victor Kovda had become head of UNESCO's Science Sector in 1958.[50] In the decade-long project, UNESCO and the UN Food and Agriculture Organization (FAO) collaborated to make a map of the world's top soil. As historian Perrin Selcer demonstrates, the project co-produced global environmental knowledge as well as a transnational community of soil experts.[51] Colonial expertise was central in the making of the World Soil Map and the undertaking proved to be particularly attractive to the Dutch and the Belgians. Stripped of their tropical colonies, these small countries were left with little tropical soil but a surplus of soil expertise.

The reliance on colonial expertise in the World Soil Map project was part of a general trend. Indeed, UN development and research projects during the 1960s functioned, partly, as job programmes for former colonial powers. A study of the postcolonial careering of 90 British former colonial officers with expertise in the field of agriculture and natural resources shows that just over half found employment in international organisations after the end of colonial rule.[52] Postcolonial careering in the UN was clearly a general European phenomenon.[53]

In the case of the World Soil Map several imperial links have been teased out expertly by Selcer. UNESCO engaged in numerous other scientific mapmaking exercises—including oceanic maps, climatic maps, vegetation maps, and geological maps among others—which have yet to be explored in detail. The World Soil Map project also highlights the fact that many of the large-scale initiatives were joint projects shared between different UN agencies and NGOs.

Figure 26.2 UNESCO Arid Zone research in colonial Kenya: deep gully and sheet erosion from over-grassing in Suk, Western Kenya, early 1960's. J.H. Blower, UNESCO collection, Paris.

In the environmental sciences and in other fields, the relations between the international organisations, colonial bodies, and national institutions were often fraught with conflict. The contestations in the 1950s and 1960s have thus been referred to as a new imperial scramble for disciplinary territory.[54] Environmental sciences and conservation were particularly intense areas at least until the establishment in 1972 of the UN Environment Programme (UNEP).[55]

Inventing science policy and debunking race science

UNESCO stands out in the family of UN organisations because of the organisation's (partly self-proclaimed) status as the intellectual branch of the UN system. This status profoundly influenced the outlook and priorities of UNESCO in the sciences. Therefore, UNESCO offers a window to explore the place of science in what historians Jansen and Osterhammel call "the intellectual history of decolonization."[56] People like Huxley intended UNESCO to be a forum for the scientific study of a modern world in which imperialism and internationalism could be regarded as complementary forces. Two decades later, the rostrums of UNESCO served as platforms for independence leaders like India's Jawaharlal Nehru to stress the international nature of modern scientific endeavour as well as the contributions and leadership of the formerly colonised in this endeavour.[57]

Below the level of high politics, the intellectual outlook of UNESCO translated into a wide range of activities which UNESCO reports often grouped under the heading Science and Society. One of the most extensive and long-lasting undertakings was a science policy programme formally set up in 1960. Its aim was to provide a uniform global structure for national science policies among UNESCO member states.[58] In a seminal study of UNESCO's science policy programmes Martha Finnemore argues persuasively that UNESCO in this case pushed extensive and expensive science policy institutions and norms on to the former colonial world through highly unequal partnership.[59]

Other Science and Society activities were directed at wider audiences. For prospective historians, journals such as *Impact of Science on Society* (est. 1950), the widely circulated magazine the *UNESCO Courier* (est. 1948), and later *Nature and Resources* (est. 1965) provide unique avenues into discussions of international science in a decolonising world. Science exhibitions, widely distributed textbooks, and large research projects in scientific history writing such as the UNESCO General History of Mankind addressed the connections between colonialism, science, and modernity.[60] From the early 1960s much of this work aimed explicitly to challenge Eurocentric ideas and diffusion models as explanations for the development of modern science.

UNESCO was particularly active and influential in relation to race questions and anti-racism. Like many of the Science and Society activities in UNESCO, race issues were dealt with by UNESCO's Social Sciences Department. The cornerstone was constituted by the UNESCO *Statements on Race* issued in 1950, 1951, 1964 and 1967.[61] UNESCO gathered groups of international scientific experts to debunk myths about race and publish authorised scientific views on human difference. The statements aimed to dismantle racial typologies and replace them with population-based conceptions of human variation and the documents are rightly regarded as testament to the dual ascendance of population genetics and anti-racist social norms in the international community.[62]

The race statements were probably what most people associated UNESCO science with during the organisation's first decades. For historians they have proven a rich subject for investigations at the intersection of science and politics in contexts tied intimately to questions of decolonisation. The statements issued in the 1950s thus had to reconcile the anti-racist ideals of a liberal post-war world order with the realities of European colonialism. The statement from 1964 was issued after the new independent nations had joined the UN, and in the context of the non-aligned movement in which solidarity among members was based on a shared colonial history and often expressed in racial terms.[63]

The drafting and receptions of the statements were influenced by national and regional concerns.[64] South Africa is a case in point. In 1948, South Africa's Nationalist Party came to power and introduced the apartheid system based on racial segregation. UNESCO's race statements proved highly controversial for South Africa's white-minority rulers. For a few years, UNESCO's officials and South African diplomats managed to postpone the inescapable breakdown of relations which finally happened in 1956 when South African withdrew from UNESCO.[65] The parting of the ways is striking. South Africa had been a highly active founding member of UNESCO in 1945. The withdrawal of the country from the organisation—the first non-communist country to do so—only a decade after its founding is testament to the fact that, inside UN organisations, increasing anticolonial tenor and mounting criticisms of imperialism quickly created a widening institutional gap between empire and internationalism. National developments such as those in South Africa during apartheid widened them even further.

Yet, also with respect to race there are notable continuities and more subtle changes below the level of formal institutional ruptures. An important study by Gil-Riaño uses UNESCO's *Statements on Race* to relocate anti-racism in science to the Global South. He demonstrates that many of the experts who participated in the drafting of the first statements also played important roles in late-colonial, postcolonial, and international projects designed to "modernize and improve" so-called backward communities. For Gil-Riaño the statements on race reveal not the disappearance but rather the transformation of racialised imageries: "by critiquing the biological determinism, the UNESCO statements confronted scientific racism yet also legitimized relations of rescue between northern experts, southern elites and so-called backwards peoples."[66] The connection between anti-racism, late-colonial expertise, and the rise of international development was surprisingly close.

Conclusion

During the post-war decades, UNESCO operated within the tripartite structure of national, imperial, and international science. Within this structure processes of decolonisation profoundly influenced UNESCO science. This was the case inside the walls of UNESCO and in both ends of the dissolving imperial axes when projects left the headquarters and the UNESCO Field Science Offices. While essential scholarly work continues to demonstrate beyond doubt the influence of the Cold War dynamics in UNESCO, decolonisation may well be the most important structural factor in the development of science within UNESCO and the UN system in general.[67]

In the course of the decades charted in this essay, flags of empire were lowered in the research institutions created as part of colonial infrastructures of science. However, it remains disputed to what extent science *de facto* was decolonised as the colonial empires disintegrated. This is, perhaps, unsurprising given the uneven distribution of science resources between Global North and Global South which persists to this day—a disparity UNESCO has the capability to document but not to change in any significant way.[68] According to some scholars, science remains a hegemonic, imperial structure inherently geared to suit the interests of the Global North.[69] This debate continues, as does the more specific discussions concerning the role of

UN and UNESCO in international science. However, even those who view UNESCO's influence in a positive light rightly point out that decline in financial resources and prestige prevents UNESCO from keeping the level of influence in international science it once enjoyed, let alone expand the influence.

Notes

1 Mark Mazower, *No Enchanted Place: The End of Empire and the Ideological Origins of the United Nations* (Princeton, NJ: Princeton University Press, 2009), 17. For UN internationalism as a decolonising factor see Akira Iriye, *Cultural Internationalism and World Order* (Baltimore, MD: Johns Hopkins University Press, 1997). A well-balanced interpretation is Glenda Sluga, *Internationalism in the Age of Nationalism* (Philadelphia: University of Pennsylvania Press, 2013).

2 Helen Tilley, *Africa as a Living Laboratory: Empire, Development and the Problem of Scientific Knowledge 1870–1950* (Chicago: Chicago University Press, 2011), 9–10.

3 UNESCO has been involved in an overwhelming number of science activities and projects—and probably more diverse in kind than any other international organisation given UNESCO's broad mandate in science, education, and culture. Today, UNESCO's online repository, UNESDOC, makes much material on these activities readily available but no less overwhelming. A great service was done with the publication in 2006 of the massive volume *Sixty Years of Science at UNESCO 1945–2005* which is essential for locating projects, key people, themes and UNESCO's connections with other institutions. It contains also the best analytical essay charting overarching themes and trends in the history of UNECO's Science Sector. See Malcolm Hadley and Lotta Nuotio, "Partnership in Science. Cross-cutting Issues in UNESCO's Natural Sciences Programme," in *Sixty Years of Science at UNESCO 1945–2005*, ed. P. Petitjean, V. Zharov, G. Glaser, J. Richardson, B. de Padirac, and G. Archibald (Paris: UNESCO Publishing, 2006), 507–71. As one perceptive reviewer of the *Sixty Years of Science at UNESCO* observes, "the lack of depth in description of work 'in the field' may accurately reflect the organization's program that attempted to 'encourage' and 'coordinate action' with limited resources." Perrin Selcer, "Review *Sixty Years of Science at UNESCO*," *Isis* 99, no. 1 (2008): 219.

4 Geert J. Somsen, "A History of Universalism: Conceptions of the Internationality of Science from the Enlightenment to the Cold War," *Minerva* 46, no. 3 (2008): 361–79.

5 Aant Elzinga, "UNESCO and the Politics of Scientific Internationalism," in *Science and Internationalism*, ed. Aant Elzinga and Catharina Landström (London: Taylor Graham, 1996), 164–99.

6 Chloé Maurel, "Huxley, Sir Julian Sorell," in *IO BIO, Biographical Dictionary of Secretaries-General of International Organizations*, ed. Bob Reinalda, Kent J. Kille, and Jaci Eisenberg, 2012, www.ru.nl/fm/iobio.

7 Chinua Achebe, *The Education of a British-Protected Child: Essays* (London: Penguin Group, 2009), 21.

8 Roy MacLeod, "Passages in Imperial Science: From Empire to Commonwealth," *Journal of World History* 4, no.1 (1993), 146–7; John Toye and Richard Toye, "One World, Two Cultures? Alfred Zimmern, Julian Huxley and the Ideological Origins of UNESCO," *History* 95, no. 319 (2010): 308–31.

9 Glenda Sluga, "UNESCO and the (One) World of Julian Huxley," *Journal of World History* 22, no. 3 (2010): 397.

10 Kenneth C. Waters and Albert Van Helden, eds., *Julian Huxley, Biologist and Statesman of Science: Proceedings of a Conference Held at Rice University, 25–27 September 1987* (Houston, TX: Rice University Press, 1992).

11 Julian Huxley, *UNESCO: Its Purpose and Its Philosophy* (London: Public Affairs Press, 1946).

12 Julian Huxley, *The Conservation of Wild Life and Natural Habitats in Central and East Africa. Report on a Mission Accomplished for UNESCO, July–September 1960* (Paris: UNESCO Publishing, 1961). The IUPN was since renamed IUCN (International Union for the Conservation of Nature and Natural Resources) and is now known as the World Conservation Union.

13 Gail Archibald, "How the 'S' came to be in UNESCO," in Petitjean et al., *Science at UNESCO*, 36–40; Casper Andersen, "The Zero Hours of Technology and the Founding of UNESCO," in *Zero Hours*, ed. Hagen Schulz-Forberg, vol. 2 (London: Palgrave Macmillan, forthcoming), Chapter 6.

14 Gary Wersky, *The Visible College: The Collective Biography of British Scientific Socialists of the 1930s* (London: Allen Lane, 1978).

15 Thomas Mougey, "Needham at the Crossroads: History, Politics and International Science in Wartime China (1942–1946)," *The British Journal for the History of Science* 50, no. 1 (2017): 83–109.

16 Joseph Needham, "The Place of Science and International Science Cooperation in the Postwar World Organization," *Nature,* no. 3987 (1945): 558–61; Joseph Needham, "Natural Sciences: Practical Steps for International Co-operation among Scientists," *UNESCO Courier 1,* no. 1 (1948): 2.

17 Thomas Mougey. *Enlightening the Dark Zone: UNESCO, Science and the Technocratic Reordering of the World in the Global South, 1937–1959* (doctoral diss., Maastricht University, 2018), doi.org/10.26481/dis.20181219tm.

18 For one example see Julian Huxley, "Colonies in a Changing World," *Free World* 315 (1942): 314–23.

19 Patrick Petitjean, "The Joint Establishment of the World Federation of Scientific Workers and of UNESCO after World War II," *Minerva* 46, no 2 (2008): 269.

20 Lynn Meskell, *A Future in Ruins: UNESCO, World Heritage, and the Dream of Peace* (Oxford: Oxford University Press, 2018), 24.

21 Frank Greenaway, *Science International: A History of the International Council of Scientific Unions* (Cambridge: Cambridge University Press, 1996).

22 Casper Andersen, "Internationalism and Engineering in UNESCO during the End Game of Empire, 1943–68," *Technology and Culture* 58, no. 3 (2017): 650–77.

23 Andersen, "Internationalism and Engineering," 676.

24 UNESCO, "Special Resolution G", *General Conference 1946, First Session* (Paris: UNESCO Publishing, 1946), 232; Jürgen Hillig, "Going Global. UNESCO's Field Science Offices," in Petitjean et al., *Science at UNESCO,* 72–6.

25 Marcel Florkin, "Ten Years of Science at UNESCO," *Impact of Science on Society* 7, no. 3 (1956): 121–47.

26 UNESCO, *The Field Scientific Liaison Work of UNESCO* (Paris: UNESCO Publishing, 1950), 18.

27 Hillig, "Going Global," 74.

28 Malcolm Hadley, "Nature to the Fore. The Early Years of UNESCO's Environmental Programme," in Petitjean et al., *Science at UNESCO,* 205–9.

29 Heloisa Maria Bertol Domingues and Patrick Petitjean, "International Science, Brazil and Diplomacy in UNESCO (1946–50)," *Science, Technology and Society* 9, no. 1 (2004): 36.

30 Mougey, *Enlightening the Dark Zone.*

31 Poul Duedahl, ed., *A History of UNESCO: Global Actions and Impacts* (London: Palgrave Macmillan, 2016); Ivan Lind Christensen and Christian Ydesen, "Routes of Knowledge: Toward a Methodological Framework for Tracing the Historical Impact of International Organizations," *European Education* 47, no. 3 (2015): 274–88.

32 John Lonsdale and D.A. Low, "Towards the New Order 1945–1963," in *History of East Africa, III,* ed. D.A. Low and A. Smith (Oxford: Clarendon Press, 1976), 1–64; Frederick Cooper, *Africa since 1940: The Past of the Present* (Cambridge: Cambridge University Press, 2002), chapter 2.

33 UNESCO, Meeting between UNESCO and the British East African Groundnut Corporation, 24 September, 1947, UNESDOCS EDUC/52, Paris.

34 Isibill Gruhn, "The Commission for Technical Co-Operation in Africa 1950–1965," *Journal of Modern African Studies* 9, no. 3 (1971): 459–69; Casper Andersen, "'Scientific Independence,' Capacity Building, and the Development of UNESCO's Science and Technology Agenda for Africa," *Canadian Journal of African Studies / Revue canadienne des études africaines* 50, no. 3 (2016): 379–94.

35 R. Maybury, "Meanwhile in the Motherland: Regional Office for Science and Technology in Africa," in Petitjean et al., *Science at UNESCO,* 75–6.

36 J. Gaillard and R. Waast, *Science in Africa at the Dawn of the 21st Century* (Paris: IRD Publishing, 2000); P.W. Geissler, "Parasite Lost: Remembering Modern Times with Kenyan Government Medical Scientists," in *Evidence, Ethos and Experiment: The Anthropology of Medical Research in Africa,* ed. P.W. Geissler and Catherine Molyneux (New York: Berghahn Books, 2011), 297–333.

37 Klaus Heinrich Standke, "Science and Technology in Global Cooperation: The Case of the United Nations and UNESCO," *Science and Public Policy* 33, no. 9 (2006): 627–46.

38 Caroline Ford, "Nature, Culture and Conservation in France and Her Colonies 1840–1940," *Past & Present* 183 (2004), 173–98. William Adams, *Against Extinction: The Story of Conservation* (London: Routledge, 2013); William Mark Adams, William Mark, and Martin Mulligan, eds., *Decolonizing Nature: Strategies for Conservation in a Post-Colonial Era* (London: Earthscan, 2003)

39 Hadley, "Nature to the Fore"; Martin Holdgate, *The Green Web: A Union for World Conservation* (Oxford: Routledge, 2014).

40 Roderick P. Neumann, "The Postwar Conservation Boom in British colonial Africa," *Environmental History* 7, no. 1 (2002): 22–47.

41 Anna-Katharina Wöbse, " 'The World After All Was One': The International Environmental Network of UNESCO and IUPN, 1945–1950," *Contemporary European History* 20, No. 3 (2011): 331–48.

42 International Union for the Protection of Nature, ed., *International Technical Conference on the Protection of Nature, Lake Success, 22–28 Aug. Proceedings and Papers* (Paris: UNESCO Publishing, 1950).

43 Daniel Speich, "Der Blick von Lake Success: Das Entwicklungsdenken der frühen UNO als lokales Wissen," *Entwicklungswelten* 6 (2009): 143–74; Thomas Jundt, "Dueling Visions for the Postwar World: The UN and UNESCO 1949 Conferences on Resources and Nature, and the Origins of Environmentalism," *Journal of American History* 101, no. 1 (2014): 44–70.

44 "A Tribute to Two of our Own: Michel Batisse and Yvan de Hemptinne," in Petitjean et al., *Science at UNESCO*, 671–2; Andersen, "Scientific Independence."

45 Peder Anker, *Imperial Ecology: Environmental Order in the British Empire, 1893–1945* (Cambridge, MA.: Harvard University Press, 2001); Helen Tilley " 'African Environments and Environmental Sciences: The African Research Survey, Ecological Paradigms, and British Colonial Development, 1920–1940," in *Social History and African Environments*, ed. William Beinart and Jo Ann McGregor (Oxford: Heinemann/James Currey Press 2003), 109–30; Libby Robin, " 'Ecology: A Science of Empire?," in *Ecology and Empire: Environmental History of Settler Societies*, ed. Tom Griffiths and Libby Robin (Edinburgh: Keele University Press), 63–75.

46 E. Barton Worthington, *Science in Africa: A Review of Scientific Research Relating to Tropical and Southern Africa* (London: Oxford University Press, 1938); UNESCO, *International Conference on the Organization of Research and Training in Africa in Relation to the Study, Conservation and Utilization of Natural Resources; Lagos; 1964. Selected Documents* (Paris: UNESCO Publishing 1965), 9–20.

47 E. Barton Worthington, *The Ecological Century: A Personal Appraisal* (Oxford: Clarendon Press, 1983).

48 Hadley, "Nature to the Fore," 211–20.

49 Peter Bridgewater, "The Man and Biosphere Programme of UNESCO: Rambunctious Child of the Sixties, But Was the Promise Fulfilled?," *Current Opinion in Environmental Sustainability* 19 (2016), 1–6; Ruida Pool-Stanvliet, "A History of the UNESCO Man and the Biosphere Programme in South Africa," *South African Journal of Science* 109, no. 9–10 (2013): 1–6.

50 Elzinga, *Unesco and the Politics*, 183–4. The Soviet Union joined UNESCO in 1955.

51 Perrin Selcer, "Fabricating Unity: the FAO-UNESCO Soil Map of the World," *Historical Social Research* 40, no. 2 (2015): 174–201.

52 Joseph M. Hodge, "British Colonial Expertise, Post-colonial Careering and the Early History of International Development," *Journal of Modern European History* 8, no. 1 (2010): 24–44.

53 Véronique Dimier, "For a New Start: Resettling French Colonial Administrators in the Prefectoral Corps," *Itinerario* 28, no.1 (2004): 49–66.

54 Neumann, "The Postwar Conservation."

55 Iris Borowy, "Before UNEP: Who Was in Charge of the Global Environment? The Struggle for Institutional Responsibility 1968–72," *Journal of Global History* 14, no. 1 (2019): 87–106.

56 Jan Jansen and Jürgen Osterhammel, *Decolonization: A Short History* (Princeton, NJ: Princeton University Press, 2017), 168.

57 David Arnold, "Nehruvian Science and Postcolonial India," *Isis 104*, no. 2 (2013): 360–70.

58 Yvan de Hemptinne, "The Science Policy of States in Course of Independent Development," *Impact of Science on Society* 12, no. 3 (1960): 233–47.

59 Martha Finnemore, "International Organizations as the Teachers of Norms: UNESCO and Science Policy," *International Organizations* 47 (1993): 565–98.

60 Poul Duedahl, "Selling Mankind: UNESCO and the Invention of Global History, 1945–1976," *Journal of World History* 22, no. 1 (2011): 101–33; Paul Betts, "Humanity's New Heritage: Unesco and the Rewriting of World History," *Past & Present* 228, no. 1 (2015): 249–85.

61 Ashly Montagu, *Statement on Race: An Annotated Elaboration and Exposition of the Four Statements on Race Issued by UNESCO* (Oxford: Oxford University Press, 1972).

62 Jenny Bangham, "What Is Race? UNESCO, Mass Communication and Human Genetics in the Early 1950s," *History of the Human Sciences* 28, no. 5 (2015): 80–107; Elazar Barkan, "The Politics of the Science of Race: Ashley Montagu and UNESCO's Anti-Racist Declarations," in *Race and Other Misadventures: Essays in Honor of Ashley Montagu in his Ninetieth year*, ed. Larry T. Reynolds and Leonard Liberman (New York: Rowman and Littlefield, 1996), 96–105.

63 Perrin Selcer, "Beyond the Cephalic Index: Negotiating Politics to Produce UNESCO's Scientific Statements on Race," *Current Anthropology* 53, no. 5 (2011): 173–84.

64 Bram Harkema and Fenneke Sysling, "Dutch Scientists and the UNESCO Statement on Race," *Studium* 11, no. 4 (2019), doi.org/10.18352/studium.10181; Michelle Brattain, "Race, Racism, and Antiracism: UNESCO and the Politics of Presenting Science to the Postwar Public," *The American Historical Review* 112, no. 5 (2007): 1386–413.

65 Michelle Brattain, "South Africa's 'Strange' Relations with UNESCO: Antiracism Versus Apartheid," in *UNESCO Without Borders: Educational Campaigns for International Understanding*, ed. Aigul Kulnazarova and Christian Ydesen (London: Routledge, 2016), 220–37.

66 Sebastian Gil-Riaño, "Relocating Anti-Racist Science: The 1950 UNESCO Statement on Race and Economic Development in the Global South," *The British Journal for the History of Science* 51, no 2 (2018): 303.

67 Louis H. Porter, "Cold War Internationalisms: The USSR in UNESCO 1945–1967" (PhD diss., University of North Carolina at Chapel Hill, 2018), https://cdr.lib.unc.edu/concern/dissertations/f7623d391?locale=en; Perrin Selcer, *The Postwar Origins of the Global Environment: How the United Nations Built Spaceship Earth* (New York: Columbia University Press, 2018).

68 UNESCO, *UNESCO World Science Report. Towards 2030*, https://en.unesco.org/unesco_science_report.

69 For insightful contributions to this debate see F. Collyer, R. Connell, J. Maia, and R. Morell, *Knowledge and Global Power: Making New Sciences in the South* (Clayton: Monash University Publishing, 2019); Paulin Hountondji, "Producing Knowledge in Africa Today: The Second Bashorun MKO Abiola Distinguished Lecture," *African Studies Review* 38, no 3 (1995): 1–10.

27

DECOLONISING SCIENCE AND MEDICINE IN INDONESIA

Hans Pols

On 17 August 1945, two days after Japan capitulated, Sukarno and Mohammad Hatta declared Indonesia's independence. A small number of Indonesian scientists and politicians already had started formulating plans for higher education and scientific research in the newly independent nation. They faced the challenge of rebuilding and transforming the remaining colonial academic institutions in a new social and political environment. By having Indonesians occupy most positions in universities and research institutions, by designing an Indonesian scientific and medical vocabulary, and by aligning teaching and research to the needs of the developing nation, they sought to decolonise science and higher education. After independence, there was a continuing need to negotiate the competing demands of furthering free scientific inquiry and assisting the nation's development. Recently, the Indonesian scientific community has encountered difficulties in producing internationally recognised research because of limited access to funds and resources. To protect the Indonesian scientific community, controversial new conditions have been legislated for international scientists conducting research in Indonesia. Nonetheless, unequal access to resources necessary for conducting high-quality scientific research forces Indonesian scientists and physicians to conduct research under neocolonial conditions.

Proposals for decolonising science and medicine had already been articulated by Indonesian physicians in the Dutch East Indies. In this chapter, I analyse their efforts to counter the colonial nature of science and medicine, which had relegated them to secondary positions. Their desire for emancipation focused on gaining equality with their European colleagues with respect to education, income, and placement within the colonial public health service. Outside of medicine, the number of Indonesians with advanced scientific qualifications remained negligible. I continue by reviewing plans for higher education and research formulated by the Ministry of Higher Education and Culture after independence, which aimed to fulfil the ideals articulated during the colonial era. Because there were sufficient university-educated physicians to staff medical schools, it was possible to fully Indonesianise medical education and research. Unfortunately, this was not the case in other scientific disciplines. I conclude by discussing the current aims of the Indonesian government to address the unequal access to resources for scientific research by regulating the activities of international scientists in the country—attempting to ensure that research activities include Indonesian scientists and benefit the nation.

Agitating against colonial medicine: Indonesian physicians in the Dutch East Indies

In the Dutch East Indies, medical students were the only Indonesians who received a scientific training including the application of experimental methods. Despite the length and rigour of their education, Indonesian physicians received very modest salaries in comparison to their European colleagues. In 1909, a number of leading Indonesian physicians founded the Association of Native Physicians (Vereeniging van Inlandsche Artsen) to improve the social position *vis-à-vis* their European colleagues.[1] All of them were graduates of the Batavia medical college (School tot Opleiding van Inlandsche Artsen, STOVIA; or School for the Education of Native Physicians).[2] In 1913, after a medical college opened in Surabaya, from that year on, both colleges granted the degree "Indisch Arts" (Indies Physician).[3] Most medical students were enthralled by the promises of modern science and medicine to improve conditions in the Indies. Upon graduation, they soon learned that the organisation of medical care in the colonies had a dual or two-tier nature: European physicians held all leading positions while their Indonesian counterparts were placed in subordinate positions, earning significantly less. The degrees awarded by the Batavia and Surabaya medical colleges were considered inferior to those awarded by Dutch university-affiliated medical schools. In the colonies, medical education consisted of a tightly scheduled course with compulsory class attendance, supervised study, and regular examinations—all characteristics of Dutch professional schools. Dutch medical schools, in contrast, were research-oriented, emphasising independent study. Examinations were conducted annually on an individual basis. The higher appraisal of Dutch medical degrees thereby reinforced the derogatory views held by colonial Dutchmen about their Indonesian colleagues.

Most European physicians working in the Dutch East Indies disparaged their Indonesian colleagues, claiming that they could only work as supervised assistants. In 1908, the director of the Indies Health Service presented a novel justification for the latter's inferior position. While he complimented Indies physicians by calling them the cornerstone of the colonial health service since they were trusted by the Indigenous population, he argued that it was essential that they did not become alienated from their ethnic communities. Providing them with a simple and mostly practical medical education and keeping their salaries modest could accomplish this.[4] Most Indonesian physicians, as well as the director of the Batavia medical college, vehemently opposed these ideas.[5] The desire of Indies physicians to gain equality with their European colleagues required expanding the medical curriculum and earning the same salary.

In response to the dualism in colonial medicine, several Indonesian physicians enrolled in university-affiliated medical schools in the Netherlands to attain Dutch medical degrees. Abdul Rivai and the Tehupeiory brothers were among the first to do so.[6] However, the completion of a Dutch medical degree did not ensure their status as "full" physicians: upon their return to the Indies, they were treated with ambivalence by their European colleagues. To prevent affronting Dutch physicians, W.K. Tehupeiory, for example, was placed in a remote village on the island of Bangka; his closest Dutch colleague practised more than 100 kilometres away, ensuring that they rarely encountered each other. Many other Indonesians with Dutch medical degrees were deeply disappointed by the lack of respect accorded to them after they had returned to the Indies.[7] The Association of Indies Physicians nevertheless continued to advocate for scholarships to enable the most able Indies physicians to attain Dutch medical degrees. It hoped that increasing their number would eventually improve their position in the colonial medical service. Several Indonesian physicians and medical students became involved in what has been called the Indonesian nationalist movement; many of them advocated for more health care facilities and improvements in the position of Indies physicians.[8]

A second strategy employed by the Association of Indies Physicians involved advocating for the establishment of a full-fledged medical school in the Indies, which would award degrees equivalent to those at European universities. By 1910, several progressive intellectuals established a committee supporting the formation of a university in the colonies.[9] The Dutch government responded by organising a number of official meetings to discuss these proposals.[10] In 1918, Rivai, then a member of the colonial parliament (Volksraad), promoted establishing university-level medical education in the colonies.[11] He presented an elaborate proposal on how to accomplish this, which was co-authored by the directors of the Batavia and Surabaya medical colleges and published by the Association of Indies Physicians.[12] When full medical degrees could be obtained in the Indies, Rivai thought, the number of fully qualified Indonesian physicians would increase, which in turn would improve their professional status, realising their emancipation.

It was not until 1927 that the Batavia Medical School (Geneeskundige Hoogeschool) opened, accepting European and Indigenous students. Teaching followed the Dutch academic model and the course awarded the same degree as those of Dutch medical schools. Unfortunately, tuition fees were high and stipends that had until then been awarded to medical students were discontinued. Consequently, only a few Indonesians were able to enrol.[13] Despite this, the number of Indonesian physicians with university medical degrees increased slowly but steadily. After 1925, the Association of Indies Physicians showcased their research in its *Bulletin* to demonstrate that they were able to conduct high-quality research equal to that of their European colleagues.[14] In the 1930s, these physicians also started to publish in the *Medical Journal of the Dutch Indies* (*Geneeskundig Tijdschrift voor Nederlandsch-Indië*), the journal

Figure 27.1 Surabaya medical college (NIAS) students in class with microscopes, c. 1930. From a photobook by Dr Soenarjo. Hans Pols collection.

of the elite Association for the Advancement of Medical Science in the Dutch Indies. Until the 1920s, this journal had exclusively published the research of European physicians.[15]

In 1939, Indies physician Abdul Rasjid, a member of the colonial parliament from 1931 to 1942 and, after 1938, president of the then renamed Association of Indonesian Physicians, presented a radical approach to decolonise the Indies Public Health Service. In a speech to the colonial parliament, he argued that Indies physicians had a superior ability to gain the trust of the population of the archipelago. The colonial health service, he continued, would function more effectively if it was fully Indonesianised. European physicians, in the meantime, could focus on private practice, which was much more lucrative anyway.[16] In a fractious manner, Rasjid subverted the arguments that had been used to curtail the medical curriculum at the Indies medical colleges and limit the remuneration of its graduates. Not surprisingly, the head of the colonial Public Health Service disparaged Rasjid's suggestions, praising the many contributions European physicians had made to the health of the Indonesian population thus far.

Rasjid's zeal to decolonise colonial medicine motivated his interest in traditional herbal medicine or *jamu*. According to him, pharmaceuticals imported from Germany and other Western countries were expensive, artificial, and, above all, foreign. The archipelago's traditional herbal medicine, which had been prepared and sold by many generations of women and traditional healers, could potentially provide an affordable alternative. In the colonial parliament, Rasjid suggested establishing a committee for the scientific investigation of traditional herbal medicine.[17] During the second congress of the Association of Indonesian Physicians, two leading physicians gave presentations on traditional herbal medicine—in Malay, marking the first time that Indonesian physicians used this language in an official setting.[18] In making his arguments, Rasjid continued to uphold European medical research methods as the ultimate arbiter of efficacy. During the Japanese occupation, interest in traditional herbal medicine intensified as standard pharmaceuticals were no longer available. Traditional herbal medicine was presented with an increased nationalistic ardour.[19] Since independence, *jamu* has at times been seen as an alternative to Western pharmaceutical products; more generally, it is served as a supplement to these.[20] From the 1920s on, it has been industrially produced and is still widely consumed.[21] They are generally seen as supplementing medications rather than constituting an alternative.

During colonial times, Indonesian physicians repeatedly protested against the colonial nature of medical education and the oppressive organisation of the Indies health service. They were steadfast in countering the dual nature of medicine, which accorded them a secondary role, by requesting scholarships for Indies physicians to attain full medical qualifications in the Netherlands, and by advocating for a university-level medical school in the Dutch East Indies. In the 1930s, some of them made more radical demands to Indonesianise the colonial health service and to investigate traditional herbal medicine. At the same time, they established contacts with Indigenous physicians in other Asian countries, which assisted them in formulating alternatives to medical care in the Indies.

The Japanese occupation and the war of independence

The Japanese occupation played a significant role in the decolonisation of medicine in Indonesia—paradoxically, even as the archipelago became part of the Japanese colonial empire. A few months after the Dutch capitulation, most Dutch physicians were interned, and all institutions of higher education, including both the Batavia Medical School and Surabaya Medical College, were closed. The Japanese military administration later favourably responded to requests to re-open the Batavia Medical School, which was renamed Ika Daigaku and enrolled students from 29 April 1943.[22] Officially, a Japanese dean and a small number of

Japanese professors directed operations. Several leading Indonesian physicians received official appointments in the medical school, the public health service, and research institutions. For the first time, Indonesians were in charge of medical teaching, research, and health care. The Japanese had outlawed the use of the Dutch language in public life; medical teaching therefore had to be conducted in Malay. Indonesian physicians established a committee to formulate Malay counterparts of Dutch and Latin medical terms to make that language suitable for medicine.[23] In the meantime, the Association of Indonesian Physicians was revived. Thus, under the much harsher colonial regime of the Japanese, Indonesian medicine undertook successful efforts to decolonise itself from the former Dutch colonial administration.

On 17 August 1945, two days after the Japan imperial army capitulated, Sukarno and Mohammad Hatta declared Indonesia's independence during a modest ceremony in front of Sukarno's residence close to the medical school. A group of leading Indonesian physicians began formulating plans to revive medical education and re-establish health services two days later. Three weeks later, the Japanese dean of the Ika Daigaku symbolically handed the keys of the building to Professor Sarwono Prariwohardjo, his assistant during the previous three years.[24] The next day, teaching resumed. Indonesian independence was not recognised by the Netherlands, which quickly sought to re-establish control over their former colonies, leading to prolonged and violent conflicts during the next four and a half years. With increasing Dutch military presence in the capital, most medical education moved to Central Java. In January 1946, the Dutch military administration established the Emergency University in Jakarta.[25] Research and higher education remained important to the benign image it wanted to convey to both Indonesians and the international world. The magazines of the Dutch Red Cross assisted in disseminating the benevolent nature of Dutch medical work in the areas it had occupied.[26]

During Indonesia's war of independence (1945–9), the Netherlands Indies Civil Administration re-established higher education and several pre-war research institutions. They organised these according to Dutch pre-war, colonial ideas about the dual nature of colonial science: higher education and research were essentially a privilege for the Dutch while Indonesian technicians could assist and apply the results.[27] In both politics and science, the Dutch were convinced that Indonesians continued to need their guidance and that the dual organisation of science and medicine needed to be maintained. In 1947, a group of 62 former colonial scientists expressed their concerns to the Minister of Colonial Affairs about the future of scientific research in the archipelago if the Indonesian Republic assumed power—they were convinced that all scientific infrastructure would be destroyed.[28] At the same time, many Dutch physicians derided their Indonesian colleagues as lacking initiative, originality, motivation, intelligence, and the ability to think independently and logically.[29] In the last few years of the Dutch Empire in the Indonesian archipelago, derogatory impulses were still common among Dutch scientists and physicians. Their colonial mindset was unusually persistent.

Decolonising science and medicine in independent Indonesia

Anticipating the imminent demise of the Dutch East Indies, the European scientific elite celebrated their scientific accomplishments.[30] Yet despite this comforting nostalgia, it soon became clear that the Dutch colonial administration had taken little initiative in providing any form of education to Indonesians, apart from a small group who occupied essential positions in the colonial bureaucracy. In 1942, over 90 per cent of Indonesians were illiterate and there were approximately 500 Indies physicians and 30 physicians with academic qualifications, a paltry result despite the fact that the colonial administration had invested more in medical education than in all other forms of higher education. In 1924, a Law School had opened in Batavia and a Technical School in Bandung, which had graduated several dozen Indonesians by 1942. Only a handful of Indonesians had received higher education in any other field. There were several research facilities that had been led by Dutch scientists, some of which had stellar international reputations.[31] With a few exceptions, scientific research in the Dutch East Indies had entirely remained a Dutch affair.

After the transfer of sovereignty on 27 December 1949, the political and organisational context for science, medicine, and higher education changed dramatically. Under the direction of the Indonesian government, colonial institutions for higher education were transformed to educate professionals to staff the government bureaucracy, schools, and businesses of the newly independent state.[32] For a few years, most Dutch scientific institutions continued to function relatively unchanged, although scientists felt pressured by the Indonesian government to hire more Indonesians. This was difficult to achieve, because few had the required qualifications. Scientific organisations founded by the Dutch were either reorganised or disbanded. The elite Association for the Advancement of Medical Research in the Dutch Indies (Vereeniging tot Bevordering der Geneeskundige Wetenschappen in Nederlandsch Indië), which had been founded in 1851, was liquidated after the Indonesian Medical Association (Ikatan Dokter Indonesia) was established.[33] The (Dutch) Organisation for Scientific Research, which had been founded in 1949 to coordinate all scientific research in the colonies, was slowly transformed as it was staffed by an increasing number of Indonesians.[34] In 1956, for example, the former Dean of Medicine, Sarwono Prawirohardjo, became the chairman of the Indonesian Council of Sciences (today's Lembaga Ilmu Pengetahuan Indonesia or LIPI).

After 1950, Indonesians faced a daunting task in building viable government institutions, an infrastructure, and an economy despite the severe economic problems, political instability, and violent insurrections. Transforming what had remained of colonial institutions for research and higher education into ones suitable for their developing nation was equally challenging.[35] Indonesian scientists aimed to decolonise science, build an Indonesian scientific community, conduct research, and shape higher education to meet the demands of the newly independent republic. In 1951, the Ministry of Education and Culture inaugurated a programme to develop higher education and define the parameters for scientific research.[36] First, the demands of the newly independent state took precedence over those of pure research—or, as it was phrased at the time, research and higher education had to develop in harmony with the state. Second, to foster the growth of the Indonesian scientific community, Indonesians had precedence when hiring staff at universities and research centres. To prepare young academics to embark on their career in science, they were sent abroad for advanced study. Third, using the Dutch language was outlawed and Indonesian preferred; English was tolerated but few instructors and students had mastered that language. Finally, instead of graduating a small number of highly skilled researchers, universities were tasked with producing large numbers of teachers, physicians, engineers, and lawyers. At this time, decolonising science and higher education was defined as Indonesianising both. Physicians, long active in the Indonesian nationalist movement in colonial times, took the lead.[37]

After the Japanese capitulated, most Dutch citizens living in Indonesia opted to migrate to the Netherlands. Several Dutch scientists, however, remained in Indonesia out of idealistic motives, because their research was based there, or in response to a variety of practical considerations. Most did not stay long because of "matters of prestige, privilege, and salary."[38] In addition, many disagreed with the science policies of the Indonesian government, opportunities to conduct research declined, and salaries were decimated as a consequence of economic problems.[39] When the conflict between Indonesia and the Netherlands over West Papua heightened in 1958, all Dutch citizens were expelled—except for those whose services were considered indispensable.[40] The exodus of Dutch scientific personnel led to skilled-labour shortages, negatively affecting research and higher education in Indonesia.

The decolonisation of medicine turned out not to be a difficult undertaking because there were enough university-educated physicians and medical researchers to take the place of their former European colleagues. During the Japanese occupation, these Indonesian physicians had already been actively organising higher education and health services. Initially, medical education was based on the so-called "free study" model common at Dutch universities and the Batavia Medical School. To meet the nation's urgent need for large numbers of physicians, however, teachers at the Faculty of Medicine at the University of Indonesia decided to reform medical education in line with the American example, which they labelled "guided study."[41] Education became cohort-based, attendance at lectures and practicals compulsory, and examinations were held regularly to check each student's progress. The University of Indonesia formed an alliance with the University of California at San Francisco (UCSF) in 1952 to realise this transformation. American lecturers spent several months every year in Jakarta while Indonesian medical staff travelled to the United States for advanced instruction. The number of medical graduates steadily increased in the early 1960s.[42]

Medicine was more readily decolonised in independent Indonesia than other sciences and academic disciplines.[43] There were few university-educated researchers in biology, botany, chemistry, physics, or any other field. It is instructive to compare the case of medicine with the research conducted at the botanical gardens (Kebun Raya Indonesia)—the flagship scientific institution during colonial times—after 1950. Indonesian policy makers supported the botanical gardens because of their international prestige. As Andrew Goss has demonstrated, its new Indonesian director "plotted how to build Indonesian science without Indonesian scientists" because there were hardly any sufficiently qualified Indonesians to receive appointments.[44] To remedy this untenable situation he decided to incorporate higher education into the core tasks of the botanical gardens, which would educate Indonesian botanists who could then conduct research there.[45]

To decolonise science and medicine in Indonesia effectively, it was necessary to further loosen the connections to Dutch scientists and to strengthen those with Asian ones. The realisation that the Dutch scientific community was only a small part of its international English-speaking counterpart aided these efforts. In 1958, the first National Science Congress (now Kongres Ilmu Pengetahuan Nasional) was held in Malang, followed by the second one in Yogyakarta in 1962. As Adrian B. Lapian, the coordinator of the second congress, stated: "It was the first time that Southeast Asian scholars had worked together. Before independence, each of them was oriented towards their respective colonial metropoles."[46] Starting during the 1960s, Indonesian scientists increasingly associated themselves with the cosmopolitan world of science. However, it remained difficult for them to conduct research that was recognised by their colleagues from the West. If, on the one hand, they addressed local problems, scientists outside Indonesia generally were not interested. If, on the other hand, they conducted fundamental scientific research, they often lacked the resources to do so in a way that impressed their colleagues from the West. International economic disparity adversely affected, and continues to affect, the research that scientists in lower- and middle-income countries can accomplish.

Indonesian science in the global world: recent developments

Since independence, Indonesian scientists and physicians have made great strides in decolonising Indonesian science and medicine. Unfortunately, lack of access to resources continues to impede progress. Subscriptions to academic journals are notoriously expensive, as

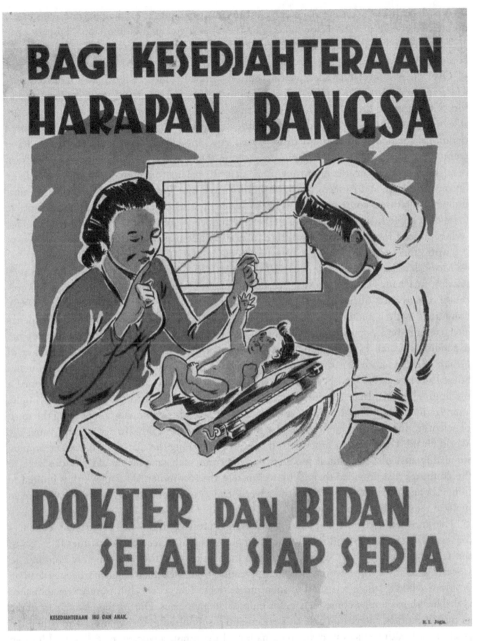

Figure 27.2   Indonesian public health education poster on maternal and child health. The text reads: "For the nation's welfare and hope: physicians and midwives are always ready and available," c. 1960. Hans Pols collection.

are laboratory equipment, biological and medical solutions and samples, and other research materials. Since around 1980, the context in which scientific research is conducted changed significantly, exacerbating this predicament. Encouraged by American president Ronald Reagan, the number of patents claimed by scientists, business, and universities has steadily

307

increased, facilitating commercialisation. At the same time, public funding for research steadily diminished, while private-sector funding has become increasingly important.[47] Disparities in access to resources for scientific research between wealthy, and lower- and middle-income countries have consequently increased.

Over the past several decades, pharmaceutical companies increasingly have conducted clinical trials in lower- and middle-income countries, partly because of the friendlier regulatory environment there.[48] In many cases, this is the only way patients in such locales are able to obtain the latest medications.[49] Other companies have appropriated and explored traditional medicines in attempts to isolate effective ingredients that could be patented and commercialised, a practice called bio-prospecting. During the 1990s, for example, Western companies and universities took out more than 70 patents on ingredients and uses of the neem tree, its oil, and its seeds, used for centuries as a source of traditional medicine in India. Indigenous farmers consequently lost control over their seeds, and they became unaffordable for most people in India.[50] Patents, intellectual property law, and bio-prospecting advantage well-resourced researchers from wealthier nations, while relegating scientists from the rest of the world to a secondary position.

Medical researchers in Indonesia have experienced this new regime of knowledge production first-hand. During the first decade of the twenty-first century, Asia became the epicentre of outbreaks of avian influenza (H5N1) and severe acute respiratory syndrome (SARS), conditions with potential for disastrous global consequences. In particular, avian influenza affected large bird populations in Indonesia, leading to intensified containment strategies and expanding research by international health organisations.[51] In 2007, Indonesia's Minister of Health became aware that the samples of flu strains it had provided to the Global Influenza Surveillance Network had been shared with a pharmaceutical company, which developed a medication unaffordable to most Indonesians. Indonesia refused to provide further samples after this incident.[52] This decision was denounced by scientists worldwide as opposing the spirit of scientific inquiry and contrary to the interests of global health. To resolve this issue, the World Health Organization proposed a framework regulating the sharing of influenza samples, stipulating that pharmaceutical products derived from such samples should be affordable to the countries that provided them.[53] Unfortunately, the recommended framework is limited to influenza and does little to counter the adverse effects of patenting and commercialisation in medical research into other diseases.

The knowledge regime emphasising intellectual property rights has adverse consequences for researchers in lower- and middle-income countries as it exacerbates disparities in access to the resources necessary for scientific inquiry. Indonesian scientists consequently continue to work in a dual scientific world, in which scientists from wealthier countries occupy leading positions while they are relegated to subordinate ones. The recent shift from international to global health has similar consequences for Indonesian physicians. During the second part of the twentieth century, international health organisations supported governments to strengthen the position of health services in developing nations. In contrast, advocates of global health often have stimulated private initiative, working through non-government organisations.[54] Indonesian physicians, like those in other developing nations, generally are employed by the state. They therefore risk being relegated to subordinate positions in global health.[55] Indonesian academics are required to publish in international journals indexed in the Scopus database to be promoted, which generally means they have to publish in English. Consequently, articles addressing local and national topics and published in Indonesian journals are considered less prestigious. Because of their limited access to essential resources, Indonesian researchers unfortunately work under conditions that resemble those of colonial times.[56]

In 2019, scientists inside and outside the archipelago were involved in discussions about a new research law which regulates all scientific research conducted in Indonesia as well as the conditions under which scientists from abroad can conduct research there. The new research law prohibits scientists from taking biological samples out of the country without a materials transfer permit and allows immigration to blacklist foreign scientists, preventing them from conducting research in Indonesia without a research permit.[57] In addition, foreign scientists are obliged to submit their raw data to the Ministry of Research, Technology, and Higher Education; involve Indonesian researchers as equal partners in research projects; and list all involved Indonesian scientists as co-authors in publications.[58] Scientists must also receive permission to conduct potentially dangerous research, which could include, critics assert, research that might threaten social harmony or national security.[59] Through these regulations, the Indonesian government seeks ways to balance its desire to support international scientists conducting research inside Indonesia with both its national interests and those of its own scientific community.

Conclusion

During the first part of the twentieth century, Indonesian physicians received their education within a colonial framework, which relegated most of them to a secondary position as assistants to their European colleagues who had enjoyed superior training at European universities. A small number of physicians were able to emancipate themselves from this secondary position by obtaining academic qualifications in Europe or graduating from the Batavia Medical School. This Indonesian medical elite took charge of medical education, research, and health care during the Japanese occupation and in independent Indonesia. Consequently, the efforts to Indonesianise and decolonise medical education and research, even though difficult and at times challenging, proved to be the most straightforward as compared to other academic disciplines.

After independence, Indonesian scientists took charge of higher education and research by rebuilding and transforming the colonial institutions they had inherited from the Dutch. Unfortunately, there were very few Indonesians with advanced scientific training outside medicine, which hampered efforts to decolonise all scientific research institutions. Indonesia aimed to Indonesianise higher education and research by making use of the Indonesian language compulsory, sending promising young staff members abroad for advanced education, and requiring higher education to develop in harmony with the state. The Indonesian government demanded the assistance of scientists, physicians, and medical researchers in addressing the multitude of problems that beset the newly independent nation. During the early 1950s, Dutch academics and those from other European countries continued to occupy leading positions at Indonesian universities. Economic problems, political instability, and an increasingly hostile attitude towards the Netherlands changed this situation toward the end of the 1950s, leading to severe staffing shortages outside medicine. Scientific research undoubtedly suffered as a result.

Scientists from lower- and middle-income countries face the challenge of gaining international recognition for their research while having limited access to necessary resources and crucial networks. This predicament has been exacerbated during the last 40 years, as scientists in wealthy countries rely on patents, intellectual property rights, and other legal measures, which place resources essential for conducting scientific research out of reach of their less-favoured colleagues. The response by the Republic of Indonesia to bolster its system of research permits could be interpreted as a modest attempt to correct this imbalance. Still, scientists from most decolonised nations struggle to be competitive on an international scale. In many respects, conditions that existed during colonial times, such as vast differences in access to resources and funding, persist in the modern global world.

Notes

1 Hans Pols, *Nurturing Indonesia: Medicine and Decolonisation in the Dutch East Indies* (Cambridge: Cambridge University Press, 2018), 102–15.

2 The Batavia medical college was founded in 1851 as the Dokter Djawa School (School for Javanese Physicians). In 1913 it was renamed School for the Education of Indies Physicians (School tot Opleiding van Indische Artsen; the acronym remained STOVIA). The medical college in Surabaya was named the Dutch Indies Physicians School (Nederlandsch-Indische Artsen School; NIAS). For a history see Liesbeth Hesselink, *Healers on the Colonial Market: Native Doctors and Midwives in the Dutch East Indies* (Leiden: KITLV Press, 2011).

3 In the remainder of this essay, I will use "Indies physician" to refer to someone who held a degree from either the Batavia or the Surabaya Medical College, and "Indonesian physician" to refer to a physician from the Indonesian archipelago, irrespective of the medical degree held. I am aware that the appellation "Indonesian" is anachronistic until the mid-1920s, but I use it here for clarity.

4 This view was forcefully presented in a report on the reorganisation of the colonial health service: J. Bijker, *Rapport der Commissie tot Voorbereiding eener Reorganisatie van den Burgerlijken Geneeskundigen Dienst* [Report of the Commission to Prepare a Reorganisation of the Civil Medical Service] (Batavia: Landsdrukkerij, 1908). See also the views of J. Haga, a member of this committee and the former director of the Military Health Service, about the Medical Service in the Netherlands Indies: J. Haga, "Geneeskundige Dienst in Ned.-Indië [Medical Service in the Dutch Indies]," *Indische Gids* 30 (1908): 25–32; J. Haga, "Reorganisatie van den Geneeskundigen Dienst in Ned.-Indië [Reorganisation of the Medical Service in the Dutch Indies]," *Indische Gids* 31, no. 1 (1909): 457–64.

5 See, for example: W.K. Tehupeiory, "Iets over de Inlandsche Geneeskundigen," [About Native Physicians], *Indisch Genootschap: Verslagen der Vergaderingen* (1908): 101–34; W.K. Tehupeiory, "Reorganisatie van het Onderwijs aan de School tot Opleiding van Inlandsche Artsen te Weltevreden" [Reorganisation of the Education at the Stovia], *Indische Gids* 31 (1909): 922–8; H.F. Roll, *Is Reorganisatie van de School tot Opleiding van Inlandsche Artsen te Weltevreden Nogmaals Nodig?* [Is a Reorganisation of Medical Instruction of Native Physicians in Weltevreden Again Necessary?] (Dordrecht: Morks & Geuze, 1909).

6 Abdul Rivai moved to the Netherlands in 1899, started his medical studies in 1902, and graduated in 1908. The Tehupeiory brothers moved to the Netherlands in 1907 and graduated in 1909. Unfortunately, Johannes Everhardus Tehupeiory died by asphyxiation in a hotel room because of a leaking gas lamp. Willem Karel Tehupeiory returned on his own to the Dutch East Indies in 1909. For Rivai see Pols, *Nurturing Indonesia*, 21–45; and Harry Poeze, "Early Indonesian Emancipation: Abdul Rivai, Van Heutsz and the Bintang Hindia," *BKI: Bijdragen tot de Taal-, Land- en Volkenkunde* 145, no. 1 (1989): 87–106; for the Tehupeiory brothers see Pols, *Nurturing Indonesia*, 40–2, 102–5.

7 Pols, *Nurturing Indonesia*, 39–43.

8 The phrase "Indonesian nationalist (or national) movement" or at times just "movement" has been widely used by historians to describe the multifaceted organisational activities among Indonesians in the first part of the twentieth century. See, for example, Takashi Shiraishi, *An Age in Motion: Popular Radicalism in Java, 1912–1926* (Ithaca, NY: Cornell University Press, 1990). For the political and organisational activities of Indonesian physicians and medical students, see Pols, *Nurturing Indonesia*.

9 This group was led by radical Indo-European politician E.F.E. Douwes Dekker, who also led the Indies Party (Indische Partij), the first group to advocate independence for the Indies.

10 In 1916, 1919, and 1924 the first, second, and third congress on colonial education were held in The Hague. See, for example, *Prae-Adviezen van het Eerste Koloniaal* [Recommendations for the first Conference on Colonial Education] (The Hague: Drukkerij Korthals, 1916). In the 1930s, the issue of higher education was studied extensively as well: J.W. Meyer Ranneft, *Rapport van de Commissie tot Bestudeering van de Toekomstige Ontwikkeling van het Hooger Onderwijs in Nederlandsch Indië* [Report from the Commission to Study the Future Development of Higher Education in the Dutch Indies] (Bandoeng: A.C. Nix, 1932).

11 Abdul Rivai, "Voorstel Betreffende de Voorbereiding der Stichting van een Universiteit in Indië, 25ᵉ Vergadering, 9 Dec 1918," [Proposal to Start Preparations for Founding a University in the Indies], in *Handelingen van den Volksraad, 2ᵉ Gewone Zitting* (Batavia: Volksraad van Nederlandsch-Indië, 1918), 570–7.

12 Abdul Rivai, J.T. Terburgh, A.E. Sitsen, and A. de Waart, *Het Hooger Onderwijs Vraagstuk, speciaal in verband met het Geneeskundig Onderwijs in Nederlandsch-Indië* [The Issue of Higher Education, in

Particular with Respect to Medical Education in the Dutch Indies], *Orgaan van Indische Artsen*, special issue (1919).

13 Pols, *Nurturing Indonesia*, 139–41.

14 Mohamad Amir proposed to use the Bulletin for this purpose; see Mohamad Amir, "Prae-Advies Dr. Amir: Een Eigen Vaktijdschrift voor den Indonesischen Arts" [Brief by Dr. Amir: A Professional Journal for the Indonesian Physician], *Bulletin van den Bond van Indische Geneeskundigen* 14, no. 3 (1925): 133–5.

15 For a quantitative analysis of the contributions of Indonesian physicians to the *Medical Journal of the Dutch Indies* see Liesbeth Hesselink, "Indigenous Authors," in *The Medical Journal of the Dutch Indies, 1852–1942: A Platform for Medical Research*, ed. Leo van Bergen, Liesbeth Hesselink, and Jan Peter Verhave (Jakarta: Indonesian Academy of Sciences (AIPI), 2018), 113–44.

16 Abdul Rasjid, "Dienst der Volksgezondheid, 25 Juli" [Public Health Service], in *Handelingen van den Volksraad, Zittingsjaar 1939–40* (Batavia: Landsdrukkerij, 1939), 418–24.

17 Abdul Rasjid, "Dienst der Volksgezondheid, 18 November 1940," in *Handelingen van den Volksraad, Zittingsjaar 1940–41* (Batavia: Landsdrukkerij, 1940), 999–1003.

18 Goelarso Astrohadikoesoemo, "Sedikit Pemandangan tentang Obat-Obat Kita oentoek Diselidiki," [Short Overview of Our Medicines to Be Explored], in *Het Tweede Congres van de Vereeniging van Indonesische Geneeskundigen* (Batavia: Kenanga, 1940), 200–21; and K.W.T. Saleh Mangoendiningrat, "Ketabiban didalam Bangsa Indonesia ditanah Djawa" [Indonesian Traditional Medicine from Java's Gardens], in *Het Tweede Congres van de Vereeniging van Indonesische Geneeskundigen* (Batavia: Kenanga, 1940), 222–47. For reflections on traditional herbal medicine after independence see Seno Sastroamidjojo, *Obat Asli Indonesia* [Indonesian Indigenous Medicine] (Djakarta: Penerbit Kebangsaan Pustaka Rakjat, 1948).

19 See, for example, Yakusyo Katuyo Iinkai-Iin (Committee for the Study of Indigenous Medicine), and Raden Mochtar, *Obat Obat dari Bahan-Bahan Negeri Sendiri* [Medicines from Our Own Country's Ingredients] (Djakarta: Balai Poestaka, 1945).

20 C. Antons, and R. Antons-Sutanto, "Traditional Medicine and Intellectual Property Rights: A Case Study of Indonesian Jamu Industry," in *Traditional Knowledge, Traditional Cultural Expressions and Intellectual Property Law on the Asia-Pacific Region*, ed. C. Antons (Alphen aan den Rijn: Wolters Kluwer, 2009), 363–84.

21 See, for example, Asih Sumardono and Mark Hanusz, *Family Business: A Case Study of Nyonya Meneer, One of Indonesia's Most Successful Traditional Medicine Companies* (Singapore: Equinox, 2007). See also Susan-Jane Beers, *Jamu: The Ancient Indonesian Art of Herbal Healing* (Singapore: Periplus, 2001).

22 Soejono Martosewojo, "Risalah Pembentukan Djakarta Ika Dai Gaku [About the Opening of the *Ika Dai Gaku* in Jakarta]," in *125 Tahun Pendidikan Dokter di Indonesia 1851–1976*, ed. M.A. Hanafia, Bahder Djohan, and Surono (Jakarta: FKUI, 1976), 33–4; T. Karimoeddin, "Pendidikan Dokter Jaman Pendudukan Jepang (Ika Dai Gaku) [Medical Education during the Japanese Occupation]," in Hanafia, Djohan, and Surono, *125 Tahun Pendidikan Dokter di Indonesia 1851–1976*, 26–32.

23 Achmad Ramali was associated with the committee to find equivalent Malay words for Dutch medical terms. The results were published as Ahmad Ramali and K. St. Pamoentjak, *Kamus Kedoktoran: Arti dan Keterangan Istilah* [Medical Dictionary] (Djakarta: Djambatan, 1953). The most recent edition is: Ahmad Ramali, K. St. Pamoentjak, and Hendra T. Laksman, *Kamus Kedoktoran: Arti dan Keterangan Istilah*, 25th ed. (Djakarta: Djambatan, 2003).

24 For a biography see Asvi Warman Adam, *Sarwono Prawirohardjo: Pembangun Institusi Ilmu Pengetahuan di Indonesia* [Sarwono Prawirohardjo: The Development of Scientific Institutions in Indonesia] (Jakarta: Lembaga Ilmu Pengetahuan Indonesia, 2009).

25 P.M. van Wulfften Palthe and P.A. Kerstens, *Opening Nood-Universiteit: Redevoeringen Uitgesproken door Prof. Dr. P.M. van Wulfften Palthe, President der Nood-Universiteit en P.A. Kerstens, wd. Directeur van Onderwijs & Eeredienst te Batavia op 21 Januari 1946* [Addresses at the Opening of the Emergency University] (Groningen en Batavia: J.B. Wolters, 1946).

26 Leo van Bergen, *The Dutch East Indies Red Cross, 1870–1950: On Humanitarianism and Colonialism* (Lanham, UK: Lexington Books, 2019).

27 Andrew Goss, "Reinventing the *Kebun Raya* in the New Republic: Scientific Research at the Bogor Botanical Gardens in the Age of Decolonization," *Studium* 11, no. 3 (2018): 206–19; Andrew Goss, *The Floracrats: State-Sponsored Science and the Failure of the Enlightenment in Indonesia* (Madison: University of Wisconsin Press, 2011), 132–6.

28 Goss, "Reinventing the *Kebun Raya*," 213.

29 E. Kits van Waveren, "De Artsenpositie in Indonesië," *Nederlands Tijdschrift voor Geneeskunde* 91 (1947): 3430–69.

30 W.H. van Helsdingen, *Daar Wèrd Wat Groots Verricht: Nederlandsch-Indië in de Twintigste Eeuw* [Something Immense Has Been Accomplished There: The Dutch Indies During the Twentieth Century] (Amsterdam: Elsevier, 1941); Pieter Honig and Frans Verdoorn, *Science and Scientists in the Netherlands Indies* (New York: Board for the Netherlands Indies, Surinam and Curaçao, 1945). For medicine see the series of articles in the journal *Quarterly Journal of Tropical Medicine and Hygiene*: N.H. Swellengrebel, "Indonesia before the War: Introduction," *Documenta Neerlandica et Indonesica de Morbis Tropicis; Quarterly Journal of Tropical Medicine and Hygiene* 1, no. 3 (1949): 193–4.

31 Peter Boomgaard, "The Making and Unmaking of Tropical Science: Dutch Research in Indonesia, 1600–2000," *BKI: Bijdragen tot de Taal-, Land- en Volkenkunde* 162, no. 2–3 (2006): 191–217; Lewis Pyenson, *Empire of Reason: Exact Sciences in Indonesia, 1840–1940* (Leiden: Brill, 1989); Dirk Schoute, *Occidental Therapeutics in the Netherlands East Indies During Three Centuries of Netherlands Settlement (1600–1900)* (Batavia: Netherlands Indies Public Health Service, 1937).

32 Critics have argued that too many elements from colonial times have been retained. See, for example, Denys Lombard, *Nusa Jawa, Silang Budaya: Kajian Sejarah Terpadu* [Java's Homeland, Cultural Crossings: Integrated Historical Studies], trans. Winarsih Partaningrat Arifin, Rahayu S. Hidayat, and Nini Hidayati Yusuf, *Batas-batas Pembaratan* [Limitations to westernisation] (Jakarta: Gramedia, 1990), 1: 121–4.

33 R. Gispen, "Tot Besluit" [To Conclude], *Medisch Maandblad* 3, no. 12 (1950): 420. The modest *Medical Monthly* replaced the pre-war *Medical Journal of the Dutch Indies* (*Geneeskundig Tijdschrift voor Nederlandsch-Indië*). See also R. Gispen, "Termination of the Activities of the Society for the Furthering of Medical Sciences in Indonesia," *OSR News* 2, no. 11 (1950): 142–3. For the establishment of the *Ikatan Dokter Indonesia*, see "Announcement [On the Founding IDI]," *OSR News* 2, no. 11 (1950): 141–2. The previous year, Dr Sardjito, a bacteriologist and the founding president of Gadjah Mada University, provided an overview of the medical research conducted by Indonesian physicians. See Sardjito, "The Development of Medical Science in Indonesia," *OSR News* 2, no. 10 (1950): 125–9.

34 Adam Messer, "Effects of the Indonesian National Revolution and Transfer of Power on the Scientific Establishment," *Indonesia* 58 (1994): 41–68.

35 For a report on the success of Indonesianising the University of Indonesia see Willard A. Hanna, "From Universiteit to Universitas: The 'Indonesiatization' of a Dutch University," *American Universities Field Staff Reports, Southeast Asia Series* 4, no. 17 (1956): 1–17.

36 See R. Thomas Murray, *A Chronicle of Indonesian Higher Education: The First Half Century, 1920–1970* (Singapore: Chopmen, 1973), 40–172; William K. Cummings and Salman Kasenda, "The Origins of Modern Indonesian Higher Education," in *From Dependence to Autonomy: The Development of Asian Universities*, ed. Philip G. Altbach and Viswanathan Selvaratnam (Dordrecht: Kluwer, 1989), 143–66; R.M. Koentjaraningrat and Harsja W. Bachtiar, "Higher Education in the Social Sciences in Indonesia," in *The Social Sciences in Indonesia*, ed. R.M. Koentjaraningrat (Jakarta: LIPI, 1975), 1–42; Bachtiar Rifai and Koesnadi Hardjasoemantri, *Perguruan Tinggi di Indonesia* (Jakarta: Departemen Perguruan Tinggi dan Ilmu Pengetahuan, 1965).

37 Warwick Anderson and Hans Pols, "Scientific Patriotism: Medical Science and National Self-Fashioning in Southeast Asia," *Comparative Studies in Society and History* 54, no. 1 (2012): 93–113; Pols, *Nurturing Indonesia*.

38 Hanna, "From Universiteit to Universitas," 8.

39 Messer, "Effects," 41–68.

40 Messer, "Effects."

41 Vivek Neelakantan, *Science, Public Health and Nation-Building in Soekarno-Era Indonesia* (Newcastle upon Tyne: Cambridge Scholars, 2017), 142–55. See also Francis Scott Smyth, "University of California Medical Science Teaching in Indonesia," *Journal of Medical Education* 32, no. 5 (1957): 344–9; Francis Scott Smyth, "Health and Medicine in Indonesia," *Journal of Medical Education* 38, no. 8 (1963): 693–6; John S. Wellington, "Medical Science and Technology," in *Indonesia: Resource and Their Technological Development*, ed. Howard M. Beers (Lexington: University Press of Kentucky, 1970), 165–73. See also Willard A. Hanna, "A Binational Project in Medical Education," *American Universities Field Staff, Southeast Asia Series* 4, no. 19 (1956): 1–10.

42 During the 1950s, there was an extreme shortage of physicians in Indonesia. In 1957, for example, there were 1,450 physicians working there, which is roughly equal to one physician for every 55,000 persons. See Willard A. Hanna, "No Place for Hypochondriacs: Medical Services in Indonesia," *American Universities Field Staff Reports, Southeast Asia Series* 5, no. 5 (1957): 1–5.

43 See Goss, *Floracrats*, 145–52; Goss, "Reinventing the *Kebun Raya*," 206–19.

44 Goss, "Reinventing the *Kebun Raya*," 214.

45 Goss, "Reinventing the *Kebun Raya*," 218–19.

46 Hendrik E. Niemeijer, "A Sea of Histories, A History of the Seas: An Interview with Adrian B. Lapian," *Itinerario* 28, no. 1 (2004): 10.

47 Shobita Parthasarathy, *Life Forms, Markets, and the Public Interest in the United States and Europe* (Chicago: University of Chicago Press, 2017); Rebecca Lave, Philip Mirowski, and Samuel Randalls, "Introduction: STS and Neoliberal Science," *Social Studies of Science* 40, no. 5 (2010): 659–75.

48 Adriana Petryna, Andrew Lakoff, and Arthur Kleinman, eds., *Global Pharmaceuticals: Ethics, Markets, Practices* (Durham, NC: Duke University Press, 2006); Adriana Petryna, *When Experiments Travel: Clinical Trials and the Global Search for Human Subjects* (Princeton, NJ: Princeton University Press, 2009).

49 See Andrew Lakoff, *Pharmaceutical Reason: Knowledge and Value in Global Psychiatry* (New York: Cambridge University Press, 2006); Janis H. Jenkins, ed. *Pharmaceutical Self: The Global Shaping of Experience in an Age of Psychopharmacology* (Santa Fe, NM: SAR Press, 2010).

50 Lori Andrews and Dorothy Nelkin, *Body Bazaar: The Market for Human Tissue in the Biotechnology Age* (New York: Crown, 2001), 71; Emily Marden, "The Neem Tree Patent: International Conflict over the Commodification of Life," *Boston College International and Comparative Law Review* 22, no. 2 (1999): 279–95. For a reaction from India see Vandana Shiva, "The Neem Tree: A Case History of Biopiracy," *TWN: Third World Network*, https://twn.my/title/pir-ch.htm; Vandana Shiva, *Biopiracy: The Plunder of Nature and Knowledge* (Boston: South End Press, 1997).

51 Celia Lowe, "Viral Clouds: Becoming H5N1 in Indonesia," *Cultural Anthropology* 25, no. 4 (2010): 625–49; Celia Low, "Preparing Indonesia: H5N1 Influenza through the Lens of Global Health," *Indonesia* 90 (2010): 147–70.

52 Kennas Mullis, "Playing Chicken with Bird Flu: 'Viral Sovereignty,' The Right to Exploit Natural Genetic Resources, and the Potential Human Rights Ramifications," *American University International Law Review* 24, no. 5 (2009): 943–67.

53 World Health Organization, *World Health Organization, Pandemic Influenza Preparedness Framework for the Sharing of Influenza Viruses and Access to Vaccines and Other Benefits* (Geneva: World Health Organization, 2011).

54 Andrew Lakoff, "Two Regimes of Public Health," *Humanity: An International Journal of Human Rights, Humanitarianism, and Development* 1, no. 1 (2010): 59–79; Theodore M. Brown, Marcos Cueto, and Elizabeth Fee, "The World Health Organization and the Transition from 'International' to 'Global' Public Health," *American Journal of Public Health* 96, no. 1 (2004): 62–72.

55 Pols, *Nurturing Indonesia*, 233–9.

56 An exception is the Eijkman Institute for Molecular Biology, which was established on the initiative of Minister of Research and Technology, and later President, B.J. Habibie, an engineer, in 1992, with Professor Sangkot Marzuki as director. Under Habibie, the Indonesian Academy of Sciences (Akademi Ilmu Pengetahuan Indonesia) was established as well.

57 The law itself is named Undang-undang Republik Indonesia Nomor 11 Tahun 2019 Tentang Sistem Ilmu Pengetahuan dan Teknologi (Law no. 11 of the year 2019 of the Republic of Indonesia Pertaining to Science and Technology). See also Luthfi T. Dzulfikar, "After Science Law, Indonesian Government Must Streamline Its Research Permits or Risk Scaring Away Foreign Scientists: Scholars," *The Conversation*, 11 September 2019, https://theconversation.com/after-science-law-indonesian-govt-must-streamline-its-research-permits-or-risk-scaring-away-foreign-scientists-scholars-122947.

58 Dyna Rochmyaningsih, "Indonesian Plan to Clamp Down on Foreign Scientists Draws Protest," *Nature* 557 (May 22, 2018): 476, doi.org/10.1038/d41586-018-05001-7.

59 Luthfi T. Dzulfikar, "Indonesia's Science Law Brings Back Memories of New Order Restrictions on Academic Freedom, Scholars Say," *The Conversation*, 12 October 2019, https://theconversation.com/indonesias-science-law-brings-back-memories-of-new-order-era-restrictions-on-academic-freedom-scholars-say-125111.

INDEX

Printed in the United States
by Baker & Taylor Publisher Services

Printed in the United States
by Baker & Taylor Publisher Services